Studies in Systems, Decision and Control

Volume 398

The series "Studies in Systems, Decision and Control" (SSDC) covers both new developments and advances, as well as the state of the art, in the various areas of broadly perceived systems, decision making and control–quickly, up to date and with a high quality. The intent is to cover the theory, applications, and perspectives on the state of the art and future developments relevant to systems, decision making, control, complex processes and related areas, as embedded in the fields of engineering, computer science, physics, economics, social and life sciences, as well as the paradigms and methodologies behind them. The series contains monographs, textbooks, lecture notes and edited volumes in systems, decision making and control spanning the areas of Cyber-Physical Systems, Autonomous Systems, Sensor Networks, Control Systems, Energy Systems, Automotive Systems, Biological Systems, Vehicular Networking and Connected Vehicles, Aerospace Systems, Automation, Manufacturing, Smart Grids, Nonlinear Systems, Power Systems, Robotics, Social Systems, Economic Systems and other. Of particular value to both the contributors and the readership are the short publication timeframe and the world-wide distribution and exposure which enable both a wide and rapid dissemination of research output.

Indexed by SCOPUS, DBLP, WTI Frankfurt eG, zbMATH, SCImago.

All books published in the series are submitted for consideration in Web of Science.

More information about this series at https://link.springer.com/bookseries/13304

George A. Anastassiou

Unification of Fractional Calculi with Applications

Springer

George A. Anastassiou
Department of Mathematical Sciences
University of Memphis
Memphis, TN, USA

ISSN 2198-4182 ISSN 2198-4190 (electronic)
Studies in Systems, Decision and Control
ISBN 978-3-030-86919-9 ISBN 978-3-030-86920-5 (eBook)
https://doi.org/10.1007/978-3-030-86920-5

This Springer imprint is published by the registered company Springer Nature Switzerland AG
The registered company address is: Gewerbestrasse 11, 6330 Cham, Switzerland

Dedicated to the truth fighting and searching mathematicians under harsh circumstances and difficult conditions and eventually achieving important results

Preface

Fractional calculus and its long history started in 1695 by L'Hospital and Leibnitz. Over the years, many different kinds of it emerged according to the needs of science given by many famous mathematicians. In the last 50 years, fractional calculus due to its wide applications to many applied sciences has become very popular. Its predominant kinds are the old Riemann–Liouville fractional calculus and the newer one of Caputo type.

Around these two trends have evolved many other variants, and all of these involve singular kernels. More recently, researchers presented also new fractional calculi of non-singular kernels. Almost all fractional calculi engage memory in their kernels, and the two divisions of singular and non-singular kernel have their own advantages and disadvantages, but both have their strong merits.

The recent Hilfer fractional calculus unifies the Riemann–Liouville and Caputo fractional calculi, and the Prabhakar fractional calculus combines both singular and non-singular kernel fractional calculi.

Finally, the newer Hilfer–Prabhakar fractional calculus is the most general one unifying all trends, and for different values of its parameters, we get the particular fractional calculi.

In this monograph, we present and employ advanced and generalized versions of Hilfer, Prabhakar and Hilfer–Prabhakar fractional calculi, and we produce very interesting fractional integral inequalities.

So in this monograph, all presented is original work by the author given at a very general level to cover a maximum number of cases in various applications.

All in all, this monograph is the natural and expected evolution of recent author's research work put in a book form for the first time. The presented approaches are original, and chapters are self-contained and can be read independently. This monograph is suitable to be used in related graduate classes and research projects. We exhibit to the maximum our computational and estimating methods to all possible directions.

The motivation to write this monograph came by the following: Various issues related to the modeling and analysis of ordinary and fractional-order systems have gained an increased popularity, as witnessed by many books and volumes in Springer's program:

http://www.springer.com/gp/search?query=fractional&submit=Prze%C5%9Blij

and the purpose of our book is to capture at a very general level a deeper formal analysis on some issues that are relevant to many areas for instance: decision making, complex processes, systems modeling and control, and related areas.

The above are deeply embedded in the fields of mathematics, engineering, computer science, physics, economics, social and life sciences.

The complete list of presented topics follows:

- Progress on generalized Hilfer fractional calculus and fractional integral inequalities,
- Landau-type advanced Hilfer fractional inequalities,
- Advanced Iyengar–Hilfer fractional inequalities,
- Generalized Hilfer–Polya, Hilfer Ostrowski and Hilfer–Hilbert–Pachpatte fractional inequalities,
- Generalized Hilfer fractional approximations of Csiszar's f-Divergence,
- Advanced Hilfer fractional self-adjoint operator analytic inequalities,
- Basic forward and reverse generalized Hilfer–Hardy fractional inequalities,
- Foundations of generalized Prabhakar–Hilfer fractional calculus with applications,
- Advanced Hilfer–Prabhakar–Hardy-type fractional inequalities,
- Vectorial Hilfer–Prabhakar–Hardy-type fractional inequalities,
- Vectorial Prabhakar–Hardy-type generalized fractional inequalities under convexity,
- Generalized multivariate Prabhakar-type fractional integrals and inequalities,
- Multiparameter fractional differentiation with non-singular kernel,
- Hilfer fractional Opial inequalities,
- Exotic fractional inequalities revisited,
- Fuzzy fractional calculus revisited.

The book's results are expected to find applications in many areas of pure and applied mathematics, especially in inequalities and differential equations as they connect to fractionality. Other possible applications can be in applied sciences like geophysics, physics, chemistry, economics, engineering, etc. Thus, this monograph is suitable for researchers, graduate students, practitioners and seminars of the above disciplines, also to be in all science and engineering libraries.

The preparation of the book took place during 2021 at the University of Memphis, at stay home of author during the COVID-19 outbreak keeping him alive, happy and in control!

The author likes to thank Prof. Alina Lupas of University of Oradea, Romania, for checking and reading the manuscript.

Memphis, TN, USA George A. Anastassiou
October 2021

Contents

Chapter 1
Progress on Generalized Hilfer Fractional Calculus and Fractional Integral Inequalities

After motivation we give a complete background on needed ψ-Hilfer fractional Calculus. Then we produce ψ-Hilfer fractional left and right Taylor formulae. We give also important ψ-Hilfer fractional left and right representation integral formulae regarding ψ-Hilfer left and right fractional derivatives. Then we give extensive applications of our ψ-Hilfer fractional results to left and right ψ-Hilfer fractional Ostrowski, Opial and Poincaré type integral inequalities. We create the space for more future forthcoming results. It follows [7].

1.1 Introduction

We are inspired by the following famous Ostrowski inequality [10]:

Theorem 1.1 *Let $f \in C^1([a,b])$, $x \in [a,b]$. Then*

$$\left| \frac{1}{b-a} \int_a^b f(y)\, dy - f(x) \right| \leq \left(\frac{1}{4} + \frac{\left(x - \frac{a+b}{2}\right)^2}{(b-a)^2} \right) (b-a) \left\| f' \right\|_\infty . \tag{1.1}$$

Notice that

$$\left(\frac{1}{4} + \frac{\left(x - \frac{a+b}{2}\right)^2}{(b-a)^2} \right) (b-a) = \frac{(x-a)^2 + (b-x)^2}{2(b-a)}. \tag{1.2}$$

Inequality (1.1) is sharp, the optimal function is $f^*(y) = |y-x|^\alpha (b-a)$, $\alpha > 1$, see [2].

We are also inspired by [13], which has great ideas but contains several mistakes.

G. A. Anastassiou, *Unification of Fractional Calculi with Applications*,
Studies in Systems, Decision and Control 398,
https://doi.org/10.1007/978-3-030-86920-5_1

We will present some necessary background from Hilfer fractional calculus, then we will develop our Hilfer type fractional Taylor formulae and we will exhibit our fractional integral representation formulae. Finally all these will be applied to establishing Hilfer fractional Ostrowski, Opial and Poincaré type inequalities.

1.2 Background

Let $-\infty < a < b < \infty$, the left and right Riemann-Liouville fractional integrals of order $\alpha \in \mathbb{C}$ ($\mathcal{R}(\alpha) > 0$) are defined by

$$\left(I_{a+}^{\alpha} f\right)(x) = \frac{1}{\Gamma(\alpha)} \int_a^x (x-t)^{\alpha-1} f(t)\, dt, \quad (1.3)$$

$x > a$; where Γ stands for the gamma function,

and

$$\left(I_{b-}^{\alpha} f\right)(x) = \frac{1}{\Gamma(\alpha)} \int_x^b (t-x)^{\alpha-1} f(t)\, dt, \quad (1.4)$$

$x < b$.

The Riemann-Liouville left and right fractional derivatives of order $\alpha \in \mathbb{C}$ ($\mathcal{R}(\alpha) \geq 0$) are defined by

$$\left(\Delta_{a+}^{\alpha} y\right)(x) = \left(\frac{d}{dx}\right)^n \left(I_{a+}^{n-\alpha} y\right)(x) = \frac{1}{\Gamma(n-\alpha)} \left(\frac{d}{dx}\right)^n \int_a^x (x-t)^{n-\alpha-1} y(t)\, dt \quad (1.5)$$

($n = \lceil \mathcal{R}(\alpha) \rceil$, $\lceil \cdot \rceil$ means ceiling of the number; $x > a$)

$$\left(\Delta_{b-}^{\alpha} y\right)(x) = (-1)^n \left(\frac{d}{dx}\right)^n \left(I_{b-}^{n-\alpha} y\right)(x) =$$

$$\frac{(-1)^n}{\Gamma(n-\alpha)} \left(\frac{d}{dx}\right)^n \int_x^b (t-x)^{n-\alpha-1} y(t)\, dt \quad (1.6)$$

($n = \lceil \mathcal{R}(\alpha) \rceil$; $x < b$), respectively, where $\mathcal{R}(\alpha)$ is the real part of α.

In particular, when $\alpha = n \in \mathbb{Z}_+$, then

$$\left(\Delta_{a+}^{0} y\right)(x) = \left(\Delta_{b-}^{0} y\right)(x) = y(x);$$

$$\left(\Delta_{a+}^{n} y\right)(x) = y^{(n)}(x), \text{ and } \left(\Delta_{b-}^{n} y\right)(x) = (-1)^n y^{(n)}(x), \ n \in \mathbb{N},$$

see [11].

Based on [11], pp. 44–45, and [8], pp. 74–76 we have the following left and right basic Riemann-Liouville fractional Taylor formulae:

Theorem 1.2 *Let* $\nu > 0$, $\nu \notin \mathbb{N}$ *and* $m = \lceil \nu \rceil$; φ, $\Delta_{a+}^{\nu}\varphi \in C([a,b])$. *Then*

$$\varphi(x) = \sum_{j=1}^{m} \frac{\Delta_{a+}^{\nu-j}\varphi(a)}{\Gamma(\nu-j+1)}(x-a)^{\nu-j} + \frac{1}{\Gamma(\nu)}\int_a^x (x-t)^{\nu-1}\Delta_{a+}^{\nu}\varphi(t)\,dt, \quad (1.7)$$

$x \in [a,b]$.

Theorem 1.3 *Let* $\nu > 0$, $\nu \notin \mathbb{N}$ *and* $m = \lceil \nu \rceil$; φ, $\Delta_{b-}^{\nu}\varphi \in C([a,b])$. *Then*

$$\varphi(x) = \sum_{j=1}^{m} \frac{\Delta_{b-}^{\nu-j}\varphi(b)}{\Gamma(\nu-j+1)}(b-x)^{\nu-j} + \frac{1}{\Gamma(\nu)}\int_x^b (t-x)^{\nu-1}\Delta_{b-}^{\nu}\varphi(t)\,dt, \quad (1.8)$$

$x \in [a,b]$.

Let $\alpha > 0$, $I = [a,b] \subset \mathbb{R}$, f an integrable function defined on I and $\psi \in C^1(I)$ an increasing function such that $\psi'(x) \neq 0$, for all $x \in I$. Left fractional integrals and left Riemann-Liouville fractional derivatives of a function f with respect to another function ψ are defined as [8, 11]

$$I_{a+}^{\alpha,\psi}f(x) = \frac{1}{\Gamma(\alpha)}\int_a^x \psi'(t)(\psi(x)-\psi(t))^{\alpha-1}f(t)\,dt, \quad (1.9)$$

and

$$\Delta_{a+}^{\alpha,\psi}f(x) = \left(\frac{1}{\psi'(x)}\frac{d}{dx}\right)^n I_{a+}^{n-\alpha,\psi}f(x) = \quad (1.10)$$

$$\frac{1}{\Gamma(n-\alpha)}\left(\frac{1}{\psi'(x)}\frac{d}{dx}\right)^n \int_a^x \psi'(t)(\psi(x)-\psi(t))^{n-\alpha-1}f(t)\,dt,$$

respectively, where $n = \lceil \alpha \rceil$.

Similarly, we define the right ones:

$$I_{b-}^{\alpha,\psi}f(x) = \frac{1}{\Gamma(\alpha)}\int_x^b \psi'(t)(\psi(t)-\psi(x))^{\alpha-1}f(t)\,dt, \quad (1.11)$$

and

$$\Delta_{b-}^{\alpha,\psi}f(x) = \left(-\frac{1}{\psi'(x)}\frac{d}{dx}\right)^n I_{b-}^{n-\alpha,\psi}f(x) =$$

$$\frac{1}{\Gamma(n-\alpha)}\left(-\frac{1}{\psi'(x)}\frac{d}{dx}\right)^n \int_x^b \psi'(t)(\psi(t)-\psi(x))^{n-\alpha-1}f(t)\,dt. \quad (1.12)$$

The following semigroup property holds; if $\alpha, \beta > 0$, $f \in C(I)$, then

$$I_{a+}^{\alpha,\psi}I_{a+}^{\beta,\psi}f = I_{a+}^{\alpha+\beta,\psi}f \quad \text{and} \quad I_{b-}^{\alpha,\psi}I_{b-}^{\beta,\psi}f = I_{b-}^{\alpha+\beta,\psi}f.$$

Next let again $\alpha > 0$, $n = \lceil \alpha \rceil$, $I = [a, b]$, $f, \psi \in C^n(I) : \psi$ is increasing and $\psi'(x) \neq 0$, for all $x \in I$. The left ψ-Caputo fractional derivative of f of order α is given by [1]

$$^{C}D_{a+}^{\alpha,\psi} f(x) = I_{a+}^{n-\alpha,\psi} \left(\frac{1}{\psi'(x)} \frac{d}{dx} \right)^n f(x), \tag{1.13}$$

and the right ψ-Caputo fractional derivative [1]

$$^{C}D_{b-}^{\alpha,\psi} f(x) = I_{b-}^{n-\alpha,\psi} \left(-\frac{1}{\psi'(x)} \frac{d}{dx} \right)^n f(x). \tag{1.14}$$

We set

$$f_{\psi}^{[n]}(x) := f_{\psi}^{(n)} f(x) := \left(\frac{1}{\psi'(x)} \frac{d}{dx} \right)^n f(x). \tag{1.15}$$

Clearly, when $\alpha = m \in \mathbb{N}$ we have

$$^{C}D_{a+}^{\alpha,\psi} f(x) = f_{\psi}^{[m]}(x) \quad \text{and} \quad ^{C}D_{b-}^{\alpha,\psi} f(x) = (-1)^m f_{\psi}^{[m]}(x),$$

and if $\alpha \notin \mathbb{N}$, then

$$^{C}D_{a+}^{\alpha,\psi} f(x) = \frac{1}{\Gamma(n-\alpha)} \int_a^x \psi'(t) (\psi(x) - \psi(t))^{n-\alpha-1} f_{\psi}^{[n]}(t)\, dt, \tag{1.16}$$

and

$$^{C}D_{b-}^{\alpha,\psi} f(x) = \frac{(-1)^n}{\Gamma(n-\alpha)} \int_x^b \psi'(t) (\psi(t) - \psi(x))^{n-\alpha-1} f_{\psi}^{[n]}(t)\, dt. \tag{1.17}$$

If $\psi(x) = x$, then we get the usual left and right Caputo fractional derivatives

$$^{C}D_{a+}^{m} f(x) = f^{(m)}(x), \quad ^{C}D_{b-}^{m} f(x) = (-1)^m f^{(m)}(x),$$

for $m \in \mathbb{N}$, and ($\alpha \notin \mathbb{N}$)

$$D_{*a}^{\alpha} f(x) = {}^{C}D_{a+}^{\alpha} f(x) = \frac{1}{\Gamma(n-\alpha)} \int_a^x (x-t)^{n-\alpha-1} f^{(n)}(t)\, dt, \tag{1.18}$$

$$D_{b-}^{\alpha}(x) = {}^{C}D_{b-}^{\alpha} f(x) = \frac{(-1)^n}{\Gamma(n-\alpha)} \int_x^b (t-x)^{n-\alpha-1} f^{(n)}(t)\, dt. \tag{1.19}$$

Also we set

$$^{C}D_{a+}^{0,\psi} f(x) = {}^{C}D_{b-}^{0,\psi} f(x) = f(x).$$

We need the following left and right ψ-Caputo fractional Taylor formulae:

Theorem 1.4 ([1]) *Let $\alpha > 0$, $n = \lceil \alpha \rceil$, $f, \psi \in C^n([a, b])$, ψ is increasing and $\psi'(x) \neq 0$ over $[a, b]$. Then*

$$I_{a+}^{\alpha,\psi} \, {}^{C}D_{a+}^{\alpha,\psi} f(x) = f(x) - \sum_{k=0}^{n-1} \frac{f_{\psi}^{[k]}(a)}{k!} (\psi(x) - \psi(a))^k , \tag{1.20}$$

and

$$I_{b-}^{\alpha,\psi} \, {}^{C}D_{b-}^{\alpha,\psi} f(x) = f(x) - \sum_{k=0}^{n-1} (-1)^k \frac{f_{\psi}^{[k]}(b)}{k!} (\psi(b) - \psi(x))^k , \tag{1.21}$$

$x \in [a, b]$.

In particular, given $\alpha \in (0, 1)$, we have

$$I_{a+}^{\alpha,\psi} \, {}^{C}D_{a+}^{\alpha,\psi} f(x) = f(x) - f(a)$$

and

$$I_{b-}^{\alpha,\psi} \, {}^{C}D_{b-}^{\alpha,\psi} f(x) = f(x) - f(b).$$

Denote by ($k \in \mathbb{Z}_+$) the sequential ψ-fractional derivatives:

$$\begin{aligned} {}^{C}D_{a+}^{k\alpha,\psi} &= {}^{C}D_{a+}^{\alpha,\psi} \, {}^{C}D_{a+}^{\alpha,\psi} \ldots {}^{C}D_{a+}^{\alpha,\psi} \\ & \qquad\qquad\qquad\qquad (k\text{-times}) \\ {}^{C}D_{b-}^{k\alpha,\psi} &= {}^{C}D_{b-}^{\alpha,\psi} \, {}^{C}D_{b-}^{\alpha,\psi} \ldots {}^{C}D_{b-}^{\alpha,\psi} \end{aligned}$$

We also mention the following left and right ψ-Caputo sequential fractional Taylor formulae (they derive from [1]).

Theorem 1.5 *Let $\alpha \in (0, 1)$, $f, \psi \in C^1([a, b])$, ψ is increasing and $\psi'(x) \neq 0$ over $[a, b]$. Let also $n \in \mathbb{N}$ and f is such that ${}^{C}D_{a+}^{k\alpha,\psi} f$ and ${}^{C}D_{b-}^{k\alpha,\psi} f$ exist and are continuous, for all $k = 0, 1, ..., n + 1$, $x \in [a, b]$. Then*

$$f(x) = \sum_{k=0}^{n} \frac{(\psi(x) - \psi(a))^{k\alpha}}{\Gamma(k\alpha + 1)} \, {}^{C}D_{a+}^{k\alpha,\psi} f(a) +$$

$$\frac{1}{\Gamma((n+1)\alpha)} \int_a^x \psi'(t) (\psi(x) - \psi(t))^{(n+1)\alpha - 1} \, {}^{C}D_{a+}^{(n+1)\alpha,\psi} f(t) \, dt, \tag{1.22}$$

and

$$f(x) = \sum_{k=0}^{n} \frac{(\psi(b) - \psi(x))^{k\alpha}}{\Gamma(k\alpha + 1)} \, {}^{C}D_{b-}^{k\alpha,\psi} f(b) +$$

$$\frac{1}{\Gamma((n+1)\alpha)}\int_x^b \psi'(t)(\psi(t)-\psi(x))^{(n+1)\alpha-1}\ {}^{C}D_{b-}^{(n+1)\alpha,\psi}f(t)\,dt. \tag{1.23}$$

Next we will deal with the ψ-Hilfer fractional derivative.

Definition 1.6 ([13]) Let $n-1<\alpha<n$, $n\in\mathbb{N}$, $I=[a,b]\subset\mathbb{R}$ and $f,\psi\in C^n([a,b])$, ψ is increasing and $\psi'(x)\neq 0$, for all $x\in I$. The ψ-Hilfer fractional derivative (left-sided and right-sided) ${}^{H}\mathbb{D}_{a+(b-)}^{\alpha,\beta;\psi}f$ of order α and type $0\le\beta\le 1$, are defined by

$${}^{H}\mathbb{D}_{a+}^{\alpha,\beta;\psi}f(x)=I_{a+}^{\beta(n-\alpha);\psi}\left(\frac{1}{\psi'(x)}\frac{d}{dx}\right)^n I_{a+}^{(1-\beta)(n-\alpha);\psi}f(x), \tag{1.24}$$

and

$${}^{H}\mathbb{D}_{b-}^{\alpha,\beta;\psi}f(x)=I_{b-}^{\beta(n-\alpha);\psi}\left(-\frac{1}{\psi'(x)}\frac{d}{dx}\right)^n I_{b-}^{(1-\beta)(n-\alpha);\psi}f(x),\quad x\in[a,b]. \tag{1.25}$$

The original Hilfer fractional derivatives [12] come from $\psi(x)=x$, and are denoted by ${}^{H}\mathbb{D}_{a+}^{\alpha,\beta}f(x)$ and ${}^{H}\mathbb{D}_{b-}^{\alpha,\beta}f(x)$.

When $\beta=0$, we get Riemann-Liouville fractional derivatives, while when $\beta=1$ we have Caputo type fractional derivatives.

We define $\gamma=\alpha+\beta(n-\alpha)$. We notice that $n-1<\alpha\le\alpha+\beta(n-\alpha)\le\alpha+n-\alpha=n$, hence $\lceil\gamma\rceil=n$. We can easily write that ([13])

$${}^{H}\mathbb{D}_{a+}^{\alpha,\beta;\psi}f(x)=I_{a+}^{\gamma-\alpha;\psi}\Delta_{a+}^{\gamma;\psi}f(x), \tag{1.26}$$

and

$${}^{H}\mathbb{D}_{b-}^{\alpha,\beta;\psi}f(x)=I_{b-}^{\gamma-\alpha;\psi}\Delta_{b-}^{\gamma;\psi}f(x),\quad x\in[a,b]. \tag{1.27}$$

We have that [13]

$$\Delta_{a+}^{\gamma,\psi}f(x)=\left(\frac{1}{\psi'(x)}\frac{d}{dx}\right)^n I_{a+}^{(1-\beta)(n-\alpha);\psi}f(x), \tag{1.28}$$

and

$$\Delta_{b-}^{\gamma,\psi}f(x)=\left(-\frac{1}{\psi'(x)}\frac{d}{dx}\right)^n I_{b-}^{(1-\beta)(n-\alpha);\psi}f(x). \tag{1.29}$$

In particular, when $0<\alpha<1$ and $0\le\beta\le 1$; $\gamma=\alpha+\beta(1-\alpha)$, we have that

$${}^{H}\mathbb{D}_{a+}^{\alpha,\beta;\psi}f(x)=\frac{1}{\Gamma(\gamma-\alpha)}\int_a^x \psi'(t)(\psi(x)-\psi(t))^{\gamma-\alpha-1}\Delta_{a+}^{\gamma;\psi}f(t)\,dt, \tag{1.30}$$

and

$$ {}^{H}\mathbb{D}_{b-}^{\alpha,\beta;\psi} f(x) = \frac{1}{\Gamma(\gamma-\alpha)} \int_x^b \psi'(t)(\psi(t)-\psi(x))^{\gamma-\alpha-1} \Delta_{b-}^{\gamma;\psi} f(t)\, dt, \quad (1.31) $$

$x \in [a, b]$.

We mention

Definition 1.7 For $n-1 < \alpha < n, n \in \mathbb{N}, 0 \leq \beta \leq 1, \gamma = \alpha + \beta(n-\alpha)$ we define the following spaces of functions:

$$ C_{\gamma,\psi}^{a}([a,b]) = \{f : [a,b] \to \mathbb{R} : (\psi(t)-\psi(a))^{\gamma} f(t) \in C([a,b])\}, \quad (1.32) $$

with norm

$$ \|f\|_{C_{\gamma,\psi}^{a}([a,b])} = \|(\psi(t)-\psi(a))^{\gamma} f(t)\|_{C([a,b])}, $$

and

$$ C_{\gamma,\psi}^{b}([a,b]) = \left\{f : [a,b] \to \mathbb{R} : (\psi(b)-\psi(t))^{\gamma} f(t) \in C([a,b])\right\}, \quad (1.33) $$

with norm

$$ \|f\|_{C_{\gamma,\psi}^{b}([a,b])} = \left\|(\psi(b)-\psi(t))^{\gamma} f(t)\right\|_{C([a,b])}, $$

and

$$ C_{\gamma,\psi}^{n,a}([a,b]) = \left\{f : [a,b] \to \mathbb{R} : f \in C^{n-1}([a,b]),\ f^{(n)} \in C_{\gamma;\psi}^{a}([a,b])\right\}, \quad (1.34) $$

with norm

$$ \|f\|_{C_{\gamma,\psi}^{n,a}([a,b])} = \sum_{k=0}^{n-1} \left\|f^{(k)}\right\|_{C([a,b])} + \left\|f^{(n)}\right\|_{C_{\gamma;\psi}^{a}([a,b])}, \quad (1.35) $$

and

$$ C_{\gamma,\psi}^{n,b}([a,b]) = \left\{f : [a,b] \to \mathbb{R} : f \in C^{n-1}([a,b]),\ f^{(n)} \in C_{\gamma;\psi}^{b}([a,b])\right\}, \quad (1.36) $$

with norm

$$ \|f\|_{C_{\gamma,\psi}^{n,b}([a,b])} = \sum_{k=0}^{n-1} \left\|f^{(k)}\right\|_{C([a,b])} + \left\|f^{(n)}\right\|_{C_{\gamma;\psi}^{b}([a,b])}. \quad (1.37) $$

We mention

Theorem 1.8 *(see also [13]) The ψ-Hilfer fractional derivatives are bounded operators for $n-1 < \alpha < n$, $n \in \mathbb{N}$, $0 \leq \beta \leq 1$, $\gamma = \alpha + \beta(n-\alpha)$. We have*

$$\left\| {}^{H}\mathbb{D}_{a+}^{\alpha,\beta;\psi} f \right\|_{C^{a}_{\gamma;\psi}} \leq K \left\| f \right\|_{C^{n,a}_{\gamma;\psi}}, \tag{1.38}$$

and

$$\left\| {}^{H}\mathbb{D}_{b-}^{\alpha,\beta;\psi} f \right\|_{C^{b}_{\gamma;\psi}} \leq K \left\| f \right\|_{C^{n,b}_{\gamma;\psi}}, \tag{1.39}$$

where

$$K = \frac{(\psi(b) - \psi(a))^{n-\alpha}}{\Gamma(n-\gamma+1)\,\Gamma(\gamma-\alpha+1)}.$$

We make

Remark 1.9 Let $\mu = n(1-\beta) + \beta\alpha$. If $\beta = 0$, then $\mu = n$, and if $\beta = 1$, then $\mu = \alpha$.

Let $0 < \beta < 1$, we have $n - 1 < \alpha < n$, then $\beta(n-1) < \beta\alpha < \beta n$ and $\beta(n-1) + n(1-\beta) < \beta\alpha + n(1-\beta) < \beta n + n(1-\beta)$. That is $n - \beta < \mu < n$. Since $\beta < 1$, then $-\beta > -1$ and $n - \beta > n - 1$. Thus $n - 1 < \mu < n$, i.e. $\lceil \mu \rceil = n$.

Assume that $g(x) = I_{a+}^{(1-\beta)(n-\alpha);\psi} f(x) \in C^n([a,b])$, we have that

$${}^{H}\mathbb{D}_{a+}^{\alpha,\beta;\psi} f(x) = I_{a+}^{n-\mu;\psi} \left(\frac{1}{\psi'(x)} \frac{d}{dx} \right)^{n} g(x). \tag{1.40}$$

Thus

$${}^{H}\mathbb{D}_{a+}^{\alpha,\beta;\psi} f = {}^{C}D_{a+}^{\mu;\psi} g(x) = {}^{C}D_{a+}^{\mu;\psi} \left[I_{a+}^{(1-\beta)(n-\alpha);\psi} f(x) \right]. \tag{1.41}$$

Assume that $w(x) = I_{b-}^{(1-\beta)(n-\alpha);\psi} f(x) \in C^n([a,b])$. Hence

$${}^{H}\mathbb{D}_{b-}^{\alpha,\beta;\psi} f(x) = I_{b-}^{\beta(n-\alpha);\psi} \left(-\frac{1}{\psi'(x)} \frac{d}{dx} \right)^{n} w(x) = I_{b-}^{n-\mu;\psi} \left(-\frac{1}{\psi'(x)} \frac{d}{dx} \right)^{n} w(x). \tag{1.42}$$

Thus

$${}^{H}\mathbb{D}_{b-}^{\alpha,\beta;\psi} f = {}^{C}D_{b-}^{\mu;\psi} w(x) = {}^{C}D_{b-}^{\mu;\psi} \left(I_{b-}^{(1-\beta)(n-\alpha);\psi} f(x) \right). \tag{1.43}$$

We also need the following ψ-Hilfer left and right fractional Taylor formulae:

Theorem 1.10 ([13]) *Let $\psi, f \in C^n([a,b])$, with ψ being increasing such that $\psi'(x) \neq 0$ over $[a,b]$, where $n - 1 < \alpha < n$, $0 \leq \beta \leq 1$, and $\gamma = \alpha + \beta(n-\alpha)$, $x \in [a,b]$. Then*

$$I_{a+}^{\alpha;\psi}\, {}^{H}\mathbb{D}_{a+}^{\alpha,\beta;\psi} f(x) = f(x) - \sum_{k=1}^{n} \frac{(\psi(x) - \psi(a))^{\gamma-k}}{\Gamma(\gamma-k+1)} f_{\psi}^{[n-k]} \left(I_{a+}^{(1-\beta)(n-\alpha);\psi} f \right)(a), \tag{1.44}$$

and

$$I_{b-}^{\alpha;\psi}\,{}^{H}\mathbb{D}_{b-}^{\alpha,\beta;\psi}f(x)=f(x)-\sum_{k=1}^{n}\frac{(-1)^{k}(\psi(b)-\psi(x))^{\gamma-k}}{\Gamma(\gamma-k+1)}f_{\psi}^{[n-k]}\left(I_{b-}^{(1-\beta)(n-\alpha);\psi}f\right)(b). \tag{1.45}$$

1.3 Main Results

1.3.1 New Advancements to Hilfer Fractional Calculus

We give the following ψ-Hilfer fractional Taylor formulae.

Theorem 1.11 *Let $f,\psi\in C^{n}([a,b])$, with ψ being increasing, $\psi'(x)\neq 0$ over $[a,b]\subset\mathbb{R}$, $\alpha>0:\lceil\alpha\rceil=n$, $0\leq\beta\leq 1$, $\mu=n(1-\beta)+\beta\alpha$. Assume that $g(x)=I_{a+}^{(1-\beta)(n-\alpha);\psi}f(x)$, $w(x)=I_{b-}^{(1-\beta)(n-\alpha);\psi}f(x)\in C^{n}([a,b])$.*

Then

(1)

$$I_{a+}^{\mu;\psi}\,{}^{H}\mathbb{D}_{a+}^{\alpha,\beta;\psi}f(x)=g(x)-\sum_{k=0}^{n-1}\frac{g_{\psi}^{[k]}(a)}{k!}(\psi(x)-\psi(a))^{k}, \tag{1.46}$$

where

$$g_{\psi}^{[k]}(x)=\left(\frac{1}{\psi'(x)}\frac{d}{dx}\right)^{k}g(x),\quad k=0,1,...,n-1,$$

and

(2)

$$I_{b-}^{\mu;\psi}\,{}^{H}\mathbb{D}_{b-}^{\alpha,\beta;\psi}f(x)=w(x)-\sum_{k=0}^{n-1}\frac{(-1)^{k}w_{\psi}^{[k]}(b)}{k!}(\psi(b)-\psi(x))^{k}, \tag{1.47}$$

where

$$w_{\psi}^{[k]}(x)=\left(\frac{1}{\psi'(x)}\frac{d}{dx}\right)^{k}w(x),\quad k=0,1,...,n-1;\ x\in[a,b].$$

Proof (1) We have $\lceil\mu\rceil=n$. We observe that

$$I_{a+}^{\mu;\psi}\,{}^{H}\mathbb{D}_{a+}^{\alpha,\beta;\psi}f(x)\overset{(1.41)}{=}I_{a+}^{\mu;\psi}\,{}^{C}D_{a+}^{\mu;\psi}g(x)\overset{(1.20)}{=}g(x)-\sum_{k=0}^{n-1}\frac{g_{\psi}^{[k]}(a)}{k!}(\psi(x)-\psi(a))^{k},$$

proving (1.46).

(2) We also have

$$I_{b-}^{\mu;\psi}\ {}^{H}\mathbb{D}_{b-}^{\alpha,\beta;\psi} f(x) \overset{(1.43)}{=} I_{b-}^{\mu;\psi}\ {}^{C}D_{b-}^{\mu;\psi} w(x) \overset{(1.21)}{=} w(x) - \sum_{k=0}^{n-1} (-1)^k \frac{w_{\psi}^{[k]}(b)}{k!} (\psi(b) - \psi(x))^k ,$$

proving (1.47). ■

Next we present some Hilfer fractional derivatives representation formulae.

Theorem 1.12 *Let* $\alpha > 0$, $\alpha \notin \mathbb{N}$, $\lceil \alpha \rceil = n$, $0 < \beta < 1$; $f \in C^n([a,b])$, $[a,b] \subset \mathbb{R}$; *and set* $\gamma = \alpha + \beta(n - \alpha)$. *Assume further that* $\Delta_{a+}^{\gamma} f \in C([a,b]) : \Delta_{a+}^{\gamma-j} f(a) = 0$, *for* $j = 1, ..., n$. *Let also* $\overline{\alpha} > 0 : \lceil \overline{\alpha} \rceil = \overline{n}$, *with* $\overline{\gamma} = \overline{\alpha} + \beta(\overline{n} - \overline{\alpha})$, *and assume that* $\alpha > \overline{\alpha}$ *and* $\gamma > \overline{\gamma}$. *Then*

$$ {}^{H}\mathbb{D}_{a+}^{\overline{\alpha},\beta} f(x) = \frac{1}{\Gamma(\alpha - \overline{\alpha})} \int_a^x (x - t)^{\alpha - \overline{\alpha} - 1}\ {}^{H}\mathbb{D}_{a+}^{\alpha,\beta} f(t)\, dt, \tag{1.48}$$

$\forall\, x \in [a,b]$,

furthermore ${}^{H}\mathbb{D}_{a+}^{\overline{\alpha},\beta} f \in AC([a,b])$ *(absolutely continuous functions) if* $\alpha - \overline{\alpha} \geq 1$ *and* ${}^{H}\mathbb{D}_{a+}^{\overline{\alpha},\beta} f \in C([a,b])$ *if* $\alpha - \overline{\alpha} \in (0,1)$.

Proof We have that $n - 1 < \alpha \leq \alpha + \beta(n - \alpha) < \alpha + (n - \alpha) = n$, that is $\lceil \gamma \rceil = n$ and $\gamma \notin \mathbb{N}$. For $\psi(x) = x$, from (1.26) we get that

$$ {}^{H}\mathbb{D}_{a+}^{\alpha,\beta} f(x) = I_{a+}^{\gamma-\alpha} \Delta_{a+}^{\gamma} f(x), \ x \in [a,b]. \tag{1.49}$$

By Theorem 1.2 and the assumptions we get that

$$f(x) = \frac{1}{\Gamma(\gamma)} \int_a^x (x - t)^{\gamma - 1} \Delta_{a+}^{\gamma} f(t)\, dt = I_{a+}^{\gamma} \Delta_{a+}^{\gamma} f(x), \ \forall\, x \in [a,b]. \tag{1.50}$$

Also by [3], p. 388, we get that ${}^{H}\mathbb{D}_{a+}^{\alpha,\beta} f \in AC([a,b])$ when $\gamma - \alpha \geq 1$, and ${}^{H}\mathbb{D}_{a+}^{\alpha,\beta} f \in C([a,b])$ when $\gamma - \alpha \in (0,1)$.

We also have that

$$\Delta_{a+}^{\overline{\gamma}} f(x) = D^{\overline{n}} I_{a+}^{\overline{n}-\overline{\gamma}} f(x) \overset{(1.50)}{=} D^{\overline{n}} I_{a+}^{\overline{n}-\overline{\gamma}} I_{a+}^{\gamma} \Delta_{a+}^{\gamma} f(x) =$$

$$D^{\overline{n}} I_{a+}^{\gamma-\overline{\gamma}+\overline{n}} \Delta_{a+}^{\gamma} f(x) \text{ (by [3], p. 117)} = I_{a+}^{\gamma-\overline{\gamma}} \Delta_{a+}^{\gamma} f(x).$$

That is

$$\Delta_{a+}^{\overline{\gamma}} f(x) = I_{a+}^{\gamma-\overline{\gamma}} \Delta_{a+}^{\gamma} f(x), \ \ \forall\, x \in [a,b]. \tag{1.51}$$

Furthermore we observe that

$$ {}^{H}\mathbb{D}_{a+}^{\overline{\alpha},\beta} f(x) = I_{a+}^{\overline{\gamma}-\overline{\alpha}} \Delta_{a+}^{\overline{\gamma}} f(x) = I_{a+}^{\overline{\gamma}-\overline{\alpha}} I_{a+}^{\gamma-\overline{\gamma}} \Delta_{a+}^{\gamma} f(x) = I_{a+}^{\gamma-\overline{\alpha}} \Delta_{a+}^{\gamma} f(x) =$$

$$I_{a+}^{\gamma-\alpha+\alpha-\overline{\alpha}} \Delta_{a+}^{\gamma} f(x) = I_{a+}^{\gamma-\alpha} I_{a+}^{\alpha-\overline{\alpha}} \Delta_{a+}^{\gamma} f(x) = I_{a+}^{\alpha-\overline{\alpha}} I_{a+}^{\gamma-\alpha} \Delta_{a+}^{\gamma} f(x) = I_{a+}^{\alpha-\overline{\alpha}}\ {}^{H}\mathbb{D}_{a+}^{\alpha,\beta} f(x).$$

That is

$$^{H}\mathbb{D}_{a+}^{\overline{\alpha},\beta} f(x) = I_{a+}^{\alpha-\overline{\alpha}} \, {}^{H}\mathbb{D}_{a+}^{\alpha,\beta} f(x), \quad \forall\, x \in [a,b]. \tag{1.52}$$

We have proved (1.48). By [3], p. 388, we get that ${}^{H}\mathbb{D}_{a+}^{\overline{\alpha},\beta} f \in AC([a,b])$ when $\alpha - \overline{\alpha} \geq 1$, and ${}^{H}\mathbb{D}_{a+}^{\overline{\alpha},\beta} f \in C([a,b])$ when $\alpha - \overline{\alpha} \in (0,1)$.

The theorem is proved. ■

We continue with

Theorem 1.13 *Let $\alpha > 0$, $\alpha \notin \mathbb{N}$, $\lceil \alpha \rceil = n$, $0 < \beta < 1$; $f \in C^n([a,b])$, $[a,b] \subset \mathbb{R}$; and set $\gamma = \alpha + \beta(n-\alpha)$. Assume further that $\Delta_{b-}^{\gamma} f \in C([a,b]) : \Delta_{b-}^{\gamma-j} f(b) = 0$, $j = 1, ..., n$. Let also $\overline{\alpha} > 0 : \lceil \overline{\alpha} \rceil = \overline{n}$, with $\overline{\gamma} = \overline{\alpha} + \beta(\overline{n} - \overline{\alpha})$, and assume that $\alpha > \overline{\alpha}$ and $\gamma > \overline{\gamma}$. Then*

$$^{H}\mathbb{D}_{b-}^{\overline{\alpha},\beta} f(x) = \frac{1}{\Gamma(\alpha-\overline{\alpha})} \int_x^b (t-x)^{\alpha-\overline{\alpha}-1} \, {}^{H}\mathbb{D}_{b-}^{\alpha,\beta} f(t)\, dt, \tag{1.53}$$

$\forall\, x \in [a,b]$,

furthermore ${}^{H}\mathbb{D}_{b-}^{\overline{\alpha},\beta} f \in AC([a,b])$ if $\alpha - \overline{\alpha} \geq 1$ and ${}^{H}\mathbb{D}_{b-}^{\overline{\alpha},\beta} f \in C([a,b])$ if $\alpha - \overline{\alpha} \in (0,1)$.

Proof Here $\lceil \gamma \rceil = n$, $\gamma \notin \mathbb{N}$, and $\lceil \overline{\gamma} \rceil = \overline{n}$. For $\psi(x) = x$, by (1.27) we get

$$^{H}\mathbb{D}_{b-}^{\alpha,\beta} f(x) = I_{b-}^{\gamma-\alpha} \Delta_{b-}^{\gamma} f(x), \quad \forall\, x \in [a,b]. \tag{1.54}$$

By Theorem 1.3 we obtain

$$f(x) = \frac{1}{\Gamma(\gamma)} \int_x^b (t-x)^{\gamma-1} \Delta_{b-}^{\gamma} f(t)\, dt = I_{b-}^{\gamma} \Delta_{b-}^{\gamma} f(x), \quad \forall\, x \in [a,b].$$

That is

$$f(x) = I_{b-}^{\gamma} \Delta_{b-}^{\gamma} f(x), \quad \forall\, x \in [a,b]. \tag{1.55}$$

Also by Anastassiou [6], p. 101, we get that ${}^{H}\mathbb{D}_{b-}^{\alpha,\beta} f \in AC([a,b])$ when $\gamma - \alpha \geq 1$, and ${}^{H}\mathbb{D}_{b-}^{\alpha,\beta} f \in C([a,b])$ when $\gamma - \alpha \in (0,1)$.

We observe that

$$\Delta_{b-}^{\overline{\gamma}} f(x) = (-1)^{\overline{n}} D^{\overline{n}} I_{b-}^{\overline{n}-\overline{\gamma}} f(x) \overset{(1.55)}{=} (-1)^{\overline{n}} D^{\overline{n}} I_{b-}^{\overline{n}-\overline{\gamma}} I_{b-}^{\gamma} \Delta_{b-}^{\gamma} f(x) =$$

$$(-1)^{\overline{n}} D^{\overline{n}} I_{b-}^{\gamma-\overline{\gamma}+\overline{n}} \Delta_{b-}^{\gamma} f(x) \text{ (by [4], p. 6)} = (-1)^{\overline{n}} (-1)^{\overline{n}} I_{b-}^{\gamma-\overline{\gamma}} \Delta_{b-}^{\gamma} f(x)$$

$$= I_{b-}^{\gamma-\overline{\gamma}} \Delta_{b-}^{\gamma} f(x).$$

That is

$$\Delta_{b-}^{\overline{\gamma}} f(x) = I_{b-}^{\gamma-\overline{\gamma}} \Delta_{b-}^{\gamma} f(x), \quad \forall x \in [a, b]. \tag{1.56}$$

Furthermore we have

$$^{H}\mathbb{D}_{b-}^{\overline{\alpha},\beta} f(x) = I_{b-}^{\overline{\gamma}-\overline{\alpha}} \Delta_{b-}^{\overline{\gamma}} f(x) = I_{b-}^{\overline{\gamma}-\overline{\alpha}} I_{b-}^{\gamma-\overline{\gamma}} \Delta_{b-}^{\gamma} f(x) = I_{b-}^{\gamma-\overline{\alpha}} \Delta_{b-}^{\gamma} f(x) =$$

$$I_{b-}^{\gamma-\alpha+\alpha-\overline{\alpha}} \Delta_{b-}^{\gamma} f(x) = I_{b-}^{\gamma-\alpha} I_{b-}^{\alpha-\overline{\alpha}} \Delta_{b-}^{\gamma} f(x) = I_{b-}^{\alpha-\overline{\alpha}} I_{b-}^{\gamma-\alpha} \Delta_{b-}^{\gamma} f(x) = I_{b-}^{\alpha-\overline{\alpha}} \, {}^{H}\mathbb{D}_{b-}^{\alpha,\beta} f(x).$$

We have proved that

$$^{H}\mathbb{D}_{b-}^{\overline{\alpha},\beta} f(x) = I_{b-}^{\alpha-\overline{\alpha}} \, {}^{H}\mathbb{D}_{b-}^{\alpha,\beta} f(x), \quad \forall x \in [a, b], \tag{1.57}$$

which is establishing (1.53).

By Anastassiou [6], p. 101, we get that ${}^{H}\mathbb{D}_{b-}^{\overline{\alpha},\beta} f \in AC([a, b])$ when $\alpha - \overline{\alpha} \geq 1$, and ${}^{H}\mathbb{D}_{b-}^{\overline{\alpha},\beta} f \in C([a, b])$ when $\alpha - \overline{\alpha} \in (0, 1)$.

The proof of the theorem is complete. ∎

We continue with two extreme cases.

Theorem 1.14 *When $\beta = 0$, Theorems 1.12, 1.13 are still valid. That is*

$$\Delta_{a+}^{\overline{\alpha}} f(x) = I_{a+}^{\alpha-\overline{\alpha}} \Delta_{a+}^{\alpha} f(x), \quad \Delta_{b-}^{\overline{\alpha}} f(x) = I_{b-}^{\alpha-\overline{\alpha}} \Delta_{b-}^{\alpha} f(x), \tag{1.58}$$

all $x \in [a, b]$, respectively.

Proof Let $\beta = 0$, then $\gamma = \alpha$, and

$$^{H}\mathbb{D}_{a+}^{\alpha,0} f(x) = \Delta_{a+}^{\alpha} f(x), \tag{1.59}$$

and

$$^{H}\mathbb{D}_{b-}^{\alpha,0} f(x) = \Delta_{b-}^{\alpha} f(x). \tag{1.60}$$

We assume $\Delta_{a+}^{\alpha} f, \Delta_{b-}^{\alpha} f \in C([a, b])$.

(1) Assume further that $\Delta_{a+}^{\alpha-j} f(a) = 0, j = 1, ..., n$. Hence $f(x) = I_{a+}^{\alpha} \Delta_{a+}^{\alpha} f(x)$, all $x \in [a, b]$. Then

$$\Delta_{a+}^{\overline{\alpha}} f(x) = D^{\overline{n}} I_{a+}^{\overline{n}-\overline{\alpha}} f(x) = D^{\overline{n}} I_{a+}^{\overline{n}-\overline{\alpha}} I_{a+}^{\alpha} \Delta_{a+}^{\alpha} f(x) =$$

$$D^{\overline{n}} I_{a+}^{\alpha-\overline{\alpha}+\overline{n}} \Delta_{a+}^{\alpha} f(x) = I_{a+}^{\alpha-\overline{\alpha}} \Delta_{a+}^{\alpha} f(x).$$

That is

$$\Delta_{a+}^{\overline{\alpha}} f(x) = I_{a+}^{\alpha-\overline{\alpha}} \Delta_{a+}^{\alpha} f(x), \quad x \in [a, b]. \tag{1.61}$$

(2) Assume further that $\Delta_{b-}^{\alpha-j} f(b) = 0, j = 1, ..., n$. Then $f(x) = I_{b-}^{\alpha} \Delta_{b-}^{\alpha} f(x)$, all $x \in [a, b]$. Furthermore it holds

$$\Delta_{b-}^{\overline{\alpha}}f(x) = (-1)^{\overline{n}} D^{\overline{n}} I_{b-}^{\overline{n}-\overline{\alpha}} f(x) = (-1)^{\overline{n}} D^{\overline{n}} I_{b-}^{\overline{n}-\overline{\alpha}} I_{b-}^{\alpha} \Delta_{b-}^{\alpha} f(x) =$$

$$(-1)^{\overline{n}} D^{\overline{n}} I_{b-}^{\alpha-\overline{\alpha}+\overline{n}} \Delta_{b-}^{\alpha} f(x) = (-1)^{\overline{n}} (-1)^{\overline{n}} I_{b-}^{\alpha-\overline{\alpha}} \Delta_{b-}^{\alpha} f(x) = I_{b-}^{\alpha-\overline{\alpha}} \Delta_{b-}^{\alpha} f(x).$$

That is

$$\Delta_{b-}^{\overline{\alpha}} f(x) = I_{b-}^{\alpha-\overline{\alpha}} \Delta_{b-}^{\alpha} f(x), \quad \text{all } x \in [a, b]. \tag{1.62}$$

The claim is proved. ■

It follows the case $\beta = 1$.

Proposition 1.15 *Let $\alpha > 0$, $\alpha \notin \mathbb{N}$, $\lceil \alpha \rceil = n$, $f \in C^n([a, b])$, $[a, b] \subset \mathbb{R}$. Assume further that $f^{(k)}(a) = 0$, $k = 0, 1, ..., n-1$. Let also $\overline{\alpha} > 0 : \lceil \overline{\alpha} \rceil = \overline{n}$ and $\alpha > \overline{n}$. Then*

$$D_{*a}^{\overline{\alpha}} f(x) = \frac{1}{\Gamma(\alpha - \overline{\alpha})} \int_a^x (x-t)^{\alpha-\overline{\alpha}-1} D_{*a}^{\alpha} f(t)\, dt, \tag{1.63}$$

$\forall\, x \in [a, b]$.

If $\alpha - \overline{\alpha} \geq 1$ then $D_{*a}^{\overline{\alpha}} f \in AC([a, b])$, and if $\alpha - \overline{\alpha} \in (0, 1)$ then $D_{*a}^{\overline{\alpha}} f \in C([a, b])$.

Proof When $\beta = 1$, then $\gamma = n$, and ${}^{H}\mathbb{D}_{a+}^{\alpha,1} f(x) = I_{a+}^{n-\alpha} f^{(n)}(x) = D_{*a}^{\alpha} f(x) \in C([a, b])$, the left Caputo fractional derivative. From [3], p. 395, we have

$$f(x) = \sum_{k=0}^{n-1} \frac{f^{(k)}(a)}{k!} (x-a)^k + I_{a+}^{\alpha} D_{*a}^{\alpha} f(x), \tag{1.64}$$

$\forall\, x \in [a, b]$.

Since $f^{(k)}(a) = 0$, $k = 0, 1, ..., n-1$, we get that

$$f(x) = I_{a+}^{\alpha} D_{*a}^{\alpha} f(x), \quad \forall\, x \in [a, b]. \tag{1.65}$$

Furthermore we observe that

$$D_{*a}^{\overline{\alpha}} f(x) = I_{a+}^{\overline{n}-\overline{\alpha}} f^{(\overline{n})}(x) = I_{a+}^{\overline{n}-\overline{\alpha}} I_{a+}^{\alpha-\overline{n}} D_{*a}^{\alpha} f(x) = I_{a+}^{\alpha-\overline{\alpha}} D_{*a}^{\alpha} f(x), \tag{1.66}$$

proving the claim. ■

Proposition 1.16 *Let $\alpha > 0$, $\alpha \notin \mathbb{N}$, $\lceil \alpha \rceil = n$, $f \in C^n([a, b])$, $[a, b] \subset \mathbb{R}$. Assume further that $f^{(k)}(b) = 0$, $k = 0, 1, ..., n-1$. Let also $\overline{\alpha} > 0 : \lceil \overline{\alpha} \rceil = \overline{n}$ and $\alpha > \overline{n}$. Then*

$$D_{b-}^{\overline{\alpha}} f(x) = \frac{1}{\Gamma(\alpha - \overline{\alpha})} \int_x^b (t-x)^{\alpha-\overline{\alpha}-1} D_{b-}^{\alpha} f(t)\, dt, \tag{1.67}$$

$\forall\, x \in [a, b]$.

If $\alpha - \overline{\alpha} \geq 1$ then $D_{b-}^{\overline{\alpha}} f \in AC([a, b])$, and if $\alpha - \overline{\alpha} \in (0, 1)$ then $D_{b-}^{\overline{\alpha}} f \in C([a, b])$.

Proof When $\beta = 1$, then $\gamma = n$, and

$$ {}^{H}\mathbb{D}_{b-}^{\alpha,1} f(x) = (-1)^n I_{b-}^{n-\alpha} f^{(n)}(x) = D_{b-}^{\alpha} f(x) \in C([a,b]), \quad \forall\, x \in [a,b], \tag{1.68} $$

the right Caputo fractional derivative. From [5], p. 341, we have

$$ f(x) = \sum_{k=0}^{n-1} \frac{f^{(k)}(b)}{k!} (x-b)^k + I_{b-}^{\alpha} D_{b-}^{\alpha} f(x), \tag{1.69} $$

$\forall\, x \in [a,b]$.

Since $f^{(k)}(b) = 0$, $k = 0, 1, \ldots, n-1$, we get that

$$ f(x) = I_{b-}^{\alpha} D_{b-}^{\alpha} f(x), \quad \forall\, x \in [a,b]. $$

Furthermore we observe that

$$ D_{b-}^{\overline{\alpha}} f(x) = (-1)^{\overline{n}} I_{b-}^{\overline{n}-\overline{\alpha}} f^{(\overline{n})}(x) = (-1)^{\overline{n}} (-1)^{\overline{n}} I_{b-}^{\overline{n}-\overline{\alpha}} I_{b-}^{\alpha-\overline{n}} D_{b-}^{\alpha} f(x) = I_{b-}^{\alpha-\overline{\alpha}} D_{b-}^{\alpha} f(x), \tag{1.70} $$

proving the claim. ■

Corollary 1.17 (to Theorem 1.10) *All as in Theorem 1.10.*

(1) If additionally $f_{\psi}^{[k]}\left(I_{a+}^{(1-\beta)(n-\alpha);\psi} f\right)(a) = 0$, $k = 0, 1, \ldots, n-1$, *then we have*

$$ f(x) = I_{a+}^{\alpha;\psi}\, {}^{H}\mathbb{D}_{a+}^{\alpha,\beta;\psi} f(x), \quad \forall\, x \in [a,b], \tag{1.71} $$

and

(2) if additionally $f_{\psi}^{[k]}\left(I_{b-}^{(1-\beta)(n-\alpha);\psi} f\right)(b) = 0$, $k = 0, 1, \ldots, n-1$, *then we have*

$$ f(x) = I_{b-}^{\alpha;\psi}\, {}^{H}\mathbb{D}_{b-}^{\alpha,\beta;\psi} f(x), \quad \forall\, x \in [a,b]. \tag{1.72} $$

Next we deal with case $\psi(x) = x$.

Corollary 1.18 (to Theorem 1.10) *Let* $f \in C^n([a,b])$, $\alpha > 0$, $\alpha \notin \mathbb{N}$ *with* $\lceil \alpha \rceil = n$, $0 \le \beta \le 1$ *and* $\gamma = \alpha + \beta(n-\alpha)$, $x \in [a,b]$.

(1) If additionally $\left(I_{a+}^{(1-\beta)(n-\alpha)} f\right)^{(k)}(a) = 0$, $k = 0, 1, \ldots, n-1$, *then*

$$ f(x) = I_{a+}^{\alpha}\, {}^{H}\mathbb{D}_{a+}^{\alpha,\beta} f(x), \quad \forall\, x \in [a,b], \tag{1.73} $$

and

(2) if additionally $\left(I_{b-}^{(1-\beta)(n-\alpha)} f\right)^{(k)}(b) = 0$, $k = 0, 1, \ldots, n-1$, *then*

$$ f(x) = I_{b-}^{\alpha}\, {}^{H}\mathbb{D}_{b-}^{\alpha,\beta} f(x), \quad \forall\, x \in [a,b]. \tag{1.74} $$

1.3.2 Basic Integral Inequalities Involving Hilfer Fractional Calculus

We start with a left ψ-Hilfer fractional Ostrowski type inequality.

Theorem 1.19 *Here all are as in Theorem 1.11, with* $g_{\psi}^{[k]}(a) = 0$, $k = 1, ..., n-1$. *Then*

$$\left| \frac{1}{\psi(b) - \psi(a)} \int_a^b \left(I_{a+}^{(1-\beta)(n-\alpha);\psi} f(x) \right) \psi'(x)\, dx - I_{a+}^{(1-\beta)(n-\alpha);\psi} f(a) \right|$$

$$\leq \frac{\left\| {}^H\mathbb{D}_{a+}^{\alpha,\beta;\psi} f \right\|_{\infty,[a,b]}}{\Gamma(\mu+2)} (\psi(b) - \psi(a))^{\mu}. \tag{1.75}$$

Proof By (1.46) we can write

$$g(x) = I_{a+}^{(1-\beta)(n-\alpha);\psi} f(x) = \sum_{k=0}^{n-1} \frac{g_{\psi}^{[k]}(a)}{k!} (\psi(x) - \psi(a))^k + \tag{1.76}$$

$$\frac{1}{\Gamma(\mu)} \int_a^x \psi'(t) (\psi(x) - \psi(t))^{\mu-1}\, {}^H\mathbb{D}_{a+}^{\alpha,\beta;\psi} f(t)\, dt,$$

$\forall\, x \in [a,b]$. Here it holds $\lceil \mu \rceil = n$.

By the assumption $g_{\psi}^{[k]}(a) = 0$, $k = 1, ..., n-1$, we have that

$$g(x) - g(a) = \frac{1}{\Gamma(\mu)} \int_a^x \psi'(t) (\psi(x) - \psi(t))^{\mu-1}\, {}^H\mathbb{D}_{a+}^{\alpha,\beta;\psi} f(t)\, dt, \tag{1.77}$$

$\forall\, x \in [a,b]$.

Hence we derive

$$|g(x) - g(a)| \leq \frac{1}{\Gamma(\mu)} \int_a^x \psi'(t) (\psi(x) - \psi(t))^{\mu-1} \left| {}^H\mathbb{D}_{a+}^{\alpha,\beta;\psi} f(t) \right| dt \leq$$

$$\left\| {}^H\mathbb{D}_{a+}^{\alpha,\beta;\psi} f \right\|_{\infty,[a,b]} \frac{1}{\Gamma(\mu)} \int_a^x \psi'(t) (\psi(x) - \psi(t))^{\mu-1}\, dt =$$

$$\frac{\left\| {}^H\mathbb{D}_{a+}^{\alpha,\beta;\psi} f \right\|_{\infty,[a,b]}}{\Gamma(\mu+1)} (\psi(x) - \psi(a))^{\mu}. \tag{1.78}$$

That is

$$|g(x) - g(a)| \leq \frac{\left\| {}^H\mathbb{D}_{a+}^{\alpha,\beta;\psi} f \right\|_{\infty,[a,b]}}{\Gamma(\mu+1)} (\psi(x) - \psi(a))^{\mu}, \tag{1.79}$$

$\forall\, x \in [a, b]$.

Next we observe that

$$\left|\frac{1}{\psi(b)-\psi(a)}\int_a^b g(x)\,\psi'(x)\,dx - g(a)\right| =$$

$$\left|\frac{1}{\psi(b)-\psi(a)}\left[\int_a^b g(x)\,\psi'(x)\,dx - g(a)\,(\psi(b)-\psi(a))\right]\right| =$$

$$\left|\frac{1}{\psi(b)-\psi(a)}\left[\int_a^b g(x)\,\psi'(x)\,dx - \int_a^b g(a)\,\psi'(x)\,dx\right]\right| = \tag{1.80}$$

$$\left|\frac{1}{\psi(b)-\psi(a)}\left[\int_a^b (g(x)-g(a))\,\psi'(x)\,dx\right]\right| \leq$$

$$\frac{1}{\psi(b)-\psi(a)}\int_a^b |g(x)-g(a)|\,\psi'(x)\,dx \leq$$

$$\frac{\left\| {}^{H}\mathbb{D}_{a+}^{\alpha,\beta;\psi} f\right\|_{\infty,[a,b]}}{(\psi(b)-\psi(a))\,\Gamma(\mu+1)}\left(\int_a^b (\psi(x)-\psi(a))^{\mu}\,\psi'(x)\,dx\right) =$$

$$\frac{\left\| {}^{H}\mathbb{D}_{a+}^{\alpha,\beta;\psi} f\right\|_{\infty,[a,b]}}{(\psi(b)-\psi(a))\,\Gamma(\mu+2)}(\psi(b)-\psi(a))^{\mu+1} = \frac{\left\| {}^{H}\mathbb{D}_{a+}^{\alpha,\beta;\psi} f\right\|_{\infty,[a,b]}}{\Gamma(\mu+2)}(\psi(b)-\psi(a))^{\mu}, \tag{1.81}$$

proving the claim. ■

We continue with a right similar inequality.

Theorem 1.20 *Here again all as in Theorem 1.11, with* $w_{\psi}^{[k]}(b) = 0,\ k = 1, ..., n-1$. *Then*

$$\left|\frac{1}{\psi(b)-\psi(a)}\int_a^b \left(I_{b-}^{(1-\beta)(n-\alpha);\psi} f(x)\right)\psi'(x)\,dx - I_{b-}^{(1-\beta)(n-\alpha);\psi} f(b)\right|$$

$$\leq \frac{\left\| {}^{H}\mathbb{D}_{b-}^{\alpha,\beta;\psi} f\right\|_{\infty,[a,b]}}{\Gamma(\mu+2)}(\psi(b)-\psi(a))^{\mu}. \tag{1.82}$$

Proof From (1.47) we obtain

$$w(x) = I_{b-}^{(1-\beta)(n-\alpha);\psi} f(x) = \sum_{k=0}^{n-1}\frac{(-1)^k\, w_{\psi}^{[k]}(b)}{k!}(\psi(b)-\psi(x))^k + \tag{1.83}$$

$$\frac{1}{\Gamma(\mu)} \int_x^b \psi'(t) (\psi(t) - \psi(x))^{\mu-1} \, {}^H\mathbb{D}_{b-}^{\alpha,\beta;\psi} f(t) \, dt,$$

$\forall x \in [a, b]$.

Since we assumed $w_{\psi}^{[k]}(b) = 0$, $k = 1, ..., n-1$, we have that

$$w(x) - w(b) = \frac{1}{\Gamma(\mu)} \int_x^b \psi'(t) (\psi(t) - \psi(x))^{\mu-1} \, {}^H\mathbb{D}_{b-}^{\alpha,\beta;\psi} f(t) \, dt, \tag{1.84}$$

$\forall x \in [a, b]$.

Hence it holds

$$|w(x) - w(b)| \leq \frac{1}{\Gamma(\mu)} \int_x^b \psi'(t) (\psi(t) - \psi(x))^{\mu-1} \left| {}^H\mathbb{D}_{b-}^{\alpha,\beta;\psi} f(t) \right| dt \leq$$

$$\left\| {}^H\mathbb{D}_{b-}^{\alpha,\beta;\psi} f \right\|_{\infty,[a,b]} \frac{1}{\Gamma(\mu)} \int_x^b \psi'(t) (\psi(t) - \psi(x))^{\mu-1} \, dt =$$

$$\frac{\left\| {}^H\mathbb{D}_{b-}^{\alpha,\beta;\psi} f \right\|_{\infty,[a,b]}}{\Gamma(\mu+1)} (\psi(b) - \psi(x))^{\mu}. \tag{1.85}$$

That is

$$|w(x) - w(b)| \leq \frac{\left\| {}^H\mathbb{D}_{b-}^{\alpha,\beta;\psi} f \right\|_{\infty,[a,b]}}{\Gamma(\mu+1)} (\psi(b) - \psi(x))^{\mu}, \tag{1.86}$$

$\forall x \in [a, b]$.

We observe that

$$\left| \frac{1}{\psi(b) - \psi(a)} \int_a^b w(x) \psi'(x) \, dx \quad w(b) \right| =$$

$$\left| \frac{1}{\psi(b) - \psi(a)} \left[\int_a^b w(x) \psi'(x) \, dx - w(b) (\psi(b) - \psi(a)) \right] \right| =$$

$$\left| \frac{1}{\psi(b) - \psi(a)} \left[\int_a^b w(x) \psi'(x) \, dx - \int_a^b w(b) \psi'(x) \, dx \right] \right| =$$

$$\left| \frac{1}{\psi(b) - \psi(a)} \left[\int_a^b (w(x) - w(b)) \psi'(x) \, dx \right] \right| \leq$$

$$\frac{1}{\psi(b) - \psi(a)} \int_a^b |w(x) - w(b)| \psi'(x) \, dx \leq \tag{1.87}$$

$$\frac{\left\| {}^{H}\mathbb{D}_{b-}^{\alpha,\beta;\psi} f \right\|_{\infty,[a,b]}}{(\psi(b) - \psi(a))\,\Gamma(\mu+1)} \int_a^b (\psi(b) - \psi(x))^{\mu}\, \psi'(x)\, dx =$$

$$\frac{\left\| {}^{H}\mathbb{D}_{b-}^{\alpha,\beta;\psi} f \right\|_{\infty,[a,b]}}{(\psi(b) - \psi(a))\,\Gamma(\mu+2)} (\psi(b) - \psi(a))^{\mu+1} = \frac{\left\| {}^{H}\mathbb{D}_{b-}^{\alpha,\beta;\psi} f \right\|_{\infty,[a,b]}}{\Gamma(\mu+2)} (\psi(b) - \psi(a))^{\mu}, \tag{1.88}$$

proving the claim. ∎

We continue with a special case

Theorem 1.21 *All as in Theorem 1.11, with* $w_2 = I_{x_0-}^{(1-\beta)(n-\alpha);\psi} f \in C^n([a, x_0])$ *and* $g_2 = I_{x_0+}^{(1-\beta)(n-\alpha);\psi} f \in C^n([x_0, b])$*, instead of* $g, w \in C^n([a, b])$*. Assume also* $g_{2\psi}^{[k]}(x_0) = 0$*,* $w_{2\psi}^{[k]}(x_0) = 0$*,* $k = 1, ..., n-1$*, where* $x_0 \in [a, b] : \psi(x_0) = \frac{\psi(a)+\psi(b)}{2}$*. Then*

$$\Bigg| \frac{2}{\psi(b) - \psi(a)} \left[\int_a^{x_0} \left(I_{x_0-}^{(1-\beta)(n-\alpha);\psi} f(x) \right) \psi'(x)\, dx + \int_{x_0}^b \left(I_{x_0+}^{(1-\beta)(n-\alpha);\psi} f(x) \right) \psi'(x)\, dx \right]$$

$$- \left[I_{x_0-}^{(1-\beta)(n-\alpha);\psi} f(x_0) + I_{x_0+}^{(1-\beta)(n-\alpha);\psi} f(x_0) \right] \Bigg| \leq$$

$$\frac{\left(\left\| {}^{H}\mathbb{D}_{x_0-}^{\alpha,\beta;\psi} f \right\|_{\infty,[a,x_0]} + \left\| {}^{H}\mathbb{D}_{x_0+}^{\alpha,\beta;\psi} f \right\|_{\infty,[x_0,b]} \right)}{\Gamma(\mu+2)} \left(\frac{\psi(b) - \psi(a)}{2} \right)^{\mu}. \tag{1.89}$$

Proof We can choose $x_0 \in [a, b] : \psi(x_0) - \psi(a) = \psi(b) - \psi(x_0)$, iff $\psi(x_0) = \frac{\psi(a)+\psi(b)}{2}$.

Of course $\psi(b) - \psi(x_0) = \psi(x_0) - \psi(a) = \frac{\psi(b)-\psi(a)}{2}$.

Inequality (1.89) comes by triangle inequality and (1.90), (1.91) next.

By (1.75) for $a = x_0$ we can write

$$\left| \frac{1}{\psi(b) - \psi(x_0)} \int_{x_0}^b \left(I_{x_0+}^{(1-\beta)(n-\alpha);\psi} f(x) \right) \psi'(x)\, dx - I_{x_0+}^{(1-\beta)(n-\alpha);\psi} f(x_0) \right|$$

$$\leq \frac{\left\| {}^{H}\mathbb{D}_{x_0+}^{\alpha,\beta;\psi} f \right\|_{\infty,[x_0,b]}}{\Gamma(\mu+2)} (\psi(b) - \psi(x_0))^{\mu}. \tag{1.90}$$

By (1.82) for $b = x_0$ we can write

$$\left| \frac{1}{\psi(x_0) - \psi(a)} \int_a^{x_0} \left(I_{x_0-}^{(1-\beta)(n-\alpha);\psi} f(x) \right) \psi'(x)\, dx - I_{x_0-}^{(1-\beta)(n-\alpha);\psi} f(x_0) \right|$$

$$\leq \frac{\left\| {}^{H}\mathbb{D}_{x_0-}^{\alpha,\beta;\psi} f \right\|_{\infty,[a,x_0]}}{\Gamma(\mu+2)} (\psi(x_0) - \psi(a))^{\mu}. \tag{1.91}$$

■

Corollary 1.22 *All as in Theorem 1.21 with $x_0 \in [a, b]$ being arbitrary. Then*

$$\left\| \left[\frac{1}{\psi(x_0) - \psi(a)} \int_a^{x_0} \left(I_{x_0-}^{(1-\beta)(n-\alpha);\psi} f(x) \right) \psi'(x)\, dx + \right. \right.$$

$$\left. \frac{1}{\psi(b) - \psi(x_0)} \int_{x_0}^{b} \left(I_{x_0+}^{(1-\beta)(n-\alpha);\psi} f(x) \right) \psi'(x)\, dx \right] -$$

$$\left. \left[I_{x_0-}^{(1-\beta)(n-\alpha);\psi} f(x_0) + I_{x_0+}^{(1-\beta)(n-\alpha);\psi} f(x_0) \right] \right\| \leq \frac{1}{\Gamma(\mu+2)}$$

$$\left[\left\| {}^{H}\mathbb{D}_{x_0-}^{\alpha,\beta;\psi} f \right\|_{\infty,[a,x_0]} (\psi(x_0) - \psi(a))^{\mu} + \left\| {}^{H}\mathbb{D}_{x_0+}^{\alpha,\beta;\psi} f \right\|_{\infty,[x_0,b]} (\psi(b) - \psi(x_0))^{\mu} \right]. \tag{1.92}$$

We can prove L_1, L_p versions of the above Ostrowski inequalities but we omit this similar task, see [3]. Of interest will be also to write down such inequalities for various $\psi(x)$, such as $\psi(x) = e^x$, $\ln x$, etc., we skip that too.

Next we give a series of Hilfer fractional left and right Opial type inequalities (see [3, 9]).

Theorem 1.23 *Let all as in Theorem 1.12, $0 \leq \beta < 1$. Let also $p, q > 1 : \frac{1}{p} + \frac{1}{q} = 1$, and assume that $\alpha - \overline{\alpha} > \frac{1}{q}$. Then*

$$\int_a^x \left| {}^{H}\mathbb{D}_{a+}^{\overline{\alpha},\beta} f(w) \right| \left| {}^{H}\mathbb{D}_{a+}^{\alpha,\beta} f(w) \right| dw \leq$$

$$\frac{2^{-\frac{1}{q}} (x-a)^{(\alpha-\overline{\alpha})+\left(\frac{1}{p}-\frac{1}{q}\right)}}{\Gamma(\alpha-\overline{\alpha}) \left[(p(\alpha-\overline{\alpha}-1)+1)(p(\alpha-\overline{\alpha}-1)+2) \right]^{\frac{1}{p}}} \left(\int_a^x \left| {}^{H}\mathbb{D}_{a+}^{\alpha,\beta} f(w) \right|^q dw \right)^{\frac{2}{q}}, \tag{1.93}$$

$\forall\, x \in [a, b]$.

Proof By (1.48) we have

$$\left| {}^{H}\mathbb{D}_{a+}^{\overline{\alpha},\beta} f(x) \right| \leq \frac{1}{\Gamma(\alpha-\overline{\alpha})} \int_a^x (x-t)^{\alpha-\overline{\alpha}-1} \left| {}^{H}\mathbb{D}_{a+}^{\alpha,\beta} f(t) \right| dt \leq$$

$$\frac{1}{\Gamma(\alpha-\overline{\alpha})} \left(\int_a^x (x-t)^{(\alpha-\overline{\alpha}-1)p} dt \right)^{\frac{1}{p}} \left(\int_a^x \left| {}^{H}\mathbb{D}_{a+}^{\alpha,\beta} f(t) \right|^q dt \right)^{\frac{1}{q}} = \tag{1.94}$$

$$\frac{1}{\Gamma(\alpha-\overline{\alpha})}\frac{(x-a)^{\frac{(p(\alpha-\overline{\alpha}-1)+1)}{p}}}{(p(\alpha-\overline{\alpha}-1)+1)^{\frac{1}{p}}}(z(x))^{\frac{1}{q}}.$$

Here we set

$$z(x)=\int_a^x\left|{}^H\mathbb{D}_{a+}^{\alpha,\beta}f(t)\right|^q dt\geq 0, \tag{1.95}$$

with $z(a)=0$.

Thus

$$z'(x)=\left|{}^H\mathbb{D}_{a+}^{\alpha,\beta}f(x)\right|^q, \tag{1.96}$$

and

$$\left(z'(x)\right)^{\frac{1}{q}}=\left|{}^H\mathbb{D}_{a+}^{\alpha,\beta}f(x)\right|\geq 0,\ \ \forall\, x\in[a,b].$$

Hence it holds

$$\left|{}^H\mathbb{D}_{a+}^{\overline{\alpha},\beta}f(w)\right|\left|{}^H\mathbb{D}_{a+}^{\alpha,\beta}f(w)\right|\leq\frac{1}{\Gamma(\alpha-\overline{\alpha})}\frac{(w-a)^{\frac{(p(\alpha-\overline{\alpha}-1)+1)}{p}}}{(p(\alpha-\overline{\alpha}-1)+1)^{\frac{1}{p}}}\left(z'(w)\,z(w)\right)^{\frac{1}{q}}, \tag{1.97}$$

$\forall\, w\in[a,b]$.

Therefore we obtain

$$\int_a^x\left|{}^H\mathbb{D}_{a+}^{\overline{\alpha},\beta}f(w)\right|\left|{}^H\mathbb{D}_{a+}^{\alpha,\beta}f(w)\right|dw\leq$$

$$\frac{1}{\Gamma(\alpha-\overline{\alpha})(p(\alpha-\overline{\alpha}-1)+1)^{\frac{1}{p}}}\int_a^x(w-a)^{\frac{(p(\alpha-\overline{\alpha}-1)+1)}{p}}\left(z'(w)\,z(w)\right)^{\frac{1}{q}}dw\leq$$

$$\frac{1}{\Gamma(\alpha-\overline{\alpha})(p(\alpha-\overline{\alpha}-1)+1)^{\frac{1}{p}}}\left(\int_a^x(w-a)^{(p(\alpha-\overline{\alpha}-1)+1)}dw\right)^{\frac{1}{p}}\left(\int_a^x z(w)\,dz(w)\right)^{\frac{1}{q}}=$$

$$\frac{1}{\Gamma(\alpha-\overline{\alpha})(p(\alpha-\overline{\alpha}-1)+1)^{\frac{1}{p}}}\frac{(x-a)^{\frac{(p(\alpha-\overline{\alpha}-1)+2)}{p}}}{(p(\alpha-\overline{\alpha}-1)+2)^{\frac{1}{p}}}\left(\frac{z^2(x)}{2}\right)^{\frac{1}{q}}=$$

$$\frac{2^{-\frac{1}{q}}(x-a)^{(\alpha-\overline{\alpha}-1)+\frac{2}{p}}}{\Gamma(\alpha-\overline{\alpha})\left[(p(\alpha-\overline{\alpha}-1)+1)(p(\alpha-\overline{\alpha}-1)+2)\right]^{\frac{1}{p}}}\left(\int_a^x\left|{}^H\mathbb{D}_{a+}^{\alpha,\beta}f(w)\right|^q dw\right)^{\frac{2}{q}},$$

proving the claim. ■

We continue with

Theorem 1.24 *Let all as in Theorem 1.13,* $0\leq\beta<1$. *Let also* $p,q>1:\frac{1}{p}+\frac{1}{q}=1$, *and assume that* $\alpha-\overline{\alpha}>\frac{1}{q}$. *Then*

$$\int_x^b \left| {}^H\mathbb{D}_{b-}^{\overline{\alpha},\beta} f(w) \right| \left| {}^H\mathbb{D}_{b-}^{\alpha,\beta} f(w) \right| dw \leq$$

$$\frac{2^{-\frac{1}{q}} (b-x)^{(\alpha-\overline{\alpha})+\left(\frac{1}{p}-\frac{1}{q}\right)}}{\Gamma(\alpha-\overline{\alpha}) \left[(p(\alpha-\overline{\alpha}-1)+1)(p(\alpha-\overline{\alpha}-1)+2) \right]^{\frac{1}{p}}} \left(\int_x^b \left| {}^H\mathbb{D}_{b-}^{\alpha,\beta} f(w) \right|^q dw \right)^{\frac{2}{q}}, \quad (1.98)$$

$\forall\, x \in [a,b]$.

Proof By (1.53) we have

$$\left| {}^H\mathbb{D}_{b-}^{\overline{\alpha},\beta} f(x) \right| \leq \frac{1}{\Gamma(\alpha-\overline{\alpha})} \int_x^b (t-x)^{\alpha-\overline{\alpha}-1} \left| {}^H\mathbb{D}_{b-}^{\alpha,\beta} f(t) \right| dt \leq$$

$$\frac{1}{\Gamma(\alpha-\overline{\alpha})} \left(\int_x^b (t-x)^{(\alpha-\overline{\alpha}-1)p} dt \right)^{\frac{1}{p}} \left(\int_x^b \left| {}^H\mathbb{D}_{b-}^{\alpha,\beta} f(t) \right|^q dt \right)^{\frac{1}{q}} =$$

$$\frac{1}{\Gamma(\alpha-\overline{\alpha})} \frac{(b-x)^{\frac{((\alpha-\overline{\alpha}-1)p+1)}{p}}}{((\alpha-\overline{\alpha}-1)p+1)^{\frac{1}{p}}} (z(x))^{\frac{1}{q}}. \quad (1.99)$$

Here we set

$$z(x) = \int_x^b \left| {}^H\mathbb{D}_{b-}^{\alpha,\beta} f(t) \right|^q dt \geq 0, \quad (1.100)$$

with $z(b) = 0$.

Thus

$$-z'(x) = \left| {}^H\mathbb{D}_{b-}^{\alpha,\beta} f(x) \right|^q,$$

and

$$\left(-z'(x)\right)^{\frac{1}{q}} = \left| {}^H\mathbb{D}_{b-}^{\alpha,\beta} f(x) \right| \geq 0, \quad \forall\, x \in [a,b]. \quad (1.101)$$

Hence it holds

$$\left| {}^H\mathbb{D}_{b-}^{\overline{\alpha},\beta} f(w) \right| \left| {}^H\mathbb{D}_{b-}^{\alpha,\beta} f(w) \right| \leq \frac{1}{\Gamma(\alpha-\overline{\alpha})} \frac{(b-w)^{\frac{((\alpha-\overline{\alpha}-1)p+1)}{p}}}{((\alpha-\overline{\alpha}-1)p+1)^{\frac{1}{p}}} \left(z(w) \left(-z'(w)\right) \right)^{\frac{1}{q}}, \quad (1.102)$$

$\forall\, w \in [a,b]$.

Therefore we obtain

$$\int_x^b \left| {}^H\mathbb{D}_{b-}^{\overline{\alpha},\beta} f(w) \right| \left| {}^H\mathbb{D}_{b-}^{\alpha,\beta} f(w) \right| dw \leq$$

$$\frac{1}{\Gamma(\alpha-\overline{\alpha})((\alpha-\overline{\alpha}-1)p+1)^{\frac{1}{p}}} \left(\int_x^b (b-w)^{((\alpha-\overline{\alpha}-1)p+1)} dw \right)^{\frac{1}{p}}$$

$$\left(\int_x^b z(w)\left(-z'(w)\right)dw\right)^{\frac{1}{q}} =$$

$$\frac{(b-x)^{\frac{((\alpha-\overline{\alpha}-1)p+2)}{p}}}{\Gamma(\alpha-\overline{\alpha})\left[((\alpha-\overline{\alpha}-1)p+1)((\alpha-\overline{\alpha}-1)p+2)\right]^{\frac{1}{p}}}\left(\frac{z^2(x)}{2}\right)^{\frac{1}{q}} =$$

$$\frac{2^{-\frac{1}{q}}(b-x)^{(\alpha-\overline{\alpha}-1)+\frac{2}{p}}}{\Gamma(\alpha-\overline{\alpha})\left[(p(\alpha-\overline{\alpha}-1)+1)(p(\alpha-\overline{\alpha}-1)+2)\right]^{\frac{1}{p}}}\left(\int_x^b \left|{}^H\mathbb{D}_{b-}^{\alpha,\beta}f(w)\right|^q dw\right)^{\frac{2}{q}}, \quad (1.103)$$

proving the claim. ■

We give further left and right ψ-Hilfer fractional Opial type inequalities.

Theorem 1.25 *Let $\alpha > 1$, $\alpha \notin \mathbb{N}$, $\lceil\alpha\rceil = n$, $0 \leq \beta \leq 1$, $x \in [a,b]$, $g = I_{a+}^{(1-\beta)(n-\alpha);\psi} f \in C^n([a,b])$. Here $\psi, f \in C^n([a,b])$, with ψ being increasing such that $\psi'(x) \neq 0$ over $[a,b]$. Assume that $f_{\psi}^{[k]}\left(I_{a+}^{(1-\beta)(n-\alpha);\psi}f\right)(a) = 0$, $k = 0, 1, ..., n-1$, and let $p, q > 1 : \frac{1}{p} + \frac{1}{q} = 1$. Then*

$$\int_a^x |f(w)|\left|\left({}^H\mathbb{D}_{a+}^{\alpha,\beta;\psi}f\right)(w)\right|\psi'(w)\,dw \leq$$

$$\frac{1}{2^{\frac{1}{q}}\Gamma(\alpha)}\left(\int_a^x\left(\int_a^w(\psi(w)-\psi(t))^{p(\alpha-1)}dt\right)dw\right)^{\frac{1}{p}}$$

$$\left(\int_a^x(\psi'(w))^q\left|{}^H\mathbb{D}_{a+}^{\alpha,\beta;\psi}f(w)\right|^q dw\right)^{\frac{2}{q}}. \quad (1.104)$$

Proof By Theorem 1.10 and Corollary 1.17 (1) we get

$$f(x) \overset{(1.71)}{=} I_{a+}^{\alpha;\psi}\,{}^H\mathbb{D}_{a+}^{\alpha,\beta;\psi}f(x) = \frac{1}{\Gamma(\alpha)}\int_a^x \psi'(t)(\psi(x)-\psi(t))^{\alpha-1}\,{}^H\mathbb{D}_{a+}^{\alpha,\beta;\psi}f(t)\,dt. \quad (1.105)$$

We have that

$$|f(x)| \leq \frac{1}{\Gamma(\alpha)}\int_a^x \psi'(t)(\psi(x)-\psi(t))^{\alpha-1}\left|{}^H\mathbb{D}_{a+}^{\alpha,\beta;\psi}f(t)\right|dt \leq$$

$$\frac{1}{\Gamma(\alpha)}\left(\int_a^x(\psi(x)-\psi(t))^{p(\alpha-1)}dt\right)^{\frac{1}{p}}\left(\int_a^x(\psi'(t))^q\left|{}^H\mathbb{D}_{a+}^{\alpha,\beta;\psi}f(t)\right|^q dt\right)^{\frac{1}{q}}. \quad (1.106)$$

Since $g = I_{a+}^{(1-\beta)(n-\alpha);\psi}f \in C^n([a,b])$, by (1.40), we have that ${}^H\mathbb{D}_{a+}^{\alpha,\beta;\psi}f \in C([a,b])$.

Call

$$z(x) = \int_a^x(\psi'(t))^q\left|{}^H\mathbb{D}_{a+}^{\alpha,\beta;\psi}f(t)\right|^q dt, \quad z(a) = 0.$$

Thus

$$z'(x) = (\psi'(x))^q \left| {}^H\mathbb{D}_{a+}^{\alpha,\beta;\psi} f(t) \right|^q \geq 0, \quad (1.107)$$

and

$$(z'(x))^{\frac{1}{q}} = \psi'(x) \left| {}^H\mathbb{D}_{a+}^{\alpha,\beta;\psi} f(t) \right| \geq 0, \quad \forall\, x \in [a,b].$$

Therefore it holds

$$|f(w)|\, \psi'(w) \left| {}^H\mathbb{D}_{a+}^{\alpha,\beta;\psi} f(w) \right| \leq$$

$$\frac{1}{\Gamma(\alpha)} \left(\int_a^w (\psi(w) - \psi(t))^{p(\alpha-1)} dt \right)^{\frac{1}{p}} (z(w)\, z'(w))^{\frac{1}{q}}, \quad \forall\, w \in [a,b]. \quad (1.108)$$

Consequently we get

$$\int_a^x |f(w)| \left| \left({}^H\mathbb{D}_{a+}^{\alpha,\beta;\psi} f \right)(w) \right| \psi'(w)\, dw \leq$$

$$\frac{1}{\Gamma(\alpha)} \int_a^x \left(\int_a^w (\psi(w) - \psi(t))^{p(\alpha-1)} dt \right)^{\frac{1}{p}} (z(w)\, z'(w))^{\frac{1}{q}} dw \leq$$

$$\frac{1}{\Gamma(\alpha)} \left(\int_a^x \left(\int_a^w (\psi(w) - \psi(t))^{p(\alpha-1)} dt \right) dw \right)^{\frac{1}{p}} \left(\int_a^x z(w)\, dz(w) \right)^{\frac{1}{q}} =$$

$$(1.109)$$

$$\frac{1}{\Gamma(\alpha)} \left(\int_a^x \left(\int_a^w (\psi(w) - \psi(t))^{p(\alpha-1)} dt \right) dw \right)^{\frac{1}{p}} \left(\frac{z^2(x)}{2} \right)^{\frac{1}{q}} =$$

$$\frac{1}{\Gamma(\alpha)} \left(\int_a^x \left(\int_a^w (\psi(w) - \psi(t))^{p(\alpha-1)} dt \right) dw \right)^{\frac{1}{p}}$$

$$\left(\int_a^x (\psi'(w))^q \left| {}^H\mathbb{D}_{a+}^{\alpha,\beta;\psi} f(w) \right|^q dw \right)^{\frac{2}{q}} 2^{-\frac{1}{q}},$$

proving the claim. ∎

It follows

Theorem 1.26 *Let* $\alpha > 1$, $\alpha \notin \mathbb{N}$, $\lceil \alpha \rceil = n$, $0 \leq \beta \leq 1$, $x \in [a,b]$, $w = I_{b-}^{(1-\beta)(n-\alpha);\psi} f \in C^n([a,b])$. *Here* $\psi, f \in C^n([a,b])$, *with* ψ *being increasing such that* $\psi'(x) \neq 0$ *over* $[a,b]$. *Assume that* $f_{\psi}^{[k]}\left(I_{b-}^{(1-\beta)(n-\alpha);\psi} f \right)(b) = 0$, $k = 0, 1, ..., n-1$, *and let* $p, q > 1 : \frac{1}{p} + \frac{1}{q} = 1$. *Then*

$$\int_x^b |f(w)| \left| \left({}^H\mathbb{D}_{b-}^{\alpha,\beta;\psi} f \right)(w) \right| \psi'(w)\, dw \leq \quad (1.110)$$

$$\frac{1}{2^{\frac{1}{q}}\Gamma(\alpha)}\left(\int_x^b\left(\int_w^b(\psi(t)-\psi(w))^{p(\alpha-1)}dt\right)dw\right)^{\frac{1}{p}}$$

$$\left(\int_x^b(\psi'(w))^q\left|{}^H\mathbb{D}_{b-}^{\alpha,\beta;\psi}f(w)\right|^q dw\right)^{\frac{2}{q}}.$$

Proof As similar to the proof of Theorem 1.25 is omitted. Use of Theorem 1.10 and Corollary 1.17 (2), etc. ■

One can find many variants of the above fractional Opial type inequalities. We skip this task here.

Next we give a series of ψ-Hilfer and Hilfer fractional left and right Poincaré type inequalities [3].

Theorem 1.27 *All as in Theorem 1.11. Let* $p, q > 1 : \frac{1}{p}+\frac{1}{q}=1$. *Assume* $\mu > 1$ *and* $g_{\psi}^{[k]}(a)=0$, $k=0,1,...,n-1$. *Then*

$$\left\|I_{a+}^{(1-\beta)(n-\alpha);\psi}f\right\|_{q,[a,b]}\leq\frac{1}{\Gamma(\mu)}\left(\int_a^b\left(\int_a^x(\psi(x)-\psi(t))^{p(\mu-1)}dt\right)^{\frac{q}{p}}dx\right)^{\frac{1}{q}} \tag{1.111}$$

$$\left\|\psi'\,{}^H\mathbb{D}_{a+}^{\alpha,\beta;\psi}f\right\|_{q,[a,b]}.$$

Proof By (1.46) we have ($x\in[a,b]$)

$$|g(x)|=\left|I_{a+}^{\mu,\psi}\,{}^H\mathbb{D}_{a+}^{\alpha,\beta;\psi}f(x)\right|=$$

$$\frac{1}{\Gamma(\mu)}\left|\int_a^x\psi'(t)(\psi(x)-\psi(t))^{\mu-1}\,{}^H\mathbb{D}_{a+}^{\alpha,\beta;\psi}f(t)\,dt\right|\leq \tag{1.112}$$

$$\frac{1}{\Gamma(\mu)}\int_a^x\psi'(t)(\psi(x)-\psi(t))^{\mu-1}\left|{}^H\mathbb{D}_{a+}^{\alpha,\beta;\psi}f(t)\right|dt\leq$$

$$\frac{1}{\Gamma(\mu)}\left(\int_a^x(\psi(x)-\psi(t))^{(\mu-1)p}dt\right)^{\frac{1}{p}}\left(\int_a^x(\psi'(t))^q\left|{}^H\mathbb{D}_{a+}^{\alpha,\beta;\psi}f(t)\right|^q dt\right)^{\frac{1}{q}}\leq$$

$$\frac{1}{\Gamma(\mu)}\left(\int_a^x(\psi(x)-\psi(t))^{(\mu-1)p}dt\right)^{\frac{1}{p}}\left(\int_a^b(\psi'(t))^q\left|{}^H\mathbb{D}_{a+}^{\alpha,\beta;\psi}f(t)\right|^q dt\right)^{\frac{1}{q}}.$$

Hence it holds

$$|g(x)|^q\leq\frac{1}{(\Gamma(\mu))^q}\left(\int_a^x(\psi(x)-\psi(t))^{(\mu-1)p}dt\right)^{\frac{q}{p}} \tag{1.113}$$

$$\left(\int_a^b (\psi'(t))^q \left|{}^H\mathbb{D}_{a+}^{\alpha,\beta;\psi} f(t)\right|^q dt\right),$$

$\forall\, x \in [a, b]$.

Therefore we get

$$\int_a^b |g(x)|^q\, dx \leq \frac{1}{(\Gamma(\mu))^q}\left(\int_a^b \left(\int_a^x (\psi(x) - \psi(t))^{(\mu-1)p}\, dt\right)^{\frac{q}{p}} dx\right)$$

$$\left(\int_a^b (\psi'(t))^q \left|{}^H\mathbb{D}_{a+}^{\alpha,\beta;\psi} f(t)\right|^q dt\right), \tag{1.114}$$

proving the claim. ■

We continue with

Theorem 1.28 *All as in Theorem 1.11. Let* $p, q > 1 : \frac{1}{p} + \frac{1}{q} = 1$. *Assume* $\mu > 1$ *and* $w_{\psi}^{[k]}(b) = 0$, $k = 0, 1, ..., n-1$. *Then*

$$\left\| I_{b-}^{(1-\beta)(n-\alpha);\psi} f \right\|_{q,[a,b]} \leq \frac{1}{\Gamma(\mu)}\left(\int_a^b \left(\int_x^b (\psi(t) - \psi(x))^{p(\mu-1)}\, dt\right)^{\frac{q}{p}} dx\right)^{\frac{1}{q}} \tag{1.115}$$

$$\left\| \psi'\, {}^H\mathbb{D}_{b-}^{\alpha,\beta;\psi} f \right\|_{q,[a,b]}.$$

Proof Similar to the proof of Theorem 1.27, now by using (1.47). ■

It follows

Theorem 1.29 *All as in Theorem 1.12,* $0 \leq \beta < 1$. *Let also* $p, q > 1 : \frac{1}{p} + \frac{1}{q} = 1$ *and assume that* $\alpha - \overline{\alpha} > \frac{1}{q}$. *Then*

$$\left\| {}^H\mathbb{D}_{a+}^{\overline{\alpha},\beta} f \right\|_{q,[a,b]} \leq \frac{(b-a)^{(\alpha-\overline{\alpha})}}{\Gamma(\alpha-\overline{\alpha})\,(q(\alpha-\overline{\alpha}))^{\frac{1}{q}}\,(p(\alpha-\overline{\alpha}-1)+1)^{\frac{1}{p}}} \left\| {}^H\mathbb{D}_{a+}^{\alpha,\beta} f \right\|_{q,[a,b]}. \tag{1.116}$$

Proof By (1.48) we get

$$\left|{}^H\mathbb{D}_{a+}^{\overline{\alpha},\beta} f(x)\right| \leq \frac{1}{\Gamma(\alpha-\overline{\alpha})}\left(\int_a^x (x-t)^{p(\alpha-\overline{\alpha}-1)}\, dt\right)^{\frac{1}{p}} \left(\int_a^b \left|{}^H\mathbb{D}_{a+}^{\alpha,\beta} f(t)\right|^q dt\right)^{\frac{1}{q}} = \tag{1.117}$$

$$\frac{(x-a)^{\frac{(p(\alpha-\overline{\alpha}-1)+1)}{p}}}{\Gamma(\alpha-\overline{\alpha})\,(p(\alpha-\overline{\alpha}-1)+1)^{\frac{1}{p}}}\left(\int_a^b \left|{}^H\mathbb{D}_{a+}^{\alpha,\beta} f(t)\right|^q dt\right)^{\frac{1}{q}}.$$

Hence it holds

$$\left|{}^{H}\mathbb{D}_{a+}^{\overline{\alpha},\beta}f(x)\right|^{q} \leq \frac{(x-a)^{\frac{(p(\alpha-\overline{\alpha}-1)+1)q}{p}}}{(\Gamma(\alpha-\overline{\alpha}))^{q}(p(\alpha-\overline{\alpha}-1)+1)^{\frac{q}{p}}}\left(\int_{a}^{b}\left|{}^{H}\mathbb{D}_{a+}^{\alpha,\beta}f(t)\right|^{q}dt\right), \tag{1.118}$$

$\forall\, x \in [a,b]$.

Therefore we get that

$$\int_{a}^{b}\left|{}^{H}\mathbb{D}_{a+}^{\overline{\alpha},\beta}f(x)\right|^{q}dx \leq \frac{1}{(\Gamma(\alpha-\overline{\alpha}))^{q}(p(\alpha-\overline{\alpha}-1)+1)^{\frac{q}{p}}}$$

$$\frac{(b-a)^{(p(\alpha-\overline{\alpha}-1)+1)\frac{q}{p}+1}}{\left((p(\alpha-\overline{\alpha}-1)+1)\frac{q}{p}+1\right)}\left(\int_{a}^{b}\left|{}^{H}\mathbb{D}_{a+}^{\alpha,\beta}f(t)\right|^{q}dt\right) = \tag{1.119}$$

$$\frac{(b-a)^{q(\alpha-\overline{\alpha})}}{q(\alpha-\overline{\alpha})(\Gamma(\alpha-\overline{\alpha}))^{q}(p(\alpha-\overline{\alpha}-1)+1)^{\frac{q}{p}}}\left(\int_{a}^{b}\left|{}^{H}\mathbb{D}_{a+}^{\alpha,\beta}f(t)\right|^{q}dt\right),$$

proving the claim. ■

We continue with

Theorem 1.30 *All as in Theorem 1.13,* $0 \leq \beta < 1$. *Let also* $p, q > 1 : \frac{1}{p} + \frac{1}{q} = 1$ *and assume that* $\alpha - \overline{\alpha} > \frac{1}{q}$. *Then*

$$\left\|{}^{H}\mathbb{D}_{b-}^{\overline{\alpha},\beta}f\right\|_{q,[a,b]} \leq \frac{(b-a)^{(\alpha-\overline{\alpha})}}{\Gamma(\alpha-\overline{\alpha})(q(\alpha-\overline{\alpha}))^{\frac{1}{q}}(p(\alpha-\overline{\alpha}-1)+1)^{\frac{1}{p}}}\left\|{}^{H}\mathbb{D}_{b-}^{\alpha,\beta}f\right\|_{q,[a,b]}. \tag{1.120}$$

Proof As similar to Theorem 1.29 is omitted. Just use (1.53), etc. ■

We also present

Theorem 1.31 *All as in Theorem 1.10. Let* $p, q > 1 : \frac{1}{p} + \frac{1}{q} = 1$. *Assume* $\alpha > 1$ *and* $f_{\psi}^{[k]}\left(I_{a+}^{(1-\beta)(n-\alpha);\psi}f\right)(a) = 0$, $k = 0, 1, ..., n-1$. *Then*

$$\|f\|_{q,[a,b]} \leq \frac{1}{\Gamma(\alpha)}\left(\int_{a}^{b}\left(\int_{a}^{x}(\psi(x)-\psi(t))^{p(\alpha-1)}dt\right)^{\frac{q}{p}}dx\right)^{\frac{1}{q}}\left\|\psi'\,{}^{H}\mathbb{D}_{a+}^{\alpha,\beta;\psi}f\right\|_{q,[a,b]}. \tag{1.121}$$

Proof It is omitted. It is similar to Theorem 1.27, use Theorem 1.10 (1.44) and Corollary 1.17 (1) (1.71). ■

The counterpart of the last theorem follows.

Theorem 1.32 *All as in Theorem 1.10. Let* $p, q > 1 : \frac{1}{p} + \frac{1}{q} = 1$. *Assume* $\alpha > 1$ *and* $f_{\psi}^{[k]}\left(I_{b-}^{(1-\beta)(n-\alpha);\psi}f\right)(b) = 0$, $k = 0, 1, ..., n-1$. *Then*

$$\|f\|_{q,[a,b]} \leq \frac{1}{\Gamma(\alpha)} \left(\int_a^b \left(\int_x^b (\psi(t) - \psi(x))^{p(\alpha-1)} dt\right)^{\frac{q}{p}} dx\right)^{\frac{1}{q}} \left\|\psi' \, {}^H\mathbb{D}_{b-}^{\alpha,\beta;\psi} f\right\|_{q,[a,b]}. \tag{1.122}$$

Proof It is skipped. It is similar to Theorem 1.27, use of Theorem 1.10 (1.45) and Corollary 1.17 (2) (1.72). ■

Many other variants of Poincaré type Hilfer fractional inequalities are possible. We skip this task here.

We finish with

Note 1.33 In this chapter the fixed fractional order $\alpha > 0$, $\lceil \alpha \rceil = n$, could be replaced by a variable order $\alpha(z) > 0$ as long as $\lceil \alpha(z) \rceil = n$, for all z in an interval $I \subseteq \mathbb{R}$.

References

1. Almeida, R.: A Caputo fractional derivative of a function with respect to another function. Commun. Nonlinear Sci. Numer. Simulat. **44**, 460–481 (2017)
2. Anastassiou, G.A.: Ostrowski type inequalities. Proc. AMS **123**, 3775–3781 (1995)
3. Anastassiou, G.A.: Fractional Differentiation Inequalities. Springer, Heidelberg (2009)
4. Anastassiou, G.A.: Advances on Fractional Inequalities. Springer, New York (2011)
5. Anastassiou, G.A.: Intelligent Mathematics: Computational Analysis. Springer, Heidelberg (2011)
6. Anastassiou, G.A.: Intelligent Computations: Abstract Fractional Calculus. Inequalities, Approximations. Springer, Heidelberg (2018)
7. Anastassiou, G.A.: Advancements on ψ-Hilfer fractional calculus and fractional integral inequalities. Nonlinearity and Complexity, accepted for publication, Discontinuity (2021)
8. Kilbas, A.A., Srivastava, H.M., Trujillo, J.J.: Theory and Applications of Fractional Differentiation Equations. North Holland, Amsterdam (2006)
9. Opial, Z.: Sur une inégalité. Ann. Polon. Math. **8**, 29–32 (1960)
10. Ostrowski, A.: Über die Absolutabweichung einer differtentiebaren Funktion von ihrem Integralmittelwert. Comment. Math. Helv. **10**, 226–227 (1938)
11. Samko, S.G., Kilbas, A.A., Marichev, O.I.: Fractional Integrals and Derivatives: Theory and Applications. Gordon and Breach Science Publishers, Switzerland (1993)
12. Tomovski, Ž, Hilfer, R., Srivastava, H.M.: Fractional and operational calculus with generalized fractional derivative operators and Mittag-Leffler type functions. Integr. Transforms Spec. Funct. **21**(11), 797–814 (2010)
13. Vanterler da, J., Sousa, C., Capelas de Oliveira, E.: On the ψ-Hilfer fractional derivative. Commun. Nonlinear Sci. Numer. Simul. **60**, 72–91 (2018)

Chapter 2
Landau Generalized Hilfer Fractional Inequalities

Here we present a series of Landau type inequalities related to left and right ψ-Hilfer fractional derivatives. It follows [5].

2.1 Introduction

Let $p \in [1, \infty]$, $I = \mathbb{R}_+$ or $I = \mathbb{R}$ and $f : I \to \mathbb{R}$ is twice differentiable with $f, f'' \in L_p(I)$, then $f' \in L_p(I)$. Moreover, there exists a constant $C_p(I) > 0$ independent of f, such that

$$\left\| f' \right\|_{p,I} \le C_p(I) \left\| f \right\|_{p,I}^{\frac{1}{2}} \left\| f'' \right\|_{p,I}^{\frac{1}{2}}, \tag{2.1}$$

where $\|\cdot\|_{p,I}$ is the p-norm on the interval I, see [1, 6].

The research on these inequalities started by Landau [12] in 1913. For the case of $p = \infty$ he proved that

$$C_\infty(\mathbb{R}_+) = 2 \text{ and } C_\infty(\mathbb{R}) = \sqrt{2}, \tag{2.2}$$

are the best constants in (2.1).

In 1932, Hardy and Littlewood [8] proved (2.1) for $p = 2$, with the best constants

$$C_2(\mathbb{R}_+) = \sqrt{2} \text{ and } C_2(\mathbb{R}) = 1. \tag{2.3}$$

In 1935, Hardy et al. [9] showed that the best constants $C_p(\mathbb{R}_+)$ in (2.1) satisfies the estimate

$$C_p(\mathbb{R}_+) \le 2, \text{ for } p \in [1, \infty), \tag{2.4}$$

which yields $C_p(\mathbb{R}) \le 2$ for $p \in [1, \infty)$.

G. A. Anastassiou, *Unification of Fractional Calculi with Applications*,
Studies in Systems, Decision and Control 398,
https://doi.org/10.1007/978-3-030-86920-5_2

In fact, in [7, 10] was shown that $C_p(\mathbb{R}) \le \sqrt{2}$.
In this chapter we present Landau type ψ-Hilfer fractional inequalities.

2.2 Background

Let $-\infty < a < b < \infty$, the left and right Riemann-Liouville fractional integrals of order $\alpha \in \mathbb{C}$ ($\mathcal{R}(\alpha) > 0$) are defined by

$$\left(I_{a+}^{\alpha} f\right)(x) = \frac{1}{\Gamma(\alpha)} \int_a^x (x-t)^{\alpha-1} f(t)\, dt, \tag{2.5}$$

$x > a$; where Γ stands for the gamma function,
and

$$\left(I_{b-}^{\alpha} f\right)(x) = \frac{1}{\Gamma(\alpha)} \int_x^b (t-x)^{\alpha-1} f(t)\, dt, \tag{2.6}$$

$x < b$.
The Riemann-Liouville left and right fractional derivatives of order $\alpha \in \mathbb{C}$ ($\mathcal{R}(\alpha) \ge 0$) are defined by

$$\left(\Delta_{a+}^{\alpha} y\right)(x) = \left(\frac{d}{dx}\right)^n \left(I_{a+}^{n-\alpha} y\right)(x) = \frac{1}{\Gamma(n-\alpha)} \left(\frac{d}{dx}\right)^n \int_a^x (x-t)^{n-\alpha-1} y(t)\, dt \tag{2.7}$$

($n = \lceil \mathcal{R}(\alpha) \rceil$, $\lceil \cdot \rceil$ means ceiling of the number; $x > a$)

$$\left(\Delta_{b-}^{\alpha} y\right)(x) = (-1)^n \left(\frac{d}{dx}\right)^n \left(I_{b-}^{n-\alpha} y\right)(x) =$$

$$\frac{(-1)^n}{\Gamma(n-\alpha)} \left(\frac{d}{dx}\right)^n \int_x^b (t-x)^{n-\alpha-1} y(t)\, dt \tag{2.8}$$

($n = \lceil \mathcal{R}(\alpha) \rceil$; $x < b$), respectively, where $\mathcal{R}(\alpha)$ is the real part of α.
In particular, when $\alpha = n \in \mathbb{Z}_+$, then

$$\left(\Delta_{a+}^{0} y\right)(x) = \left(\Delta_{b-}^{0} y\right)(x) = y(x);$$

$$\left(\Delta_{a+}^{n} y\right)(x) = y^{(n)}(x), \text{ and } \left(\Delta_{b-}^{n} y\right)(x) = (-1)^n y^{(n)}(x), \; n \in \mathbb{N},$$

see [13].
Let $\alpha > 0$, $I = [a, b] \subset \mathbb{R}$, f an integrable function defined on I and $\psi \in C^1(I)$ an increasing function such that $\psi'(x) \ne 0$, for all $x \in I$. Left fractional integrals and left Riemann-Liouville fractional derivatives of a function f with respect to another function ψ are defined as [11, 13]

$$I_{a+}^{\alpha,\psi} f(x) = \frac{1}{\Gamma(\alpha)} \int_a^x \psi'(t) (\psi(x) - \psi(t))^{\alpha-1} f(t) \, dt, \tag{2.9}$$

and

$$\Delta_{a+}^{\alpha,\psi} f(x) = \left(\frac{1}{\psi'(x)} \frac{d}{dx} \right)^n I_{a+}^{n-\alpha,\psi} f(x) = \tag{2.10}$$

$$\frac{1}{\Gamma(n-\alpha)} \left(\frac{1}{\psi'(x)} \frac{d}{dx} \right)^n \int_a^x \psi'(t) (\psi(x) - \psi(t))^{n-\alpha-1} f(t) \, dt,$$

respectively, where $n = \lceil \alpha \rceil$.

Similarly, we define the right ones:

$$I_{b-}^{\alpha,\psi} f(x) = \frac{1}{\Gamma(\alpha)} \int_x^b \psi'(t) (\psi(t) - \psi(x))^{\alpha-1} f(t) \, dt, \tag{2.11}$$

and

$$\Delta_{b-}^{\alpha,\psi} f(x) = \left(-\frac{1}{\psi'(x)} \frac{d}{dx} \right)^n I_{b-}^{n-\alpha,\psi} f(x) =$$

$$\frac{1}{\Gamma(n-\alpha)} \left(-\frac{1}{\psi'(x)} \frac{d}{dx} \right)^n \int_x^b \psi'(t) (\psi(t) - \psi(x))^{n-\alpha-1} f(t) \, dt. \tag{2.12}$$

The following semigroup property holds; if $\alpha, \beta > 0$, $f \in C(I)$, then

$$I_{a+}^{\alpha,\psi} I_{a+}^{\beta,\psi} f = I_{a+}^{\alpha+\beta,\psi} f \quad \text{and} \quad I_{b-}^{\alpha,\psi} I_{b-}^{\beta,\psi} f = I_{b-}^{\alpha+\beta,\psi} f.$$

Next let again $\alpha > 0$, $n = \lceil \alpha \rceil$, $I = [a, b]$, $f, \psi \in C^n(I) : \psi$ is increasing and $\psi'(x) \neq 0$, for all $x \in I$. The left ψ -Caputo fractional derivative of f of order α is given by Almeida [2]

$${}^C D_{a+}^{\alpha,\psi} f(x) = I_{a+}^{n-\alpha,\psi} \left(\frac{1}{\psi'(x)} \frac{d}{dx} \right)^n f(x), \tag{2.13}$$

and the right ψ-Caputo fractional derivative [2]

$${}^C D_{b-}^{\alpha,\psi} f(x) = I_{b-}^{n-\alpha,\psi} \left(-\frac{1}{\psi'(x)} \frac{d}{dx} \right)^n f(x). \tag{2.14}$$

We set

$$f_{\psi}^{[n]}(x) := f_{\psi}^{(n)} f(x) := \left(\frac{1}{\psi'(x)} \frac{d}{dx} \right)^n f(x). \tag{2.15}$$

Clearly, when $\alpha = m \in \mathbb{N}$ we have

$$^{C}D_{a+}^{\alpha,\psi}f(x)=f_{\psi}^{[m]}(x)\quad\text{and}\quad ^{C}D_{b-}^{\alpha,\psi}f(x)=(-1)^{m}f_{\psi}^{[m]}(x),$$

and if $\alpha\notin\mathbb{N}$, then

$$^{C}D_{a+}^{\alpha,\psi}f(x)=\frac{1}{\Gamma(n-\alpha)}\int_{a}^{x}\psi'(t)(\psi(x)-\psi(t))^{n-\alpha-1}f_{\psi}^{[n]}(t)\,dt,\tag{2.16}$$

and

$$^{C}D_{b-}^{\alpha,\psi}f(x)=\frac{(-1)^{n}}{\Gamma(n-\alpha)}\int_{x}^{b}\psi'(t)(\psi(t)-\psi(x))^{n-\alpha-1}f_{\psi}^{[n]}(t)\,dt.\tag{2.17}$$

If $\psi(x)=x$, then we get the usual left and right Caputo fractional derivatives

$$^{C}D_{a+}^{m}f(x)=f^{(m)}(x),\ ^{C}D_{b-}^{m}f(x)=(-1)^{m}f^{(m)}(x),$$

for $m\in\mathbb{N}$, and ($\alpha\notin\mathbb{N}$)

$$D_{*a}^{\alpha}f(x)={}^{C}D_{a+}^{\alpha}f(x)=\frac{1}{\Gamma(n-\alpha)}\int_{a}^{x}(x-t)^{n-\alpha-1}f^{(n)}(t)\,dt,\tag{2.18}$$

$$D_{b-}^{\alpha}(x)={}^{C}D_{b-}^{\alpha}f(x)=\frac{(-1)^{n}}{\Gamma(n-\alpha)}\int_{x}^{b}(t-x)^{n-\alpha-1}f^{(n)}(t)\,dt.\tag{2.19}$$

Also we set

$$^{C}D_{a+}^{0,\psi}f(x)={}^{C}D_{b-}^{0,\psi}f(x)=f(x).$$

Next we will deal with the ψ-Hilfer fractional derivative.

Definition 2.1 ([15]) Let $n-1<\alpha<n$, $n\in\mathbb{N}$, $I=[a,b]\subset\mathbb{R}$ and $f,\psi\in C^{n}([a,b])$, ψ is increasing and $\psi'(x)\neq0$, for all $x\in I$. The ψ-Hilfer fractional derivative (left-sided and right-sided) $^{H}\mathbb{D}_{a+(b-)}^{\alpha,\beta;\psi}f$ of order α and type $0\leq\beta\leq1$, respectively, are defined by

$$^{H}\mathbb{D}_{a+}^{\alpha,\beta;\psi}f(x)=I_{a+}^{\beta(n-\alpha);\psi}\left(\frac{1}{\psi'(x)}\frac{d}{dx}\right)^{n}I_{a+}^{(1-\beta)(n-\alpha);\psi}f(x),\tag{2.20}$$

and

$$^{H}\mathbb{D}_{b-}^{\alpha,\beta;\psi}f(x)=I_{b-}^{\beta(n-\alpha);\psi}\left(-\frac{1}{\psi'(x)}\frac{d}{dx}\right)^{n}I_{b-}^{(1-\beta)(n-\alpha);\psi}f(x),\quad x\in[a,b].\tag{2.21}$$

The original Hilfer fractional derivatives [14] come from $\psi(x)=x$, and are denoted by $^{H}\mathbb{D}_{a+}^{\alpha,\beta}f(x)$ and $^{H}\mathbb{D}_{b-}^{\alpha,\beta}f(x)$.

When $\beta = 0$, we get Riemann-Liouville fractional derivatives, while when $\beta = 1$ we have Caputo type fractional derivatives.

We define $\gamma = \alpha + \beta(n - \alpha)$. We notice that $n - 1 < \alpha \leq \alpha + \beta(n - \alpha) \leq \alpha + n - \alpha = n$, hence $\lceil \gamma \rceil = n$. We can easily write that [15]

$$^{H}\mathbb{D}_{a+}^{\alpha,\beta;\psi} f(x) = I_{a+}^{\gamma-\alpha;\psi} \Delta_{a+}^{\gamma;\psi} f(x), \tag{2.22}$$

and

$$^{H}\mathbb{D}_{b-}^{\alpha,\beta;\psi} f(x) = I_{b-}^{\gamma-\alpha;\psi} \Delta_{b-}^{\gamma;\psi} f(x), \quad x \in [a,b]. \tag{2.23}$$

We have that [15]

$$\Delta_{a+}^{\gamma,\psi} f(x) = \left(\frac{1}{\psi'(x)}\frac{d}{dx}\right)^n I_{a+}^{(1-\beta)(n-\alpha);\psi} f(x), \tag{2.24}$$

and

$$\Delta_{b-}^{\gamma,\psi} f(x) = \left(-\frac{1}{\psi'(x)}\frac{d}{dx}\right)^n I_{b-}^{(1-\beta)(n-\alpha);\psi} f(x). \tag{2.25}$$

In particular, when $0 < \alpha < 1$ and $0 \leq \beta \leq 1$; $\gamma = \alpha + \beta(1 - \alpha)$, we have that

$$^{H}\mathbb{D}_{a+}^{\alpha,\beta;\psi} f(x) = \frac{1}{\Gamma(\gamma-\alpha)} \int_a^x \psi'(t)(\psi(x) - \psi(t))^{\gamma-\alpha-1} \Delta_{a+}^{\gamma;\psi} f(t)\,dt, \tag{2.26}$$

and

$$^{H}\mathbb{D}_{b-}^{\alpha,\beta;\psi} f(x) = \frac{1}{\Gamma(\gamma-\alpha)} \int_x^b \psi'(t)(\psi(t) - \psi(x))^{\gamma-\alpha-1} \Delta_{b-}^{\gamma;\psi} f(t)\,dt, \tag{2.27}$$

$x \in [a,b]$.

Remark 2.2 ([4, 15]) Let $\mu = n(1-\beta) + \beta\alpha$, then $\lceil \mu \rceil = n$.

Assume that $g(x) = I_{a+}^{(1-\beta)(n-\alpha);\psi} f(x) \in C^n([a,b])$, we have that

$$^{H}\mathbb{D}_{a+}^{\alpha,\beta;\psi} f(x) = I_{a+}^{n-\mu;\psi} \left(\frac{1}{\psi'(x)}\frac{d}{dx}\right)^n g(x). \tag{2.28}$$

Thus

$$^{H}\mathbb{D}_{a+}^{\alpha,\beta;\psi} f = {}^{C}D_{a+}^{\mu;\psi} g(x) = {}^{C}D_{a+}^{\mu;\psi}\left[I_{a+}^{(1-\beta)(n-\alpha);\psi} f(x)\right]. \tag{2.29}$$

Assume that $w(x) = I_{b-}^{(1-\beta)(n-\alpha);\psi} f(x) \in C^n([a,b])$. Hence

$$ {}^{H}\mathbb{D}_{b-}^{\alpha,\beta;\psi} f(x) = I_{b-}^{\beta(n-\alpha);\psi} \left(-\frac{1}{\psi'(x)} \frac{d}{dx} \right)^{n} w(x) = I_{b-}^{n-\mu;\psi} \left(-\frac{1}{\psi'(x)} \frac{d}{dx} \right)^{n} w(x). \tag{2.30} $$

Thus

$$ {}^{H}\mathbb{D}_{b-}^{\alpha,\beta;\psi} f = {}^{C}D_{b-}^{\mu;\psi} w(x) = {}^{C}D_{b-}^{\mu;\psi} \left(I_{b-}^{(1-\beta)(n-\alpha);\psi} f(x) \right). \tag{2.31} $$

We need the following ψ-Hilfer fractional Taylor formulae.

Theorem 2.3 ([4]) *Let $f, \psi \in C^{n}([a,b])$, with ψ being increasing, $\psi'(x) \neq 0$ over $[a,b] \subset \mathbb{R}$, $\alpha > 0 : \lceil \alpha \rceil = n$, $0 \leq \beta \leq 1$, $\mu = n(1-\beta) + \alpha\beta$. Assume that $g_a(x) = I_{a+}^{(1-\beta)(n-\alpha);\psi} f(x)$, $w_b(x) = I_{b-}^{(1-\beta)(n-\alpha);\psi} f(x) \in C^{n}([a,b])$. Then*
(1)

$$ I_{a+}^{\mu;\psi}\, {}^{H}\mathbb{D}_{a+}^{\alpha,\beta;\psi} f(x) = g_a(x) - \sum_{k=0}^{n-1} \frac{g_{a\psi}^{[k]}(a)}{k!} (\psi(x) - \psi(a))^{k}, \tag{2.32} $$

where

$$ g_{a\psi}^{[k]}(x) = \left(\frac{1}{\psi'(x)} \frac{d}{dx} \right)^{k} g_a(x), \quad k = 0, 1, ..., n-1, $$

and
(2)

$$ I_{b-}^{\mu;\psi}\, {}^{H}\mathbb{D}_{b-}^{\alpha,\beta;\psi} f(x) = w_b(x) - \sum_{k=0}^{n-1} \frac{w_{b\psi}^{[k]}(b)}{k!} (\psi(x) - \psi(b))^{k}, \tag{2.33} $$

where

$$ w_{b\psi}^{[k]}(x) = \left(\frac{1}{\psi'(x)} \frac{d}{dx} \right)^{k} w_b(x), \quad k = 0, 1, ..., n-1; $$

$x \in [a,b]$.

We also need

Theorem 2.4 ([3]) *Let $\overline{\alpha}, \overline{\beta}, A, B > 0$ and the function $y(h) = \frac{A}{h^{\overline{\alpha}}} + Bh^{\overline{\beta}}$, $h > 0$. It has only one critical number $h_0 = \left(\frac{\overline{\alpha} A}{\overline{\beta} B} \right)^{\frac{1}{(\overline{\alpha}+\overline{\beta})}}$, such that $y'(h_0) = 0$. Furthermore y has a global minimum which is*

$$ y(h_0) = \frac{(\overline{\alpha} + \overline{\beta})}{\overline{\alpha}^{\frac{\overline{\alpha}}{\overline{\alpha}+\overline{\beta}}} \overline{\beta}^{\frac{\overline{\beta}}{\overline{\alpha}+\overline{\beta}}}} A^{\frac{\overline{\beta}}{\overline{\alpha}+\overline{\beta}}} B^{\frac{\overline{\alpha}}{\overline{\alpha}+\overline{\beta}}}. \tag{2.34} $$

2.3 Main Results

We make

Convention 2.5 *Let* $f : \mathbb{R} \to \mathbb{R}$. *Define* ${}^{H}\mathbb{D}_{a+}^{\alpha,\beta;\psi} f(x) = 0$, *for* $x < a$, *and* ${}^{H}\mathbb{D}_{b-}^{\alpha,\beta;\psi} f(x) = 0$, *for* $x > b$, *for any* $\alpha > 0$, $\alpha \notin \mathbb{N}$; $a, b \in \mathbb{R}$.

Here we prove several Landau type ψ-Hilfer fractional inequalities.
We present the following theorem.

Theorem 2.6 *Let* $f, \psi \in C^2(\mathbb{R}_+)$, *with* ψ *being strictly increasing,* $1 < \alpha < 2$, $0 \le \beta \le 1$, $\mu = 2(1-\beta) + \beta\alpha$. *Assume further that* $g_a := I_{a+}^{(1-\beta)(2-\alpha);\psi} f \in C^2([a, +\infty))$, $\forall\ a \in \mathbb{R}_+$; $f \in C_B(\mathbb{R}_+)$ *(continuous and bounded functions) and* $\sup_{a \in \mathbb{R}_+} \left\| {}^{H}\mathbb{D}_{a+}^{\alpha,\beta;\psi} f \right\|_{\infty,\mathbb{R}_+} < +\infty$.

Then

$$\sup_{a \in \mathbb{R}_+} \left| g_{a\psi}^{[1]}(a) \right| \le$$

$$\left[\frac{(\mu - (1-\beta)(2-\alpha))}{[\Gamma(\mu+1)(1-(1-\beta)(2-\alpha))]^{\frac{(1-(1-\beta)(2-\alpha))}{(\mu-(1-\beta)(2-\alpha))}} [(\mu-1)\Gamma((1-\beta)(2-\alpha)+1)]^{\frac{(\mu-1)}{(\mu-(1-\beta)(2-\alpha))}}} \right]$$

$$\|f\|_{\infty,\mathbb{R}_+}^{\frac{(\mu-1)}{(\mu-(1-\beta)(2-\alpha))}} \left(\sup_{a \in \mathbb{R}_+} \left\| {}^{H}\mathbb{D}_{a+}^{\alpha,\beta;\psi} f \right\|_{\infty,\mathbb{R}_+} \right)^{\frac{(1-(1-\beta)(2-\alpha))}{(\mu-(1-\beta)(2-\alpha))}} < +\infty. \quad (2.35)$$

Proof We have that

$$g_a(x) = I_{a+}^{(1-\beta)(2-\alpha);\psi} f(x) =$$

$$\frac{1}{\Gamma((1-\beta)(2-\alpha))} \int_a^x \psi'(t)(\psi(x) - \psi(t))^{(1-\beta)(2-\alpha)-1} f(t)\, dt,$$

for $x \ge a \in \mathbb{R}_+$.

Hence

$$|g_a(x)| \le \frac{1}{\Gamma((1-\beta)(2-\alpha))} \int_a^x \psi'(t)(\psi(x) - \psi(t))^{(1-\beta)(2-\alpha)-1} |f(t)|\, dt \le$$

$$\frac{\|f\|_{\infty,\mathbb{R}_+}}{\Gamma((1-\beta)(2-\alpha)+1)} (\psi(x) - \psi(a))^{(1-\beta)(2-\alpha)}, \quad \forall\, x \ge a. \quad (2.36)$$

Clearly it holds $g_a(a) = 0$.

By Theorem 2.3 (2.32), we have

$$g_a(x) - g_{a\psi}^{[1]}(a)(\psi(x) - \psi(a)) = I_{a+}^{\mu;\psi}\, {}^{H}\mathbb{D}_{a+}^{\alpha,\beta;\psi} f(x), \forall x \ge a. \quad (2.37)$$

That is

$$g_{a\psi}^{[1]}(a)(\psi(x)-\psi(a)) = g_a(x) - I_{a+}^{\mu;\psi}\ {}^{H}\mathbb{D}_{a+}^{\alpha,\beta;\psi} f(x), \quad \forall\, x \geq a. \tag{2.38}$$

Let $\psi(x_1) - \psi(a) = h$ (where $x_1 > a$), then

$$g_{a\psi}^{[1]}(a)\, h = g_a(x_1) - I_{a+}^{\mu;\psi}\ {}^{H}\mathbb{D}_{a+}^{\alpha,\beta;\psi} f(x_1),$$

$$g_{a\psi}^{[1]}(a) = \frac{1}{h}\left(g_a(x_1) - I_{a+}^{\mu;\psi}\ {}^{H}\mathbb{D}_{a+}^{\alpha,\beta;\psi} f(x_1)\right) \tag{2.39}$$

and

$$\left|g_{a\psi}^{[1]}(a)\right| = \frac{1}{h}\left|g_a(x_1) - I_{a+}^{\mu;\psi}\ {}^{H}\mathbb{D}_{a+}^{\alpha,\beta;\psi} f(x_1)\right| \leq$$

$$\frac{1}{h}\left[|g_a(x_1)| + \frac{1}{\Gamma(\mu)}\int_a^{x_1}\psi'(t)(\psi(x_1)-\psi(t))^{\mu-1}\left|{}^{H}\mathbb{D}_{a+}^{\alpha,\beta;\psi} f(t)\right| dt\right] \leq$$

$$\frac{1}{h}\left[\frac{1}{\Gamma((1-\beta)(2-\alpha))}\int_a^{x_1}\psi'(t)(\psi(x_1)-\psi(t))^{(1-\beta)(2-\alpha)-1}|f(t)|\, dt +\right.$$

$$\left.\left\|{}^{H}\mathbb{D}_{a+}^{\alpha,\beta;\psi} f\right\|_{\infty,[a,+\infty)}\frac{(\psi(x_1)-\psi(a))^{\mu}}{\Gamma(\mu+1)}\right] \leq \tag{2.40}$$

$$\frac{1}{h}\left[\frac{\|f\|_{\infty,\mathbb{R}_+}}{\Gamma((1-\beta)(2-\alpha)+1)}h^{(1-\beta)(2-\alpha)} + \frac{\left\|{}^{H}\mathbb{D}_{a+}^{\alpha,\beta;\psi} f\right\|_{\infty,[a,+\infty)}h^{\mu}}{\Gamma(\mu+1)}\right].$$

Therefore it holds

$$\left|g_{a\psi}^{[1]}(a)\right| \leq \frac{\|f\|_{\infty,\mathbb{R}_+}}{\Gamma((1-\beta)(2-\alpha)+1)}h^{(1-\beta)(2-\alpha)-1} + \frac{\sup_{a\in\mathbb{R}_+}\left\|{}^{H}\mathbb{D}_{a+}^{\alpha,\beta;\psi} f\right\|_{\infty,[a,+\infty)}}{\Gamma(\mu+1)}h^{\mu-1}, \tag{2.41}$$

$\forall\, h > 0$ and $\forall\, a \in \mathbb{R}_+$.

Thus we derive

$$\sup_{a\in\mathbb{R}_+}\left|g_{a\psi}^{[1]}(a)\right| \leq \frac{\|f\|_{\infty,\mathbb{R}_+}}{\Gamma((1-\beta)(2-\alpha)+1)\, h^{1-(1-\beta)(2-\alpha)}} + \tag{2.42}$$

$$\frac{\sup_{a\in\mathbb{R}_+}\left\|{}^{H}\mathbb{D}_{a+}^{\alpha,\beta;\psi} f\right\|_{\infty,\mathbb{R}_+}}{\Gamma(\mu+1)}h^{\mu-1}, \quad \forall\, h > 0.$$

Here $\mu - 1 = 1 - 2\beta + \beta\alpha$. Since $\alpha > 1$ we have $\frac{1}{2-\alpha} > 1$. But $0 \leq \beta \leq 1 < \frac{1}{2-\alpha}$, hence $\beta < \frac{1}{2-\alpha}$, that is $\beta(\alpha - 2) + 1 > 0$, giving $\mu - 1 > 0$.

Next we apply Theorem 2.4.

Let

$$A := \frac{\|f\|_{\infty,\mathbb{R}_+}}{\Gamma((1-\beta)(2-\alpha)+1)}, \quad \overline{\alpha} = 1 - (1-\beta)(2-\alpha),$$

and

$$B := \frac{\sup_{a\in\mathbb{R}_+} \left\| {}^{H}\mathbb{D}_{a+}^{\alpha,\beta;\psi} f \right\|_{\infty,\mathbb{R}_+}}{\Gamma(\mu+1)}, \quad \overline{\beta} = \mu - 1.$$

All $A, B, \overline{\alpha}, \overline{\beta} > 0$.

The critical number is $h_0 = \left(\frac{\overline{\alpha}A}{\overline{\beta}B}\right)^{\frac{1}{\overline{\alpha}+\overline{\beta}}} : y'(h_0) = 0$, where

$$y(h) := \frac{A}{h^{\overline{\alpha}}} + Bh^{\overline{\beta}}, \ h > 0. \tag{2.43}$$

The global minimum of y is given by

$$y(h_0) = \frac{(\overline{\alpha}+\overline{\beta})}{\overline{\alpha}^{\frac{\overline{\alpha}}{\overline{\alpha}+\overline{\beta}}}\overline{\beta}^{\frac{\overline{\beta}}{\overline{\alpha}+\overline{\beta}}}} A^{\frac{\overline{\beta}}{\overline{\alpha}+\overline{\beta}}} B^{\frac{\overline{\alpha}}{\overline{\alpha}+\overline{\beta}}}. \tag{2.44}$$

Consequently it holds

$$\sup_{a\in\mathbb{R}_+} \left| g_{a\psi}^{[1]}(a) \right| \leq \frac{(\mu - (1-\beta)(2-\alpha))}{(1-(1-\beta)(2-\alpha))^{\frac{(1-(1-\beta)(2-\alpha))}{(\mu-(1-\beta)(2-\alpha))}} (\mu-1)^{\frac{(\mu-1)}{(\mu-(1-\beta)(2-\alpha))}}} \tag{2.45}$$

$$\left(\frac{\|f\|_{\infty,\mathbb{R}_+}}{\Gamma((1-\beta)(2-\alpha)+1)}\right)^{\frac{(\mu-1)}{(\mu-(1-\beta)(2-\alpha))}} \left(\frac{\sup_{a\in\mathbb{R}_+} \left\| {}^{H}\mathbb{D}_{a+}^{\alpha,\beta;\psi} f \right\|_{\infty,\mathbb{R}_+}}{\Gamma(\mu+1)}\right)^{\frac{(1-(1-\beta)(2-\alpha))}{(\mu-(1-\beta)(2-\alpha))}}.$$

The claim is proved. ■

We also give the following theorem.

Theorem 2.7 *Let $f, \psi \in C^3(\mathbb{R}_+)$ with ψ being strictly increasing, $2 < \alpha < 3$, $0 \leq \beta \leq 1$, $\mu = 3(1-\beta) + \beta\alpha$. Assume further that $g_a := I_{a+}^{(1-\beta)(3-\alpha);\psi} f \in C^3([a, +\infty))$, $\forall\, a \in \mathbb{R}_+$; $f \in C_B(\mathbb{R}_+)$ and $\sup_{a\in\mathbb{R}_+} \left\| {}^{H}\mathbb{D}_{a+}^{\alpha,\beta;\psi} f \right\|_{\infty,\mathbb{R}_+} < +\infty$.*

Then

(1)

$$\sup_{a\in\mathbb{R}_+} \left| g_{a\psi}^{[1]}(a) \right| \leq$$

$$\left[\frac{(\mu-(1-\beta)(3-\alpha))}{[2\Gamma(\mu+1)(1-(1-\beta)(3-\alpha))]^{\frac{(1-(1-\beta)(3-\alpha))}{(\mu-(1-\beta)(3-\alpha))}}[2\Gamma((1-\beta)(3-\alpha)+1)(\mu-1)]^{\frac{(\mu-1)}{(\mu-(1-\beta)(3-\alpha))}}}\right]$$

$$\left(4+2^{(1-\beta)(3-\alpha)}\right)^{\frac{(\mu-1)}{(\mu-(1-\beta)(3-\alpha))}}\left(4+2^{\mu}\right)^{\frac{(1-(1-\beta)(3-\alpha))}{(\mu-(1-\beta)(3-\alpha))}}$$

$$\|f\|_{\infty,\mathbb{R}_+}^{\frac{(\mu-1)}{(\mu-(1-\beta)(3-\alpha))}}\left(\sup_{a\in\mathbb{R}_+}\left\|{}^{H}\mathbb{D}_{a+}^{\alpha,\beta;\psi}f\right\|_{\infty,\mathbb{R}_+}\right)^{\frac{(1-(1-\beta)(3-\alpha))}{(\mu-(1-\beta)(3-\alpha))}}<+\infty, \tag{2.46}$$

and

(2)

$$\sup_{a\in\mathbb{R}_+}\left|g_{a\psi}^{[2]}(a)\right|\leq$$

$$\left[\frac{(\mu-(1-\beta)(3-\alpha))}{[\Gamma(\mu+1)(2-(1-\beta)(3-\alpha))]^{\frac{(2-(1-\beta)(3-\alpha))}{(\mu-(1-\beta)(3-\alpha))}}[\Gamma((1-\beta)(3-\alpha)+1)(\mu-2)]^{\frac{(\mu-2)}{(\mu-(1-\beta)(3-\alpha))}}}\right]$$

$$\left(2+2^{(1-\beta)(3-\alpha)}\right)^{\frac{(\mu-2)}{(\mu-(1-\beta)(3-\alpha))}}\left(2+2^{\mu}\right)^{\frac{(2-(1-\beta)(3-\alpha))}{(\mu-(1-\beta)(3-\alpha))}}$$

$$\|f\|_{\infty,\mathbb{R}_+}^{\frac{(\mu-2)}{(\mu-(1-\beta)(3-\alpha))}}\left(\sup_{a\in\mathbb{R}_+}\left\|{}^{H}\mathbb{D}_{a+}^{\alpha,\beta;\psi}f\right\|_{\infty,\mathbb{R}_+}\right)^{\frac{(2-(1-\beta)(3-\alpha))}{(\mu-(1-\beta)(3-\alpha))}}<+\infty. \tag{2.47}$$

Proof Similarly, here it holds $g_a(a)=0$. By Theorem 2.3 (2.32), we have

$$g_a(x)-g_{a\psi}^{[1]}(a)(\psi(x)-\psi(a))-g_{a\psi}^{[2]}(a)\frac{(\psi(x)-\psi(a))^2}{2}=I_{a+}^{\mu;\psi}\,{}^{H}\mathbb{D}_{a+}^{\alpha,\beta;\psi}f(x), \tag{2.48}$$

$\forall x\geq a\in\mathbb{R}_+$.

That is

$$g_{a\psi}^{[1]}(a)(\psi(x)-\psi(a))+g_{a\psi}^{[2]}(a)\frac{(\psi(x)-\psi(a))^2}{2}=g_a(x)-I_{a+}^{\mu;\psi}\,{}^{H}\mathbb{D}_{a+}^{\alpha,\beta;\psi}f(x), \tag{2.49}$$

$\forall x\geq a$.

Let $\psi(x_1)-\psi(a)=h>0$ (where $x_1>a$), then

$$g_{a\psi}^{[1]}(a)\,h+g_{a\psi}^{[2]}(a)\frac{h^2}{2}=g_a(x_1)-I_{a+}^{\mu;\psi}\,{}^{H}\mathbb{D}_{a+}^{\alpha,\beta;\psi}f(x_1)=:A_1. \tag{2.50}$$

Let $\psi(x_2)-\psi(a)=2h$ (where $x_2>a$), then

$$g_{a\psi}^{[1]}(a)\,2h+g_{a\psi}^{[2]}(a)\,2h^2=g_a(x_2)-I_{a+}^{\mu;\psi}\,{}^{H}\mathbb{D}_{a+}^{\alpha,\beta;\psi}f(x_2)=:A_2. \tag{2.51}$$

So we solve the system

$$2g_{a\psi}^{[1]}(a)\,h+g_{a\psi}^{[2]}(a)\,h^2=2A_1, \tag{2.52}$$

$$2g_{a\psi}^{[1]}(a)\,h + 2g_{a\psi}^{[2]}(a)\,h^2 = A_2. \tag{2.53}$$

We get that

$$g_{a\psi}^{[2]}(a)\,h^2 = A_2 - 2A_1,$$

i.e.

$$g_{a\psi}^{[2]}(a) = \frac{A_2 - 2A_1}{h^2}, \tag{2.54}$$

and

$$g_{a\psi}^{[1]}(a) = \frac{4A_1 - A_2}{2h}. \tag{2.55}$$

We estimate

$$\left|g_{a\psi}^{[1]}(a)\right| = \frac{1}{2h}\left|4A_1 - A_2\right| \leq \frac{1}{2h}\left(4\left|A_1\right| + \left|A_2\right|\right) =$$

$$\frac{1}{2h}\left[4\left|g_a(x_1) - I_{a+}^{\mu;\psi}\ {}^H\mathbb{D}_{a+}^{\alpha,\beta;\psi} f(x_1)\right| + \left|g_a(x_2) - I_{a+}^{\mu;\psi}\ {}^H\mathbb{D}_{a+}^{\alpha,\beta;\psi} f(x_2)\right|\right] \leq \tag{2.56}$$

$$\frac{1}{2h}\left[4\left[\left|g_a(x_1)\right| + I_{a+}^{\mu;\psi}\left|{}^H\mathbb{D}_{a+}^{\alpha,\beta;\psi} f\right|(x_1)\right] + \left[\left|g_a(x_2)\right| + I_{a+}^{\mu;\psi}\left|{}^H\mathbb{D}_{a+}^{\alpha,\beta;\psi} f\right|(x_2)\right]\right] \leq$$

$$\frac{1}{2h}\left[4\left[\frac{1}{\Gamma((1-\beta)(3-\alpha))}\int_a^{x_1}\psi'(t)\left(\psi(x_1)-\psi(a)\right)^{(1-\beta)(3-\alpha)-1}\left|f(t)\right|dt+\right.\right.$$

$$\left.\frac{1}{\Gamma(\mu)}\int_a^{x_1}\psi'(t)\left(\psi(x_1)-\psi(a)\right)^{\mu-1}\left|{}^H\mathbb{D}_{a+}^{\alpha,\beta;\psi} f(t)\right|dt\right]+$$

$$\left[\frac{1}{\Gamma((1-\beta)(3-\alpha))}\int_a^{x_2}\psi'(t)\left(\psi(x_2)-\psi(a)\right)^{(1-\beta)(3-\alpha)-1}\left|f(t)\right|dt+\right.$$

$$\left.\left.\frac{1}{\Gamma(\mu)}\int_a^{x_2}\psi'(t)\left(\psi(x_2)-\psi(a)\right)^{\mu-1}\left|{}^H\mathbb{D}_{a+}^{\alpha,\beta;\psi} f(t)\right|dt\right]\right] \leq$$

$$\frac{1}{2h}\left[4\left[\frac{\|f\|_{\infty,\mathbb{R}_+}}{\Gamma((1-\beta)(3-\alpha)+1)}h^{(1-\beta)(3-\alpha)} + \frac{\left\|{}^H\mathbb{D}_{a+}^{\alpha,\beta;\psi} f\right\|_{\infty,[a,+\infty)}}{\Gamma(\mu+1)}h^{\mu}\right]+\right. \tag{2.57}$$

$$\left.\left[\frac{\|f\|_{\infty,\mathbb{R}_+}}{\Gamma((1-\beta)(3-\alpha)+1)}(2h)^{(1-\beta)(3-\alpha)} + \frac{\left\|{}^H\mathbb{D}_{a+}^{\alpha,\beta;\psi} f\right\|_{\infty,[a,+\infty)}}{\Gamma(\mu+1)}(2h)^{\mu}\right]\right] =$$

$$\frac{1}{2h}\left[\frac{\|f\|_{\infty,\mathbb{R}_+}}{\Gamma\left(\left(1-\beta\right)\left(3-\alpha\right)+1\right)}\left(4+2^{(1-\beta)(3-\alpha)}\right)h^{(1-\beta)(3-\alpha)}+\right. \tag{2.58}$$

$$\left.\frac{\left\|{}^{H}\mathbb{D}_{a+}^{\alpha,\beta;\psi}f\right\|_{\infty,[a,+\infty)}}{\Gamma\left(\mu+1\right)}\left(4+2^{\mu}\right)h^{\mu}\right]=$$

$$\frac{\|f\|_{\infty,\mathbb{R}_+}\left(4+2^{(1-\beta)(3-\alpha)}\right)}{2\Gamma\left(\left(1-\beta\right)\left(3-\alpha\right)+1\right)h^{1-(1-\beta)(3-\alpha)}}+\left(\frac{\left\|{}^{H}\mathbb{D}_{a+}^{\alpha,\beta;\psi}f\right\|_{\infty,[a,+\infty)}\left(4+2^{\mu}\right)}{2\Gamma\left(\mu+1\right)}\right)h^{\mu-1},$$

$\forall a \in \mathbb{R}_+, \forall h > 0$.

Consequently, we get

$$\sup_{a\in\mathbb{R}_+}\left|g_{a\psi}^{[1]}\left(a\right)\right|\leq\left(\frac{\|f\|_{\infty,\mathbb{R}_+}\left(4+2^{(1-\beta)(3-\alpha)}\right)}{2\Gamma\left(\left(1-\beta\right)\left(3-\alpha\right)+1\right)}\right)\frac{1}{h^{1-(1-\beta)(3-\alpha)}} \tag{2.59}$$

$$+\left(\frac{\sup_{a\in\mathbb{R}_+}\left\|{}^{H}\mathbb{D}_{a+}^{\alpha,\beta;\psi}f\right\|_{\infty,[a,+\infty)}\left(4+2^{\mu}\right)}{2\Gamma\left(\mu+1\right)}\right)h^{\mu-1},\quad\forall\, h>0.$$

Notice here that $1-(1-\beta)(3-\alpha)>0$, and since $\alpha>1$, then $2>3-\alpha$, $\frac{2}{3-\alpha}>1$, $\beta<\frac{2}{3-\alpha}$, $2>\beta(3-\alpha)$, and finally we have $\mu-1>0$.

Next we apply Theorem 2.4 to (2.59). Here it is

$$A:=\left(\frac{\|f\|_{\infty,\mathbb{R}_+}\left(4+2^{(1-\beta)(3-\alpha)}\right)}{2\Gamma((1-\beta)(3-\alpha)+1)}\right),\ \overline{\alpha}=1-\left(1-\beta\right)\left(3-\alpha\right),$$

and

$$B:=\left(\frac{\sup_{a\in\mathbb{R}_+}\left\|{}^{H}\mathbb{D}_{a+}^{\alpha,\beta;\psi}f\right\|_{\infty,\mathbb{R}_+}(4+2^{\mu})}{2\Gamma(\mu+1)}\right),\quad\overline{\beta}=\mu-1. \tag{2.60}$$

It is all $A, B, \overline{\alpha}, \overline{\beta} > 0$.

Therefore we get

$$\sup_{a\in\mathbb{R}_+}\left|g_{a\psi}^{[1]}\left(a\right)\right|\leq\frac{\left(\mu-\left(1-\beta\right)\left(3-\alpha\right)\right)}{\left(1-\left(1-\beta\right)\left(3-\alpha\right)\right)^{\frac{(1-(1-\beta)(3-\alpha))}{(\mu-(1-\beta)(3-\alpha))}}\left(\mu-1\right)^{\frac{(\mu-1)}{(\mu-(1-\beta)(3-\alpha))}}} \tag{2.61}$$

$$\left(\frac{\|f\|_{\infty,\mathbb{R}_+}\left(4+2^{(1-\beta)(3-\alpha)}\right)}{2\Gamma\left(\left(1-\beta\right)\left(3-\alpha\right)+1\right)}\right)^{\frac{(\mu-1)}{(\mu-(1-\beta)(3-\alpha))}}$$

$$\left(\frac{\sup_{a \in \mathbb{R}_+} \left\| {}^H\mathbb{D}_{a+}^{\alpha,\beta;\psi} f \right\|_{\infty,\mathbb{R}_+} (4 + 2^\mu)}{2\Gamma(\mu+1)} \right)^{\frac{(1-(1-\beta)(3-\alpha))}{(\mu-(1-\beta)(3-\alpha))}},$$

proving (2.46).

We also have

$$\left| g_{a\psi}^{[2]}(a) \right| \leq \frac{1}{h^2} \left(|A_2| + 2|A_1| \right) =$$

$$\frac{1}{h^2} \left[\left[|g_a(x_2)| + I_{a+}^{\mu;\psi} \left| {}^H\mathbb{D}_{a+}^{\alpha,\beta;\psi} f \right| (x_2) \right] + 2 \left[|g_a(x_1)| + I_{a+}^{\mu;\psi} \left| {}^H\mathbb{D}_{a+}^{\alpha,\beta;\psi} f \right| (x_1) \right] \right] \leq$$

$$\frac{1}{h^2} \left[\left[\frac{\|f\|_{\infty,\mathbb{R}_+}}{\Gamma((1-\beta)(3-\alpha)+1)} (2h)^{(1-\beta)(3-\alpha)} + \frac{\left\| {}^H\mathbb{D}_{a+}^{\alpha,\beta;\psi} f \right\|_{\infty,[a,+\infty)}}{\Gamma(\mu+1)} (2h)^\mu \right] + \right. \tag{2.62}$$

$$\left. 2 \left[\frac{\|f\|_{\infty,\mathbb{R}_+}}{\Gamma((1-\beta)(3-\alpha)+1)} h^{(1-\beta)(3-\alpha)} + \frac{\left\| {}^H\mathbb{D}_{a+}^{\alpha,\beta;\psi} f \right\|_{\infty,[a,+\infty)}}{\Gamma(\mu+1)} h^\mu \right] \right] =$$

$$\frac{1}{h^2} \left[\frac{\|f\|_{\infty,\mathbb{R}_+}}{\Gamma((1-\beta)(3-\alpha)+1)} \left(2 + 2^{(1-\beta)(3-\alpha)}\right) h^{(1-\beta)(3-\alpha)} + \right.$$

$$\left. \frac{\left\| {}^H\mathbb{D}_{a+}^{\alpha,\beta;\psi} f \right\|_{\infty,[a,+\infty)}}{\Gamma(\mu+1)} (2 + 2^\mu) h^\mu \right] \leq$$

$$\left(\frac{\|f\|_{\infty,\mathbb{R}_+} \left(2 + 2^{(1-\beta)(3-\alpha)}\right)}{\Gamma((1-\beta)(3-\alpha)+1)} \right) \frac{1}{h^{2-(1-\beta)(3-\alpha)}} + \tag{2.63}$$

$$\left(\frac{\sup_{a \in \mathbb{R}_+} \left\| {}^H\mathbb{D}_{a+}^{\alpha,\beta;\psi} f \right\|_{\infty,[a,+\infty)} (2 + 2^\mu)}{2\Gamma(\mu+1)} \right) h^{\mu-2},$$

$\forall\, a \in \mathbb{R}_+, \forall\, h > 0$.

That is

$$\sup_{a \in \mathbb{R}_+} \left| g_{a\psi}^{[2]}(a) \right| \leq \left(\frac{\|f\|_{\infty,\mathbb{R}_+} \left(2 + 2^{(1-\beta)(3-\alpha)}\right)}{\Gamma((1-\beta)(3-\alpha)+1)} \right) \frac{1}{h^{2-(1-\beta)(3-\alpha)}} \tag{2.64}$$

$$+\left(\frac{\sup_{a\in\mathbb{R}_+}\left\|{}^{H}\mathbb{D}_{a+}^{\alpha,\beta;\psi}f\right\|_{\infty,\mathbb{R}_+}\left(2+2^{\mu}\right)}{\Gamma\left(\mu+1\right)}\right)h^{\mu-2},\ \ \forall\, h>0.$$

Notice here that $2-(1-\beta)(3-\alpha)>0$, and for $\beta\neq 0$ we have $\mu=3(1-\beta)+\beta\alpha>3(1-\beta)+2\beta=3-\beta$, and $\mu-2>3-\beta-2=1-\beta\geq 0$, hence $\mu-2>0$; if $\beta=0$, $\mu-2=1>0$.

So we can apply Theorem 2.4 to (2.64). Call

$$A^{*}:=\left(\frac{\|f\|_{\infty,\mathbb{R}_+}\left(2+2^{(1-\beta)(3-\alpha)}\right)}{\Gamma((1-\beta)(3-\alpha)+1)}\right),\ \alpha^{*}=2-(1-\beta)(3-\alpha),$$

and

$$B^{*}:=\left(\frac{\sup_{a\in\mathbb{R}_+}\left\|{}^{H}\mathbb{D}_{a+}^{\alpha,\beta;\psi}f\right\|_{\infty,\mathbb{R}_+}(2+2^{\mu})}{\Gamma(\mu+1)}\right),\quad \beta^{*}=\mu-2. \tag{2.65}$$

All $A^{*},B^{*},\alpha^{*},\beta^{*}>0$.

We will have that

$$\sup_{a\in\mathbb{R}_+}\left|g_{a\psi}^{[2]}(a)\right|\leq\frac{(\mu-(1-\beta)(3-\alpha))}{(2-(1-\beta)(3-\alpha))^{\frac{(2-(1-\beta)(3-\alpha))}{(\mu-(1-\beta)(3-\alpha))}}(\mu-2)^{\frac{(\mu-2)}{(\mu-(1-\beta)(3-\alpha))}}} \tag{2.66}$$

$$\left(\frac{\|f\|_{\infty,\mathbb{R}_+}\left(2+2^{(1-\beta)(3-\alpha)}\right)}{\Gamma\left((1-\beta)(3-\alpha)+1\right)}\right)^{\frac{(\mu-2)}{(\mu-(1-\beta)(3-\alpha))}}$$

$$\left(\frac{\sup_{a\in\mathbb{R}_+}\left\|{}^{H}\mathbb{D}_{a+}^{\alpha,\beta;\psi}f\right\|_{\infty,\mathbb{R}_+}\left(2+2^{\mu}\right)}{\Gamma\left(\mu+1\right)}\right)^{\frac{(2-(1-\beta)(3-\alpha))}{(\mu-(1-\beta)(3-\alpha))}},$$

proving (2.47). ■

We present

Theorem 2.8 *Let $f,\psi\in C^{2}(\mathbb{R}_-)$, with ψ being strictly increasing, $1<\alpha<2$, $0\leq\beta\leq 1$, $\mu=2(1-\beta)+\beta\alpha$. Assume further that $w_b:=I_{b-}^{(1-\beta)(2-\alpha);\psi}f\in C^{2}((-\infty,b])$, $\forall\, b\in\mathbb{R}_-$; $f\in C_B(\mathbb{R}_-)$ and $\sup_{b\in\mathbb{R}_-}\left\|{}^{H}\mathbb{D}_{b-}^{\alpha,\beta;\psi}f\right\|_{\infty,\mathbb{R}_-}<+\infty$.*

Then

$$\sup_{b\in\mathbb{R}_-}\left|w_{b\psi}^{[1]}(b)\right|\leq$$

$$\left[\frac{(\mu-(1-\beta)(2-\alpha))}{\left[\Gamma(\mu+1)(1-(1-\beta)(2-\alpha))\right]^{\frac{(1-(1-\beta)(2-\alpha))}{(\mu-(1-\beta)(2-\alpha))}}\left[(\mu-1)\Gamma((1-\beta)(2-\alpha)+1)\right]^{\frac{(\mu-1)}{(\mu-(1-\beta)(2-\alpha))}}}\right]$$

$$\|f\|_{\infty,\mathbb{R}_-}^{\frac{(\mu-1)}{(\mu-(1-\beta)(2-\alpha))}}\left(\sup_{b\in\mathbb{R}_-}\left\|{}^{H}\mathbb{D}_{b-}^{\alpha,\beta;\psi}f\right\|_{\infty,\mathbb{R}_-}\right)^{\frac{(1-(1-\beta)(2-\alpha))}{(\mu-(1-\beta)(2-\alpha))}}<+\infty. \quad (2.67)$$

Proof We have that

$$w_b(x)=I_{b-}^{(1-\beta)(2-\alpha);\psi}f(x)=$$

$$\frac{1}{\Gamma((1-\beta)(2-\alpha))}\int_x^b\psi'(t)(\psi(t)-\psi(x))^{(1-\beta)(2-\alpha)-1}f(t)\,dt,$$

for $x\leq b\in\mathbb{R}_-$.

Hence

$$|w_b(x)|\leq\frac{1}{\Gamma((1-\beta)(2-\alpha))}\int_x^b\psi'(t)(\psi(t)-\psi(x))^{(1-\beta)(2-\alpha)-1}|f(t)|\,dt\leq$$

$$\frac{\|f\|_{\infty,\mathbb{R}_-}}{\Gamma((1-\beta)(2-\alpha)+1)}(\psi(b)-\psi(x))^{(1-\beta)(2-\alpha)},\quad\forall\,x\leq b. \quad (2.68)$$

Clearly it holds $w_b(b)=0$.

By Theorem 2.3 (2.33), we have

$$w_b(x)-w_{b\psi}^{[1]}(b)(\psi(x)-\psi(b))=I_{b-}^{\mu;\psi}\,{}^{H}\mathbb{D}_{b-}^{\alpha,\beta;\psi}f(x),\quad\forall\,x\leq b. \quad (2.69)$$

That is

$$w_{b\psi}^{[1]}(b)(\psi(x)-\psi(b))=w_b(x)-I_{b-}^{\mu;\psi}\,{}^{H}\mathbb{D}_{b-}^{\alpha,\beta;\psi}f(x),\ \forall x\leq b. \quad (2.70)$$

Let $\psi(b)-\psi(x_1)=h>0(x_1<b)$, then

$$-w_{b\psi}^{[1]}(b)\,h=w_b(x_1)-I_{b-}^{\mu;\psi}\,{}^{H}\mathbb{D}_{b-}^{\alpha,\beta;\psi}f(x_1), \quad (2.71)$$

and

$$\left|w_{b\psi}^{[1]}(b)\right|=\frac{1}{h}\left|w_b(x_1)-I_{b-}^{\mu;\psi}\,{}^{H}\mathbb{D}_{b-}^{\alpha,\beta;\psi}f(x_1)\right|\leq$$

$$\frac{1}{h}\left[|w_b(x_1)|+I_{b-}^{\mu;\psi}\left|{}^{H}\mathbb{D}_{b-}^{\alpha,\beta;\psi}f\right|(x_1)\right]\leq$$

$$\frac{1}{h}\left[\frac{1}{\Gamma((1-\beta)(2-\alpha))}\int_{x_1}^b\psi'(t)(\psi(t)-\psi(x_1))^{(1-\beta)(2-\alpha)-1}|f(t)|\,dt+\right.$$

$$\left.\frac{1}{\Gamma(\mu)}\int_{x_1}^b\psi'(t)(\psi(t)-\psi(x_1))^{\mu-1}\left|{}^{H}\mathbb{D}_{b-}^{\alpha,\beta;\psi}f(t)\right|dt\right]\leq$$

$$\frac{1}{h}\left[\frac{\|f\|_{\infty,\mathbb{R}_-}}{\Gamma((1-\beta)(2-\alpha)+1)}(\psi(b)-\psi(x_1))^{(1-\beta)(2-\alpha)}+\right.$$

$$\left.\frac{\left\|{}^{H}\mathbb{D}_{b-}^{\alpha,\beta;\psi}f\right\|_{\infty,(-\infty,b]}}{\Gamma(\mu+1)}(\psi(b)-\psi(x_1))^{\mu}\right]= \tag{2.72}$$

$$\frac{1}{h}\left[\frac{\|f\|_{\infty,\mathbb{R}_-}}{\Gamma((1-\beta)(2-\alpha)+1)}h^{(1-\beta)(2-\alpha)}+\frac{\left\|{}^{H}\mathbb{D}_{b-}^{\alpha,\beta;\psi}f\right\|_{\infty,(-\infty,b]}h^{\mu}}{\Gamma(\mu+1)}\right].$$

Therefore it holds

$$\left|w_{b\psi}^{[1]}(b)\right|\leq\frac{\|f\|_{\infty,\mathbb{R}_-}}{\Gamma((1-\beta)(2-\alpha)+1)}h^{(1-\beta)(2-\alpha)-1}+\frac{\sup_{b\in\mathbb{R}_-}\left\|{}^{H}\mathbb{D}_{b-}^{\alpha,\beta;\psi}f\right\|_{\infty,(-\infty,b]}}{\Gamma(\mu+1)}h^{\mu-1}, \tag{2.73}$$

$\forall\, h>0$ and $\forall\, b\in\mathbb{R}_-$.

Thus we derive

$$\sup_{b\in\mathbb{R}_-}\left|w_{b\psi}^{[1]}(b)\right|\leq\left(\frac{\|f\|_{\infty,\mathbb{R}_-}}{\Gamma((1-\beta)(2-\alpha)+1)}\right)\frac{1}{h^{1-(1-\beta)(2-\alpha)}}+$$

$$\left(\frac{\sup_{b\in\mathbb{R}_-}\left\|{}^{H}\mathbb{D}_{b-}^{\alpha,\beta;\psi}f\right\|_{\infty,\mathbb{R}_-}}{\Gamma(\mu+1)}\right)h^{\mu-1},\quad\forall\, h>0. \tag{2.74}$$

The rest of the proof comes by application of Theorem 2.4, as in the proof of Theorem 2.6. ■

We finish with

Theorem 2.9 *Let* $f,\psi\in C^3(\mathbb{R}_-)$, *with* ψ *being strictly increasing,* $2<\alpha<3$, $0\leq\beta\leq1$, $\mu=3(1-\beta)+\beta\alpha$. *Assume further that* $w_b:=I_{b-}^{(1-\beta)(3-\alpha);\psi}f\in C^3((-\infty,b])$, $\forall\, b\in\mathbb{R}_-$; $f\in C_B(\mathbb{R}_-)$ *and* $\sup_{b\in\mathbb{R}_-}\left\|{}^{H}\mathbb{D}_{b-}^{\alpha,\beta;\psi}f\right\|_{\infty,\mathbb{R}_-}<+\infty$.

Then

(1)

$$\sup_{b\in\mathbb{R}-}\left|w_{b\psi}^{[1]}(b)\right|\leq$$

$$\left[\frac{(\mu-(1-\beta)(3-\alpha))}{[2\Gamma(\mu+1)(1-(1-\beta)(3-\alpha))]^{\frac{(1-(1-\beta)(3-\alpha))}{(\mu-(1-\beta)(3-\alpha))}}[2\Gamma((1-\beta)(3-\alpha)+1)(\mu-1)]^{\frac{(\mu-1)}{(\mu-(1-\beta)(3-\alpha))}}}\right]$$

$$\left(4+2^{(1-\beta)(3-\alpha)}\right)^{\frac{(\mu-1)}{(\mu-(1-\beta)(3-\alpha))}}\left(4+2^{\mu}\right)^{\frac{(1-(1-\beta)(3-\alpha))}{(\mu-(1-\beta)(3-\alpha))}}$$

$$\|f\|_{\infty,\mathbb{R}_-}^{\frac{(\mu-1)}{(\mu-(1-\beta)(3-\alpha))}}\left(\sup_{b\in\mathbb{R}_-}\left\|{}^{H}\mathbb{D}_{b-}^{\alpha,\beta;\psi}f\right\|_{\infty,\mathbb{R}-}\right)^{\frac{(1-(1-\beta)(3-\alpha))}{(\mu-(1-\beta)(3-\alpha))}}<+\infty, \tag{2.75}$$

and

(2)

$$\sup_{b\in\mathbb{R}-}\left|w_{b\psi}^{[2]}(b)\right|\leq$$

$$\left[\frac{(\mu-(1-\beta)(3-\alpha))}{[\Gamma(\mu+1)(2-(1-\beta)(3-\alpha))]^{\frac{(2-(1-\beta)(3-\alpha))}{(\mu-(1-\beta)(3-\alpha))}}[\Gamma((1-\beta)(3-\alpha)+1)(\mu-2)]^{\frac{(\mu-2)}{(\mu-(1-\beta)(3-\alpha))}}}\right]$$

$$\left(2+2^{(1-\beta)(3-\alpha)}\right)^{\frac{(\mu-2)}{(\mu-(1-\beta)(3-\alpha))}}\left(2+2^{\mu}\right)^{\frac{(2-(1-\beta)(3-\alpha))}{(\mu-(1-\beta)(3-\alpha))}}$$

$$\|f\|_{\infty,\mathbb{R}_-}^{\frac{(\mu-2)}{(\mu-(1-\beta)(3-\alpha))}}\left(\sup_{b\in\mathbb{R}_-}\left\|{}^{H}\mathbb{D}_{b-}^{\alpha,\beta;\psi}f\right\|_{\infty,\mathbb{R}_-}\right)^{\frac{(2-(1-\beta)(3-\alpha))}{(\mu-(1-\beta)(3-\alpha))}}<+\infty. \tag{2.76}$$

Proof Similarly, here it holds $w_b(b)=0$. By Theorem 2.3 (2.33), we have

$$w_b(x)-w_{b\psi}^{[1]}(b)(\psi(x)-\psi(b))-w_{b\psi}^{[2]}(b)\frac{(\psi(x)-\psi(b))^2}{2}=I_{b-}^{\mu;\psi}\,{}^{H}\mathbb{D}_{b-}^{\alpha,\beta;\psi}f(x), \tag{2.77}$$

$\forall\, x\leq b\in\mathbb{R}_-$.

That is

$$w_{b\psi}^{[1]}(b)(\psi(x)-\psi(b))+w_{b\psi}^{[2]}(b)\frac{(\psi(x)-\psi(b))^2}{2}=w_b(x)-I_{b-}^{\mu;\psi}\,{}^{H}\mathbb{D}_{b-}^{\alpha,\beta;\psi}f(x), \tag{2.78}$$

$\forall\, x\leq b$.

Let $\psi(b)-\psi(x_1)=h>0$ $(x_1<b)$, then

$$\left(-w_{b\psi}^{[1]}(b)\right)h+w_{b\psi}^{[2]}(b)\frac{h^2}{2}=w_b(x_1)-I_{b-}^{\mu;\psi}\,{}^{H}\mathbb{D}_{b-}^{\alpha,\beta;\psi}f(x_1)=:B_1. \tag{2.79}$$

Let also $\psi(b)-\psi(x_2)=2h$ $(x_2<b)$, then

$$2\left(-w_{b\psi}^{[1]}(b)\right)h+2w_{b\psi}^{[2]}(b)h^2=w_b(x_2)-I_{b-}^{\mu;\psi}\,{}^{H}\mathbb{D}_{b-}^{\alpha,\beta;\psi}f(x_2)=:B_2. \tag{2.80}$$

So we solve the system

$$\begin{aligned}2\left(-w_{b\psi}^{[1]}(b)\right)h+w_{b\psi}^{[2]}(b)h^2&=2B_1,\\2\left(-w_{b\psi}^{[1]}(b)\right)h+2w_{b\psi}^{[2]}(b)h^2&=B_2.\end{aligned} \tag{2.81}$$

We get that

$$w_{b\psi}^{[2]}(b)\,h^2 = B_2 - 2B_1,$$

i.e.

$$w_{b\psi}^{[2]}(b) = \frac{B_2 - 2B_1}{h^2}, \tag{2.82}$$

and

$$-w_{b\psi}^{[1]}(b) = \frac{4B_1 - B_2}{2h}. \tag{2.83}$$

Consequently, we get that

$$\left|w_{b\psi}^{[1]}(b)\right| \leq \frac{1}{2h}\left(4|B_1| + |B_2|\right), \tag{2.84}$$

and

$$\left|w_{b\psi}^{[2]}(b)\right| \leq \frac{1}{h^2}\left(|B_2| + 2|B_1|\right). \tag{2.85}$$

Therefore we have

$$\left|w_{b\psi}^{[1]}(b)\right| \leq$$

$$\frac{1}{2h}\left[4\left[\frac{1}{\Gamma((1-\beta)(3-\alpha))}\int_{x_1}^{b}\psi'(t)\left(\psi(t)-\psi(x_1)\right)^{(1-\beta)(3-\alpha)-1}|f(t)|\,dt+\right.\right.$$

$$\left.\frac{1}{\Gamma(\mu)}\int_{x_1}^{b}\psi'(t)\left(\psi(t)-\psi(x_1)\right)^{\mu-1}\left|{}^{H}\mathbb{D}_{b-}^{\alpha,\beta;\psi}f(t)\right|dt\right]+$$

$$\left[\frac{1}{\Gamma((1-\beta)(3-\alpha))}\int_{x_2}^{b}\psi'(t)\left(\psi(t)-\psi(x_2)\right)^{(1-\beta)(3-\alpha)-1}|f(t)|\,dt+\right.$$

$$\left.\left.\frac{1}{\Gamma(\mu)}\int_{x_2}^{b}\psi'(t)\left(\psi(t)-\psi(x_2)\right)^{\mu-1}\left|{}^{H}\mathbb{D}_{b-}^{\alpha,\beta;\psi}f(t)\right|dt\right]\right] \leq \tag{2.86}$$

$$\frac{1}{2h}\left[4\left[\frac{\|f\|_{\infty,\mathbb{R}_-}}{\Gamma((1-\beta)(3-\alpha)+1)}h^{(1-\beta)(3-\alpha)} + \frac{\left\|{}^{H}\mathbb{D}_{b-}^{\alpha,\beta;\psi}f\right\|_{\infty,\mathbb{R}_-}}{\Gamma(\mu+1)}h^{\mu}\right]+\right.$$

$$\left.\left[\frac{\|f\|_{\infty,\mathbb{R}_-}}{\Gamma((1-\beta)(3-\alpha)+1)}(2h)^{(1-\beta)(3-\alpha)} + \frac{\left\|{}^{H}\mathbb{D}_{b-}^{\alpha,\beta;\psi}f\right\|_{\infty,\mathbb{R}_-}}{\Gamma(\mu+1)}(2h)^{\mu}\right]\right] =$$

$$\frac{1}{2h}\left[\frac{\|f\|_{\infty,\mathbb{R}_-}}{\Gamma((1-\beta)(3-\alpha)+1)}\left(4+2^{(1-\beta)(3-\alpha)}\right)h^{(1-\beta)(3-\alpha)}+\right.$$

$$\left.\frac{\left\| {}^{H}\mathbb{D}_{b-}^{\alpha,\beta;\psi} f \right\|_{\infty,\mathbb{R}_-}}{\Gamma(\mu+1)} (4+2^{\mu}) h^{\mu} \right] \leq$$

$$\frac{\|f\|_{\infty,\mathbb{R}_-}\left(4+2^{(1-\beta)(3-\alpha)}\right)}{2\Gamma((1-\beta)(3-\alpha)+1)} \frac{1}{h^{1-(1-\beta)(3-\alpha)}} + \frac{\sup_{b\in\mathbb{R}_-} \left\| {}^{H}\mathbb{D}_{b-}^{\alpha,\beta;\psi} f \right\|_{\infty,\mathbb{R}_-} (4+2^{\mu})}{2\Gamma(\mu+1)} h^{\mu-1},$$

$\forall\, b \in \mathbb{R}_-, \forall\, h > 0.$

Therefore it holds

$$\sup_{b\in\mathbb{R}_-} \left| w_{b\psi}^{[1]}(b) \right| \leq \left(\frac{\|f\|_{\infty,\mathbb{R}_-}\left(4+2^{(1-\beta)(3-\alpha)}\right)}{2\Gamma((1-\beta)(3-\alpha)+1)} \right) \frac{1}{h^{1-(1-\beta)(3-\alpha)}} \tag{2.87}$$

$$+ \left(\frac{\sup_{b\in\mathbb{R}_-} \left\| {}^{H}\mathbb{D}_{b-}^{\alpha,\beta;\psi} f \right\|_{\infty,\mathbb{R}_-} (4+2^{\mu})}{2\Gamma(\mu+1)} \right) h^{\mu-1}, \quad \forall\, h > 0.$$

Next we apply Theorem 2.4 to (2.87), to derive (2.75).

Finally, we estimate

$$\left| w_{b\psi}^{[2]}(b) \right| \leq \frac{1}{h^2} \left[\left[\frac{\|f\|_{\infty,\mathbb{R}_-}}{\Gamma((1-\beta)(3-\alpha)+1)} (2h)^{(1-\beta)(3-\alpha)} + \frac{\left\| {}^{H}\mathbb{D}_{b-}^{\alpha,\beta;\psi} f \right\|_{\infty,\mathbb{R}_-}}{\Gamma(\mu+1)} (2h)^{\mu} \right] + \right.$$

$$\left. 2 \left[\frac{\|f\|_{\infty,\mathbb{R}_-}}{\Gamma((1-\beta)(3-\alpha)+1)} h^{(1-\beta)(3-\alpha)} + \frac{\left\| {}^{H}\mathbb{D}_{b-}^{\alpha,\beta;\psi} f \right\|_{\infty,\mathbb{R}_-}}{\Gamma(\mu+1)} h^{\mu} \right] \right] =$$

$$\left(\frac{\|f\|_{\infty,\mathbb{R}_-}\left(2+2^{(1-\beta)(3-\alpha)}\right)}{\Gamma((1-\beta)(3-\alpha)+1)} \right) \frac{1}{h^{2-(1-\beta)(3-\alpha)}} +$$

$$\left(\frac{\sup_{b\in\mathbb{R}_-} \left\| {}^{H}\mathbb{D}_{b-}^{\alpha,\beta;\psi} f \right\|_{\infty,\mathbb{R}_-} (2+2^{\mu})}{\Gamma(\mu+1)} \right) h^{\mu-2}, \tag{2.88}$$

$\forall\, b \in \mathbb{R}_-, \forall\, h > 0.$

Therefore it holds

$$\sup_{b\in\mathbb{R}_-}\left|w_{b\psi}^{[2]}(b)\right| \le \left(\frac{\|f\|_{\infty,\mathbb{R}_-}\left(2+2^{(1-\beta)(3-\alpha)}\right)}{\Gamma((1-\beta)(3-\alpha)+1)}\right)\frac{1}{h^{2-(1-\beta)(3-\alpha)}} \tag{2.89}$$

$$+\left(\frac{\sup_{b\in\mathbb{R}-}\left\|{}^{H}\mathbb{D}_{b-}^{\alpha,\beta;\psi}f\right\|_{\infty,\mathbb{R}_-}(2+2^{\mu})}{\Gamma(\mu+1)}\right)h^{\mu-2},\ \ \forall\, h>0.$$

Next by applying Theorem 2.4 to (2.89), we obtain (2.76).

The theorem is proved. ■

References

1. Aglic Aljinovic, A., Marangunic, L., Pecaric, J.: On Landau type inequalities via Ostrowski inequalities. Nonlinear Funct. Anal. Appl. **10**(4), 565–579 (2005)
2. Almeida, R.: A Caputo fractional derivative of a function with respect to another function. Commun. Nonlinear Sci. Numer. Simulat. **44**, 460–481 (2017)
3. Anastassiou, G.: High order abstract generalized Landau inequalities. J. Appl. Pure Math. accepted for publication (2020)
4. Anastassiou, G.: Advancements on ψ-Hilfer fractional calculus and fractional integral inequalities. Nonlinearity and Complexity, accepted, Discontinuity (2021)
5. Anastassiou, G.: Landau type ψ-Hilfer fractional inequalities, submitted (2021)
6. Barnett, N.S., Dragomir, S.S.: Some Landau type inequalities for functions whose derivatives are of locally bounded variation. Tamkang J. Math. **37**(4), 301–308 (2006)
7. Ditzian, Z.: Remarks, questions and conjectures on Landau-Kolmogorov-type inequalities. Math. Inequal. Appl. **3**, 15–24 (2000)
8. Hardy, G.H., Littlewood, J.E.: Some integral inequalities connected with the calculus of variations. Quart. J. Math. Oxford Ser. **3**, 241–252 (1932)
9. Hardy, G.H., Landau, E., Littlewood, J.E.: Some inequalities satisfied by the integrals or derivatives of real or analytic functions. Math. Z. **39**, 677–695 (1935)
10. Kallman, R.R., Rota, G.C.: On the inequality $\left|\left|f'\right|\right|^2 \le 4\,||f||\cdot\left|\left|f''\right|\right|$. In: Shisha, O. (ed.) Inequalities, Vol. II, 187–192. Academic Press, New York (1970)
11. Kilbas, A.A., Srivastava, H.M., Trujillo, J.J.: Theory and Applications of Fractional Differentiation Equations. North Holland, Amsterdam, New York (2006)
12. Landau, E.: Einige Ungleichungen für zweimal differentzierban funktionen. Proc. London Math. Soc. **13**, 43–49 (1913)
13. Samko, S.G., Kilbas, A.A., Marichev, O.I.: Fractional Integrals and Derivatives: Theory and Applications. Gordon and Breach Science Publishers, Switzerland (1993)
14. Tomovski, Ž, Hilfer, R., Srivastava, H.M.: Fractional and Operational Calculus with Generalized Fractional Derivative Operators and Mittag-Leffler Type Functions. Integral Transforms Spec Funct. **21**(11), 797–814 (2010)
15. Vanterler da, J., Sousa, C., Capelas de Oliveira, E.: On the ψ-Hilfer fractional derivative. Commun. Nonlinear Sci. Numer. Simul. **60**, 72–91 (2018)

Chapter 3
Iyengar-Hilfer Generalized Fractional Inequalities

Here we present Iyengar type integral inequalities. At the univariate level they involve ψ-Hilfer left and right fractional derivatives. At the multivariate level they involve Hilfer left and right fractional derivatives, and they deal with radial and non-radial functions on the ball and spherical shell. All estimates are with respect to norms $\|\cdot\|_p$, $1 \leq p \leq \infty$. At the end we provide an application. It follows [2].

3.1 Introduction

We are motivated by the following famous Iyengar inequality (1938), [4].

Theorem 3.1 *Let f be a differentiable function on $[a, b]$ and $|f'(x)| \leq M$. Then*

$$\left| \int_a^b f(x)\,dx - \frac{1}{2}(b-a)(f(a)+f(b)) \right| \leq \frac{M(b-a)^2}{4} - \frac{(f(b)-f(a))^2}{4M}, \tag{3.1}$$

Iyengar type inequalities provide tight estimates to the integral of a function. In this chapter we present Iyengar type integral inequalities involving the ψ-Hilfer and Hilfer left and right fractional derivatives. These are with respect to norms $\|\cdot\|_p$, $1 \leq p \leq \infty$. We treat also the multivariate cases of radial and non-radial approaches over the shell and ball.

3.2 Background

Let $-\infty < a < b < \infty$, the left and right Riemann-Liouville fractional integrals of order $\alpha \in \mathbb{C}$ $(\mathcal{R}(\alpha) > 0)$ are defined by

G. A. Anastassiou, *Unification of Fractional Calculi with Applications*,
Studies in Systems, Decision and Control 398,
https://doi.org/10.1007/978-3-030-86920-5_3

$$\left(I_{a+}^{\alpha} f\right)(x) = \frac{1}{\Gamma(\alpha)} \int_a^x (x-t)^{\alpha-1} f(t)\, dt, \tag{3.2}$$

$x > a$; where Γ stands for the gamma function,

and

$$\left(I_{b-}^{\alpha} f\right)(x) = \frac{1}{\Gamma(\alpha)} \int_x^b (t-x)^{\alpha-1} f(t)\, dt, \tag{3.3}$$

$x < b$.

The Riemann-Liouville left and right fractional derivatives of order $\alpha \in \mathbb{C}$ ($\mathcal{R}(\alpha) \geq 0$) are defined by

$$\left(\Delta_{a+}^{\alpha} y\right)(x) = \left(\frac{d}{dx}\right)^n \left(I_{a+}^{n-\alpha} y\right)(x) = \frac{1}{\Gamma(n-\alpha)} \left(\frac{d}{dx}\right)^n \int_a^x (x-t)^{n-\alpha-1} y(t)\, dt \tag{3.4}$$

($n = \lceil \mathcal{R}(\alpha) \rceil$, $\lceil \cdot \rceil$ means ceiling of the number; $x > a$)

$$\left(\Delta_{b-}^{\alpha} y\right)(x) = (-1)^n \left(\frac{d}{dx}\right)^n \left(I_{b-}^{n-\alpha} y\right)(x) =$$

$$\frac{(-1)^n}{\Gamma(n-\alpha)} \left(\frac{d}{dx}\right)^n \int_x^b (t-x)^{n-\alpha-1} y(t)\, dt \tag{3.5}$$

($n = \lceil \mathcal{R}(\alpha) \rceil$; $x < b$), respectively, where $\mathcal{R}(\alpha)$ is the real part of α.

In particular, when $\alpha = n \in \mathbb{Z}_+$, then

$$\left(\Delta_{a+}^{0} y\right)(x) = \left(\Delta_{b-}^{0} y\right)(x) = y(x);$$

$$\left(\Delta_{a+}^{n} y\right)(x) = y^{(n)}(x), \text{ and } \left(\Delta_{b-}^{n} y\right)(x) = (-1)^n y^{(n)}(x), \ n \in \mathbb{N},$$

see [7].

Let $\alpha > 0$, $I = [a, b] \subset \mathbb{R}$, f an integrable function defined on I and $\psi \in C^1(I)$ an increasing function such that $\psi'(x) \neq 0$, for all $x \in I$. Left fractional integrals and left Riemann-Liouville fractional derivatives of a function f with respect to another function ψ are defined as [5, 7]

$$I_{a+}^{\alpha,\psi} f(x) = \frac{1}{\Gamma(\alpha)} \int_a^x \psi'(t) (\psi(x) - \psi(t))^{\alpha-1} f(t)\, dt, \tag{3.6}$$

and

$$\Delta_{a+}^{\alpha,\psi} f(x) = \left(\frac{1}{\psi'(x)} \frac{d}{dx}\right)^n I_{a+}^{n-\alpha,\psi} f(x) = \tag{3.7}$$

$$\frac{1}{\Gamma(n-\alpha)} \left(\frac{1}{\psi'(x)} \frac{d}{dx}\right)^n \int_a^x \psi'(t) (\psi(x) - \psi(t))^{n-\alpha-1} f(t)\, dt,$$

respectively, where $n = \lceil \alpha \rceil$.

Similarly, we define the right ones:

$$I_{b-}^{\alpha,\psi} f(x) = \frac{1}{\Gamma(\alpha)} \int_x^b \psi'(t) (\psi(t) - \psi(x))^{\alpha-1} f(t)\, dt, \tag{3.8}$$

and

$$\Delta_{b-}^{\alpha,\psi} f(x) = \left(-\frac{1}{\psi'(x)} \frac{d}{dx} \right)^n I_{b-}^{n-\alpha,\psi} f(x) =$$

$$\frac{1}{\Gamma(n-\alpha)} \left(-\frac{1}{\psi'(x)} \frac{d}{dx} \right)^n \int_x^b \psi'(t) (\psi(t) - \psi(x))^{n-\alpha-1} f(t)\, dt. \tag{3.9}$$

The following semigroup property holds; if $\alpha, \beta > 0$, $f \in C(I)$, then

$$I_{a+}^{\alpha,\psi} I_{a+}^{\beta,\psi} f = I_{a+}^{\alpha+\beta,\psi} f \quad \text{and} \quad I_{b-}^{\alpha,\psi} I_{b-}^{\beta,\psi} f = I_{b-}^{\alpha+\beta,\psi} f.$$

Next let again $\alpha > 0$, $n = \lceil \alpha \rceil$, $I = [a, b]$, $f, \psi \in C^n(I) : \psi$ is increasing and $\psi'(x) \neq 0$, for all $x \in I$. The left ψ-Caputo fractional derivative of f of order α is given by Almeida [1]

$$^C D_{a+}^{\alpha,\psi} f(x) = I_{a+}^{n-\alpha,\psi} \left(\frac{1}{\psi'(x)} \frac{d}{dx} \right)^n f(x), \tag{3.10}$$

and the right ψ-Caputo fractional derivative [1]

$$^C D_{b-}^{\alpha,\psi} f(x) = I_{b-}^{n-\alpha,\psi} \left(-\frac{1}{\psi'(x)} \frac{d}{dx} \right)^n f(x). \tag{3.11}$$

We set

$$f_\psi^{[n]}(x) := f_\psi^{(n)} f(x) := \left(\frac{1}{\psi'(x)} \frac{d}{dx} \right)^n f(x). \tag{3.12}$$

Clearly, when $\alpha = m \in \mathbb{N}$ we have

$$^C D_{a+}^{\alpha,\psi} f(x) = f_\psi^{[m]}(x) \quad \text{and} \quad {}^C D_{b-}^{\alpha,\psi} f(x) = (-1)^m f_\psi^{[m]}(x),$$

and if $\alpha \notin \mathbb{N}$, then

$$^C D_{a+}^{\alpha,\psi} f(x) = \frac{1}{\Gamma(n-\alpha)} \int_a^x \psi'(t) (\psi(x) - \psi(t))^{n-\alpha-1} f_\psi^{[n]}(t)\, dt, \tag{3.13}$$

and

$$ {}^{C}D_{b-}^{\alpha,\psi} f(x) = \frac{(-1)^n}{\Gamma(n-\alpha)} \int_x^b \psi'(t) (\psi(t) - \psi(x))^{n-\alpha-1} f_{\psi}^{[n]}(t) \, dt. \tag{3.14} $$

If $\psi(x) = x$, then we get the usual left and right Caputo fractional derivatives

$$ {}^{C}D_{a+}^{m} f(x) = f^{(m)}(x), \quad {}^{C}D_{b-}^{m} f(x) = (-1)^m f^{(m)}(x), $$

for $m \in \mathbb{N}$, and ($\alpha \notin \mathbb{N}$)

$$ D_{*a}^{\alpha} f(x) = {}^{C}D_{a+}^{\alpha} f(x) = \frac{1}{\Gamma(n-\alpha)} \int_a^x (x-t)^{n-\alpha-1} f^{(n)}(t) \, dt, \tag{3.15} $$

$$ D_{b-}^{\alpha}(x) = {}^{C}D_{b-}^{\alpha} f(x) = \frac{(-1)^n}{\Gamma(n-\alpha)} \int_x^b (t-x)^{n-\alpha-1} f^{(n)}(t) \, dt. \tag{3.16} $$

Also we set

$$ {}^{C}D_{a+}^{0,\psi} f(x) = {}^{C}D_{b-}^{0,\psi} f(x) = f(x). $$

Next we will deal with the ψ-Hilfer fractional derivative.

Definition 3.2 ([10]) Let $n-1 < \alpha < n$, $n \in \mathbb{N}$, $I = [a,b] \subset \mathbb{R}$ and $f, \psi \in C^n([a,b])$, ψ is increasing and $\psi'(x) \neq 0$, for all $x \in I$. The ψ-Hilfer fractional derivative (left-sided and right-sided) ${}^{H}\mathbb{D}_{a+(b-)}^{\alpha,\beta;\psi} f$ of order α and type $0 \leq \beta \leq 1$, respectively, are defined by

$$ {}^{H}\mathbb{D}_{a+}^{\alpha,\beta;\psi} f(x) = I_{a+}^{\beta(n-\alpha);\psi} \left(\frac{1}{\psi'(x)} \frac{d}{dx} \right)^n I_{a+}^{(1-\beta)(n-\alpha);\psi} f(x), \tag{3.17} $$

and

$$ {}^{H}\mathbb{D}_{b-}^{\alpha,\beta;\psi} f(x) = I_{b-}^{\beta(n-\alpha);\psi} \left(-\frac{1}{\psi'(x)} \frac{d}{dx} \right)^n I_{b-}^{(1-\beta)(n-\alpha);\psi} f(x), \quad x \in [a,b]. \tag{3.18} $$

The original Hilfer fractional derivatives [9] come from $\psi(x) = x$, and are denoted by ${}^{H}\mathbb{D}_{a+}^{\alpha,\beta} f(x)$ and ${}^{H}\mathbb{D}_{b-}^{\alpha,\beta} f(x)$.

When $\beta = 0$, we get Riemann-Liouville fractional derivatives, while when $\beta = 1$ we have Caputo type fractional derivatives.

We define $\gamma = \alpha + \beta(n-\alpha)$. We notice that $n-1 < \alpha \leq \alpha + \beta(n-\alpha) \leq \alpha + n - \alpha = n$, hence $\lceil \gamma \rceil = n$. We can easily write that [10]

$$ {}^{H}\mathbb{D}_{a+}^{\alpha,\beta;\psi} f(x) = I_{a+}^{\gamma-\alpha;\psi} \Delta_{a+}^{\gamma;\psi} f(x), \tag{3.19} $$

and

$$ {}^{H}\mathbb{D}_{b-}^{\alpha,\beta;\psi} f(x) = I_{b-}^{\gamma-\alpha;\psi} \Delta_{b-}^{\gamma;\psi} f(x), \quad x \in [a,b]. \tag{3.20} $$

We have that [10]

$$\Delta_{a+}^{\gamma,\psi} f(x) = \left(\frac{1}{\psi'(x)}\frac{d}{dx}\right)^n I_{a+}^{(1-\beta)(n-\alpha);\psi} f(x), \tag{3.21}$$

and

$$\Delta_{b-}^{\gamma,\psi} f(x) = \left(-\frac{1}{\psi'(x)}\frac{d}{dx}\right)^n I_{b-}^{(1-\beta)(n-\alpha);\psi} f(x). \tag{3.22}$$

In particular, when $0 < \alpha < 1$ and $0 \leq \beta \leq 1$; $\gamma = \alpha + \beta(1-\alpha)$, we have that

$${}^{H}\mathbb{D}_{a+}^{\alpha,\beta;\psi} f(x) = \frac{1}{\Gamma(\gamma-\alpha)} \int_a^x \psi'(t)(\psi(x)-\psi(t))^{\gamma-\alpha-1} \Delta_{a+}^{\gamma;\psi} f(t)\, dt, \tag{3.23}$$

and

$${}^{H}\mathbb{D}_{b-}^{\alpha,\beta;\psi} f(x) = \frac{1}{\Gamma(\gamma-\alpha)} \int_x^b \psi'(t)(\psi(t)-\psi(x))^{\gamma-\alpha-1} \Delta_{b-}^{\gamma;\psi} f(t)\, dt, \tag{3.24}$$

$x \in [a,b]$.

Remark 3.3 ([10]) Let $\mu = n(1-\beta) + \beta\alpha$, then $\lceil \mu \rceil = n$.
Assume that $g(x) = I_{a+}^{(1-\beta)(n-\alpha);\psi} f(x) \in C^n([a,b])$, we have that

$${}^{H}\mathbb{D}_{a+}^{\alpha,\beta;\psi} f(x) = I_{a+}^{n-\mu;\psi}\left(\frac{1}{\psi'(x)}\frac{d}{dx}\right)^n g(x). \tag{3.25}$$

Thus

$${}^{H}\mathbb{D}_{a+}^{\alpha,\beta;\psi} f = {}^{C}D_{a+}^{\mu;\psi} g(x) = {}^{C}D_{a+}^{\mu;\psi}\left[I_{a+}^{(1-\beta)(n-\alpha);\psi} f(x)\right]. \tag{3.26}$$

Assume that $w(x) = I_{b-}^{(1-\beta)(n-\alpha);\psi} f(x) \in C^n([a,b])$. Hence

$${}^{H}\mathbb{D}_{b-}^{\alpha,\beta;\psi} f(x) = I_{b-}^{\beta(n-\alpha);\psi}\left(-\frac{1}{\psi'(x)}\frac{d}{dx}\right)^n w(x) = I_{b-}^{n-\mu;\psi}\left(-\frac{1}{\psi'(x)}\frac{d}{dx}\right)^n w(x). \tag{3.27}$$

Thus

$${}^{H}\mathbb{D}_{b-}^{\alpha,\beta;\psi} f = {}^{C}D_{b-}^{\mu;\psi} w(x) = {}^{C}D_{b-}^{\mu;\psi}\left(I_{b-}^{(1-\beta)(n-\alpha);\psi} f(x)\right). \tag{3.28}$$

We also need the following ψ-Hilfer left and right fractional Taylor formulae:

Theorem 3.4 ([10]) *Let $\psi, f \in C^n([a,b])$, with ψ being increasing such that $\psi'(x) \neq 0$ over $[a,b]$, where $n-1 < \alpha < n$, $0 \leq \beta \leq 1$, and $\gamma = \alpha + \beta(n-\alpha)$, $x \in [a,b]$. Then*

$$I_{a+}^{\alpha;\psi}\ {}^{H}\mathbb{D}_{a+}^{\alpha,\beta;\psi} f(x) = f(x) - \sum_{k=1}^{n} \frac{(\psi(x) - \psi(a))^{\gamma-k}}{\Gamma(\gamma-k+1)} f_{\psi}^{[n-k]} \left(I_{a+}^{(1-\beta)(n-\alpha);\psi} f \right)(a), \quad (3.29)$$

and

$$I_{b-}^{\alpha;\psi}\ {}^{H}\mathbb{D}_{b-}^{\alpha,\beta;\psi} f(x) = f(x) - \sum_{k=1}^{n} \frac{(-1)^k (\psi(b) - \psi(x))^{\gamma-k}}{\Gamma(\gamma-k+1)} f_{\psi}^{[n-k]} \left(I_{b-}^{(1-\beta)(n-\alpha);\psi} f \right)(b). \quad (3.30)$$

A simplification to Theorem 3.4 follows:

Theorem 3.5 *Let $\psi, f \in C^n([a,b])$, with ψ being increasing such that $\psi'(x) \neq 0$ over $[a,b]$, where $n-1 < \alpha < n$, $0 \leq \beta \leq 1$, and $\gamma = \alpha + \beta(n-\alpha)$, $x \in [a,b]$. Then*

$$f(x) - \sum_{k=1}^{n-1} \frac{(\psi(x) - \psi(a))^{\gamma-k}}{\Gamma(\gamma-k+1)} f_{\psi}^{[n-k]} \left(I_{a+}^{(1-\beta)(n-\alpha);\psi} f \right)(a) =$$

$$\frac{1}{\Gamma(\alpha)} \int_a^x \psi'(t) (\psi(x) - \psi(t))^{\alpha-1}\ {}^{H}\mathbb{D}_{a+}^{\alpha,\beta;\psi} f(t)\, dt, \quad (3.31)$$

and

$$f(x) - \sum_{k=1}^{n-1} \frac{(-1)^k (\psi(b) - \psi(x))^{\gamma-k}}{\Gamma(\gamma-k+1)} f_{\psi}^{[n-k]} \left(I_{b-}^{(1-\beta)(n-\alpha);\psi} f \right)(b) =$$

$$\frac{1}{\Gamma(\alpha)} \int_x^b \psi'(t) (\psi(t) - \psi(x))^{\alpha-1}\ {}^{H}\mathbb{D}_{b-}^{\alpha,\beta;\psi} f(t)\, dt. \quad (3.32)$$

Proof From Theorem 3.4. Also notice that

$$f_{\psi}^{[n-n]} \left(I_{a+}^{(1-\beta)(n-\alpha);\psi} f \right)(a) = f_{\psi}^{[0]} \left(I_{a+}^{(1-\beta)(n-\alpha);\psi} f \right)(a) =$$

$$\left(I_{a+}^{(1-\beta)(n-\alpha);\psi} f \right)(a) = 0.$$

The last is true from the following:

$$\left| \left(I_{a+}^{(1-\beta)(n-\alpha);\psi} f \right)(x) \right| \leq$$

$$\frac{1}{\Gamma((1-\beta)(n-\alpha))} \int_a^x \psi'(t) (\psi(x) - \psi(t))^{(1-\beta)(n-\alpha)-1} |f(t)|\, dt \leq$$

$$\frac{\|f\|_{\infty}}{\Gamma((1-\beta)(n-\alpha)+1)} (\psi(x) - \psi(a))^{(1-\beta)(n-\alpha)}, \quad \forall\, x \in [a,b]. \quad (3.33)$$

Clearly it holds $\left(I_{a+}^{(1-\beta)(n-\alpha);\psi} f\right)(a) = 0$.

Furthermore we have

$$f_{\psi}^{[n-n]}\left(I_{b-}^{(1-\beta)(n-\alpha);\psi} f\right)(b) = \left(I_{b-}^{(1-\beta)(n-\alpha);\psi} f\right)(b) = 0.$$

The last follows from

$$\left|\left(I_{b-}^{(1-\beta)(n-\alpha);\psi} f\right)(x)\right| \leq$$

$$\frac{1}{\Gamma((1-\beta)(n-\alpha))} \int_x^b \psi'(t)(\psi(t) - \psi(x))^{(1-\beta)(n-\alpha)-1} |f(t)| \, dt \leq$$

$$\frac{\|f\|_{\infty}}{\Gamma((1-\beta)(n-\alpha)+1)} (\psi(b) - \psi(x))^{(1-\beta)(n-\alpha)}, \quad \forall\, x \in [a,b]. \tag{3.34}$$

Hence $\left(I_{b-}^{(1-\beta)(n-\alpha);\psi} f\right)(b) = 0$. ■

3.3 Main Results

3.3.1 Univariate Results

We present

Theorem 3.6 *Let $\psi, f \in C^n([a,b])$, with ψ being increasing such that $\psi'(x) \neq 0$ over $[a,b]$, where $n-1 < \alpha < n \in \mathbb{N}$, $0 \leq \beta \leq 1$, and $\gamma = \alpha + \beta(n-\alpha)$. Call $\rho_1 := \left\| {}^H\mathbb{D}_{a+}^{\alpha,\beta;\psi} f \right\|_{\infty}$, $\rho_2 := \left\| {}^H\mathbb{D}_{b-}^{\alpha,\beta;\psi} f \right\|_{\infty}$, and set $\rho := \max(\rho_1, \rho_2)$. Then*

(1)

$$\left| \int_a^b f(x)\, d\psi(x) - \left\{ \sum_{k=1}^{n-1} \frac{1}{\Gamma(\gamma - k + 2)} \left[f_{\psi}^{[n-k]}\left(I_{a+}^{(1-\beta)(n-\alpha);\psi} f\right)(a)(\psi(t) - \psi(a))^{\gamma-k+1} \right.\right.\right.$$

$$\left.\left.\left. + (-1)^k f_{\psi}^{[n-k]}\left(I_{b-}^{(1-\beta)(n-\alpha);\psi} f\right)(b)(\psi(b) - \psi(t))^{\gamma-k+1} \right]\right\}\right| \leq$$

$$\frac{\rho}{\Gamma(\alpha+2)} \left[(\psi(b) - \psi(t))^{\alpha+1} + (\psi(t) - \psi(a))^{\alpha+1}\right], \tag{3.35}$$

$\forall\, t \in [a,b]$,

(2) the right hand side (R.H.S.) of (3.35) is minimized at $t \in [a,b]$, such that $\psi(t) = \frac{\psi(a)+\psi(b)}{2}$, with a minimum value of $\frac{\rho}{\Gamma(\alpha+2)} \frac{(\psi(b)-\psi(a))^{\alpha+1}}{2^{\alpha}}$, and we have

$$\left| \int_a^b f(x)\, d\psi(x) - \left\{ \sum_{k=1}^{n-1} \frac{1}{\Gamma(\gamma-k+2)} \left(\frac{\psi(b)-\psi(a)}{2} \right)^{\gamma-k+1} \right. \right.$$

$$\left. \left. \left[f_\psi^{[n-k]} \left(I_{a+}^{(1-\beta)(n-\alpha);\psi} f \right)(a) + (-1)^k f_\psi^{[n-k]} \left(I_{b-}^{(1-\beta)(n-\alpha);\psi} f \right)(b) \right] \right\} \right| \leq$$

$$\frac{\rho (\psi(b)-\psi(a))^{\alpha+1}}{\Gamma(\alpha+2)\, 2^\alpha}, \tag{3.36}$$

(3) assuming $f_\psi^{[n-k]} \left(I_{a+}^{(1-\beta)(n-\alpha);\psi} f \right)(a) = f_\psi^{[n-k]} \left(I_{b-}^{(1-\beta)(n-\alpha);\psi} f \right)(b) = 0$, *for* $k = 1, ..., n-1$, *we get*

$$\left| \int_a^b f(x)\, d\psi(x) \right| \leq \frac{\rho (\psi(b)-\psi(a))^{\alpha+1}}{\Gamma(\alpha+2)\, 2^\alpha}, \tag{3.37}$$

a sharp inequality,

(4) additionally, assume that ψ *is strictly increasing,* $j = 0, 1, 2, ..., N \in \mathbb{N}$, *it holds*

$$\left| \int_a^b f(x)\, d\psi(x) - \left\{ \sum_{k=1}^{n-1} \frac{1}{\Gamma(\gamma-k+2)} \left(\frac{\psi(b)-\psi(a)}{N} \right)^{\gamma-k+1} \right. \right.$$

$$\left[f_\psi^{[n-k]} \left(I_{a+}^{(1-\beta)(n-\alpha);\psi} f \right)(a)\, j^{\gamma-k+1} + \right.$$

$$\left. \left. \left. (-1)^k f_\psi^{[n-k]} \left(I_{b-}^{(1-\beta)(n-\alpha);\psi} f \right)(b)\, (N-j)^{\gamma-k+1} \right] \right\} \right| \leq$$

$$\frac{\rho}{\Gamma(\alpha+2)} \left(\frac{\psi(b)-\psi(a)}{N} \right)^{\alpha+1} \left(j^{\alpha+1} + (N-j)^{\alpha+1} \right), \tag{3.38}$$

(5) if $f_\psi^{[n-k]} \left(I_{a+}^{(1-\beta)(n-\alpha);\psi} f \right)(a) = f_\psi^{[n-k]} \left(I_{b-}^{(1-\beta)(n-\alpha);\psi} f \right)(b) = 0$, $k = 1, ..., n-1$, *from (3.38) we derive*

$$\left| \int_a^b f(x)\, d\psi(x) \right| \leq$$

$$\frac{\rho}{\Gamma(\alpha+2)} \left(\frac{\psi(b)-\psi(a)}{N} \right)^{\alpha+1} \min_{j \in \{0,1,...,N\}} \left(j^{\alpha+1} + (N-j)^{\alpha+1} \right), \tag{3.39}$$

(6) when $N = 2$ *and* $j = 1$, *from (3.39) we get*

$$\left|\int_a^b f(x)\,d\psi(x)\right| \leq \frac{\rho}{\Gamma(\alpha+2)} \frac{(\psi(b)-\psi(a))^{\alpha+1}}{2^\alpha}, \tag{3.40}$$

(7) if $0 < \alpha < 1$, *we have (3.40) valid without any boundary conditions.*

Proof From (3.31) we have

$$\left| f(x) - \sum_{k=1}^{n-1} \frac{(\psi(x)-\psi(a))^{\gamma-k}}{\Gamma(\gamma-k+1)} f_\psi^{[n-k]} \left(I_{a+}^{(1-\beta)(n-\alpha);\psi} f \right)(a) \right| \leq$$

$$\left\| {}^H\mathbb{D}_{a+}^{\alpha,\beta;\psi} f \right\|_\infty \frac{(\psi(x)-\psi(a))^\alpha}{\Gamma(\alpha+1)}, \quad \forall\, x \in [a,b]. \tag{3.41}$$

And by (3.32) we derive

$$\left| f(x) - \sum_{k=1}^{n-1} \frac{(-1)^k (\psi(b)-\psi(x))^{\gamma-k}}{\Gamma(\gamma-k+1)} f_\psi^{[n-k]} \left(I_{b-}^{(1-\beta)(n-\alpha);\psi} f \right)(b) \right| \leq$$

$$\left\| {}^H\mathbb{D}_{b-}^{\alpha,\beta;\psi} f \right\|_\infty \frac{(\psi(b)-\psi(x))^\alpha}{\Gamma(\alpha+1)}, \quad \forall\, x \in [a,b]. \tag{3.42}$$

That is

$$\left| f(x) - \sum_{k=1}^{n-1} \frac{(\psi(x)-\psi(a))^{\gamma-k}}{\Gamma(\gamma-k+1)} f_\psi^{[n-k]} \left(I_{a+}^{(1-\beta)(n-\alpha);\psi} f \right)(a) \right| \leq$$

$$\frac{\rho}{\Gamma(\alpha+1)} (\psi(x)-\psi(a))^\alpha, \tag{3.43}$$

and

$$\left| f(x) - \sum_{k=1}^{n-1} \frac{(-1)^k (\psi(b)-\psi(x))^{\gamma-k}}{\Gamma(\gamma-k+1)} f_\psi^{[n-k]} \left(I_{b-}^{(1-\beta)(n-\alpha);\psi} f \right)(b) \right| \leq$$

$$\frac{\rho}{\Gamma(\alpha+1)} (\psi(b)-\psi(x))^\alpha, \quad \forall\, x \in [a,b]. \tag{3.44}$$

Equivalently, we have

$$\sum_{k=1}^{n-1} \frac{(\psi(x)-\psi(a))^{\gamma-k}}{\Gamma(\gamma-k+1)} f_\psi^{[n-k]} \left(I_{a+}^{(1-\beta)(n-\alpha);\psi} f \right)(a) - \frac{\rho}{\Gamma(\alpha+1)} (\psi(x)-\psi(a))^\alpha$$

$$\leq f(x) \leq$$

$$\sum_{k=1}^{n-1} \frac{(\psi(x) - \psi(a))^{\gamma-k}}{\Gamma(\gamma-k+1)} f_{\psi}^{[n-k]} \left(I_{a+}^{(1-\beta)(n-\alpha);\psi} f \right)(a) + \frac{\rho}{\Gamma(\alpha+1)} (\psi(x) - \psi(a))^{\alpha}, \quad (3.45)$$

and

$$\sum_{k=1}^{n-1} \frac{(-1)^k (\psi(b) - \psi(x))^{\gamma-k}}{\Gamma(\gamma-k+1)} f_{\psi}^{[n-k]} \left(I_{b-}^{(1-\beta)(n-\alpha);\psi} f \right)(b) - \frac{\rho}{\Gamma(\alpha+1)} (\psi(b) - \psi(x))^{\alpha}$$

$$\leq f(x) \leq$$

$$\sum_{k=1}^{n-1} \frac{(-1)^k (\psi(b) - \psi(x))^{\gamma-k}}{\Gamma(\gamma-k+1)} f_{\psi}^{[n-k]} \left(I_{b-}^{(1-\beta)(n-\alpha);\psi} f \right)(b) + \frac{\rho}{\Gamma(\alpha+1)} (\psi(b) - \psi(x))^{\alpha}, \quad (3.46)$$

$\forall\, x \in [a, b]$.

Let any $t \in [a, b]$, then by integration against ψ over $[a, t]$ and $[t, b]$, respectively, we obtain

$$\sum_{k=1}^{n-1} \frac{f_{\psi}^{[n-k]} \left(I_{a+}^{(1-\beta)(n-\alpha);\psi} f \right)(a)}{\Gamma(\gamma-k+2)} (\psi(t) - \psi(a))^{\gamma-k+1} - \frac{\rho}{\Gamma(\alpha+2)} (\psi(t) - \psi(a))^{\alpha+1}$$

$$\leq \int_a^t f(x)\, d\psi(x) \leq$$

$$\sum_{k=1}^{n-1} \frac{f_{\psi}^{[n-k]} \left(I_{a+}^{(1-\beta)(n-\alpha);\psi} f \right)(a)}{\Gamma(\gamma-k+2)} (\psi(t) - \psi(a))^{\gamma-k+1} + \frac{\rho}{\Gamma(\alpha+2)} (\psi(t) - \psi(a))^{\alpha+1}, \quad (3.47)$$

and

$$\sum_{k=1}^{n-1} \frac{(-1)^k f_{\psi}^{[n-k]} \left(I_{b-}^{(1-\beta)(n-\alpha);\psi} f \right)(b)}{\Gamma(\gamma-k+2)} (\psi(b) - \psi(t))^{\gamma-k+1} - \frac{\rho}{\Gamma(\alpha+2)} (\psi(b) - \psi(t))^{\alpha+1}$$

$$\leq \int_t^b f(x)\, d\psi(x) \leq$$

$$\sum_{k=1}^{n-1} \frac{(-1)^k f_{\psi}^{[n-k]} \left(I_{b-}^{(1-\beta)(n-\alpha);\psi} f \right)(b)}{\Gamma(\gamma-k+2)} (\psi(b) - \psi(t))^{\gamma-k+1} + \frac{\rho}{\Gamma(\alpha+2)} (\psi(b) - \psi(t))^{\alpha+1}. \quad (3.48)$$

Adding (3.47) and (3.48), we obtain

$$\left\{ \sum_{k=1}^{n-1} \frac{1}{\Gamma(\gamma-k+2)} \left[f_{\psi}^{[n-k]} \left(I_{a+}^{(1-\beta)(n-\alpha);\psi} f \right)(a) (\psi(t) - \psi(a))^{\gamma-k+1} + \right. \right.$$

$$\left. (-1)^k f_{\psi}^{[n-k]} \left(I_{b-}^{(1-\beta)(n-\alpha);\psi} f \right)(b) (\psi(b) - \psi(t))^{\gamma-k+1} \right] -$$

$$\frac{\rho}{\Gamma(\alpha+2)}\left[(\psi(t)-\psi(a))^{\alpha+1}+(\psi(b)-\psi(t))^{\alpha+1}\right]\Bigg\}$$

$$\leq \int_a^b f(x)\,d\psi(x) \leq$$

$$\Bigg\{\sum_{k=1}^{n-1}\frac{1}{\Gamma(\gamma-k+2)}\left[f_{\psi}^{[n-k]}\left(I_{a+}^{(1-\beta)(n-\alpha);\psi} f\right)(a)\,(\psi(t)-\psi(a))^{\gamma-k+1}+\right.$$

$$\left.(-1)^k f_{\psi}^{[n-k]}\left(I_{b-}^{(1-\beta)(n-\alpha);\psi} f\right)(b)\,(\psi(b)-\psi(t))^{\gamma-k+1}\right]+$$

$$\frac{\rho}{\Gamma(\alpha+2)}\left[(\psi(t)-\psi(a))^{\alpha+1}+(\psi(b)-\psi(t))^{\alpha+1}\right]\Bigg\}, \quad (3.49)$$

$\forall\, t \in [a,b]$.

Consequently, we derive

$$\left|\int_a^b f(x)\,d\psi(x)-\Bigg\{\sum_{k=1}^{n-1}\frac{1}{\Gamma(\gamma-k+2)}\left[f_{\psi}^{[n-k]}\left(I_{a+}^{(1-\beta)(n-\alpha);\psi} f\right)(a)\,(\psi(t)-\psi(a))^{\gamma-k+1}\right.\right.$$

$$\left.\left.+(-1)^k f_{\psi}^{[n-k]}\left(I_{b-}^{(1-\beta)(n-\alpha);\psi} f\right)(b)\,(\psi(b)-\psi(t))^{\gamma-k+1}\right]\Bigg\}\right| \leq$$

$$\frac{\rho}{\Gamma(\alpha+2)}\left[(\psi(b)-\psi(t))^{\alpha+1}+(\psi(t)-\psi(a))^{\alpha+1}\right], \quad (3.50)$$

$\forall\, t \in [a,b]$.

Notice that

$$(\psi(b)-\psi(a))^{\alpha+1}=((\psi(b)-\psi(t))+(\psi(t)-\psi(a)))^{\alpha+1}\leq$$

$$2^{\alpha}\left((\psi(b)-\psi(t))^{\alpha+1}+(\psi(t)-\psi(a))^{\alpha+1}\right), \quad (3.51)$$

$\forall\, t \in [a,b]$.

That is

$$\frac{(\psi(b)-\psi(a))^{\alpha+1}}{2^{\alpha}}\leq\left[(\psi(b)-\psi(t))^{\alpha+1}+(\psi(t)-\psi(a))^{\alpha+1}\right], \quad (3.52)$$

$\forall\, t \in [a,b]$.

Choosing $t \in [a,b]$ such that $\psi(t)=\frac{\psi(a)+\psi(b)}{2}$, we get the minimum of

$$\left[(\psi(b)-\psi(t))^{\alpha+1}+(\psi(t)-\psi(a))^{\alpha+1}\right]=\frac{(\psi(b)-\psi(a))^{\alpha+1}}{2^{\alpha}}. \quad (3.53)$$

Consequently (3.50) becomes

$$\left|\int_a^b f(x)\,d\psi(x) - \left\{\sum_{k=1}^{n-1} \frac{1}{\Gamma(\gamma-k+2)}\left(\frac{\psi(b)-\psi(a)}{2}\right)^{\gamma-k+1}\right.\right.$$

$$\left.\left.\left[f_{\psi}^{[n-k]}\left(I_{a+}^{(1-\beta)(n-\alpha);\psi} f\right)(a) + (-1)^k f_{\psi}^{[n-k]}\left(I_{b-}^{(1-\beta)(n-\alpha);\psi} f\right)(b)\right]\right\}\right| \leq$$

$$\frac{\rho(\psi(b)-\psi(a))^{\alpha+1}}{\Gamma(\alpha+2)\,2^{\alpha}}, \tag{3.54}$$

Assuming further $f_{\psi}^{[n-k]}\left(I_{a+}^{(1-\beta)(n-\alpha);\psi} f\right)(a) = f_{\psi}^{[n-k]}\left(I_{b-}^{(1-\beta)(n-\alpha);\psi} f\right)(b) = 0$, for $k = 1, ..., n-1$, it holds

$$\left|\int_a^b f(x)\,d\psi(x)\right| \leq \frac{\rho(\psi(b)-\psi(a))^{\alpha+1}}{\Gamma(\alpha+2)\,2^{\alpha}}, \tag{3.55}$$

which is a sharp inequality.

Next we assume that ψ is strictly increasing and let $N \in \mathbb{N}$, $j = 0, 1, 2, ..., N$ and $\psi(t_j) = \psi(a) + j\left(\frac{\psi(b)-\psi(a)}{N}\right)$, that is $\psi(t_0) = \psi(a)$, $\psi(t_1) = \psi(a) + \left(\frac{\psi(b)-\psi(a)}{N}\right)$, ..., $\psi(t_N) = \psi(b)$. Hence it holds

$$\psi(t_j) - \psi(a) = j\left(\frac{\psi(b)-\psi(a)}{N}\right), \quad \psi(b) - \psi(t_j) = (N-j)\left(\frac{\psi(b)-\psi(a)}{N}\right), \tag{3.56}$$

$j = 0, 1, 2, ..., N$.

We notice

$$(\psi(b) - \psi(t_j))^{\alpha+1} + (\psi(t_j) - \psi(a))^{\alpha+1} = \left(\frac{\psi(b)-\psi(a)}{N}\right)^{\alpha+1}\left[j^{\alpha+1} + (N-j)^{\alpha+1}\right], \tag{3.57}$$

$j = 0, 1, 2, ..., N$,

and (for $k = 1, ..., n-1$)

$$f_{\psi}^{[n-k]}\left(I_{a+}^{(1-\beta)(n-\alpha);\psi} f\right)(a)\left(\psi(t_j) - \psi(a)\right)^{\gamma-k+1} +$$

$$(-1)^k f_{\psi}^{[n-k]}\left(I_{b-}^{(1-\beta)(n-\alpha);\psi} f\right)(b)\left(\psi(b) - \psi(t_j)\right)^{\gamma-k+1} =$$

$$\left(\frac{\psi(b)-\psi(a)}{N}\right)^{\gamma-k+1}\left[f_{\psi}^{[n-k]}\left(I_{a+}^{(1-\beta)(n-\alpha);\psi} f\right)(a)\, j^{\gamma-k+1} +\right.$$

$$\left.(-1)^k f_{\psi}^{[n-k]}\left(I_{b-}^{(1-\beta)(n-\alpha);\psi} f\right)(b)\,(N-j)^{\gamma-k+1}\right], \tag{3.58}$$

$j = 0, 1, 2, ..., N$.

By (3.50) we get

$$\left| \int_a^b f(x)\, d\psi(x) - \left\{ \sum_{k=1}^{n-1} \frac{1}{\Gamma(\gamma - k + 2)} \left(\frac{\psi(b) - \psi(a)}{N} \right)^{\gamma - k + 1} \right.\right.$$

$$\left[f_\psi^{[n-k]} \left(I_{a+}^{(1-\beta)(n-\alpha);\psi} f \right)(a)\, j^{\gamma-k+1} + \right.$$

$$\left.\left.\left. (-1)^k f_\psi^{[n-k]} \left(I_{b-}^{(1-\beta)(n-\alpha);\psi} f \right)(b)\, (N - j)^{\gamma-k+1} \right] \right\} \right| \leq$$

$$\frac{\rho}{\Gamma(\alpha + 2)} \left(\frac{\psi(b) - \psi(a)}{N} \right)^{\alpha+1} \left(j^{\alpha+1} + (N - j)^{\alpha+1} \right),$$

$j = 0, 1, 2, ..., N$.

If again $f_\psi^{[n-k]} \left(I_{a+}^{(1-\beta)(n-\alpha);\psi} f \right)(a) = f_\psi^{[n-k]} \left(I_{b-}^{(1-\beta)(n-\alpha);\psi} f \right)(b) = 0$, for $k = 1, ..., n-1$, then

$$\left| \int_a^b f(x)\, d\psi(x) \right| \leq$$

$$\frac{\rho}{\Gamma(\alpha + 2)} \left(\frac{\psi(b) - \psi(a)}{N} \right)^{\alpha+1} \left(j^{\alpha+1} + (N - j)^{\alpha+1} \right), \tag{3.60}$$

$j = 0, 1, 2, ..., N$.

When $N = 2$ and $j = 1$, then (3.60) becomes

$$\left| \int_a^b f(x)\, d\psi(x) \right| \leq \frac{\rho}{\Gamma(\alpha + 2)} \frac{(\psi(b) - \psi(a))^{\alpha+1}}{2^\alpha}. \tag{3.61}$$

Let $0 < \alpha < 1$, then $n = \lceil \alpha \rceil = 1$.

Thus, without any boundary conditions (3.61) is true again.

We have proved theorem in all possible cases. ■

The L_1 variant follows:

Theorem 3.7 *Let $\alpha > 1 : \lceil \alpha \rceil = n$; $\psi, f \in C^n([a, b])$, with ψ being increasing such that $\psi'(x) \neq 0$ over $[a, b]$, where $0 \leq \beta \leq 1$ and $\gamma = \alpha + \beta(n - \alpha)$. Call $\delta := \max \left\{ \left\| {}^H\mathbb{D}_{a+}^{\alpha,\beta;\psi} f \right\|_{L_1([a,b],\psi)}, \left\| {}^H\mathbb{D}_{b-}^{\alpha,\beta;\psi} f \right\|_{L_1([a,b],\psi)} \right\}$. Then*

(1)

$$\left| \int_a^b f(x)\, d\psi(x) - \left\{ \sum_{k=1}^{n-1} \frac{1}{\Gamma(\gamma - k + 2)} \left[f_\psi^{[n-k]} \left(I_{a+}^{(1-\beta)(n-\alpha);\psi} f \right)(a)\, (\psi(t) - \psi(a))^{\gamma-k+1} \right.\right.\right.$$

$$+(-1)^k f_{\psi}^{[n-k]} \left(I_{b-}^{(1-\beta)(n-\alpha);\psi} f \right)(b) \left(\psi(b) - \psi(t) \right)^{\gamma-k+1} \Big]\Big\} \Big| \leq$$

$$\frac{\delta}{\Gamma(\alpha+1)} \left[(\psi(b) - \psi(t))^{\alpha} + (\psi(t) - \psi(a))^{\alpha} \right], \tag{3.62}$$

$\forall\, t \in [a,b]$,

(2) the R.H.S of (3.62) is minimized at $t \in [a,b]$, *such that* $\psi(t) = \frac{\psi(a)+\psi(b)}{2}$, *with a minimum value of* $\frac{\delta}{\Gamma(\alpha+1)} \frac{(\psi(b)-\psi(a))^{\alpha}}{2^{\alpha-1}}$, *and we have*

$$\Big| \int_a^b f(x)\, d\psi(x) - \Big\{ \sum_{k=1}^{n-1} \frac{1}{\Gamma(\gamma-k+2)} \left(\frac{\psi(b)-\psi(a)}{2} \right)^{\gamma-k+1}$$

$$\Big[f_{\psi}^{[n-k]} \left(I_{a+}^{(1-\beta)(n-\alpha);\psi} f \right)(a) + (-1)^k f_{\psi}^{[n-k]} \left(I_{b-}^{(1-\beta)(n-\alpha);\psi} f \right)(b) \Big]\Big\} \Big| \leq$$

$$\frac{\delta}{\Gamma(\alpha+1)} \frac{(\psi(b)-\psi(a))^{\alpha}}{2^{\alpha-1}}, \tag{3.63}$$

(3) assuming $f_{\psi}^{[n-k]} \left(I_{a+}^{(1-\beta)(n-\alpha);\psi} f \right)(a) = f_{\psi}^{[n-k]} \left(I_{b-}^{(1-\beta)(n-\alpha);\psi} f \right)(b) = 0$, *for* $k = 1, ..., n-1$, *we get*

$$\left| \int_a^b f(x)\, d\psi(x) \right| \leq \frac{\delta}{\Gamma(\alpha+1)} \frac{(\psi(b)-\psi(a))^{\alpha}}{2^{\alpha-1}}, \tag{3.64}$$

which is a sharp inequality,

(4) additionally, assume that ψ *is strictly increasing, more generally, for* $j = 0, 1, 2, ..., N \in \mathbb{N}$, *it holds*

$$\Big| \int_a^b f(x)\, d\psi(x) - \Big\{ \sum_{k=1}^{n-1} \frac{1}{\Gamma(\gamma-k+2)} \left(\frac{\psi(b)-\psi(a)}{N} \right)^{\gamma-k+1}$$

$$\Big[f_{\psi}^{[n-k]} \left(I_{a+}^{(1-\beta)(n-\alpha);\psi} f \right)(a)\, j^{\gamma-k+1} +$$

$$(-1)^k f_{\psi}^{[n-k]} \left(I_{b-}^{(1-\beta)(n-\alpha);\psi} f \right)(b)\, (N-j)^{\gamma-k+1} \Big]\Big\} \Big| \leq$$

$$\frac{\delta}{\Gamma(\alpha+1)} \left(\frac{\psi(b)-\psi(a)}{N} \right)^{\alpha} \left(j^{\alpha} + (N-j)^{\alpha} \right), \tag{3.65}$$

(5) if $f_{\psi}^{[n-k]} \left(I_{a+}^{(1-\beta)(n-\alpha);\psi} f \right)(a) = f_{\psi}^{[n-k]} \left(I_{b-}^{(1-\beta)(n-\alpha);\psi} f \right)(b) = 0$, $k = 1, ..., n-1$, *from (3.65) we derive*

$$\left|\int_a^b f(x)\,d\psi(x)\right| \le$$

$$\frac{\delta}{\Gamma(\alpha+1)}\left(\frac{\psi(b)-\psi(a)}{N}\right)^{\alpha} \min_{j\in\{0,1,\ldots,N\}}\left(j^{\alpha}+(N-j)^{\alpha}\right), \tag{3.66}$$

(6) when $N=2$ and $j=1$, from (3.66) we get

$$\left|\int_a^b f(x)\,d\psi(x)\right| \le \frac{\delta}{\Gamma(\alpha+1)}\frac{(\psi(b)-\psi(a))^{\alpha}}{2^{\alpha-1}}. \tag{3.67}$$

Proof From (3.31) we have

$$\left|f(x)-\sum_{k=1}^{n-1}\frac{(\psi(x)-\psi(a))^{\gamma-k}}{\Gamma(\gamma-k+1)} f_{\psi}^{[n-k]}\left(I_{a+}^{(1-\beta)(n-\alpha);\psi}f\right)(a)\right| \le$$

$$\frac{1}{\Gamma(\alpha)}\int_a^x \psi'(t)(\psi(x)-\psi(t))^{\alpha-1}\left|{}^{H}\mathbb{D}_{a+}^{\alpha,\beta;\psi}f(t)\right|dt \le$$

$$\frac{(\psi(x)-\psi(a))^{\alpha-1}}{\Gamma(\alpha)}\int_a^x \left|{}^{H}\mathbb{D}_{a+}^{\alpha,\beta;\psi}f(t)\right|d\psi(t) \le$$

$$\frac{(\psi(x)-\psi(a))^{\alpha-1}}{\Gamma(\alpha)}\int_a^b \left|{}^{H}\mathbb{D}_{a+}^{\alpha,\beta;\psi}f(t)\right|d\psi(t) =$$

$$\frac{\left\|{}^{H}\mathbb{D}_{a+}^{\alpha,\beta;\psi}f\right\|_{L_1([a,b],\psi)}}{\Gamma(\alpha)}(\psi(x)-\psi(a))^{\alpha-1}, \quad \forall\, x\in[a,b]. \tag{3.68}$$

Similarly, from (3.32) we obtain

$$\left|f(x)-\sum_{k=1}^{n-1}\frac{(-1)^k(\psi(b)-\psi(x))^{\gamma-k}}{\Gamma(\gamma-k+1)} f_{\psi}^{[n-k]}\left(I_{b-}^{(1-\beta)(n-\alpha);\psi}f\right)(b)\right| \le$$

$$\frac{1}{\Gamma(\alpha)}\int_x^b \psi'(t)(\psi(t)-\psi(x))^{\alpha-1}\left|{}^{H}\mathbb{D}_{b-}^{\alpha,\beta;\psi}f(t)\right|dt \le$$

$$\frac{(\psi(b)-\psi(x))^{\alpha-1}}{\Gamma(\alpha)}\int_x^b \left|{}^{H}\mathbb{D}_{b-}^{\alpha,\beta;\psi}f(t)\right|d\psi(t) \le$$

$$\frac{(\psi(b)-\psi(x))^{\alpha-1}}{\Gamma(\alpha)}\int_a^b \left|{}^{H}\mathbb{D}_{b-}^{\alpha,\beta;\psi}f(t)\right|d\psi(t) =$$

$$\frac{\left\| {}^{H}\mathbb{D}_{b-}^{\alpha,\beta;\psi} f \right\|_{L_1([a,b],\psi)}}{\Gamma(\alpha)} (\psi(b) - \psi(x))^{\alpha-1}, \quad \forall\, x \in [a,b]. \tag{3.69}$$

We have proved that

$$\left| f(x) - \sum_{k=1}^{n-1} \frac{(\psi(x) - \psi(a))^{\gamma-k}}{\Gamma(\gamma-k+1)} f_{\psi}^{[n-k]} \left(I_{a+}^{(1-\beta)(n-\alpha);\psi} f \right)(a) \right| \leq$$

$$\frac{\delta}{\Gamma(\alpha)} (\psi(x) - \psi(a))^{\alpha-1}, \tag{3.70}$$

and

$$\left| f(x) - \sum_{k=1}^{n-1} \frac{(-1)^k (\psi(b) - \psi(x))^{\gamma-k}}{\Gamma(\gamma-k+1)} f_{\psi}^{[n-k]} \left(I_{b-}^{(1-\beta)(n-\alpha);\psi} f \right)(b) \right| \leq$$

$$\frac{\delta}{\Gamma(\alpha)} (\psi(b) - \psi(x))^{\alpha-1}, \quad \forall\, x \in [a,b]. \tag{3.71}$$

The rest of the proof as in Theorem 3.6. ∎

The L_q variant follows:

Theorem 3.8 *Let* $p, q > 1 : \frac{1}{p} + \frac{1}{q} = 1,\ \frac{1}{q} < \alpha \notin \mathbb{N},\ \lceil \alpha \rceil = n,\ 0 \leq \beta \leq 1,\ \gamma = \alpha + \beta(n-\alpha);\ \psi, f \in C^n([a,b])$, *with* ψ *being increasing such that* $\psi'(x) \neq 0$ *over* $[a,b]$. *Call* $\varepsilon := \max \left\{ \left\| {}^{H}\mathbb{D}_{a+}^{\alpha,\beta;\psi} f \right\|_{L_q([a,b],\psi)}, \left\| {}^{H}\mathbb{D}_{b-}^{\alpha,\beta;\psi} f \right\|_{L_q([a,b],\psi)} \right\}$. *Then*

(1)

$$\left| \int_a^b f(x)\, d\psi(x) - \left\{ \sum_{k=1}^{n-1} \frac{1}{\Gamma(\gamma-k+2)} \left[f_{\psi}^{[n-k]} \left(I_{a+}^{(1-\beta)(n-\alpha);\psi} f \right)(a) (\psi(t) - \psi(a))^{\gamma-k+1} \right.\right.\right.$$

$$\left.\left.\left. + (-1)^k f_{\psi}^{[n-k]} \left(I_{b-}^{(1-\beta)(n-\alpha);\psi} f \right)(b) (\psi(b) - \psi(t))^{\gamma-k+1} \right] \right\} \right| \leq$$

$$\left(\frac{\varepsilon}{\Gamma(\alpha)\left(\alpha + \frac{1}{p}\right)(p(\alpha-1)+1)^{\frac{1}{p}}} \right) \left[(\psi(b) - \psi(t))^{\alpha+\frac{1}{p}} + (\psi(t) - \psi(a))^{\alpha+\frac{1}{p}} \right], \tag{3.72}$$

$\forall\, t \in [a,b]$,

(2) the R.H.S. of (3.72) is minimized at $t \in [a,b]$, *such that* $\psi(t) = \frac{\psi(a)+\psi(b)}{2}$, *with a minimum value of* $\left(\frac{\varepsilon}{\Gamma(\alpha)\left(\alpha+\frac{1}{p}\right)(p(\alpha-1)+1)^{\frac{1}{p}}} \right) \frac{(\psi(b)-\psi(a))^{\alpha+\frac{1}{p}}}{2^{\alpha-\frac{1}{q}}}$, *and we have*

$$\left| \int_a^b f(x)\, d\psi(x) - \left\{ \sum_{k=1}^{n-1} \frac{1}{\Gamma(\gamma - k + 2)} \left(\frac{\psi(b) - \psi(a)}{2} \right)^{\gamma - k + 1} \right.\right.$$

$$\left.\left. \left[f_\psi^{[n-k]} \left(I_{a+}^{(1-\beta)(n-\alpha);\psi} f \right)(a) + (-1)^k f_\psi^{[n-k]} \left(I_{b-}^{(1-\beta)(n-\alpha);\psi} f \right)(b) \right] \right\} \right| \leq$$

$$\left(\frac{\varepsilon}{\Gamma(\alpha) \left(\alpha + \frac{1}{p} \right) \left(p(\alpha - 1) + 1 \right)^{\frac{1}{p}}} \right) \frac{(\psi(b) - \psi(a))^{\alpha + \frac{1}{p}}}{2^{\alpha - \frac{1}{q}}}, \tag{3.73}$$

(3) assuming $f_\psi^{[n-k]} \left(I_{a+}^{(1-\beta)(n-\alpha);\psi} f \right)(a) = f_\psi^{[n-k]} \left(I_{b-}^{(1-\beta)(n-\alpha);\psi} f \right)(b) = 0$, *for* $k = 1, \ldots, n-1$, *we get*

$$\left| \int_a^b f(x)\, d\psi(x) \right| \leq \left(\frac{\varepsilon}{\Gamma(\alpha) \left(\alpha + \frac{1}{p} \right) \left(p(\alpha - 1) + 1 \right)^{\frac{1}{p}}} \right) \frac{(\psi(b) - \psi(a))^{\alpha + \frac{1}{p}}}{2^{\alpha - \frac{1}{q}}}, \tag{3.74}$$

which is a sharp inequality,

(4) additionally, assume that ψ *is strictly increasing,* $j = 0, 1, 2, \ldots, N \in \mathbb{N}$, *it holds*

$$\left| \int_a^b f(x)\, d\psi(x) - \left\{ \sum_{k=1}^{n-1} \frac{1}{\Gamma(\gamma - k + 2)} \left(\frac{\psi(b) - \psi(a)}{N} \right)^{\gamma - k + 1} \right.\right.$$

$$\left[f_\psi^{[n-k]} \left(I_{a+}^{(1-\beta)(n-\alpha);\psi} f \right)(a)\, j^{\gamma - k + 1} + \right.$$

$$\left.\left.\left. (-1)^k f_\psi^{[n-k]} \left(I_{b-}^{(1-\beta)(n-\alpha);\psi} f \right)(b)\, (N - j)^{\gamma - k + 1} \right] \right\} \right| \leq$$

$$\left(\frac{\varepsilon}{\Gamma(\alpha) \left(\alpha + \frac{1}{p} \right) \left(p(\alpha - 1) + 1 \right)^{\frac{1}{p}}} \right) \left(\frac{\psi(b) - \psi(a)}{N} \right)^{\alpha + \frac{1}{p}} \left[j^{\alpha + \frac{1}{p}} + (N - j)^{\alpha + \frac{1}{p}} \right], \tag{3.75}$$

(5) if again $f_\psi^{[n-k]} \left(I_{a+}^{(1-\beta)(n-\alpha);\psi} f \right)(a) = f_\psi^{[n-k]} \left(I_{b-}^{(1-\beta)(n-\alpha);\psi} f \right)(b) = 0$, $k = 1, \ldots, n-1$, *from (3.75) we derive*

$$\left| \int_a^b f(x)\, d\psi(x) \right| \leq \left(\frac{\varepsilon}{\Gamma(\alpha) \left(\alpha + \frac{1}{p} \right) \left(p(\alpha - 1) + 1 \right)^{\frac{1}{p}}} \right)$$

$$\left(\frac{\psi(b)-\psi(a)}{N}\right)^{\alpha+\frac{1}{p}} \min_{j\in\{0,1,\ldots,N\}}\left[j^{\alpha+\frac{1}{p}}+(N-j)^{\alpha+\frac{1}{p}}\right], \tag{3.76}$$

(6) when $N=2$ and $j=1$, from (3.76) we get

$$\left|\int_a^b f(x)\,d\psi(x)\right| \leq \left(\frac{\varepsilon}{\Gamma(\alpha)\left(\alpha+\frac{1}{p}\right)(p(\alpha-1)+1)^{\frac{1}{p}}}\right)\frac{(\psi(b)-\psi(a))^{\alpha+\frac{1}{p}}}{2^{\alpha-\frac{1}{q}}}, \tag{3.77}$$

(7) if $\frac{1}{q}<\alpha<1$, we have (3.77) valid without any boundary conditions.

Proof From (3.31) we have

$$\left|f(x)-\sum_{k=1}^{n-1}\frac{(\psi(x)-\psi(a))^{\gamma-k}}{\Gamma(\gamma-k+1)}f_\psi^{[n-k]}\left(I_{a+}^{(1-\beta)(n-\alpha);\psi}f\right)(a)\right| \leq$$

$$\frac{1}{\Gamma(\alpha)}\int_a^x \psi'(t)(\psi(x)-\psi(t))^{\alpha-1}\left|{}^H\mathbb{D}_{a+}^{\alpha,\beta;\psi}f(t)\right|dt =$$

$$\frac{1}{\Gamma(\alpha)}\int_a^x (\psi(x)-\psi(t))^{\alpha-1}\left|{}^H\mathbb{D}_{a+}^{\alpha,\beta;\psi}f(t)\right|d\psi(t) \leq$$

by Dragomir [3]

$$\frac{1}{\Gamma(\alpha)}\left(\int_a^x (\psi(x)-\psi(t))^{p(\alpha-1)}d\psi(t)\right)^{\frac{1}{p}}\left(\int_a^x \left|{}^H\mathbb{D}_{a+}^{\alpha,\beta;\psi}f(t)\right|^q d\psi(t)\right)^{\frac{1}{q}} \leq$$

$$\frac{(\psi(x)-\psi(a))^{\alpha-\frac{1}{q}}}{\Gamma(\alpha)(p(\alpha-1)+1)^{\frac{1}{p}}}\left\|{}^H\mathbb{D}_{a+}^{\alpha,\beta;\psi}f\right\|_{L_q([a,b],\psi)}, \tag{3.78}$$

$\forall\, x\in[a,b]$, where $\alpha>\frac{1}{q}$.

Similarly, from (3.32) we obtain

$$\left|f(x)-\sum_{k=1}^{n-1}\frac{(-1)^k(\psi(b)-\psi(x))^{\gamma-k}}{\Gamma(\gamma-k+1)}f_\psi^{[n-k]}\left(I_{b-}^{(1-\beta)(n-\alpha);\psi}f\right)(b)\right| \leq$$

$$\frac{1}{\Gamma(\alpha)}\int_x^b \psi'(t)(\psi(t)-\psi(x))^{\alpha-1}\left|{}^H\mathbb{D}_{b-}^{\alpha,\beta;\psi}f(t)\right|dt =$$

$$\frac{1}{\Gamma(\alpha)}\int_x^b (\psi(t)-\psi(x))^{\alpha-1}\left|{}^H\mathbb{D}_{b-}^{\alpha,\beta;\psi}f(t)\right|d\psi(t) \leq$$

$$\frac{1}{\Gamma(\alpha)}\left(\int_x^b (\psi(t)-\psi(x))^{p(\alpha-1)}d\psi(t)\right)^{\frac{1}{p}}\left(\int_x^b \left|{}^H\mathbb{D}_{b-}^{\alpha,\beta;\psi}f(t)\right|^q d\psi(t)\right)^{\frac{1}{q}} \le$$

$$\frac{(\psi(b)-\psi(x))^{\alpha-\frac{1}{q}}}{\Gamma(\alpha)(p(\alpha-1)+1)^{\frac{1}{p}}}\left\|{}^H\mathbb{D}_{b-}^{\alpha,\beta;\psi}f\right\|_{L_q([a,b],\psi)}, \tag{3.79}$$

$\forall\, x \in [a,b]$, where $\alpha > \frac{1}{q}$.

We have proved that

$$\left|f(x)-\sum_{k=1}^{n-1}\frac{(\psi(x)-\psi(a))^{\gamma-k}}{\Gamma(\gamma-k+1)}f_\psi^{[n-k]}\left(I_{a+}^{(1-\beta)(n-\alpha);\psi}f\right)(a)\right| \le$$

$$\frac{\varepsilon}{\Gamma(\alpha)(p(\alpha-1)+1)^{\frac{1}{p}}}(\psi(x)-\psi(a))^{\alpha-\frac{1}{q}}, \tag{3.80}$$

and

$$\left|f(x)-\sum_{k=1}^{n-1}\frac{(-1)^k(\psi(b)-\psi(x))^{\gamma-k}}{\Gamma(\gamma-k+1)}f_\psi^{[n-k]}\left(I_{b-}^{(1-\beta)(n-\alpha);\psi}f\right)(b)\right| \le$$

$$\frac{\varepsilon}{\Gamma(\alpha)(p(\alpha-1)+1)^{\frac{1}{p}}}(\psi(b)-\psi(x))^{\alpha-\frac{1}{q}}, \tag{3.81}$$

$\forall\, x \in [a,b]$, where $\alpha > \frac{1}{q}$. The rest of the proof as in Theorem 3.6. ■

3.3.2 Multivariate Results

We need

Remark 3.9 We define the ball $B(0,R) = \{x \in \mathbb{R}^N : |x| < R\} \subseteq \mathbb{R}^N$, $N \ge 2$, $R > 0$, and the sphere

$$S^{N-1} := \{x \in \mathbb{R}^N : |x| = 1\},$$

where $|\cdot|$ is the Euclidean norm. Let $d\omega$ be the element of surface measure on S^{N-1} and

$$\omega_N = \int_{S^{N-1}} d\omega = \frac{2\pi^{\frac{N}{2}}}{\Gamma\left(\frac{N}{2}\right)}$$

is the area of S^{N-1}.

For $x \in \mathbb{R}^N - \{0\}$ we can write uniquely $x = r\omega$, where $r = |x| > 0$ and $\omega = \frac{x}{r} \in S^{N-1}$, $|\omega| = 1$. Note that $\int_{B(0,R)} dy = \frac{\omega_N R^N}{N}$ is the Lebesgue measure on the ball, that is the volume of $B(0, R)$, which exactly is $Vol(B(0, R)) = \frac{\pi^{\frac{N}{2}} R^N}{\Gamma(\frac{N}{2}+1)}$.

Following [[6], pp. 149–150, exercise 6], and [8] [pp. 87–88, Theorem 5.2.2] we can write for $F : \overline{B(0, R)} \to \mathbb{R}$ a Lebesgue integrable function that

$$\int_{B(0,R)} F(x)\,dx = \int_{S^{N-1}} \left(\int_0^R F(r\omega)\, r^{N-1} dr \right) d\omega, \tag{3.82}$$

and we use this formula a lot.

Typically here the function $f : \overline{B(0, R)} \to \mathbb{R}$ is not radial. A radial function f is such that there exists a function g such that $f(x) = g(r)$, where $r = |x|, r \in [0, R]$, $\forall\, x \in \overline{B(0, R)}$.

We also need

Remark 3.10 Let the spherical shell $A := B(0, R_2) - \overline{B(0, R_1)}$, $0 < R_1 < R_2$, $A \subseteq \mathbb{R}^N$, $N \geq 2$, $x \in \overline{A}$. Consider that $f : \overline{A} \to \mathbb{R}$ is not necessarily radial. A radial function f is such that there exists a function g such that $f(x) = g(r)$, $r = |x|$, $r \in [R_1, R_2], \forall\, x \in \overline{A}$. Here x can be written uniquely as $x = r\omega$, where $r = |x| > 0$ and $\omega = \frac{x}{r} \in S^{N-1}$, $|\omega| = 1$, see [6], p. 149–150, furthermore for $F : \overline{A} \to \mathbb{R}$ a Lebesgue integrable function we have that

$$\int_A F(x)\,dx = \int_{S^{N-1}} \left(\int_{R_1}^{R_2} F(r\omega)\, r^{N-1} dr \right) d\omega. \tag{3.83}$$

Here

$$Vol(A) = \frac{\omega_N \left(R_2^N - R_1^N\right)}{N} = \frac{\pi^{\frac{N}{2}} \left(R_2^N - R_1^N\right)}{\Gamma\left(\frac{N}{2} + 1\right)}. \tag{3.84}$$

Next in this chapter we derive general multivariate fractional Hilfer-Iyengar type inequalities on the shell and ball of $\mathbb{R}^N$, $N \geq 2$, for not necessarily radial functions.

From now on our Hilfer fractional calculus so far results are applied to $f(s\omega)\, s^{N-1}$, where $s \in [R_1, R_2]$ or $s \in [0, R]$. We give Hilfer-Iyengar fractional inequalities on the shell and ball.

We present

Theorem 3.11 *Consider $f : \overline{A} \to \mathbb{R}$ be Lebesgue integrable, which is not necessarily radial. Let $\mathbb{N} \not\ni \alpha > 0 : \lceil \alpha \rceil = n, 0 \leq \beta \leq 1, \gamma = \alpha + \beta(n - \alpha)$. Assume that $f(s\omega)\, s^{N-1} \in C^n([R_1, R_2])$, $\forall$ $\omega \in S^{N-1}$, $N \geq 2$, and that*

$$\max\left\{ \left\| {}^H\mathbb{D}_{R_1+}^{\alpha,\beta} \left(f(s\omega)\, s^{N-1}\right) \right\|_{\infty,[R_1,R_2]}, \left\| {}^H\mathbb{D}_{R_2-}^{\alpha,\beta} \left(f(s\omega)\, s^{N-1}\right) \right\|_{\infty,[R_1,R_2]} \right\} \leq K_1,$$

where $K_1 > 0$, $s \in [R_1, R_2]$, $\forall\, \omega \in S^{N-1}$. Furthermore assume that $\left(I_{R_1+}^{(1-\beta)(n-\alpha)} f(s\omega)\, s^{N-1}\right)^{(n-k)}(R_1)$, $\left(I_{R_2-}^{(1-\beta)(n-\alpha)} f(s\omega)\, s^{N-1}\right)^{(n-k)}(R_2)$, $k = 1, \ldots, n-1$, are all measurable in ω over S^{N-1}.

Then
(1)

$$\left|\int_A f(y)\,dy - \left\{\sum_{k=1}^{n-1} \frac{1}{\Gamma(\gamma-k+2)}\right.\right.$$

$$\left[\left(\int_{S^{N-1}} \left(I_{R_1+}^{(1-\beta)(n-\alpha)} f(s\omega)\,s^{N-1}\right)^{(n-k)} (R_1)\,d\omega\right)(t-R_1)^{\gamma-k+1}\right.$$

$$\left.\left.\left.+(-1)^k \left(\int_{S^{n-1}} \left(I_{R_2-}^{(1-\beta)(n-\alpha)} f(s\omega)\,s^{N-1}\right)^{(n-k)} (R_2)\,d\omega\right)(R_2-t)^{\gamma-k+1}\right]\right\}\right| \le$$

$$\frac{2\pi^{\frac{N}{2}}}{\Gamma\left(\frac{N}{2}\right)} \frac{K_1}{\Gamma(\alpha+2)} \left[(R_2-t)^{\alpha+1} + (t-R_1)^{\alpha+1}\right], \tag{3.85}$$

$\forall\, t \in [R_1, R_2]$,

(2) at $t = \frac{R_1+R_2}{2}$, *the R.H.S (3.85) is minimized, and we get:*

$$\left|\int_A f(y)\,dy - \left\{\sum_{k=1}^{n-1} \frac{1}{\Gamma(\gamma-k+2)} \left(\frac{R_2-R_1}{2}\right)^{\gamma-k+1}\right.\right.$$

$$\left[\left(\int_{S^{N-1}} \left(I_{R_1+}^{(1-\beta)(n-\alpha)} f(s\omega)\,s^{N-1}\right)^{(n-k)} (R_1)\,d\omega\right)\right.$$

$$\left.\left.\left.+(-1)^k \left(\int_{S^{n-1}} \left(I_{R_2-}^{(1-\beta)(n-\alpha)} f(s\omega)\,s^{N-1}\right)^{(n-k)} (R_2)\,d\omega\right)\right]\right\}\right| \le$$

$$\frac{\pi^{\frac{N}{2}}}{\Gamma\left(\frac{N}{2}\right)} \frac{K_1}{\Gamma(\alpha+2)} \frac{(R_2-R_1)^{\alpha+1}}{2^{\alpha-1}}, \tag{3.86}$$

(3) if $\left(I_{R_1+}^{(1-\beta)(n-\alpha)} f(s\omega)\,s^{N-1}\right)^{(n-k)} (R_1) = \left(I_{R_2-}^{(1-\beta)(n-\alpha)} f(s\omega)\,s^{N-1}\right)^{(n-k)} (R_2)$ $= 0$, $\forall\, \omega \in S^{N-1}$, *(i.e. they vanish on* $\partial B(0, R_1)$ *and* $\partial B(0, R_2)$*, respectively), for all* $k = 1, ..., n-1$, *we obtain*

$$\left|\int_A f(y)\,dy\right| \le \frac{\pi^{\frac{N}{2}}}{\Gamma\left(\frac{N}{2}\right)} \frac{K_1}{\Gamma(\alpha+2)} \frac{(R_2-R_1)^{\alpha+1}}{2^{\alpha-1}}, \tag{3.87}$$

which is a sharp inequality,

(4) more generally, for $j = 0, 1, 2, ..., \overline{N} \in \mathbb{N}$, *it holds*

$$\left|\int_A f(y)\,dy - \left\{\sum_{k=1}^{n-1} \frac{1}{\Gamma(\gamma-k+2)} \left(\frac{R_2-R_1}{\overline{N}}\right)^{\gamma-k+1}\right.\right.$$

$$\left[j^{\gamma-k+1}\left(\int_{S^{N-1}}\left(I_{R_1+}^{(1-\beta)(n-\alpha)}f(s\omega)s^{N-1}\right)^{(n-k)}(R_1)\,d\omega\right)\right.$$

$$\left.\left.\left.+(-1)^k\left(\overline{N}-j\right)^{\gamma-k+1}\left(\int_{S^{n-1}}\left(I_{R_2-}^{(1-\beta)(n-\alpha)}f(s\omega)s^{N-1}\right)^{(n-k)}(R_2)\,d\omega\right)\right]\right\}\right|\leq$$

$$\frac{2\pi^{\frac{N}{2}}}{\Gamma\left(\frac{N}{2}\right)}\frac{K_1}{\Gamma(\alpha+2)}\left(\frac{R_2-R_1}{\overline{N}}\right)^{\alpha+1}\left(j^{\alpha+1}+\left(\overline{N}-j\right)^{\alpha+1}\right), \quad (3.88)$$

(5) if $\left(I_{R_1+}^{(1-\beta)(n-\alpha)}f(s\omega)s^{N-1}\right)^{(n-k)}(R_1)=\left(I_{R_2-}^{(1-\beta)(n-\alpha)}f(s\omega)s^{N-1}\right)^{(n-k)}(R_2)$ $=0$, $\forall\,\omega\in S^{N-1}$, *for all* $k=1,\dots,n-1$, *we obtain*

$$\left|\int_A f(y)\,dy\right|\leq\frac{2\pi^{\frac{N}{2}}}{\Gamma\left(\frac{N}{2}\right)}\frac{K_1}{\Gamma(\alpha+2)}\left(\frac{R_2-R_1}{\overline{N}}\right)^{\alpha+1}\min_{j\in\{0,1,2,\dots,\overline{N}\}}\left(j^{\alpha+1}+\left(\overline{N}-j\right)^{\alpha+1}\right), \quad (3.89)$$

(6) when $\overline{N}=2$ *and* $j=1$, *from (3.89) we get*

$$\left|\int_A f(y)\,dy\right|\leq\frac{\pi^{\frac{N}{2}}}{\Gamma\left(\frac{N}{2}\right)}\frac{K_1}{\Gamma(\alpha+2)}\frac{(R_2-R_1)^{\alpha+1}}{2^{\alpha-1}}, \quad (3.90)$$

(7) if $0<\alpha<1$, *we have (3.90) valid without any boundary conditions.*

Proof We use Theorem 3.6, all the way, for $f(s\omega)s^{N-1}$ and $\psi(x)=x$. First we apply (3.35):

$$\left|\int_{R_1}^{R_2}f(s\omega)s^{N-1}ds-\left\{\sum_{k=1}^{n-1}\frac{1}{\Gamma(\gamma-k+2)}\right.\right.$$

$$\left[\left(I_{R_1+}^{(1-\beta)(n-\alpha)}f(s\omega)s^{N-1}\right)^{(n-k)}(R_1)(t-R_1)^{\gamma-k+1}\right.$$

$$\left.\left.\left.+(-1)^k\left(I_{R_2-}^{(1-\beta)(n-\alpha)}f(s\omega)s^{N-1}\right)^{(n-k)}(R_2)(R_2-t)^{\gamma-k+1}\right]\right\}\right|\leq$$

$$\frac{K_1}{\Gamma(\alpha+2)}\left[(R_2-t)^{\alpha+1}+(t-R_1)^{\alpha+1}\right]=:\lambda_1(t),\quad\forall\,t\in[R_1,R_2], \quad (3.91)$$

and $\forall\,\omega\in S^{N-1}$. Equivalently, we have that

$$-\lambda_1(t)\leq\int_{R_1}^{R_2}f(s\omega)s^{N-1}ds-\left\{\sum_{k=1}^{n-1}\frac{1}{\Gamma(\gamma-k+2)}\right.$$

$$\left[\left(I_{R_1+}^{(1-\beta)(n-\alpha)} f(s\omega) s^{N-1}\right)^{(n-k)} (R_1)(t-R_1)^{\gamma-k+1}\right.$$

$$\left.\left. + (-1)^k \left(I_{R_2-}^{(1-\beta)(n-\alpha)} f(s\omega) s^{N-1}\right)^{(n-k)} (R_2)(R_2-t)^{\gamma-k+1}\right]\right\} \leq \lambda_1(t), \quad (3.92)$$

$\forall\, t \in [R_1, R_2], \forall\, \omega \in S^{N-1}$.

Therefore it holds

$$-\lambda_1(t) \int_{S^{N-1}} d\omega \leq \int_{S^{N-1}} \left(\int_{R_1}^{R_2} f(s\omega) s^{N-1} ds\right) d\omega - \left\{\sum_{k=1}^{n-1} \frac{1}{\Gamma(\gamma-k+2)}\right.$$

$$\left[\left(\int_{S^{N-1}} \left(I_{R_1+}^{(1-\beta)(n-\alpha)} f(s\omega) s^{N-1}\right)^{(n-k)} (R_1)\, d\omega\right)(t-R_1)^{\gamma-k+1}\right.$$

$$\left.\left. + (-1)^k \left(\int_{S^{N-1}} \left(I_{R_2-}^{(1-\beta)(n-\alpha)} f(s\omega) s^{N-1}\right)^{(n-k)} (R_2)\, d\omega\right)(R_2-t)^{\gamma-k+1}\right]\right\} \quad (3.93)$$

$$\leq \lambda_1(t) \int_{S^{N-1}} d\omega, \quad \forall\, t \in [R_1, R_2].$$

By (3.83) we obtain

$$-\lambda_1(t) \frac{2\pi^{\frac{N}{2}}}{\Gamma\left(\frac{N}{2}\right)} \leq \int_A f(y)\, dy - \left\{\sum_{k=1}^{n-1} \frac{1}{\Gamma(\gamma-k+2)}\right.$$

$$\left[\left(\int_{S^{N-1}} \left(I_{R_1+}^{(1-\beta)(n-\alpha)} f(s\omega) s^{N-1}\right)^{(n-k)} (R_1)\, d\omega\right)(t-R_1)^{\gamma-k+1}\right.$$

$$\left.\left.\left. + (-1)^k \left(\int_{S^{n-1}} \left(I_{R_2-}^{(1-\beta)(n-\alpha)} f(s\omega) s^{N-1}\right)^{(n-k)} (R_2)\, d\omega\right)(R_2-t)^{\gamma-k+1}\right]\right\}\right| \leq$$

$$\lambda_1(t) \frac{2\pi^{\frac{N}{2}}}{\Gamma\left(\frac{N}{2}\right)},$$

$\forall\, t \in [R_1, R_2]$, proving (3.85). The rest is obvious. ■

The L_1 variant follows

Theorem 3.12 *Consider $f : \overline{A} \to \mathbb{R}$ be Lebesgue integrable, which is not necessarily radial. Let $\mathbb{N} \not\ni \alpha > 1 : \lceil \alpha \rceil = n, 0 \leq \beta \leq 1, \gamma = \alpha + \beta(n-\alpha)$. Assume that $f(s\omega) s^{N-1} \in C^n([R_1, R_2])$, $\forall\ \omega \in S^{N-1}$, $N \geq 2$, and that* $\max\left\{\left\| {}^H\mathbb{D}_{R_1+}^{\alpha,\beta}\left(f(s\omega) s^{N-1}\right)\right\|_{L_1([R_1,R_2])}, \left\| {}^H\mathbb{D}_{R_2-}^{\alpha,\beta}\left(f(s\omega) s^{N-1}\right)\right\|_{L_1([R_1,R_2])}\right\} \leq K_2,$

where $K_2 > 0$, $s \in [R_1, R_2]$, $\forall\ \omega \in S^{N-1}$. *Furthermore assume that* $\left(I_{R_1+}^{(1-\beta)(n-\alpha)} f(s\omega)\, s^{N-1}\right)^{(n-k)}(R_1)$, $\left(I_{R_2-}^{(1-\beta)(n-\alpha)} f(s\omega)\, s^{N-1}\right)^{(n-k)}(R_2)$, $k = 1, \ldots, n-1$, *are all measurable in* ω *over* S^{N-1}.

Then

(1)

$$\left|\int_A f(y)\,dy - \left\{\sum_{k=1}^{n-1} \frac{1}{\Gamma(\gamma - k + 2)}\right.\right.$$

$$\left[\left(\int_{S^{N-1}} \left(I_{R_1+}^{(1-\beta)(n-\alpha)} f(s\omega)\, s^{N-1}\right)^{(n-k)}(R_1)\, d\omega\right)(t - R_1)^{\gamma-k+1}\right.$$

$$\left.\left.\left. + (-1)^k \left(\int_{S^{n-1}} \left(I_{R_2-}^{(1-\beta)(n-\alpha)} f(s\omega)\, s^{N-1}\right)^{(n-k)}(R_2)\, d\omega\right)(R_2 - t)^{\gamma-k+1}\right]\right\}\right| \leq$$

$$\frac{2\pi^{\frac{N}{2}}}{\Gamma\left(\frac{N}{2}\right)} \frac{K_2}{\Gamma(\alpha+1)} \left[(R_2 - t)^{\alpha} + (t - R_1)^{\alpha}\right], \tag{3.95}$$

$\forall\, t \in [R_1, R_2]$,

(2) at $t = \frac{R_1+R_2}{2}$, *the R.H.S (3.95) is minimized, and we get:*

$$\left|\int_A f(y)\,dy - \left\{\sum_{k=1}^{n-1} \frac{1}{\Gamma(\gamma - k + 2)} \left(\frac{R_2 - R_1}{2}\right)^{\gamma-k+1}\right.\right.$$

$$\left[\left(\int_{S^{N-1}} \left(I_{R_1+}^{(1-\beta)(n-\alpha)} f(s\omega)\, s^{N-1}\right)^{(n-k)}(R_1)\, d\omega\right)\right.$$

$$\left.\left.\left. + (-1)^k \left(\int_{S^{n-1}} \left(I_{R_2-}^{(1-\beta)(n-\alpha)} f(s\omega)\, s^{N-1}\right)^{(n-k)}(R_2)\, d\omega\right)\right]\right\}\right| \leq$$

$$\frac{\pi^{\frac{N}{2}}}{\Gamma\left(\frac{N}{2}\right)} \frac{K_2}{\Gamma(\alpha+1)} \frac{(R_2 - R_1)^{\alpha}}{2^{\alpha-2}}, \tag{3.96}$$

(3) if $\left(I_{R_1+}^{(1-\beta)(n-\alpha)} f(s\omega)\, s^{N-1}\right)^{(n-k)}(R_1) = \left(I_{R_2-}^{(1-\beta)(n-\alpha)} f(s\omega)\, s^{N-1}\right)^{(n-k)}(R_2) = 0$, $\forall\ \omega \in S^{N-1}$, *(i.e. they vanish on* $\partial B(0, R_1)$ *and* $\partial B(0, R_2)$*, respectively), for all* $k = 1, \ldots, n-1$, *we obtain*

$$\left|\int_A f(y)\,dy\right| \leq \frac{\pi^{\frac{N}{2}}}{\Gamma\left(\frac{N}{2}\right)} \frac{K_2}{\Gamma(\alpha+1)} \frac{(R_2 - R_1)^{\alpha}}{2^{\alpha-2}}, \tag{3.97}$$

which is a sharp inequality,

(4) more generally, for $j = 0, 1, 2, ..., \overline{N} \in \mathbb{N}$, *it holds*

$$\left| \int_A f(y)\,dy - \left\{ \sum_{k=1}^{n-1} \frac{1}{\Gamma(\gamma - k + 2)} \left(\frac{R_2 - R_1}{\overline{N}} \right)^{\gamma - k + 1} \right. \right.$$

$$\left[j^{\gamma-k+1} \left(\int_{S^{N-1}} \left(I_{R_1+}^{(1-\beta)(n-\alpha)} f(s\omega) s^{N-1} \right)^{(n-k)} (R_1)\,d\omega \right) \right.$$

$$\left. \left. \left. + (-1)^k \left(\overline{N} - j \right)^{\gamma-k+1} \left(\int_{S^{n-1}} \left(I_{R_2-}^{(1-\beta)(n-\alpha)} f(s\omega) s^{N-1} \right)^{(n-k)} (R_2)\,d\omega \right) \right] \right\} \right| \leq$$

$$\frac{2\pi^{\frac{N}{2}}}{\Gamma\left(\frac{N}{2}\right)} \frac{K_2}{\Gamma(\alpha+1)} \left(\frac{R_2 - R_1}{\overline{N}} \right)^{\alpha} \left(j^{\alpha} + \left(\overline{N} - j \right)^{\alpha} \right), \tag{3.98}$$

(5) if $\left(I_{R_1+}^{(1-\beta)(n-\alpha)} f(s\omega) s^{N-1} \right)^{(n-k)} (R_1) = \left(I_{R_2-}^{(1-\beta)(n-\alpha)} f(s\omega) s^{N-1} \right)^{(n-k)} (R_2)$ $= 0$, $\forall\, \omega \in S^{N-1}$, *for all* $k = 1, ..., n-1$, *we obtain*

$$\left| \int_A f(y)\,dy \right| \leq \frac{2\pi^{\frac{N}{2}}}{\Gamma\left(\frac{N}{2}\right)} \frac{K_2}{\Gamma(\alpha+1)} \left(\frac{R_2 - R_1}{\overline{N}} \right)^{\alpha} \min_{j \in \{0,1,2,...,\overline{N}\}} \left(j^{\alpha} + \left(\overline{N} - j \right)^{\alpha} \right), \tag{3.99}$$

(6) when $\overline{N} = 2$ *and* $j = 1$, *from (3.99) we get*

$$\left| \int_A f(y)\,dy \right| \leq \frac{\pi^{\frac{N}{2}}}{\Gamma\left(\frac{N}{2}\right)} \frac{K_2}{\Gamma(\alpha+1)} \frac{(R_2 - R_1)^{\alpha}}{2^{\alpha-2}}, \tag{3.100}$$

Proof Use of Theorem 3.7 and similar to the proof of Theorem 3.11. ■

The L_q variant follows:

Theorem 3.13 *Let* $p, q > 1 : \frac{1}{p} + \frac{1}{q} = 1$. *Consider* $f : \overline{A} \to \mathbb{R}$ *be Lebesgue integrable, which is not necessarily radial. Let* $\mathbb{N} \not\ni \alpha > \frac{1}{q} : \lceil \alpha \rceil = n$, $0 \leq \beta \leq 1$, $\gamma = \alpha + \beta(n - \alpha)$. *Assume that* $f(s\omega) s^{N-1} \in C^n([R_1, R_2])$, $\forall\, \omega \in S^{N-1}$, $N \geq 2$, *and that* $\max \left\{ \left\| {}^H\mathbb{D}_{R_1+}^{\alpha,\beta} \left(f(s\omega) s^{N-1} \right) \right\|_{L_q([R_1,R_2])}, \left\| {}^H\mathbb{D}_{R_2-}^{\alpha,\beta} \left(f(s\omega) s^{N-1} \right) \right\|_{L_q([R_1,R_2])} \right\} \leq$ K_3, *where* $K_3 > 0$, $s \in [R_1, R_2]$, $\forall\, \omega \in S^{N-1}$. *Furthermore assume that* $\left(I_{R_1+}^{(1-\beta)(n-\alpha)} f(s\omega) s^{N-1} \right)^{(n-k)} (R_1)$, $\left(I_{R_2-}^{(1-\beta)(n-\alpha)} f(s\omega) s^{N-1} \right)^{(n-k)} (R_2)$, $k = 1, ..., n-1$, *are all measurable in* ω *over* S^{N-1}.

Then

(1)

$$\left| \int_A f(y)\,dy - \left\{ \sum_{k=1}^{n-1} \frac{1}{\Gamma(\gamma - k + 2)} \right. \right.$$

$$\left[\left(\int_{S^{N-1}}\left(I_{R_1+}^{(1-\beta)(n-\alpha)}f(s\omega)s^{N-1}\right)^{(n-k)}(R_1)\,d\omega\right)(t-R_1)^{\gamma-k+1}\right.$$

$$\left.\left.\left.+(-1)^k\left(\int_{S^{n-1}}\left(I_{R_2-}^{(1-\beta)(n-\alpha)}f(s\omega)s^{N-1}\right)^{(n-k)}(R_2)\,d\omega\right)(R_2-t)^{\gamma-k+1}\right]\right\}\right|\leq$$

$$\frac{2\pi^{\frac{N}{2}}}{\Gamma\left(\frac{N}{2}\right)}\frac{K_3}{\Gamma(\alpha)\left(\alpha+\frac{1}{p}\right)(p(\alpha-1)+1)^{\frac{1}{p}}}\left[(R_2-t)^{\alpha+\frac{1}{p}}+(t-R_1)^{\alpha+\frac{1}{p}}\right], \quad (3.101)$$

$\forall\, t\in[R_1,R_2]$,

(2) at $t=\frac{R_1+R_2}{2}$*, the R.H.S (3.101) is minimized, and we get:*

$$\left|\int_A f(y)\,dy-\left\{\sum_{k=1}^{n-1}\frac{1}{\Gamma(\gamma-k+2)}\left(\frac{R_2-R_1}{2}\right)^{\gamma-k+1}\right.\right.$$

$$\left[\left(\int_{S^{N-1}}\left(I_{R_1+}^{(1-\beta)(n-\alpha)}f(s\omega)s^{N-1}\right)^{(n-k)}(R_1)\,d\omega\right)\right.$$

$$\left.\left.\left.+(-1)^k\left(\int_{S^{n-1}}\left(I_{R_2-}^{(1-\beta)(n-\alpha)}f(s\omega)s^{N-1}\right)^{(n-k)}(R_2)\,d\omega\right)\right]\right\}\right|\leq$$

$$\frac{\pi^{\frac{N}{2}}}{\Gamma\left(\frac{N}{2}\right)}\frac{K_3}{\Gamma(\alpha)\left(\alpha+\frac{1}{p}\right)(p(\alpha-1)+1)^{\frac{1}{p}}}\frac{(R_2-R_1)^{\alpha+\frac{1}{p}}}{2^{\alpha-1-\frac{1}{q}}}, \quad (3.102)$$

(3) if $\left(I_{R_1+}^{(1-\beta)(n-\alpha)}f(s\omega)s^{N-1}\right)^{(n-k)}(R_1)=\left(I_{R_2-}^{(1-\beta)(n-\alpha)}f(s\omega)s^{N-1}\right)^{(n-k)}(R_2)$ $=0$, $\forall\,\omega\in S^{N-1}$*, (i.e. they vanish on* $\partial B(0,R_1)$ *and* $\partial B(0,R_2)$*, respectively), for all* $k=1,...,n-1$*, we obtain*

$$\left|\int_A f(y)\,dy\right|\leq\frac{\pi^{\frac{N}{2}}}{\Gamma\left(\frac{N}{2}\right)}\frac{K_3}{\Gamma(\alpha)\left(\alpha+\frac{1}{p}\right)(p(\alpha-1)+1)^{\frac{1}{p}}}\frac{(R_2-R_1)^{\alpha+\frac{1}{p}}}{2^{\alpha-1-\frac{1}{q}}}, \quad (3.103)$$

which is a sharp inequality,

(4) more generally, for $j=0,1,2,...,\overline{N}\in\mathbb{N}$*, it holds*

$$\left|\int_A f(y)\,dy-\left\{\sum_{k=1}^{n-1}\frac{1}{\Gamma(\gamma-k+2)}\left(\frac{R_2-R_1}{\overline{N}}\right)^{\gamma-k+1}\right.\right.$$

$$\left[j^{\gamma-k+1}\left(\int_{S^{N-1}}\left(I_{R_1+}^{(1-\beta)(n-\alpha)}f(s\omega)s^{N-1}\right)^{(n-k)}(R_1)\,d\omega\right)\right.$$

$$+\left(-1\right)^{k}\left(\overline{N}-j\right)^{\gamma-k+1}\left(\int_{S^{n-1}}\left(I_{R_{2}-}^{(1-\beta)(n-\alpha)}f\left(s\omega\right)s^{N-1}\right)^{(n-k)}\left(R_{2}\right)d\omega\right)\Bigg]\Bigg\}\Bigg|\leq$$

$$\frac{2\pi^{\frac{N}{2}}}{\Gamma\left(\frac{N}{2}\right)}\frac{K_{3}}{\Gamma\left(\alpha\right)\left(\alpha+\frac{1}{p}\right)\left(p\left(\alpha-1\right)+1\right)^{\frac{1}{p}}}\left(\frac{R_{2}-R_{1}}{\overline{N}}\right)^{\alpha+\frac{1}{p}}\left(j^{\alpha+\frac{1}{p}}+\left(\overline{N}-j\right)^{\alpha+\frac{1}{p}}\right), \tag{3.104}$$

(5) if $\left(I_{R_{1}+}^{(1-\beta)(n-\alpha)}f\left(s\omega\right)s^{N-1}\right)^{(n-k)}\left(R_{1}\right)=\left(I_{R_{2}-}^{(1-\beta)(n-\alpha)}f\left(s\omega\right)s^{N-1}\right)^{(n-k)}\left(R_{2}\right)$ $=0$, $\forall\,\omega\in S^{N-1}$, *for all* $k=1,\ldots,n-1$, *we obtain*

$$\left|\int_{A}f\left(y\right)dy\right|\leq\frac{2\pi^{\frac{N}{2}}}{\Gamma\left(\frac{N}{2}\right)}\frac{K_{3}}{\Gamma\left(\alpha\right)\left(\alpha+\frac{1}{p}\right)\left(p\left(\alpha-1\right)+1\right)^{\frac{1}{p}}}$$

$$\left(\frac{R_{2}-R_{1}}{\overline{N}}\right)^{\alpha+\frac{1}{p}}\min_{j\in\{0,1,2,\ldots,\overline{N}\}}\left(j^{\alpha+\frac{1}{p}}+\left(\overline{N}-j\right)^{\alpha+\frac{1}{p}}\right), \tag{3.105}$$

(6) when $\overline{N}=2$ *and* $j=1$, *from (3.105) we get*

$$\left|\int_{A}f\left(y\right)dy\right|\leq\frac{\pi^{\frac{N}{2}}}{\Gamma\left(\frac{N}{2}\right)}\frac{K_{3}}{\Gamma\left(\alpha\right)\left(\alpha+\frac{1}{p}\right)\left(p\left(\alpha-1\right)+1\right)^{\frac{1}{p}}}\frac{\left(R_{2}-R_{1}\right)^{\alpha+\frac{1}{p}}}{2^{\alpha-1-\frac{1}{q}}}, \tag{3.106}$$

(7) if $\frac{1}{q}<\alpha<1$, *we have (3.106) valid without any boundary conditions.*

Proof By the use of Theorem 3.8, similar to the proof of Theorem 3.11. ■

We continue with results on the ball.

Theorem 3.14 *Let* $0<\alpha<1$, $0\leq\beta\leq1$, $\gamma=\alpha+\beta\left(1-\alpha\right)$. *Consider* $f:\overline{B\left(0,R\right)}\to\mathbb{R}$ *be Lebesgue integrable, which is not necessarily radial. Assume that* $f\left(s\omega\right)s^{N-1}\in C^{1}\left(\left[0,R\right]\right)$, $R>0$, $\forall\ \omega\in S^{N-1}$, $N\geq2$, *and that* $\max\left\{\left\|{}^{H}\mathbb{D}_{0+}^{\alpha,\beta}\left(f\left(s\omega\right)s^{N-1}\right)\right\|_{\infty,[0,R]},\left\|{}^{H}\mathbb{D}_{R-}^{\alpha,\beta}\left(f\left(s\omega\right)s^{N-1}\right)\right\|_{\infty,[0,R]}\right\}\leq M_{1}$, *where* $M_{1}>0$, $s\in\left[0,R\right]$, $\forall\,\omega\in S^{N-1}$.

Then

(1)

$$\left|\int_{B(0,R)}f\left(y\right)dy\right|\leq\frac{2\pi^{\frac{N}{2}}M_{1}}{\Gamma\left(\frac{N}{2}\right)\Gamma\left(\alpha+2\right)}\left(\left(R-t\right)^{\alpha+1}+t^{\alpha+1}\right), \tag{3.107}$$

$\forall\,t\in\left[0,R\right]$,

(2) at $t=\frac{R}{2}$, *the R.H.S (3.107) is minimized, and we get:*

$$\left|\int_{B(0,R)} f(y)\,dy\right| \leq \frac{\pi^{\frac{N}{2}} M_1}{\Gamma\left(\frac{N}{2}\right)\Gamma(\alpha+2)} \frac{R^{\alpha+1}}{2^{\alpha-1}}, \tag{3.108}$$

which is a sharp inequality.

Proof Same as in the proof of Theorem 3.11, just set there $R_1 = 0$ and $R_2 = R$ and use (3.82). ■

The L_q analog follows:

Theorem 3.15 *Let* $p, q > 1 : \frac{1}{p} + \frac{1}{q} = 1$, $\frac{1}{q} < \alpha < 1$, $0 \leq \beta \leq 1$, $\gamma = \alpha + \beta(1 - \alpha)$. *Consider* $f : \overline{B(0,R)} \to \mathbb{R}$ *be Lebesgue integrable, which is not necessarily radial. Assume that* $f(s\omega)s^{N-1} \in C^1([0,R])$, $R > 0$, $\forall\ \omega \in S^{N-1}$, $N \geq 2$, *and that* $\max\left\{\left\| {}^{H}\mathbb{D}_{0+}^{\alpha,\beta}\left(f(s\omega)s^{N-1}\right)\right\|_{L_q([0,R])}, \left\| {}^{H}\mathbb{D}_{R-}^{\alpha,\beta}\left(f(s\omega)s^{N-1}\right)\right\|_{L_q([0,R])}\right\} \leq M_2$, *where* $M_2 > 0$, $s \in [0,R]$, $\forall\ \omega \in S^{N-1}$.

Then

(1)

$$\left|\int_{B(0,R)} f(y)\,dy\right| \leq \frac{2\pi^{\frac{N}{2}} M_2}{\Gamma\left(\frac{N}{2}\right)\Gamma(\alpha)\left(\alpha+\frac{1}{p}\right)(p(\alpha-1)+1)^{\frac{1}{p}}} \left((R-t)^{\alpha+\frac{1}{p}} + t^{\alpha+\frac{1}{p}}\right), \tag{3.109}$$

$\forall\, t \in [0,R]$,

(2) at $t = \frac{R}{2}$, *the R.H.S (3.109) is minimized, and we get:*

$$\left|\int_{B(0,R)} f(y)\,dy\right| \leq \frac{\pi^{\frac{N}{2}} M_2}{\Gamma\left(\frac{N}{2}\right)\Gamma(\alpha)\left(\alpha+\frac{1}{p}\right)(p(\alpha-1)+1)^{\frac{1}{p}}} \frac{R^{\alpha+\frac{1}{p}}}{2^{\alpha-1-\frac{1}{q}}}, \tag{3.110}$$

which is a sharp inequality.

Proof As in the Theorem 3.13, set $R_1 = 0$ and $R_2 = R$ and use (3.82). ■

We continue with results on radial functions.

In the rest of this chapter we consider the functions:

(i) $f : \overline{A} \to \mathbb{R}$ which is radial, i.e. there exists g such that $f(x) = g(r), r = |x|$, $r \in [R_1, R_2]$, $\forall\ x \in \overline{A}$; where A is the spherical shell $A := B(0,R_2) - \overline{B(0,R_1)}$, $0 < R_1 < R_2$, $A \subseteq \mathbb{R}^N$, $N \geq 2$, also

(ii) $f : \overline{B(0,R)} \to \mathbb{R}$ which is radial, i.e. there exists g such that $f(x) = g(r)$, $r = |x|, r \in [0,R]$, $\forall\ x \in \overline{B(0,R)}$; where $B(0,R)$ is the ball $B(0,R) \subseteq \mathbb{R}^N$, $N \geq 2$, $R > 0$.

We will employ the related function $h(s) := g(s)s^{N-1}$, where $s \in [R_1, R_2]$ or $s \in [0,R]$.

Theorem 3.16 *Let the radial* $f : \overline{A} \to \mathbb{R}$, $n - 1 < \alpha < n \in \mathbb{N}$, $0 \leq \beta \leq 1$, $\gamma = \alpha + \beta(n - \alpha)$. *Assume that* $h \in C^n([R_1, R_2])$. *We set*

$$\max\left\{\left\|{}^{H}\mathbb{D}_{R_1+}^{\alpha,\beta}h\right\|_{\infty,[R_1,R_2]},\left\|{}^{H}\mathbb{D}_{R_2-}^{\alpha,\beta}h\right\|_{\infty,[R_1,R_2]}\right\}=:L_1.$$

Then

(1)

$$\left|\int_A f(y)\,dy-\left\{\sum_{k=1}^{n-1}\frac{1}{\Gamma(\gamma-k+2)}\left[\left(I_{R_1+}^{(1-\beta)(n-\alpha)}h\right)^{(n-k)}(R_1)(t-R_1)^{\gamma-k+1}\right.\right.\right.$$

$$\left.\left.\left.+(-1)^k\left(I_{R_2-}^{(1-\beta)(n-\alpha)}h\right)^{(n-k)}(R_2)(R_2-t)^{\gamma-k+1}\right]\right\}\frac{2\pi^{\frac{N}{2}}}{\Gamma\left(\frac{N}{2}\right)}\right|\leq$$

$$\frac{2\pi^{\frac{N}{2}}}{\Gamma\left(\frac{N}{2}\right)}\frac{L_1}{\Gamma(\alpha+2)}\left[(R_2-t)^{\alpha+1}+(t-R_1)^{\alpha+1}\right],\tag{3.111}$$

$\forall\, t\in[R_1,R_2]$,

(2) at $t=\frac{R_1+R_2}{2}$, *the R.H.S (3.111) is minimized, and we get:*

$$\left|\int_A f(y)\,dy-\left\{\sum_{k=1}^{n-1}\frac{1}{\Gamma(\gamma-k+2)}\frac{(R_2-R_1)^{\gamma-k+1}}{2^{\gamma-k}}\right.\right.$$

$$\left.\left.\left[\left(I_{R_1+}^{(1-\beta)(n-\alpha)}h\right)^{(n-k)}(R_1)+(-1)^k\left(I_{R_2-}^{(1-\beta)(n-\alpha)}h\right)^{(n-k)}(R_2)\right]\right\}\frac{\pi^{\frac{N}{2}}}{\Gamma\left(\frac{N}{2}\right)}\right|$$

$$\leq\frac{\pi^{\frac{N}{2}}}{\Gamma\left(\frac{N}{2}\right)}\frac{L_1}{\Gamma(\alpha+2)}\frac{(R_2-R_1)^{\alpha+1}}{2^{\alpha-1}},\tag{3.112}$$

(3) if $\left(I_{R_1+}^{(1-\beta)(n-\alpha)}h\right)^{(n-k)}(R_1)-\left(I_{R_2-}^{(1-\beta)(n-\alpha)}h\right)^{(n-k)}(R_2)=0$, *for* $k=1,\ldots,n-1$, *we derive that*

$$\left|\int_A f(y)\,dy\right|\leq\frac{\pi^{\frac{N}{2}}}{\Gamma\left(\frac{N}{2}\right)}\frac{L_1}{\Gamma(\alpha+2)}\frac{(R_2-R_1)^{\alpha+1}}{2^{\alpha-1}},\tag{3.113}$$

which is a sharp inequality,

(4) more generally, for $j=0,1,2,\ldots,\overline{N}\in\mathbb{N}$, *it holds*

$$\left|\int_A f(y)\,dy-\left\{\sum_{k=1}^{n-1}\frac{1}{\Gamma(\gamma-k+2)}\left(\frac{R_2-R_1}{\overline{N}}\right)^{\gamma-k+1}\right.\right.$$

$$\left[j^{\gamma-k+1}\left(I_{R_1+}^{(1-\beta)(n-\alpha)}h\right)^{(n-k)}(R_1)\right.$$

$$+(-1)^k \left(\overline{N}-j\right)^{\gamma-k+1} \left(I_{R_2-}^{(1-\beta)(n-\alpha)} h\right)^{(n-k)} (R_2)\Big]\Big\} \frac{2\pi^{\frac{N}{2}}}{\Gamma\left(\frac{N}{2}\right)}\Bigg| \leq$$

$$\frac{2\pi^{\frac{N}{2}}}{\Gamma\left(\frac{N}{2}\right)} \frac{L_1}{\Gamma(\alpha+2)} \left(\frac{R_2-R_1}{\overline{N}}\right)^{\alpha+1} \left[j^{\alpha+1}+\left(\overline{N}-j\right)^{\alpha+1}\right], \tag{3.114}$$

(5) if $\left(I_{R_1+}^{(1-\beta)(n-\alpha)} h\right)^{(n-k)} (R_1) = \left(I_{R_2-}^{(1-\beta)(n-\alpha)} h\right)^{(n-k)} (R_2) = 0,$ *for* $k = 1, ..., n-1$, *it holds*

$$\left|\int_A f(y)\, dy\right| \leq \frac{2\pi^{\frac{N}{2}}}{\Gamma\left(\frac{N}{2}\right)} \frac{L_1}{\Gamma(\alpha+2)} \left(\frac{R_2-R_1}{\overline{N}}\right)^{\alpha+1} \min_{j\in\{0,1,\ldots,\overline{N}\}} \left[j^{\alpha+1}+\left(\overline{N}-j\right)^{\alpha+1}\right], \tag{3.115}$$

(6) when $\overline{N} = 2$ *and* $j = 1$, *(3.115) turns to*

$$\left|\int_A f(y)\, dy\right| \leq \frac{\pi^{\frac{N}{2}}}{\Gamma\left(\frac{N}{2}\right)} \frac{L_1}{\Gamma(\alpha+2)} \frac{(R_2-R_1)^{\alpha+1}}{2^{\alpha-1}}, \tag{3.116}$$

(7) when $0 < \alpha < 1$, *inequality (3.116) is again valid without any boundary conditions.*

Proof By Theorem 3.6 (3.35) ($\psi(x) = x$) we have

$$\Bigg|\int_{R_1}^{R_2} h(s)\, ds - \left\{\sum_{k=1}^{n-1} \frac{1}{\Gamma(\gamma-k+2)} \left[\left(I_{R_1+}^{(1-\beta)(n-\alpha)} h\right)^{(n-k)} (R_1)(t-R_1)^{\gamma-k+1}\right.\right.$$

$$\left.\left. +(-1)^k \left(I_{R_2-}^{(1-\beta)(n-\alpha)} h\right)^{(n-k)} (R_2)(R_2-t)^{\gamma-k+1}\right]\right\}\Bigg| \leq$$

$$\frac{L_1}{\Gamma(\alpha+2)} \left[(R_2-t)^{\alpha+1}+(t-R_1)^{\alpha+1}\right] =: \Phi(t), \quad \forall\, t \in [R_1, R_2]. \tag{3.117}$$

Equivalently, we have

$$-\Phi(t) \leq \int_{R_1}^{R_2} h(s)\, ds - \left\{\sum_{k=1}^{n-1} \frac{1}{\Gamma(\gamma-k+2)}\right.$$

$$\left[\left(I_{R_1+}^{(1-\beta)(n-\alpha)} h\right)^{(n-k)} (R_1)(t-R_1)^{\gamma-k+1}\right.$$

$$\left.\left. +(-1)^k \left(I_{R_2-}^{(1-\beta)(n-\alpha)} h\right)^{(n-k)} (R_2)(R_2-t)^{\gamma-k+1}\right]\right\} \leq \Phi(t), \tag{3.118}$$

$\forall\, t \in [R_1, R_2]$.

That is

$$-\Phi(t) \leq \int_{R_1}^{R_2} f(s\omega)\, s^{N-1} ds - \left\{ \sum_{k=1}^{n-1} \frac{1}{\Gamma(\gamma - k + 2)} \right.$$

$$\left[\left(I_{R_1+}^{(1-\beta)(n-\alpha)} h \right)^{(n-k)} (R_1)(t - R_1)^{\gamma-k+1} \right.$$

$$\left.\left. + (-1)^k \left(I_{R_2-}^{(1-\beta)(n-\alpha)} h \right)^{(n-k)} (R_2)(R_2 - t)^{\gamma-k+1} \right] \right\} \leq \Phi(t), \tag{3.119}$$

$\forall\, t \in [R_1, R_2]$, $\forall\, \omega \in S^{N-1}$.

Therefore it holds

$$-\Phi(t) \int_{S^{N-1}} d\omega \leq \int_{S^{N-1}} \left(\int_{R_1}^{R_2} f(s\omega)\, s^{N-1} ds \right) d\omega - \left\{ \sum_{k=1}^{n-1} \frac{1}{\Gamma(\gamma - k + 2)} \right.$$

$$\left[\left(I_{R_1+}^{(1-\beta)(n-\alpha)} h \right)^{(n-k)} (R_1)(t - R_1)^{\gamma-k+1} \right.$$

$$\left.\left. + (-1)^k \left(I_{R_2-}^{(1-\beta)(n-\alpha)} h \right)^{(n-k)} (R_2)(R_2 - t)^{\gamma-k+1} \right] \right\} \int_{S^{N-1}} d\omega \leq \Phi(t) \int_{S^{N-1}} d\omega, \tag{3.120}$$

$\forall\, t \in [R_1, R_2]$,

which is (by 3.83)

$$-\Phi(t) \frac{2\pi^{\frac{N}{2}}}{\Gamma\left(\frac{N}{2}\right)} \leq \int_A f(y)\, dy - \left\{ \sum_{k=1}^{n-1} \frac{1}{\Gamma(\gamma - k + 2)} \right.$$

$$\left[\left(I_{R_1+}^{(1-\beta)(n-\alpha)} h \right)^{(n-k)} (R_1)(t - R_1)^{\gamma-k+1} \right.$$

$$\left.\left. + (-1)^k \left(I_{R_2-}^{(1-\beta)(n-\alpha)} h \right)^{(n-k)} (R_2)(R_2 - t)^{\gamma-k+1} \right] \right\} \frac{2\pi^{\frac{N}{2}}}{\Gamma\left(\frac{N}{2}\right)} \leq$$

$$\Phi(t) \frac{2\pi^{\frac{N}{2}}}{\Gamma\left(\frac{N}{2}\right)}, \quad \forall\, t \in [R_1, R_2]. \tag{3.121}$$

Consequently, we derive

$$\left| \int_A f(y)\,dy - \left\{ \sum_{k=1}^{n-1} \frac{1}{\Gamma(\gamma-k+2)} \left[\left(I_{R_1+}^{(1-\beta)(n-\alpha)} h \right)^{(n-k)} (R_1)(t-R_1)^{\gamma-k+1} \right.\right.\right.$$

$$\left.\left.\left. + (-1)^k \left(I_{R_2-}^{(1-\beta)(n-\alpha)} h \right)^{(n-k)} (R_2)(R_2-t)^{\gamma-k+1} \right] \right\} \frac{2\pi^{\frac{N}{2}}}{\Gamma\left(\frac{N}{2}\right)} \right| \leq$$

$$\frac{2\pi^{\frac{N}{2}}}{\Gamma\left(\frac{N}{2}\right)} \frac{L_1}{\Gamma(\alpha+2)} \left[(R_2-t)^{\alpha+1} + (t-R_1)^{\alpha+1} \right], \tag{3.122}$$

$\forall\, t \in [R_1, R_2]$, proving (3.111). The rest is obvious. ■

The L_1 variant follows:

Theorem 3.17 *Let the radial $f : \overline{A} \to \mathbb{R}$, $\alpha > 1$, $n-1 < \alpha < n \in \mathbb{N}$, $0 \leq \beta \leq 1$, $\gamma = \alpha + \beta(n-\alpha)$. Assume that $h \in C^n([R_1, R_2])$. We set $\max\left\{ \left\| {}^H\mathbb{D}_{R_1+}^{\alpha,\beta} h \right\|_{L_1([R_1,R_2])}, \left\| {}^H\mathbb{D}_{R_2-}^{\alpha,\beta} h \right\|_{L_1([R_1,R_2])} \right\} =: L_2$.*
Then
(1)

$$\left| \int_A f(y)\,dy - \left\{ \sum_{k=1}^{n-1} \frac{1}{\Gamma(\gamma-k+2)} \left[\left(I_{R_1+}^{(1-\beta)(n-\alpha)} h \right)^{(n-k)} (R_1)(t-R_1)^{\gamma-k+1} \right.\right.\right.$$

$$\left.\left.\left. + (-1)^k \left(I_{R_2-}^{(1-\beta)(n-\alpha)} h \right)^{(n-k)} (R_2)(R_2-t)^{\gamma-k+1} \right] \right\} \frac{2\pi^{\frac{N}{2}}}{\Gamma\left(\frac{N}{2}\right)} \right| \leq$$

$$\frac{2\pi^{\frac{N}{2}}}{\Gamma\left(\frac{N}{2}\right)} \frac{L_2}{\Gamma(\alpha+1)} \left[(R_2-t)^{\alpha} + (t-R_1)^{\alpha} \right], \tag{3.123}$$

$\forall\, t \in [R_1, R_2]$,
(2) at $t = \frac{R_1+R_2}{2}$, the R.H.S (3.123) is minimized, and we get:

$$\left| \int_A f(y)\,dy - \left\{ \sum_{k=1}^{n-1} \frac{1}{\Gamma(\gamma-k+2)} \frac{(R_2-R_1)^{\gamma-k+1}}{2^{\gamma-k}} \right.\right.$$

$$\left.\left. \left[\left(I_{R_1+}^{(1-\beta)(n-\alpha)} h \right)^{(n-k)} (R_1) + (-1)^k \left(I_{R_2-}^{(1-\beta)(n-\alpha)} h \right)^{(n-k)} (R_2) \right] \right\} \frac{\pi^{\frac{N}{2}}}{\Gamma\left(\frac{N}{2}\right)} \right|$$

$$\leq \frac{\pi^{\frac{N}{2}}}{\Gamma\left(\frac{N}{2}\right)} \frac{L_2}{\Gamma(\alpha+1)} \frac{(R_2-R_1)^{\alpha}}{2^{\alpha-2}}, \tag{3.124}$$

(3) if $\left(I_{R_1+}^{(1-\beta)(n-\alpha)}h\right)^{(n-k)}(R_1) = \left(I_{R_2-}^{(1-\beta)(n-\alpha)}h\right)^{(n-k)}(R_2) = 0$*, for* $k = 1, ..., n-1$*, we derive that*

$$\left|\int_A f(y)\,dy\right| \le \frac{\pi^{\frac{N}{2}}}{\Gamma\left(\frac{N}{2}\right)} \frac{L_2}{\Gamma(\alpha+1)} \frac{(R_2 - R_1)^{\alpha}}{2^{\alpha-2}}, \tag{3.125}$$

which is a sharp inequality,

(4) more generally, for $j = 0, 1, 2, ..., \overline{N} \in \mathbb{N}$*, it holds*

$$\left|\int_A f(y)\,dy - \left\{\sum_{k=1}^{n-1} \frac{1}{\Gamma(\gamma-k+2)} \left(\frac{R_2 - R_1}{\overline{N}}\right)^{\gamma-k+1}\right.\right.$$

$$\left[j^{\gamma-k+1}\left(I_{R_1+}^{(1-\beta)(n-\alpha)}h\right)^{(n-k)}(R_1)\right.$$

$$\left.\left.\left.+(-1)^k\left(\overline{N}-j\right)^{\gamma-k+1}\left(I_{R_2-}^{(1-\beta)(n-\alpha)}h\right)^{(n-k)}(R_2)\right]\right\}\frac{2\pi^{\frac{N}{2}}}{\Gamma\left(\frac{N}{2}\right)}\right| \le$$

$$\frac{2\pi^{\frac{N}{2}}}{\Gamma\left(\frac{N}{2}\right)} \frac{L_2}{\Gamma(\alpha+1)} \left(\frac{R_2 - R_1}{\overline{N}}\right)^{\alpha} \left[j^{\alpha} + \left(\overline{N}-j\right)^{\alpha}\right], \tag{3.126}$$

(5) if $\left(I_{R_1+}^{(1-\beta)(n-\alpha)}h\right)^{(n-k)}(R_1) = \left(I_{R_2-}^{(1-\beta)(n-\alpha)}h\right)^{(n-k)}(R_2) = 0$*, for* $k = 1, ..., n-1$*, it holds*

$$\left|\int_A f(y)\,dy\right| \le \frac{2\pi^{\frac{N}{2}}}{\Gamma\left(\frac{N}{2}\right)} \frac{L_2}{\Gamma(\alpha+1)} \left(\frac{R_2 - R_1}{\overline{N}}\right)^{\alpha} \min_{j\in\{0,1,...,\overline{N}\}} \left[j^{\alpha} + \left(\overline{N}-j\right)^{\alpha}\right], \tag{3.127}$$

(6) when $\overline{N} = 2$ *and* $j = 1$*, (3.127) turns to*

$$\left|\int_A f(y)\,dy\right| \le \frac{\pi^{\frac{N}{2}}}{\Gamma\left(\frac{N}{2}\right)} \frac{L_2}{\Gamma(\alpha+1)} \frac{(R_2 - R_1)^{\alpha}}{2^{\alpha-2}}, \tag{3.128}$$

Proof Use of Theorem 3.7, proof similar to the proof of Theorem 3.16. ■

The L_q analog follows:

Theorem 3.18 *Let the radial* $f : \overline{A} \to \mathbb{R}$*;* $p, q > 1 : \frac{1}{p} + \frac{1}{q} = 1$*,* $\alpha > \frac{1}{q}$*,* $n - 1 < \alpha < n \in \mathbb{N}$*,* $0 \le \beta \le 1$*,* $\gamma = \alpha + \beta(n - \alpha)$*. Assume that* $h \in C^n([R_1, R_2])$*. We set* $\max\left\{\left\|{}^{H}\mathbb{D}_{R_1+}^{\alpha,\beta}h\right\|_{L_q([R_1,R_2])}, \left\|{}^{H}\mathbb{D}_{R_2-}^{\alpha,\beta}h\right\|_{L_q([R_1,R_2])}\right\} =: L_3.$

Then
(1)

$$\left| \int_A f(y)\,dy - \left\{ \sum_{k=1}^{n-1} \frac{1}{\Gamma(\gamma-k+2)} \left[\left(I_{R_1+}^{(1-\beta)(n-\alpha)} h \right)^{(n-k)} (R_1)(t-R_1)^{\gamma-k+1} \right.\right.\right.$$

$$\left.\left.\left. + (-1)^k \left(I_{R_2-}^{(1-\beta)(n-\alpha)} h \right)^{(n-k)} (R_2)(R_2-t)^{\gamma-k+1} \right] \right\} \frac{2\pi^{\frac{N}{2}}}{\Gamma\left(\frac{N}{2}\right)} \right| \leq$$

$$\frac{2\pi^{\frac{N}{2}}}{\Gamma\left(\frac{N}{2}\right)} \frac{L_3}{\Gamma(\alpha)\left(\alpha+\frac{1}{p}\right)(p(\alpha-1)+1)^{\frac{1}{p}}} \left[(R_2-t)^{\alpha+\frac{1}{p}} + (t-R_1)^{\alpha+\frac{1}{p}} \right], \quad (3.129)$$

$\forall\, t \in [R_1, R_2]$,
(2) at $t = \frac{R_1+R_2}{2}$, *the R.H.S (3.129) is minimized, and we get:*

$$\left| \int_A f(y)\,dy - \left\{ \sum_{k=1}^{n-1} \frac{1}{\Gamma(\gamma-k+2)} \frac{(R_2-R_1)^{\gamma-k+1}}{2^{\gamma-k}} \right.\right.$$

$$\left.\left. \left[\left(I_{R_1+}^{(1-\beta)(n-\alpha)} h \right)^{(n-k)} (R_1) + (-1)^k \left(I_{R_2-}^{(1-\beta)(n-\alpha)} h \right)^{(n-k)} (R_2) \right] \right\} \frac{\pi^{\frac{N}{2}}}{\Gamma\left(\frac{N}{2}\right)} \right|$$

$$\leq \frac{\pi^{\frac{N}{2}}}{\Gamma\left(\frac{N}{2}\right)} \frac{L_3}{\Gamma(\alpha)\left(\alpha+\frac{1}{p}\right)(p(\alpha-1)+1)^{\frac{1}{p}}} \frac{(R_2-R_1)^{\alpha+\frac{1}{p}}}{2^{\alpha-1-\frac{1}{q}}}, \quad (3.130)$$

(3) if $\left(I_{R_1+}^{(1-\beta)(n-\alpha)} h \right)^{(n-k)} (R_1) = \left(I_{R_2-}^{(1-\beta)(n-\alpha)} h \right)^{(n-k)} (R_2) = 0$, *for* $k = 1, \ldots, n-1$, *we derive that*

$$\left| \int_A f(y)\,dy \right| \leq \frac{\pi^{\frac{N}{2}}}{\Gamma\left(\frac{N}{2}\right)} \frac{L_3}{\Gamma(\alpha)\left(\alpha+\frac{1}{p}\right)(p(\alpha-1)+1)^{\frac{1}{p}}} \frac{(R_2-R_1)^{\alpha+\frac{1}{p}}}{2^{\alpha-1-\frac{1}{q}}}, \quad (3.131)$$

which is a sharp inequality,
(4) more generally, for $j = 0, 1, 2, \ldots, \overline{N} \in \mathbb{N}$, *it holds*

$$\left| \int_A f(y)\,dy - \left\{ \sum_{k=1}^{n-1} \frac{1}{\Gamma(\gamma-k+2)} \left(\frac{R_2-R_1}{\overline{N}} \right)^{\gamma-k+1} \right.\right.$$

$$\left[j^{\gamma-k+1} \left(I_{R_1+}^{(1-\beta)(n-\alpha)} h \right)^{(n-k)} (R_1) \right.$$

$$+(-1)^k\left(\overline{N}-j\right)^{\gamma-k+1}\left(I_{R_2-}^{(1-\beta)(n-\alpha)}h\right)^{(n-k)}(R_2)\Bigg]\Bigg\}\frac{2\pi^{\frac{N}{2}}}{\Gamma\left(\frac{N}{2}\right)}\Bigg|\leq$$

$$\frac{2\pi^{\frac{N}{2}}}{\Gamma\left(\frac{N}{2}\right)}\frac{L_3}{\Gamma(\alpha)\left(\alpha+\frac{1}{p}\right)(p(\alpha-1)+1)^{\frac{1}{p}}}\left(\frac{R_2-R_1}{\overline{N}}\right)^{\alpha+\frac{1}{p}}\left[j^{\alpha+\frac{1}{p}}+\left(\overline{N}-j\right)^{\alpha+\frac{1}{p}}\right],\tag{3.132}$$

(5) if $\left(I_{R_1+}^{(1-\beta)(n-\alpha)}h\right)^{(n-k)}(R_1)=\left(I_{R_2-}^{(1-\beta)(n-\alpha)}h\right)^{(n-k)}(R_2)=0$*, for* $k=1,\ldots,n-1$*, it holds*

$$\left|\int_A f(y)\,dy\right|\leq\frac{2\pi^{\frac{N}{2}}}{\Gamma\left(\frac{N}{2}\right)}\frac{L_3}{\Gamma(\alpha)\left(\alpha+\frac{1}{p}\right)(p(\alpha-1)+1)^{\frac{1}{p}}}$$

$$\left(\frac{R_2-R_1}{\overline{N}}\right)^{\alpha+\frac{1}{p}}\min_{j\in\{0,1,\ldots,\overline{N}\}}\left[j^{\alpha+\frac{1}{p}}+\left(\overline{N}-j\right)^{\alpha+\frac{1}{p}}\right],\tag{3.133}$$

(6) when $\overline{N}=2$ *and* $j=1$*, (3.133) turns to*

$$\left|\int_A f(y)\,dy\right|\leq\frac{\pi^{\frac{N}{2}}}{\Gamma\left(\frac{N}{2}\right)}\frac{L_3}{\Gamma(\alpha)\left(\alpha+\frac{1}{p}\right)(p(\alpha-1)+1)^{\frac{1}{p}}}\frac{(R_2-R_1)^{\alpha+\frac{1}{p}}}{2^{\alpha-1-\frac{1}{q}}},\tag{3.134}$$

(7) when $\frac{1}{q}<\alpha<1$*, inequality (3.134) is again valid without any boundary conditions.*

Proof Use of Theorem 3.8, similar to the proof of Theorem 3.16. ■

Next we give radial results on the ball.

Wc givc

Theorem 3.19 *Let the radial* $f:\overline{B(0,R)}\to\mathbb{R}$, $n-1<\alpha<n\in\mathbb{N}$, $0\leq\beta\leq1$, $\gamma=\alpha+\beta(n-\alpha)$*. Assume that* $h\in C^n([0,R])$*. We set*
$\max\left\{\left\|{}^H\mathbb{D}_{0+}^{\alpha,\beta}h\right\|_{\infty,[0,R]},\left\|{}^H\mathbb{D}_{R-}^{\alpha,\beta}h\right\|_{\infty,[0,R]}\right\}=:T_1.$

Then

(1)

$$\left|\int_{B(0,R)}f(y)\,dy-\left\{\sum_{k=1}^{n-1}\frac{1}{\Gamma(\gamma-k+2)}\left[\left(I_{0+}^{(1-\beta)(n-\alpha)}h\right)^{(n-k)}(0)\,t^{\gamma-k+1}\right.\right.\right.$$

$$\left.\left.\left.+(-1)^k\left(I_{R-}^{(1-\beta)(n-\alpha)}h\right)^{(n-k)}(R)(R-t)^{\gamma-k+1}\right]\right\}\frac{2\pi^{\frac{N}{2}}}{\Gamma\left(\frac{N}{2}\right)}\right|\leq$$

$$\frac{2\pi^{\frac{N}{2}}}{\Gamma\left(\frac{N}{2}\right)}\frac{T_1}{\Gamma(\alpha+2)}\left[(R-t)^{\alpha+1}+t^{\alpha+1}\right], \tag{3.135}$$

$\forall\, t \in [0, R]$,

(2) at $t = \frac{R}{2}$, the R.H.S (3.135) is minimized, and we get:

$$\left|\int_{B(0,R)} f(y)\,dy - \left\{\sum_{k=1}^{n-1}\frac{1}{\Gamma(\gamma-k+2)}\frac{R^{\gamma-k+1}}{2^{\gamma-k}}\right.\right.$$

$$\left.\left.\left[\left(I_{0+}^{(1-\beta)(n-\alpha)}h\right)^{(n-k)}(0)+(-1)^k\left(I_{R-}^{(1-\beta)(n-\alpha)}h\right)^{(n-k)}(R)\right]\right\}\frac{\pi^{\frac{N}{2}}}{\Gamma\left(\frac{N}{2}\right)}\right|$$

$$\leq \frac{\pi^{\frac{N}{2}}}{\Gamma\left(\frac{N}{2}\right)}\frac{T_1}{\Gamma(\alpha+2)}\frac{R^{\alpha+1}}{2^{\alpha-1}}, \tag{3.136}$$

(3) if $\left(I_{0+}^{(1-\beta)(n-\alpha)}h\right)^{(n-k)}(0) = \left(I_{R-}^{(1-\beta)(n-\alpha)}h\right)^{(n-k)}(R) = 0$, for $k = 1, ..., n-1$, we derive that

$$\left|\int_{B(0,R)} f(y)\,dy\right| \leq \frac{\pi^{\frac{N}{2}}}{\Gamma\left(\frac{N}{2}\right)}\frac{T_1}{\Gamma(\alpha+2)}\frac{R^{\alpha+1}}{2^{\alpha-1}}, \tag{3.137}$$

which is a sharp inequality,

(4) more generally, for $j = 0, 1, 2, ..., \overline{N} \in \mathbb{N}$, it holds

$$\left|\int_{B(0,R)} f(y)\,dy - \left\{\sum_{k=1}^{n-1}\frac{1}{\Gamma(\gamma-k+2)}\left(\frac{R}{\overline{N}}\right)^{\gamma-k+1}\right.\right.$$

$$\left[j^{\gamma-k+1}\left(I_{0+}^{(1-\beta)(n-\alpha)}h\right)^{(n-k)}(0)\right.$$

$$\left.\left.\left.+(-1)^k\left(\overline{N}-j\right)^{\gamma-k+1}\left(I_{R-}^{(1-\beta)(n-\alpha)}h\right)^{(n-k)}(R)\right]\right\}\frac{2\pi^{\frac{N}{2}}}{\Gamma\left(\frac{N}{2}\right)}\right| \leq$$

$$\frac{2\pi^{\frac{N}{2}}}{\Gamma\left(\frac{N}{2}\right)}\frac{T_1}{\Gamma(\alpha+2)}\left(\frac{R}{\overline{N}}\right)^{\alpha+1}\left[j^{\alpha+1}+\left(\overline{N}-j\right)^{\alpha+1}\right], \tag{3.138}$$

(5) if $\left(I_{0+}^{(1-\beta)(n-\alpha)}h\right)^{(n-k)}(0) = \left(I_{R-}^{(1-\beta)(n-\alpha)}h\right)^{(n-k)}(R) = 0$, for $k = 1, ..., n-1$, it holds

$$\left|\int_{B(0,R)} f(y)\,dy\right| \leq \frac{2\pi^{\frac{N}{2}}}{\Gamma\left(\frac{N}{2}\right)} \frac{T_1}{\Gamma(\alpha+2)} \left(\frac{R}{\overline{N}}\right)^{\alpha+1} \min_{j\in\{0,1,\ldots,\overline{N}\}} \left[j^{\alpha+1} + \left(\overline{N}-j\right)^{\alpha+1}\right], \tag{3.139}$$

(6) when $\overline{N}=2$ and $j=1$, (3.139) turns to

$$\left|\int_{B(0,R)} f(y)\,dy\right| \leq \frac{\pi^{\frac{N}{2}}}{\Gamma\left(\frac{N}{2}\right)} \frac{T_1}{\Gamma(\alpha+2)} \frac{R^{\alpha+1}}{2^{\alpha-1}}, \tag{3.140}$$

(7) when $0<\alpha<1$, inequality (3.140) is again valid without any boundary conditions.

Proof By Theorem 3.16, just set $R_1=0$ and $R_2=R$, the assumptions now are on $B(0,R)$, and use (3.82). ■

It follows the L_1 analog.

We give

Theorem 3.20 *Let the radial $f:\overline{B(0,R)}\to\mathbb{R}$, $\alpha>1$, $n-1<\alpha<n\in\mathbb{N}$, $0\leq\beta\leq1$, $\gamma=\alpha+\beta(n-\alpha)$. Assume that $h\in C^n([0,R])$. We set* $\max\left\{\left\|{}^H\mathbb{D}_{0+}^{\alpha,\beta}h\right\|_{L_1([0,R])}, \left\|{}^H\mathbb{D}_{R-}^{\alpha,\beta}h\right\|_{L_1([0,R])}\right\} =: T_2$.

Then

(1)

$$\left|\int_{B(0,R)} f(y)\,dy - \left\{\sum_{k=1}^{n-1} \frac{1}{\Gamma(\gamma-k+2)} \left[\left(I_{0+}^{(1-\beta)(n-\alpha)}h\right)^{(n-k)}(0)\,t^{\gamma-k+1}\right.\right.\right.$$

$$\left.\left.\left. +(-1)^k \left(I_{R-}^{(1-\beta)(n-\alpha)}h\right)^{(n-k)}(R)\,(R-t)^{\gamma-k+1}\right]\right\} \frac{2\pi^{\frac{N}{2}}}{\Gamma\left(\frac{N}{2}\right)}\right| \leq$$

$$\frac{2\pi^{\frac{N}{2}}}{\Gamma\left(\frac{N}{2}\right)} \frac{T_2}{\Gamma(\alpha+1)} \left[(R-t)^{\alpha}+t^{\alpha}\right], \tag{3.141}$$

$\forall\, t\in[0,R]$,

(2) at $t=\frac{R}{2}$, the R.H.S (3.141) is minimized, and we get:

$$\left|\int_{B(0,R)} f(y)\,dy - \left\{\sum_{k=1}^{n-1} \frac{1}{\Gamma(\gamma-k+2)} \frac{R^{\gamma-k+1}}{2^{\gamma-k}}\right.\right.$$

$$\left.\left.\left[\left(I_{0+}^{(1-\beta)(n-\alpha)}h\right)^{(n-k)}(0) + (-1)^k \left(I_{R-}^{(1-\beta)(n-\alpha)}h\right)^{(n-k)}(R)\right]\right\} \frac{\pi^{\frac{N}{2}}}{\Gamma\left(\frac{N}{2}\right)}\right|$$

$$\leq \frac{\pi^{\frac{N}{2}}}{\Gamma\left(\frac{N}{2}\right)} \frac{T_2}{\Gamma(\alpha+1)} \frac{R^{\alpha}}{2^{\alpha-2}}, \tag{3.142}$$

(3) if $\left(I_{0+}^{(1-\beta)(n-\alpha)}h\right)^{(n-k)}(0) = \left(I_{R-}^{(1-\beta)(n-\alpha)}h\right)^{(n-k)}(R) = 0$, *for* $k = 1, ..., n-1$, *we derive that*

$$\left|\int_{B(0,R)} f(y)\,dy\right| \leq \frac{\pi^{\frac{N}{2}}}{\Gamma\left(\frac{N}{2}\right)} \frac{T_2}{\Gamma(\alpha+1)} \frac{R^{\alpha}}{2^{\alpha-2}}, \tag{3.143}$$

which is a sharp inequality,

(4) more generally, for $j = 0, 1, 2, ..., \overline{N} \in \mathbb{N}$, *it holds*

$$\left|\int_{B(0,R)} f(y)\,dy - \left\{\sum_{k=1}^{n-1} \frac{1}{\Gamma(\gamma-k+2)} \left(\frac{R}{\overline{N}}\right)^{\gamma-k+1}\right.\right.$$

$$\left[j^{\gamma-k+1}\left(I_{0+}^{(1-\beta)(n-\alpha)}h\right)^{(n-k)}(0)\right.$$

$$\left.\left.\left.+(-1)^k\left(\overline{N}-j\right)^{\gamma-k+1}\left(I_{R-}^{(1-\beta)(n-\alpha)}h\right)^{(n-k)}(R)\right]\right\}\frac{2\pi^{\frac{N}{2}}}{\Gamma\left(\frac{N}{2}\right)}\right| \leq$$

$$\frac{2\pi^{\frac{N}{2}}}{\Gamma\left(\frac{N}{2}\right)} \frac{T_2}{\Gamma(\alpha+1)} \left(\frac{R}{\overline{N}}\right)^{\alpha} \left[j^{\alpha} + \left(\overline{N}-j\right)^{\alpha}\right], \tag{3.144}$$

(5) if $\left(I_{0+}^{(1-\beta)(n-\alpha)}h\right)^{(n-k)}(0) = \left(I_{R-}^{(1-\beta)(n-\alpha)}h\right)^{(n-k)}(R) = 0$, *for* $k = 1, ..., n-1$, *it holds*

$$\left|\int_{B(0,R)} f(y)\,dy\right| \leq \frac{2\pi^{\frac{N}{2}}}{\Gamma\left(\frac{N}{2}\right)} \frac{T_2}{\Gamma(\alpha+1)} \left(\frac{R}{\overline{N}}\right)^{\alpha} \min_{j\in\{0,1,...,\overline{N}\}} \left[j^{\alpha} + \left(\overline{N}-j\right)^{\alpha}\right], \tag{3.145}$$

(6) when $\overline{N} = 2$ *and* $j = 1$, *(3.145) turns to*

$$\left|\int_{B(0,R)} f(y)\,dy\right| \leq \frac{\pi^{\frac{N}{2}}}{\Gamma\left(\frac{N}{2}\right)} \frac{T_2}{\Gamma(\alpha+1)} \frac{R^{\alpha}}{2^{\alpha-2}}. \tag{3.146}$$

Proof By Theorem 3.17, just set $R_1 = 0$ and $R_2 = R$, the assumptions now are on $B(0, R)$, and use (3.82). ■

It follows the L_q analog.

We give

Theorem 3.21 *Let the radial $f : \overline{B(0,R)} \to \mathbb{R}$; $p, q > 1 : \frac{1}{p} + \frac{1}{q} = 1$, $\alpha > \frac{1}{q}$, $n - 1 < \alpha < n \in \mathbb{N}$, $0 \leq \beta \leq 1$, $\gamma = \alpha + \beta(n - \alpha)$. Assume that $h \in C^n([0, R])$. We set* $\max\left\{\left\|{}^H\mathbb{D}_{0+}^{\alpha,\beta}h\right\|_{L_q([0,R])}, \left\|{}^H\mathbb{D}_{R-}^{\alpha,\beta}h\right\|_{L_q([0,R])}\right\} =: T_3$.

Then

(1)

$$\left|\int_{B(0,R)} f(y)\,dy - \left\{\sum_{k=1}^{n-1} \frac{1}{\Gamma(\gamma - k + 2)} \left[\left(I_{0+}^{(1-\beta)(n-\alpha)}h\right)^{(n-k)}(0)\,t^{\gamma-k+1}\right.\right.\right.$$

$$\left.\left.\left. + (-1)^k \left(I_{R-}^{(1-\beta)(n-\alpha)}h\right)^{(n-k)}(R)\,(R-t)^{\gamma-k+1}\right]\right\} \frac{2\pi^{\frac{N}{2}}}{\Gamma\left(\frac{N}{2}\right)}\right| \leq$$

$$\frac{2\pi^{\frac{N}{2}}}{\Gamma\left(\frac{N}{2}\right)} \frac{T_3}{\Gamma(\alpha)\left(\alpha + \frac{1}{p}\right)(p(\alpha-1)+1)^{\frac{1}{p}}} \left[(R-t)^{\alpha+\frac{1}{p}} + t^{\alpha+\frac{1}{p}}\right], \tag{3.147}$$

$\forall\, t \in [0, R]$,

(2) at $t = \frac{R}{2}$, the R.H.S (3.147) is minimized, and we get:

$$\left|\int_{B(0,R)} f(y)\,dy - \left\{\sum_{k=1}^{n-1} \frac{1}{\Gamma(\gamma - k + 2)} \frac{R^{\gamma-k+1}}{2^{\gamma-k}}\right.\right.$$

$$\left.\left.\left[\left(I_{0+}^{(1-\beta)(n-\alpha)}h\right)^{(n-k)}(0) + (-1)^k \left(I_{R-}^{(1-\beta)(n-\alpha)}h\right)^{(n-k)}(R)\right]\right\} \frac{\pi^{\frac{N}{2}}}{\Gamma\left(\frac{N}{2}\right)}\right|$$

$$\leq \frac{\pi^{\frac{N}{2}}}{\Gamma\left(\frac{N}{2}\right)} \frac{T_3}{\Gamma(\alpha)\left(\alpha + \frac{1}{p}\right)(p(\alpha-1)+1)^{\frac{1}{p}}} \frac{R^{\alpha+\frac{1}{p}}}{2^{\alpha-1-\frac{1}{q}}}, \tag{3.148}$$

(3) if $\left(I_{0+}^{(1-\beta)(n-\alpha)}h\right)^{(n-k)}(0) = \left(I_{R-}^{(1-\beta)(n-\alpha)}h\right)^{(n-k)}(R) = 0$, for $k = 1, \ldots, n-1$, we derive that

$$\left|\int_{B(0,R)} f(y)\,dy\right| \leq \frac{\pi^{\frac{N}{2}}}{\Gamma\left(\frac{N}{2}\right)} \frac{T_3}{\Gamma(\alpha)\left(\alpha + \frac{1}{p}\right)(p(\alpha-1)+1)^{\frac{1}{p}}} \frac{R^{\alpha+\frac{1}{p}}}{2^{\alpha-1-\frac{1}{q}}}, \tag{3.149}$$

which is a sharp inequality,

(4) more generally, for $j = 0, 1, 2, \ldots, \overline{N} \in \mathbb{N}$, it holds

$$\left| \int_{B(0,R)} f(y)\,dy - \left\{ \sum_{k=1}^{n-1} \frac{1}{\Gamma(\gamma - k + 2)} \left(\frac{R}{\overline{N}} \right)^{\gamma - k + 1} \right. \right.$$

$$\left[j^{\gamma - k + 1} \left(I_{0+}^{(1-\beta)(n-\alpha)} h \right)^{(n-k)} (0) \right.$$

$$\left. \left. \left. + (-1)^k \left(\overline{N} - j \right)^{\gamma - k + 1} \left(I_{R-}^{(1-\beta)(n-\alpha)} h \right)^{(n-k)} (R) \right] \right\} \frac{2\pi^{\frac{N}{2}}}{\Gamma\left(\frac{N}{2}\right)} \right| \leq$$

$$\frac{2\pi^{\frac{N}{2}}}{\Gamma\left(\frac{N}{2}\right)} \frac{T_3}{\Gamma(\alpha)\left(\alpha + \frac{1}{p}\right)(p(\alpha - 1) + 1)^{\frac{1}{p}}} \left(\frac{R}{\overline{N}} \right)^{\alpha + \frac{1}{p}} \left[j^{\alpha + \frac{1}{p}} + \left(\overline{N} - j \right)^{\alpha + \frac{1}{p}} \right], \tag{3.150}$$

(5) if $\left(I_{0+}^{(1-\beta)(n-\alpha)} h \right)^{(n-k)} (0) = \left(I_{R-}^{(1-\beta)(n-\alpha)} h \right)^{(n-k)} (R) = 0$, *for* $k = 1, \ldots, n - 1$, *it holds*

$$\left| \int_{B(0,R)} f(y)\,dy \right| \leq \frac{2\pi^{\frac{N}{2}}}{\Gamma\left(\frac{N}{2}\right)} \frac{T_3}{\Gamma(\alpha)\left(\alpha + \frac{1}{p}\right)(p(\alpha - 1) + 1)^{\frac{1}{p}}}$$

$$\left(\frac{R}{\overline{N}} \right)^{\alpha + \frac{1}{p}} \min_{j \in \{0, 1, \ldots, \overline{N}\}} \left[j^{\alpha + \frac{1}{p}} + \left(\overline{N} - j \right)^{\alpha + \frac{1}{p}} \right], \tag{3.151}$$

(6) when $\overline{N} = 2$ *and* $j = 1$, *(3.151) turns to*

$$\left| \int_{B(0,R)} f(y)\,dy \right| \leq \frac{\pi^{\frac{N}{2}}}{\Gamma\left(\frac{N}{2}\right)} \frac{T_3}{\Gamma(\alpha)\left(\alpha + \frac{1}{p}\right)(p(\alpha - 1) + 1)^{\frac{1}{p}}} \frac{R^{\alpha + \frac{1}{p}}}{2^{\alpha - 1 - \frac{1}{q}}}, \tag{3.152}$$

(7) when $\frac{1}{q} < \alpha < 1$, *inequality (3.152) is again valid without any boundary conditions.*

Proof By Theorem 3.18, just set $R_1 = 0$ and $R_2 = R$, the assumptions now are on $B(0, R)$, and use (3.82). ■

3.4 Application

When $\psi(x) = e^x$ we have:

Corollary 3.22 (to Theorem 3.6) *Let* $f \in C^n([a, b])$, $n - 1 < \alpha < n \in \mathbb{N}$, $0 \leq \beta \leq 1$, $\gamma = \alpha + \beta(n - \alpha)$. *Call* $\overline{\rho_1} := \left\| {}^H\mathbb{D}_{a+}^{\alpha, \beta; e^x} f \right\|_\infty$, $\overline{\rho_2} := \left\| {}^H\mathbb{D}_{b-}^{\alpha, \beta; e^x} f \right\|_\infty$, *and* $\overline{\rho} := \max\left(\overline{\rho_1}, \overline{\rho_2}\right)$. *Then*

(1)

$$\left|\int_a^b f(x)\, e^x dx - \left\{\sum_{k=1}^{n-1} \frac{1}{\Gamma(\gamma-k+2)} \left[f_{e^x}^{[n-k]} \left(I_{a+}^{(1-\beta)(n-\alpha);e^x} f \right)(a) \left(e^t - e^a\right)^{\gamma-k+1} \right.\right.\right.$$

$$\left.\left.\left. + (-1)^k f_{e^x}^{[n-k]} \left(I_{b-}^{(1-\beta)(n-\alpha);e^x} f \right)(b) \left(e^b - e^t\right)^{\gamma-k+1} \right]\right\}\right| \leq$$

$$\frac{\overline{\rho}}{\Gamma(\alpha+2)} \left[\left(e^b - e^t\right)^{\alpha+1} + \left(e^t - e^a\right)^{\alpha+1} \right], \tag{3.153}$$

$\forall\, t \in [a, b]$,

(2) the R.H.S (3.153) is minimized at $t \in [a, b]$, such that $e^t = \frac{e^a + e^b}{2}$, with a minimum value of $\frac{\overline{\rho}}{\Gamma(\alpha+2)} \frac{\left(e^b - e^a\right)^{\alpha+1}}{2^\alpha}$, and we have

$$\left|\int_a^b f(x)\, e^x dx - \left\{\sum_{k=1}^{n-1} \frac{1}{\Gamma(\gamma-k+2)} \left(\frac{e^b - e^a}{2}\right)^{\gamma-k+1} \right.\right.$$

$$\left.\left. \left[f_{e^x}^{[n-k]} \left(I_{a+}^{(1-\beta)(n-\alpha);e^x} f \right)(a) + (-1)^k f_{e^x}^{[n-k]} \left(I_{b-}^{(1-\beta)(n-\alpha);e^x} f \right)(b) \right]\right\}\right| \leq$$

$$\frac{\overline{\rho}\left(e^b - e^a\right)^{\alpha+1}}{\Gamma(\alpha+2)\, 2^\alpha}, \tag{3.154}$$

(3) assuming $f_{e^x}^{[n-k]} \left(I_{a+}^{(1-\beta)(n-\alpha);e^x} f \right)(a) = f_{e^x}^{[n-k]} \left(I_{b-}^{(1-\beta)(n-\alpha);e^x} f \right)(b) = 0$, for $k = 1, ..., n-1$, we get

$$\left|\int_a^b f(x)\, e^x dx\right| \leq \frac{\overline{\rho}\left(e^b - e^a\right)^{\alpha+1}}{\Gamma(\alpha+2)\, 2^\alpha}, \tag{3.155}$$

a sharp inequality,

(4) when $j = 0, 1, 2, ..., N \in \mathbb{N}$, it holds

$$\left|\int_a^b f(x)\, e^x dx - \left\{\sum_{k=1}^{n-1} \frac{1}{\Gamma(\gamma-k+2)} \left(\frac{e^b - e^a}{N}\right)^{\gamma-k+1} \right.\right.$$

$$\left[f_{e^x}^{[n-k]} \left(I_{a+}^{(1-\beta)(n-\alpha);e^x} f \right)(a)\, j^{\gamma-k+1} + \right.$$

$$\left.\left.\left. (-1)^k f_{e^x}^{[n-k]} \left(I_{b-}^{(1-\beta)(n-\alpha);e^x} f \right)(b)\, (N-j)^{\gamma-k+1} \right]\right\}\right| \leq$$

$$\frac{\overline{\rho}}{\Gamma(\alpha+2)}\left(\frac{e^b-e^a}{N}\right)^{\alpha+1}\left(j^{\alpha+1}+(N-j)^{\alpha+1}\right), \tag{3.156}$$

(5) if $f_{e^x}^{[n-k]}\left(I_{a+}^{(1-\beta)(n-\alpha);e^x}f\right)(a)=f_{e^x}^{[n-k]}\left(I_{b-}^{(1-\beta)(n-\alpha);e^x}f\right)(b)=0$, $k=1,...,n-1$, *from (3.156) we derive*

$$\left|\int_a^b f(x)\,e^x dx\right| \leq$$

$$\frac{\overline{\rho}}{\Gamma(\alpha+2)}\left(\frac{e^b-e^a}{N}\right)^{\alpha+1}\min_{j\in\{0,1,...,N\}}\left(j^{\alpha+1}+(N-j)^{\alpha+1}\right), \tag{3.157}$$

(6) when $N=2$ *and* $j=1$, *from (3.157) we get*

$$\left|\int_a^b f(x)\,e^x dx\right| \leq \frac{\overline{\rho}}{\Gamma(\alpha+2)}\frac{\left(e^b-e^a\right)^{\alpha+1}}{2^{\alpha}}, \tag{3.158}$$

(7) if $0<\alpha<1$, *(3.158) is true without boundary conditions.*

One can have many other applications but we omit this task.

References

1. Almeida, R.: A Caputo fractional derivative of a function with respect to another function. Commun. Nonlinear Sci. Numer. Simulat. **44**, 460–481 (2017)
2. Anastassiou, G.A.: Iyengar-Hilfer fractional inequalities. Mathematical Foundations of Computing, accepted (2021)
3. Dragomir, S.S.: Inequalities for the Riemann-Stieljes integral of $(p, q) - H-$Dominated integrators with applications. Appl. Math. E-Notes **15**, 243–260 (2015). http://www.math.nthu.edu.tw/~amen
4. Iyengar, K.S.K.: Note on an inequality. Math. Student **6**, 75–76 (1938)
5. Kilbas, A.A., Srivastava, H.M., Trujillo, J.J.: Theory and Applications of Fractional Differentiation Equations. North Holland, Amsterdam (2006)
6. Rudin, W.: Real and Complex Analysis, International Student Edition. Mc Graw Hill, London (1970)
7. Samko, S.G., Kilbas, A.A., Marichev, O.I.: Fractional Integrals and Derivatives: Theory and Applications. Gordon and Breach Science Publishers, Switzerland (1993)
8. Stroock, D.: A Concise Introduction to the Theory of Integration, 3rd edn. Birkhaüser, Boston, Basel, Berlin (1999)
9. Tomovski, Ž, Hilfer, R., Srivastava, H.M.: Fractional and operational calculus with generalized fractional derivative operators and Mittag-Leffler type functions. Integral Transforms Spec. Funct. **21**(11), 797–814 (2010)
10. Vanterler da, J., Sousa, C., Capelas de Oliveira, E.: On the ψ-Hilfer fractional derivative. Commun. Nonlinear Sci. Numer. Simul. **60**, 72–91 (2018)

Chapter 4
Generalized Hilfer-Polya, Hilfer-Ostrowski and Hilfer-Hilbert-Pachpatte Fractional Inequalities

Here we present Hilfer-Polya, ψ-Hilfer Ostrowski and ψ-Hilfer-Hilbert-Pachpatte types fractional inequalities. They are univariate inequalities involving left and right Hilfer and ψ-Hilfer fractional derivatives. All estimates are with respect to norms $\|\cdot\|_p$, $1 \leq p \leq \infty$. At the end we provide applications. It follows [4].

4.1 Introduction

We are motivated by the following famous Polya's integral inequality, see [6], [7, p. 62], [8] and [9, p. 83].

Theorem 4.1 *Let $f(x)$ be a differentiable and not identically a constant on $[a, b]$ with $f(a) = f(b) = 0$. Then there exists at least one point $\xi \in [a, b]$ such that*

$$\left|f'(\xi)\right| > \frac{4}{(b-a)^2} \int_a^b f(x)\, dx. \tag{4.1}$$

We are inspired also by the related first fractional Polya Inequality, see Chap. 2, p. 9, [2].

In this chapter we establish fractional integral inequalities using the Hilfer and ψ-Hilfer fractional derivatives. These are of Polya, Ostrowski and Hilbert-Pachpatte types.

4.2 Background

Let $-\infty < a < b < \infty$, the left and right Riemann-Liouville fractional integrals of order $\alpha \in \mathbb{C}$ ($\mathcal{R}(\alpha) > 0$) are defined by

G. A. Anastassiou, *Unification of Fractional Calculi with Applications*,
Studies in Systems, Decision and Control 398,
https://doi.org/10.1007/978-3-030-86920-5_4

$$\left(I_{a+}^{\alpha}f\right)(x)=\frac{1}{\Gamma(\alpha)}\int_{a}^{x}(x-t)^{\alpha-1}f(t)\,dt, \tag{4.2}$$

$x>a$; where Γ stands for the gamma function,

and

$$\left(I_{b-}^{\alpha}f\right)(x)=\frac{1}{\Gamma(\alpha)}\int_{x}^{b}(t-x)^{\alpha-1}f(t)\,dt, \tag{4.3}$$

$x<b$.

The Riemann-Liouville left and right fractional derivatives of order $\alpha\in\mathbb{C}$ ($\mathcal{R}(\alpha)\geq 0$) are defined by

$$\left(\Delta_{a+}^{\alpha}y\right)(x)=\left(\frac{d}{dx}\right)^{n}\left(I_{a+}^{n-\alpha}y\right)(x)=\frac{1}{\Gamma(n-\alpha)}\left(\frac{d}{dx}\right)^{n}\int_{a}^{x}(x-t)^{n-\alpha-1}y(t)\,dt \tag{4.4}$$

($n=\lceil\mathcal{R}(\alpha)\rceil$, $\lceil\cdot\rceil$ means ceiling of the number; $x>a$)

$$\left(\Delta_{b-}^{\alpha}y\right)(x)=(-1)^{n}\left(\frac{d}{dx}\right)^{n}\left(I_{b-}^{n-\alpha}y\right)(x)=$$

$$\frac{(-1)^{n}}{\Gamma(n-\alpha)}\left(\frac{d}{dx}\right)^{n}\int_{x}^{b}(t-x)^{n-\alpha-1}y(t)\,dt \tag{4.5}$$

($n=\lceil\mathcal{R}(\alpha)\rceil$; $x<b$), respectively, where $\mathcal{R}(\alpha)$ is the real part of α.

In particular, when $\alpha=n\in\mathbb{Z}_{+}$, then

$$\left(\Delta_{a+}^{0}y\right)(x)=\left(\Delta_{b-}^{0}y\right)(x)=y(x);$$

$$\left(\Delta_{a+}^{n}y\right)(x)=y^{(n)}(x),\text{ and }\left(\Delta_{b-}^{n}y\right)(x)=(-1)^{n}y^{(n)}(x),\ n\in\mathbb{N},$$

see [10].

Let $\alpha>0$, $I=[a,b]\subset\mathbb{R}$, f an integrable function defined on I and $\psi\in C^{1}(I)$ an increasing function such that $\psi'(x)\neq 0$, for all $x\in I$. Left fractional integrals and left Riemann-Liouville fractional derivatives of a function f with respect to another function ψ are defined as [5, 10]

$$I_{a+}^{\alpha,\psi}f(x)=\frac{1}{\Gamma(\alpha)}\int_{a}^{x}\psi'(t)(\psi(x)-\psi(t))^{\alpha-1}f(t)\,dt, \tag{4.6}$$

and

$$\Delta_{a+}^{\alpha,\psi}f(x)=\left(\frac{1}{\psi'(x)}\frac{d}{dx}\right)^{n}I_{a+}^{n-\alpha,\psi}f(x)= \tag{4.7}$$

$$\frac{1}{\Gamma(n-\alpha)}\left(\frac{1}{\psi'(x)}\frac{d}{dx}\right)^{n}\int_{a}^{x}\psi'(t)(\psi(x)-\psi(t))^{n-\alpha-1}f(t)\,dt,$$

respectively, where $n = \lceil \alpha \rceil$.

Similarly, we define the right ones:

$$I_{b-}^{\alpha,\psi} f(x) = \frac{1}{\Gamma(\alpha)} \int_x^b \psi'(t) (\psi(t) - \psi(x))^{\alpha-1} f(t)\, dt, \tag{4.8}$$

and

$$\Delta_{b-}^{\alpha,\psi} f(x) = \left(-\frac{1}{\psi'(x)} \frac{d}{dx}\right)^n I_{b-}^{n-\alpha,\psi} f(x) =$$

$$\frac{1}{\Gamma(n-\alpha)} \left(-\frac{1}{\psi'(x)} \frac{d}{dx}\right)^n \int_x^b \psi'(t) (\psi(t) - \psi(x))^{n-\alpha-1} f(t)\, dt. \tag{4.9}$$

The following semigroup property holds; if $\alpha, \beta > 0$, $f \in C(I)$, then

$$I_{a+}^{\alpha,\psi} I_{a+}^{\beta,\psi} f = I_{a+}^{\alpha+\beta,\psi} f \text{ and } I_{b-}^{\alpha,\psi} I_{b-}^{\beta,\psi} f = I_{b-}^{\alpha+\beta,\psi} f.$$

Next let again $\alpha > 0$, $n = \lceil \alpha \rceil$, $I = [a, b]$, $f, \psi \in C^n(I) : \psi$ is increasing and $\psi'(x) \neq 0$, for all $x \in I$. The left ψ-Caputo fractional derivative of f of order α is given by Almeida [1]

$${}^C D_{a+}^{\alpha,\psi} f(x) = I_{a+}^{n-\alpha,\psi} \left(\frac{1}{\psi'(x)} \frac{d}{dx}\right)^n f(x), \tag{4.10}$$

and the right ψ-Caputo fractional derivative [1]

$${}^C D_{b-}^{\alpha,\psi} f(x) = I_{b-}^{n-\alpha,\psi} \left(-\frac{1}{\psi'(x)} \frac{d}{dx}\right)^n f(x). \tag{4.11}$$

We set

$$f_{\psi}^{[n]}(x) := f_{\psi}^{(n)} f(x) := \left(\frac{1}{\psi'(x)} \frac{d}{dx}\right)^n f(x). \tag{4.12}$$

Clearly, when $\alpha = m \in \mathbb{N}$ we have

$${}^C D_{a+}^{\alpha,\psi} f(x) = f_{\psi}^{[m]}(x) \text{ and } {}^C D_{b-}^{\alpha,\psi} f(x) = (-1)^m f_{\psi}^{[m]}(x),$$

and if $\alpha \notin \mathbb{N}$, then

$${}^C D_{a+}^{\alpha,\psi} f(x) = \frac{1}{\Gamma(n-\alpha)} \int_a^x \psi'(t) (\psi(x) - \psi(t))^{n-\alpha-1} f_{\psi}^{[n]}(t)\, dt, \tag{4.13}$$

and

$$ {}^{C}D_{b-}^{\alpha,\psi} f(x) = \frac{(-1)^n}{\Gamma(n-\alpha)} \int_x^b \psi'(t) (\psi(t) - \psi(x))^{n-\alpha-1} f_{\psi}^{[n]}(t)\, dt. \tag{4.14} $$

If $\psi(x) = x$, then we get the usual left and right Caputo fractional derivatives

$$ {}^{C}D_{a+}^{m} f(x) = f^{(m)}(x), \quad {}^{C}D_{b-}^{m} f(x) = (-1)^m f^{(m)}(x), $$

for $m \in \mathbb{N}$, and ($\alpha \notin \mathbb{N}$)

$$ D_{*a}^{\alpha} f(x) = {}^{C}D_{a+}^{\alpha} f(x) = \frac{1}{\Gamma(n-\alpha)} \int_a^x (x-t)^{n-\alpha-1} f^{(n)}(t)\, dt, \tag{4.15} $$

$$ D_{b-}^{\alpha}(x) = {}^{C}D_{b-}^{\alpha} f(x) = \frac{(-1)^n}{\Gamma(n-\alpha)} \int_x^b (t-x)^{n-\alpha-1} f^{(n)}(t)\, dt. \tag{4.16} $$

Also we set

$$ {}^{C}D_{a+}^{0,\psi} f(x) = {}^{C}D_{b-}^{0,\psi} f(x) = f(x). $$

Next we will deal with the ψ-Hilfer fractional derivative.

Definition 4.2 ([12]) Let $n-1 < \alpha < n$, $n \in \mathbb{N}$, $I = [a,b] \subset \mathbb{R}$ and $f, \psi \in C^n([a,b])$, ψ is increasing and $\psi'(x) \neq 0$, for all $x \in I$. The ψ-Hilfer fractional derivative (left-sided and right-sided) ${}^{H}\mathbb{D}_{a+(b-)}^{\alpha,\beta;\psi} f$ of order α and type $0 \leq \beta \leq 1$, respectively, are defined by

$$ {}^{H}\mathbb{D}_{a+}^{\alpha,\beta;\psi} f(x) = I_{a+}^{\beta(n-\alpha);\psi} \left(\frac{1}{\psi'(x)} \frac{d}{dx} \right)^n I_{a+}^{(1-\beta)(n-\alpha);\psi} f(x), \tag{4.17} $$

and

$$ {}^{H}\mathbb{D}_{b-}^{\alpha,\beta;\psi} f(x) = I_{b-}^{\beta(n-\alpha);\psi} \left(-\frac{1}{\psi'(x)} \frac{d}{dx} \right)^n I_{b-}^{(1-\beta)(n-\alpha);\psi} f(x), \; x \in [a,b]. \tag{4.18} $$

The original Hilfer fractional derivatives [11] come from $\psi(x) = x$, and are denoted by ${}^{H}\mathbb{D}_{a+}^{\alpha,\beta} f(x)$ and ${}^{H}\mathbb{D}_{b-}^{\alpha,\beta} f(x)$.

When $\beta = 0$, we get Riemann-Liouville fractional derivatives, while when $\beta = 1$ we have Caputo type fractional derivatives.

We define $\gamma = \alpha + \beta(n-\alpha)$. We notice that $n-1 < \alpha \leq \alpha + \beta(n-\alpha) \leq \alpha + n - \alpha = n$, hence $\lceil \gamma \rceil = n$. We can easily write that ([12])

$$ {}^{H}\mathbb{D}_{a+}^{\alpha,\beta;\psi} f(x) = I_{a+}^{\gamma-\alpha;\psi} \Delta_{a+}^{\gamma;\psi} f(x), \tag{4.19} $$

and

$$ {}^{H}\mathbb{D}_{b-}^{\alpha,\beta;\psi} f(x) = I_{b-}^{\gamma-\alpha;\psi} \Delta_{b-}^{\gamma;\psi} f(x), \; x \in [a,b]. \tag{4.20} $$

We have that [12]

$$\Delta_{a+}^{\gamma,\psi} f(x) = \left(\frac{1}{\psi'(x)}\frac{d}{dx}\right)^n I_{a+}^{(1-\beta)(n-\alpha);\psi} f(x), \tag{4.21}$$

and

$$\Delta_{b-}^{\gamma,\psi} f(x) = \left(-\frac{1}{\psi'(x)}\frac{d}{dx}\right)^n I_{b-}^{(1-\beta)(n-\alpha);\psi} f(x). \tag{4.22}$$

In particular, when $0 < \alpha < 1$ and $0 \leq \beta \leq 1$; $\gamma = \alpha + \beta(1-\alpha)$, we have that

$${}^{H}\mathbb{D}_{a+}^{\alpha,\beta;\psi} f(x) = \frac{1}{\Gamma(\gamma-\alpha)} \int_a^x \psi'(t)(\psi(x)-\psi(t))^{\gamma-\alpha-1} \Delta_{a+}^{\gamma;\psi} f(t)\,dt, \tag{4.23}$$

and

$${}^{H}\mathbb{D}_{b-}^{\alpha,\beta;\psi} f(x) = \frac{1}{\Gamma(\gamma-\alpha)} \int_x^b \psi'(t)(\psi(t)-\psi(x))^{\gamma-\alpha-1} \Delta_{b-}^{\gamma;\psi} f(t)\,dt, \tag{4.24}$$

$x \in [a,b]$.

Remark 4.3 ([12]) Let $\mu = n(1-\beta) + \beta\alpha$, then $\lceil \mu \rceil = n$.

Assume that $g(x) = I_{a+}^{(1-\beta)(n-\alpha);\psi} f(x) \in C^n([a,b])$, we have that

$${}^{H}\mathbb{D}_{a+}^{\alpha,\beta;\psi} f(x) = I_{a+}^{n-\mu;\psi} \left(\frac{1}{\psi'(x)}\frac{d}{dx}\right)^n g(x). \tag{4.25}$$

Thus

$${}^{H}\mathbb{D}_{a+}^{\alpha,\beta;\psi} f = {}^{C}D_{a+}^{\mu;\psi} g(x) = {}^{C}D_{a+}^{\mu;\psi}\left[I_{a+}^{(1-\beta)(n-\alpha);\psi} f(x)\right]. \tag{4.26}$$

Assume that $w(x) = I_{b-}^{(1-\beta)(n-\alpha);\psi} f(x) \in C^n([a,b])$. Hence

$${}^{H}\mathbb{D}_{b-}^{\alpha,\beta;\psi} f(x) = I_{b-}^{\beta(n-\alpha);\psi} \left(-\frac{1}{\psi'(x)}\frac{d}{dx}\right)^n w(x) = I_{b-}^{n-\mu;\psi} \left(-\frac{1}{\psi'(x)}\frac{d}{dx}\right)^n w(x). \tag{4.27}$$

Thus

$${}^{H}\mathbb{D}_{b-}^{\alpha,\beta;\psi} f = {}^{C}D_{b-}^{\mu;\psi} w(x) = {}^{C}D_{b-}^{\mu;\psi}\left(I_{b-}^{(1-\beta)(n-\alpha);\psi} f(x)\right). \tag{4.28}$$

We mention the simplified ψ-Hilfer fractional Taylor formulae:

Theorem 4.4 (see also [12]) *Let $\psi, f \in C^n([a,b])$, with ψ being increasing such that $\psi'(x) \neq 0$ over $[a,b]$, where $n-1 < \alpha < n$, $0 \leq \beta \leq 1$, and $\gamma = \alpha + \beta(n-\alpha)$, $x \in [a,b]$. Then*

$$f(x) - \sum_{k=1}^{n-1} \frac{(\psi(x)-\psi(a))^{\gamma-k}}{\Gamma(\gamma-k+1)} f_{\psi}^{[n-k]}\left(I_{a+}^{(1-\beta)(n-\alpha);\psi} f\right)(a) =$$

$$\frac{1}{\Gamma(\alpha)} \int_a^x \psi'(t) (\psi(x) - \psi(t))^{\alpha-1} \, {}^H\mathbb{D}_{a+}^{\alpha,\beta;\psi} f(t) \, dt, \tag{4.29}$$

and

$$f(x) - \sum_{k=1}^{n-1} \frac{(-1)^k (\psi(b) - \psi(x))^{\gamma-k}}{\Gamma(\gamma - k + 1)} f_\psi^{[n-k]} \left(I_{b-}^{(1-\beta)(n-\alpha);\psi} f \right)(b) =$$

$$\frac{1}{\Gamma(\alpha)} \int_x^b \psi'(t) (\psi(t) - \psi(x))^{\alpha-1} \, {}^H\mathbb{D}_{b-}^{\alpha,\beta;\psi} f(t) \, dt. \tag{4.30}$$

Here notice that $\left(I_{a+}^{(1-\beta)(n-\alpha);\psi} f \right)(a) = \left(I_{b-}^{(1-\beta)(n-\alpha);\psi} f \right)(b) = 0.$

We also mention the following alternative ψ-Hilfer fractional Taylor formulae:

Theorem 4.5 ([3]) *Let $f, \psi \in C^n([a,b])$, with ψ being increasing, $\psi'(x) \neq 0$ over $[a,b] \subset \mathbb{R}$, $\alpha > 0 : \lceil \alpha \rceil = n$, $0 \leq \beta \leq 1$, $\mu = n(1-\beta) + \beta\alpha$. Assume that $g(x) = I_{a+}^{(1-\beta)(n-\alpha);\psi} f(x)$, $w(x) = I_{b-}^{(1-\beta)(n-\alpha);\psi} f(x) \in C^n([a,b])$.*

Then

(1)

$$I_{a+}^{\mu;\psi} \, {}^H\mathbb{D}_{a+}^{\alpha,\beta;\psi} f(x) = g(x) - \sum_{k=0}^{n-1} \frac{g_\psi^{[k]}(a)}{k!} (\psi(x) - \psi(a))^k, \tag{4.31}$$

where

$$g_\psi^{[k]}(x) = \left(\frac{1}{\psi'(x)} \frac{d}{dx} \right)^k g(x), \; k = 0, 1, ..., n-1,$$

and

(2)

$$I_{b-}^{\mu;\psi} \, {}^H\mathbb{D}_{b-}^{\alpha,\beta;\psi} f(x) = w(x) - \sum_{k=0}^{n-1} \frac{(-1)^k w_\psi^{[k]}(b)}{k!} (\psi(b) - \psi(x))^k, \tag{4.32}$$

where

$$w_\psi^{[k]}(x) = \left(\frac{1}{\psi'(x)} \frac{d}{dx} \right)^k w(x), \; k = 0, 1, ..., n-1; x \in [a,b].$$

Next we list two Hilfer fractional derivatives representation formulae:

Theorem 4.6 ([3]) *Let* $\alpha > 0$, $\alpha \notin \mathbb{N}$, $\lceil \alpha \rceil = n$, $0 < \beta < 1$; $f \in C^n([a,b])$, $[a,b] \subset \mathbb{R}$; *and set* $\gamma = \alpha + \beta(n - \alpha)$. *Assume further that* $\Delta_{a+}^{\gamma} f \in C([a,b]) : \Delta_{a+}^{\gamma - j} f(a) = 0$, *for* $j = 1, ..., n$. *Let also* $\overline{\alpha} > 0 : \lceil \overline{\alpha} \rceil = \overline{n}$, *with* $\overline{\gamma} = \overline{\alpha} + \beta(\overline{n} - \overline{\alpha})$, *and assume that* $\alpha > \overline{\alpha}$ *and* $\gamma > \overline{\gamma}$. *Then*

$$ {}^{H}\mathbb{D}_{a+}^{\overline{\alpha},\beta} f(x) = \frac{1}{\Gamma(\alpha - \overline{\alpha})} \int_a^x (x-t)^{\alpha - \overline{\alpha} - 1}\, {}^{H}\mathbb{D}_{a+}^{\alpha,\beta} f(t)\, dt, \tag{4.33}$$

$\forall\, x \in [a,b]$,

furthermore ${}^{H}\mathbb{D}_{a+}^{\overline{\alpha},\beta} f \in AC([a,b])$ *(absolutely continuous functions) if* $\alpha - \overline{\alpha} \geq 1$ *and* ${}^{H}\mathbb{D}_{a+}^{\overline{\alpha},\beta} f \in C([a,b])$ *if* $\alpha - \overline{\alpha} \in (0,1)$.

Theorem 4.7 ([3]) *Let* $\alpha > 0$, $\alpha \notin \mathbb{N}$, $\lceil \alpha \rceil = n$, $0 < \beta < 1$; $f \in C^n([a,b])$, $[a,b] \subset \mathbb{R}$; *and set* $\gamma = \alpha + \beta(n - \alpha)$. *Assume further that* $\Delta_{b-}^{\gamma} f \in C([a,b]) : \Delta_{b-}^{\gamma - j} f(b) = 0$, $j = 1, ..., n$. *Let also* $\overline{\alpha} > 0 : \lceil \overline{\alpha} \rceil = \overline{n}$, *with* $\overline{\gamma} = \overline{\alpha} + \beta(\overline{n} - \overline{\alpha})$, *and assume that* $\alpha > \overline{\alpha}$ *and* $\gamma > \overline{\gamma}$. *Then*

$$ {}^{H}\mathbb{D}_{b-}^{\overline{\alpha},\beta} f(x) = \frac{1}{\Gamma(\alpha - \overline{\alpha})} \int_x^b (t-x)^{\alpha - \overline{\alpha} - 1}\, {}^{H}\mathbb{D}_{b-}^{\alpha,\beta} f(t)\, dt, \tag{4.34}$$

$\forall\, x \in [a,b]$,

furthermore ${}^{H}\mathbb{D}_{b-}^{\overline{\alpha},\beta} f \in AC([a,b])$ *if* $\alpha - \overline{\alpha} \geq 1$ *and* ${}^{H}\mathbb{D}_{b-}^{\overline{\alpha},\beta} f \in C([a,b])$ *if* $\alpha - \overline{\alpha} \in (0,1)$.

4.3 Main Results

We present the following Hilfer-Polya type fractional inequalities:

Theorem 4.8 *Let* $\alpha > 0$, $\alpha \notin \mathbb{N}$, $\lceil \alpha \rceil = n$, $0 < \beta < 1$; $f \in C^n([a,b])$, $[a,b] \subset \mathbb{R}$; *and set* $\gamma = \alpha + \beta(n - \alpha)$. *Assume further that* $\Delta_{a+}^{\gamma} f \in C([a,b]) : \Delta_{a+}^{\gamma - j} f(a) = 0$, *for* $j = 1, ..., n$; *and* $\Delta_{b-}^{\gamma} f \in C([a,b]) : \Delta_{b-}^{\gamma - j} f(b) = 0$, $j = 1, ..., n$. *Let also* $\overline{\alpha} > 0 : \lceil \overline{\alpha} \rceil = \overline{n}$, *with* $\overline{\gamma} = \overline{\alpha} + \beta(\overline{n} - \overline{\alpha})$, *and assume that* $\alpha > \overline{\alpha}$ *and* $\gamma > \overline{\gamma}$.

Set

$$ {}^{H}\mathbb{D}^{\overline{\alpha},\beta} f(x) := \begin{cases} {}^{H}\mathbb{D}_{a+}^{\overline{\alpha},\beta} f(x), & x \in \left[a, \frac{a+b}{2}\right], \\ {}^{H}\mathbb{D}_{b-}^{\overline{\alpha},\beta} f(x), & x \in \left(\frac{a+b}{2}, b\right], \end{cases} \tag{4.35}$$

and

$$ M_1 := \max \left\{ \left\| {}^{H}\mathbb{D}_{a+}^{\alpha,\beta} f \right\|_{\infty,\left[a,\frac{a+b}{2}\right]}, \left\| {}^{H}\mathbb{D}_{b-}^{\alpha,\beta} f \right\|_{\infty,\left[\frac{a+b}{2},b\right]} \right\}. \tag{4.36}$$

Then

$$\left| \int_a^b {}^H\mathbb{D}^{\overline{\alpha},\beta} f(x)\, dx \right| \leq \int_a^b \left| {}^H\mathbb{D}^{\overline{\alpha},\beta} f(x) \right| dx \leq \frac{M_1 (b-a)^{\alpha-\overline{\alpha}+1}}{2^{\alpha-\overline{\alpha}} \Gamma(\alpha-\overline{\alpha}+2)}. \tag{4.37}$$

Proof From (4.33) we have

$${}^H\mathbb{D}_{a+}^{\overline{\alpha},\beta} f(x) = \frac{1}{\Gamma(\alpha-\overline{\alpha})} \int_a^x (x-t)^{\alpha-\overline{\alpha}-1}\, {}^H\mathbb{D}_{a+}^{\alpha,\beta} f(t)\, dt, \tag{4.38}$$

$\forall\, x \in \left[a, \frac{a+b}{2}\right]$.

By (4.34), we get

$${}^H\mathbb{D}_{b-}^{\overline{\alpha},\beta} f(x) = \frac{1}{\Gamma(\alpha-\overline{\alpha})} \int_x^b (t-x)^{\alpha-\overline{\alpha}-1}\, {}^H\mathbb{D}_{b-}^{\alpha,\beta} f(t)\, dt, \tag{4.39}$$

$\forall\, x \in \left[\frac{a+b}{2}, b\right]$.

We derive that

$$\left| {}^H\mathbb{D}_{a+}^{\overline{\alpha},\beta} f(x) \right| \leq \frac{\left\| {}^H\mathbb{D}_{a+}^{\alpha,\beta} f \right\|_{\infty,\left[a,\frac{a+b}{2}\right]}}{\Gamma(\alpha-\overline{\alpha}+1)} (x-a)^{\alpha-\overline{\alpha}}, \tag{4.40}$$

$\forall\, x \in \left[a, \frac{a+b}{2}\right]$,

and similarly,

$$\left| {}^H\mathbb{D}_{b-}^{\overline{\alpha},\beta} f(x) \right| \leq \frac{\left\| {}^H\mathbb{D}_{b-}^{\alpha,\beta} f \right\|_{\infty,\left[\frac{a+b}{2},b\right]}}{\Gamma(\alpha-\overline{\alpha}+1)} (b-x)^{\alpha-\overline{\alpha}}, \tag{4.41}$$

$\forall\, x \in \left[\frac{a+b}{2}, b\right]$.

We notice that:

$$\int_a^b {}^H\mathbb{D}^{\overline{\alpha},\beta} f(x)\, dx = \int_a^{\frac{a+b}{2}} {}^H\mathbb{D}^{\overline{\alpha},\beta} f(x)\, dx + \int_{\frac{a+b}{2}}^b {}^H\mathbb{D}^{\overline{\alpha},\beta} f(x)\, dx =$$

$$\int_a^{\frac{a+b}{2}} {}^H\mathbb{D}_{a+}^{\overline{\alpha},\beta} f(x)\, dx + \int_{\frac{a+b}{2}}^b {}^H\mathbb{D}_{b-}^{\overline{\alpha},\beta} f(x)\, dx. \tag{4.42}$$

We further derive that

$$\int_a^{\frac{a+b}{2}} \left| {}^H\mathbb{D}_{a+}^{\overline{\alpha},\beta} f(x) \right| dx \leq \frac{\left\| {}^H\mathbb{D}_{a+}^{\alpha,\beta} f \right\|_{\infty,\left[a,\frac{a+b}{2}\right]}}{\Gamma(\alpha-\overline{\alpha}+1)} \int_a^{\frac{a+b}{2}} (x-a)^{\alpha-\overline{\alpha}}\, dx = \tag{4.43}$$

$$\frac{\left\| {}^{H}\mathbb{D}_{a+}^{\alpha,\beta} f \right\|_{\infty,[a,\frac{a+b}{2}]}}{\Gamma(\alpha-\overline{\alpha}+2)} \left(\frac{a+b}{2}-a\right)^{\alpha-\overline{\alpha}+1} = \frac{\left\| {}^{H}\mathbb{D}_{a+}^{\alpha,\beta} f \right\|_{\infty,[a,\frac{a+b}{2}]}}{\Gamma(\alpha-\overline{\alpha}+2)} \left(\frac{b-a}{2}\right)^{\alpha-\overline{\alpha}+1}.$$

That is, it holds

$$\int_a^{\frac{a+b}{2}} \left| {}^{H}\mathbb{D}_{a+}^{\overline{\alpha},\beta} f(x) \right| dx \leq \frac{\left\| {}^{H}\mathbb{D}_{a+}^{\alpha,\beta} f \right\|_{\infty,[a,\frac{a+b}{2}]}}{\Gamma(\alpha-\overline{\alpha}+2)} \left(\frac{b-a}{2}\right)^{\alpha-\overline{\alpha}+1}. \tag{4.44}$$

Similarly, it holds

$$\int_{\frac{a+b}{2}}^{b} \left| {}^{H}\mathbb{D}_{b-}^{\overline{\alpha},\beta} f(x) \right| dx \leq \frac{\left\| {}^{H}\mathbb{D}_{b-}^{\alpha,\beta} f \right\|_{\infty,[\frac{a+b}{2},b]}}{\Gamma(\alpha-\overline{\alpha}+2)} \left(\frac{b-a}{2}\right)^{\alpha-\overline{\alpha}+1}. \tag{4.45}$$

Therefore, we obtain

$$\left| \int_a^b {}^{H}\mathbb{D}^{\overline{\alpha},\beta} f(x)\, dx \right| \leq \int_a^b \left| {}^{H}\mathbb{D}^{\overline{\alpha},\beta} f(x) \right| dx =$$

$$\int_a^{\frac{a+b}{2}} \left| {}^{H}\mathbb{D}_{a+}^{\overline{\alpha},\beta} f(x) \right| dx + \int_{\frac{a+b}{2}}^{b} \left| {}^{H}\mathbb{D}_{b-}^{\overline{\alpha},\beta} f(x) \right| dx \leq$$

$$\frac{\left\| {}^{H}\mathbb{D}_{a+}^{\alpha,\beta} f \right\|_{\infty,[a,\frac{a+b}{2}]}}{\Gamma(\alpha-\overline{\alpha}+2)} \left(\frac{b-a}{2}\right)^{\alpha-\overline{\alpha}+1} + \frac{\left\| {}^{H}\mathbb{D}_{b-}^{\alpha,\beta} f \right\|_{\infty,[\frac{a+b}{2},b]}}{\Gamma(\alpha-\overline{\alpha}+2)} \left(\frac{b-a}{2}\right)^{\alpha-\overline{\alpha}+1} =$$

$$\frac{\left(\frac{b-a}{2}\right)^{\alpha-\overline{\alpha}+1}}{\Gamma(\alpha-\overline{\alpha}+2)} \left[\left\| {}^{H}\mathbb{D}_{a+}^{\alpha,\beta} f \right\|_{\infty,[a,\frac{a+b}{2}]} + \left\| {}^{H}\mathbb{D}_{b-}^{\alpha,\beta} f \right\|_{\infty,[\frac{a+b}{2},b]} \right] \leq \tag{4.46}$$

$$\frac{2M_1 (b-a)^{\alpha-\overline{\alpha}+1}}{2^{\alpha-\overline{\alpha}+1}\Gamma(\alpha-\overline{\alpha}+2)} = \frac{M_1 (b-a)^{\alpha-\overline{\alpha}+1}}{2^{\alpha-\overline{\alpha}}\Gamma(\alpha-\overline{\alpha}+2)}.$$

■

We continue with the L_1-variant:

Theorem 4.9 *All as in Theorem 4.8 with* $\alpha-\overline{\alpha}>1$ *(i.e.* $\alpha>\overline{\alpha}+1$*). Call*

$$M_2 := \max\left\{ \left\| {}^{H}\mathbb{D}_{a+}^{\alpha,\beta} f \right\|_{1,[a,\frac{a+b}{2}]}, \left\| {}^{H}\mathbb{D}_{b-}^{\alpha,\beta} f \right\|_{1,[\frac{a+b}{2},b]} \right\}. \tag{4.47}$$

Then

$$\left| \int_a^b {}^{H}\mathbb{D}^{\overline{\alpha},\beta} f(x)\, dx \right| \leq \int_a^b \left| {}^{H}\mathbb{D}^{\overline{\alpha},\beta} f(x) \right| dx \leq \frac{M_2 (b-a)^{\alpha-\overline{\alpha}}}{2^{\alpha-\overline{\alpha}-1}\Gamma(\alpha-\overline{\alpha}+1)}. \tag{4.48}$$

Proof By (4.38) we have

$$\left|{}^{H}\mathbb{D}_{a+}^{\overline{\alpha},\beta} f(x)\right| \leq \frac{1}{\Gamma(\alpha-\overline{\alpha})} \int_{a}^{x} (x-t)^{\alpha-\overline{\alpha}-1} \left|{}^{H}\mathbb{D}_{a+}^{\alpha,\beta} f(t)\right| dt \leq$$

$$\frac{(x-a)^{\alpha-\overline{\alpha}-1}}{\Gamma(\alpha-\overline{\alpha})} \left\|{}^{H}\mathbb{D}_{a+}^{\alpha,\beta} f\right\|_{1,\left[a,\frac{a+b}{2}\right]}, \tag{4.49}$$

$\forall\, x \in \left[a, \frac{a+b}{2}\right]$.

Similarly, from (4.39) we find that

$$\left|{}^{H}\mathbb{D}_{b-}^{\overline{\alpha},\beta} f(x)\right| \leq \frac{(b-x)^{\alpha-\overline{\alpha}-1}}{\Gamma(\alpha-\overline{\alpha})} \left\|{}^{H}\mathbb{D}_{b-}^{\alpha,\beta} f\right\|_{1,\left[\frac{a+b}{2},b\right]}, \tag{4.50}$$

$\forall\, x \in \left[\frac{a+b}{2}, b\right]$.

Furthermore we obtain

$$\int_{a}^{\frac{a+b}{2}} \left|{}^{H}\mathbb{D}_{a+}^{\overline{\alpha},\beta} f(x)\right| dx \leq \frac{\left\|{}^{H}\mathbb{D}_{a+}^{\alpha,\beta} f\right\|_{1,\left[a,\frac{a+b}{2}\right]}}{\Gamma(\alpha-\overline{\alpha})} \int_{a}^{\frac{a+b}{2}} (x-a)^{\alpha-\overline{\alpha}-1} dx =$$

$$\frac{\left\|{}^{H}\mathbb{D}_{a+}^{\alpha,\beta} f\right\|_{1,\left[a,\frac{a+b}{2}\right]}}{\Gamma(\alpha-\overline{\alpha}+1)} \left(\frac{b-a}{2}\right)^{\alpha-\overline{\alpha}}. \tag{4.51}$$

Similarly, we derive

$$\int_{\frac{a+b}{2}}^{b} \left|{}^{H}\mathbb{D}_{b-}^{\overline{\alpha},\beta} f(x)\right| dx \leq \frac{\left\|{}^{H}\mathbb{D}_{b-}^{\alpha,\beta} f\right\|_{1,\left[\frac{a+b}{2},b\right]}}{\Gamma(\alpha-\overline{\alpha}+1)} \left(\frac{b-a}{2}\right)^{\alpha-\overline{\alpha}}. \tag{4.52}$$

Therefore we obtain

$$\left|\int_{a}^{b} {}^{H}\mathbb{D}^{\overline{\alpha},\beta} f(x)\, dx\right| \leq \int_{a}^{b} \left|{}^{H}\mathbb{D}^{\overline{\alpha},\beta} f(x)\right| dx =$$

$$\int_{a}^{\frac{a+b}{2}} \left|{}^{H}\mathbb{D}_{a+}^{\overline{\alpha},\beta} f(x)\right| dx + \int_{\frac{a+b}{2}}^{b} \left|{}^{H}\mathbb{D}_{b-}^{\overline{\alpha},\beta} f(x)\right| dx \leq$$

$$\frac{\left[\left\|{}^{H}\mathbb{D}_{a+}^{\alpha,\beta} f\right\|_{1,\left[a,\frac{a+b}{2}\right]} + \left\|{}^{H}\mathbb{D}_{b-}^{\alpha,\beta} f\right\|_{1,\left[\frac{a+b}{2},b\right]}\right]}{\Gamma(\alpha-\overline{\alpha}+1)} \left(\frac{b-a}{2}\right)^{\alpha-\overline{\alpha}} \leq \tag{4.53}$$

$$\frac{2M_2}{\Gamma(\alpha-\overline{\alpha}+1)}\frac{(b-a)^{\alpha-\overline{\alpha}}}{2^{\alpha-\overline{\alpha}}}=\frac{M_2}{\Gamma(\alpha-\overline{\alpha}+1)}\frac{(b-a)^{\alpha-\overline{\alpha}}}{2^{\alpha-\overline{\alpha}-1}}.$$

■

Next comes the L_q-variant of Hilfer-Polya fractional inequality:

Theorem 4.10 *All as in Theorem 4.8 with* $\alpha-\overline{\alpha}>\frac{1}{q}$, *where* $p,q>1:\frac{1}{p}+\frac{1}{q}=1$. *Call*

$$M_3:=\max\left\{\left\|{}^H\mathbb{D}_{a+}^{\alpha,\beta}f\right\|_{q,\left[a,\frac{a+b}{2}\right]},\left\|{}^H\mathbb{D}_{b-}^{\alpha,\beta}f\right\|_{q,\left[\frac{a+b}{2},b\right]}\right\}. \tag{4.54}$$

Then

$$\left|\int_a^b {}^H\mathbb{D}^{\overline{\alpha},\beta}f(x)\,dx\right|\leq\int_a^b\left|{}^H\mathbb{D}^{\overline{\alpha},\beta}f(x)\right|dx\leq$$

$$\frac{M_3}{\Gamma(\alpha-\overline{\alpha})\,(p(\alpha-\overline{\alpha}-1)+1)^{\frac{1}{p}}\left(\alpha-\overline{\alpha}+\frac{1}{p}\right)}\frac{(b-a)^{\alpha-\overline{\alpha}+\frac{1}{p}}}{2^{\alpha-\overline{\alpha}-\frac{1}{q}}}. \tag{4.55}$$

Proof By (4.38) we have

$$\left|{}^H\mathbb{D}_{a+}^{\overline{\alpha},\beta}f(x)\right|\leq\frac{1}{\Gamma(\alpha-\overline{\alpha})}\int_a^x(x-t)^{\alpha-\overline{\alpha}-1}\left|{}^H\mathbb{D}_{a+}^{\alpha,\beta}f(t)\right|dt\leq$$

$$\frac{1}{\Gamma(\alpha-\overline{\alpha})}\left(\int_a^x(x-t)^{p(\alpha-\overline{\alpha}-1)}dt\right)^{\frac{1}{p}}\left\|{}^H\mathbb{D}_{a+}^{\alpha,\beta}f\right\|_{q,\left[a,\frac{a+b}{2}\right]}=$$

$$\frac{(x-a)^{\left(\alpha-\overline{\alpha}-\frac{1}{q}\right)}}{\Gamma(\alpha-\overline{\alpha})\,(p(\alpha-\overline{\alpha}-1)+1)^{\frac{1}{p}}}\left\|{}^H\mathbb{D}_{a+}^{\alpha,\beta}f\right\|_{q,\left[a,\frac{a+b}{2}\right]}, \tag{4.56}$$

$\forall\, x\in\left[a,\frac{a+b}{2}\right]$, with $\alpha-\overline{\alpha}>\frac{1}{q}$.

And, by (4.39), similarly we derive

$$\left|{}^H\mathbb{D}_{b-}^{\overline{\alpha},\beta}f(x)\right|\leq\frac{(b-x)^{\left(\alpha-\overline{\alpha}-\frac{1}{q}\right)}}{\Gamma(\alpha-\overline{\alpha})\,(p(\alpha-\overline{\alpha}-1)+1)^{\frac{1}{p}}}\left\|{}^H\mathbb{D}_{b-}^{\alpha,\beta}f\right\|_{q,\left[\frac{a+b}{2},b\right]}, \tag{4.57}$$

$\forall\, x\in\left[\frac{a+b}{2},b\right]$, with $\alpha-\overline{\alpha}>\frac{1}{q}$.

Consequently, we obtain that

$$\int_a^{\frac{a+b}{2}}\left|{}^H\mathbb{D}_{a+}^{\overline{\alpha},\beta}f(x)\right|dx\leq\frac{\left\|{}^H\mathbb{D}_{a+}^{\alpha,\beta}f\right\|_{q,\left[a,\frac{a+b}{2}\right]}}{\Gamma(\alpha-\overline{\alpha})\,(p(\alpha-\overline{\alpha}-1)+1)^{\frac{1}{p}}}\int_a^{\frac{a+b}{2}}(x-a)^{\left(\alpha-\overline{\alpha}-\frac{1}{q}\right)}dx=$$

$$\frac{\left\| {}^{H}\mathbb{D}_{a+}^{\alpha,\beta} f \right\|_{q,\left[a,\frac{a+b}{2}\right]}}{\Gamma\left(\alpha-\overline{\alpha}\right)\left(p\left(\alpha-\overline{\alpha}-1\right)+1\right)^{\frac{1}{p}}\left(\alpha-\overline{\alpha}+\frac{1}{p}\right)}\left(\frac{b-a}{2}\right)^{\alpha-\overline{\alpha}+\frac{1}{p}}. \tag{4.58}$$

Similarly, we derive

$$\int_{\frac{a+b}{2}}^{b}\left| {}^{H}\mathbb{D}_{b-}^{\overline{\alpha},\beta} f(x)\right| dx \leq \frac{\left\| {}^{H}\mathbb{D}_{b-}^{\alpha,\beta} f \right\|_{q,\left[\frac{a+b}{2},b\right]}}{\Gamma\left(\alpha-\overline{\alpha}\right)\left(p\left(\alpha-\overline{\alpha}-1\right)+1\right)^{\frac{1}{p}}\left(\alpha-\overline{\alpha}+\frac{1}{p}\right)}\left(\frac{b-a}{2}\right)^{\alpha-\overline{\alpha}+\frac{1}{p}}. \tag{4.59}$$

Therefore, we obtain

$$\left|\int_a^b {}^{H}\mathbb{D}^{\overline{\alpha},\beta} f(x)\,dx\right| \leq \int_a^b \left| {}^{H}\mathbb{D}^{\overline{\alpha},\beta} f(x)\right| dx =$$

$$\frac{\left(\left\| {}^{H}\mathbb{D}_{a+}^{\alpha,\beta} f \right\|_{q,\left[a,\frac{a+b}{2}\right]} + \left\| {}^{H}\mathbb{D}_{b-}^{\alpha,\beta} f \right\|_{q,\left[\frac{a+b}{2},b\right]}\right)}{\Gamma\left(\alpha-\overline{\alpha}\right)\left(p\left(\alpha-\overline{\alpha}-1\right)+1\right)^{\frac{1}{p}}\left(\alpha-\overline{\alpha}+\frac{1}{p}\right)}\left(\frac{b-a}{2}\right)^{\alpha-\overline{\alpha}+\frac{1}{p}} \leq \tag{4.60}$$

$$\frac{M_3}{\Gamma\left(\alpha-\overline{\alpha}\right)\left(p\left(\alpha-\overline{\alpha}-1\right)+1\right)^{\frac{1}{p}}\left(\alpha-\overline{\alpha}+\frac{1}{p}\right)}\frac{(b-a)^{\alpha-\overline{\alpha}+\frac{1}{p}}}{2^{\alpha-\overline{\alpha}-\frac{1}{q}}},$$

proving the claim.

■ Next come ψ-Hilfer-Ostrowski type inequalities for several functions involved. For basic ψ-Hilfer-Ostrowski type inequalities involving one function see [3]. We make

Remark 4.11 Our setting here follows: Let $f_i \in C^n([a,b])$, $\alpha \notin \mathbb{N}$, $n = \lceil \alpha \rceil$, $\alpha > 0$; $i = 1, ..., r \in \mathbb{N} - \{1\}$, $x_0 \in [a,b]$. Assume that $g_{1i}(x) = I_{x_0+}^{(1-\beta)(n-\alpha);\psi} f_i(x) \in C^n([x_0,b])$ and $w_{1i}(x) = I_{x_0-}^{(1-\beta)(n-\alpha);\psi} f_i(x) \in C^n([a,x_0])$, for all $i = 1, ..., r$.

Define

$$\varphi_{ix_0}(x) := \left\{ \begin{array}{l} g_{1i}(x), \ x \in [x_0,b] \\ \\ w_{1i}(x), \ x \in [a,x_0) \end{array} \right\}. \tag{4.61}$$

Notice that if $\beta = 1$, we get $g_{1i}(x_0) = w_{1i}(x_0) = \varphi_{ix_0}(x_0) = f_i(x_0)$, all $i = 1, ..., r$.

In general, for $f \in C([a,b])$ we have

$$\left| I_{a+}^{\alpha,\psi} f(x)\right| \leq \frac{1}{\Gamma(\alpha)}\int_a^x \psi'(t)\left(\psi(x)-\psi(t)\right)^{\alpha-1}|f(t)|\,dt \leq$$

$$\frac{\|f\|_{\infty,[a,b]}}{\Gamma(\alpha+1)}\left(\psi(x)-\psi(a)\right)^{\alpha}, \ \forall x \in [a,b]. \tag{4.62}$$

Hence $I_{a+}^{\alpha,\psi} f(a) = 0$.

Similarly, we have

$$\left| I_{b-}^{\alpha,\psi} f(x) \right| \leq \frac{1}{\Gamma(\alpha)} \int_x^b \psi'(t) (\psi(t) - \psi(x))^{\alpha-1} |f(t)| \, dt \leq$$

$$\frac{\|f\|_{\infty,[a,b]}}{\Gamma(\alpha+1)} (\psi(b) - \psi(x))^{\alpha}, \ \forall x \in [a,b]. \tag{4.63}$$

That is $I_{b-}^{\alpha,\psi} f(b) = 0$.

So when $0 \leq \beta < 1$, by the above we obtain $g_{1i}(x_0) = w_{1i}(x_0) = \varphi_{ix_0}(x_0) = 0$, for all $i = 1, ..., r$.

Thus, it is always true that $g_{1i}(x_0) = w_{1i}(x_0)$, $i = 1, ..., r$.

We present

Theorem 4.12 *Let $\psi, f_i \in C^n([a,b])$, $\alpha \notin \mathbb{N}$, $n = \lceil \alpha \rceil$, $\alpha > 0$; $i = 1, ..., r \in \mathbb{N} - \{1\}$, $x_0 \in [a,b]$. Here ψ is increasing, $\psi'(x) \neq 0$ over $[a,b] \subset \mathbb{R}$, $0 \leq \beta \leq 1$, $\mu = n(1-\beta) + \beta\alpha$. Assume that $g_{1i}(x) = I_{x_0+}^{(1-\beta)(n-\alpha);\psi} f_i(x) \in C^n([x_0,b])$ and $w_{1i}(x) = I_{x_0-}^{(1-\beta)(n-\alpha);\psi} f_i(x) \in C^n([a,x_0])$, for all $i = 1, ..., r$, and $\varphi_{ix_0}(x)$ is as in (4.61). Assume also that $g_{1i\psi}^{[k]}(x_0) = w_{1i\psi}^{[k]}(x_0) = 0$, for all $k = 1, ..., n-1$.*

Then

(1)

$$\theta^{\psi}(f_1, ..., f_r)(x_0) := \tag{4.64}$$

$$r \int_a^b \left(\prod_{\lambda=1}^r \varphi_{\lambda x_0}(x) \right) dx - \sum_{i=1}^r \left[\varphi_{ix_0}(x_0) \int_a^b \left(\prod_{\substack{j=1 \\ j \neq i}}^r \varphi_{jx_0}(x) \right) dx \right] =$$

$$\sum_{i=1}^r \left[\left[\int_a^{x_0} \left(\prod_{\substack{j=1 \\ j \neq i}}^r \varphi_{jx_0}(x) \right) \left(I_{x_0-}^{\mu;\psi} {}^H\mathbb{D}_{x_0-}^{\alpha,\beta;\psi} f_i(x) \right) dx \right] + \right.$$

$$\left. \left[\int_{x_0}^b \left(\prod_{\substack{j=1 \\ j \neq i}}^r \varphi_{jx_0}(x) \right) \left(I_{x_0+}^{\mu;\psi} {}^H\mathbb{D}_{x_0+}^{\alpha,\beta;\psi} f_i(x) \right) dx \right] \right],$$

and in case of $0 \leq \beta < 1$, we have that

$$\theta^{\psi}(f_1, ..., f_r)(x_0) = r \int_a^b \left(\prod_{\lambda=1}^r \varphi_{\lambda x_0}(x) \right) dx, \tag{4.65}$$

(2) furthermore, it holds

$$\left|\theta^{\psi}\left(f_1, ..., f_r\right)(x_0)\right| \leq \frac{1}{\Gamma(\mu+1)}$$

$$\left\{\left(\sum_{i=1}^{r}\left\|{}^{H}\mathbb{D}_{x_0-}^{\alpha,\beta;\psi} f_i\right\|_{\infty,[a,x_0]}\left\|\prod_{\substack{j=1\\ j\neq i}}^{r}\varphi_{jx_0}\right\|_{1,[a,x_0]}\right)(\psi(x_0)-\psi(a))^{\mu}+\right.$$

$$\left.\left(\sum_{i=1}^{r}\left\|{}^{H}\mathbb{D}_{x_0+}^{\alpha,\beta;\psi} f_i\right\|_{\infty,[x_0,b]}\left\|\prod_{\substack{j=1\\ j\neq i}}^{r}\varphi_{jx_0}\right\|_{1,[x_0,b]}\right)(\psi(b)-\psi(x_0))^{\mu}\right\}. \tag{4.66}$$

It follows the L_1-variant.

Theorem 4.13 *All as in Theorem 4.12, with* $\alpha > 1$. *Then*

$$\left|\theta^{\psi}\left(f_1, ..., f_r\right)(x_0)\right| \leq \frac{1}{\Gamma(\mu)}$$

$$\left\{\left(\sum_{i=1}^{r}\left\|{}^{H}\mathbb{D}_{x_0-}^{\alpha,\beta;\psi} f_i\right\|_{L_1([a,x_0],\psi)}\left\|\prod_{\substack{j=1\\ j\neq i}}^{r}\varphi_{jx_0}\right\|_{1,[a,x_0]}\right)(\psi(x_0)-\psi(a))^{\mu-1}+\right.$$

$$\left.\left(\sum_{i=1}^{r}\left\|{}^{H}\mathbb{D}_{x_0+}^{\alpha,\beta;\psi} f_i\right\|_{L_1([x_0,b],\psi)}\left\|\prod_{\substack{j=1\\ j\neq i}}^{r}\varphi_{jx_0}\right\|_{1,[x_0,b]}\right)(\psi(b)-\psi(x_0))^{\mu-1}\right\}. \tag{4.67}$$

Next we have the L_q-variant.

Theorem 4.14 *All as in Theorem 4.12. Let also* $p, q > 1 : \frac{1}{p}+\frac{1}{q}=1$ *with* $\alpha > \frac{1}{q}$. *Then*

$$\left|\theta^{\psi}\left(f_1, ..., f_r\right)(x_0)\right| \leq \frac{1}{\Gamma(\mu)\left(p(\mu-1)+1\right)^{\frac{1}{p}}}$$

$$\left\{\left(\sum_{i=1}^{r}\left\|{}^{H}\mathbb{D}_{x_0-}^{\alpha,\beta;\psi} f_i\right\|_{L_q([a,x_0],\psi)}\left\|\prod_{\substack{j=1\\ j\neq i}}^{r}\varphi_{jx_0}\right\|_{1,[a,x_0]}\right)(\psi(x_0)-\psi(a))^{\mu-\frac{1}{q}}+\right.$$

$$\left(\sum_{i=1}^{r}\left\|{}^{H}\mathbb{D}_{x_0+}^{\alpha,\beta;\psi} f_i\right\|_{L_q([x_0,b],\psi)}\left\|\prod_{\substack{j=1\\ j\neq i}}^{r}\varphi_{jx_0}\right\|_{1,[x_0,b]}\right)(\psi(b)-\psi(x_0))^{\mu-\frac{1}{q}}\Bigg\}. \quad (4.68)$$

Proof **of Theorems** 4.12–4.14.

By Theorem 4.5 we have

$$g_{1i}(x)-g_{1i}(x_0)=I_{x_0+}^{\mu;\psi}\,{}^{H}\mathbb{D}_{x_0+}^{\alpha,\beta;\psi} f_i(x),\ \forall x\in[x_0,b],$$
$$\text{and} \quad (4.69)$$
$$w_{1i}(x)-w_{1i}(x_0)=I_{x_0-}^{\mu;\psi}\,{}^{H}\mathbb{D}_{x_0-}^{\alpha,\beta;\psi} f_i(x),\ \forall x\in[a,x_0],$$

for all $i=1,\ldots,r$.

That is

$$\varphi_{ix_0}(x)-\varphi_{ix_0}(x_0)=I_{x_0+}^{\mu;\psi}\,{}^{H}\mathbb{D}_{x_0+}^{\alpha,\beta;\psi} f_i(x),\ \forall x\in[x_0,b],$$
$$\text{and} \quad (4.70)$$
$$\varphi_{ix_0}(x)-\varphi_{ix_0}(x_0)=I_{x_0-}^{\mu;\psi}\,{}^{H}\mathbb{D}_{x_0-}^{\alpha,\beta;\psi} f_i(x),\ \forall x\in[a,x_0),$$

for all $i=1,\ldots,r$.

Multiplying (4.70) by $\left(\prod_{\substack{j=1\\ j\neq i}}^{r}\varphi_{jx_0}(x)\right)$ we get, respectively,

$$\prod_{\lambda=1}^{r}\varphi_{\lambda x_0}(x)-\left(\prod_{\substack{j=1\\ j\neq i}}^{r}\varphi_{jx_0}(x)\right)\varphi_{ix_0}(x_0)=\left(\prod_{\substack{j=1\\ j\neq i}}^{r}\varphi_{jx_0}(x)\right)I_{x_0+}^{\mu;\psi}\,{}^{H}\mathbb{D}_{x_0+}^{\alpha,\beta;\psi} f_i(x), \quad (4.71)$$

$\forall x\in[x_0,b]$,

and

$$\prod_{\lambda=1}^{r}\varphi_{\lambda x_0}(x)-\left(\prod_{\substack{j=1\\ j\neq i}}^{r}\varphi_{jx_0}(x)\right)\varphi_{ix_0}(x_0)=\left(\prod_{\substack{j=1\\ j\neq i}}^{r}\varphi_{jx_0}(x)\right)I_{x_0-}^{\mu;\psi}\,{}^{H}\mathbb{D}_{x_0-}^{\alpha,\beta;\psi} f_i(x), \quad (4.72)$$

$\forall x\in[a,x_0)$, for all $i=1,\ldots,r$.

Adding (4.71) and (4.72), separately, we obtain

$$r\left(\prod_{\lambda=1}^{r}\varphi_{\lambda x_0}(x)\right)-\sum_{i=1}^{r}\left[\left(\prod_{\substack{j=1\\ j\neq i}}^{r}\varphi_{jx_0}(x)\right)\varphi_{ix_0}(x_0)\right]= \quad (4.73)$$

$$\sum_{i=1}^{r}\left[\left(\prod_{\substack{j=1\\ j\neq i}}^{r}\varphi_{jx_0}(x)\right) I_{x_0+}^{\mu;\psi\, H}\mathbb{D}_{x_0+}^{\alpha,\beta;\psi} f_i(x)\right],$$

$\forall\, x \in [x_0, b]$,

and

$$r\left(\prod_{\lambda=1}^{r}\varphi_{\lambda x_0}(x)\right) - \sum_{i=1}^{r}\left[\left(\prod_{\substack{j=1\\ j\neq i}}^{r}\varphi_{jx_0}(x)\right)\varphi_{ix_0}(x_0)\right] = \tag{4.74}$$

$$\sum_{i=1}^{r}\left[\left(\prod_{\substack{j=1\\ j\neq i}}^{r}\varphi_{jx_0}(x)\right) I_{x_0-}^{\mu;\psi\, H}\mathbb{D}_{x_0-}^{\alpha,\beta;\psi} f_i(x)\right],$$

$\forall\, x \in [a, x_0)$.

Next integrate (4.73) and (4.74) with respect to $x \in [a, b]$. We have

$$r\int_{x_0}^{b}\left(\prod_{\lambda=1}^{r}\varphi_{\lambda x_0}(x)\right)dx - \sum_{i=1}^{r}\left[\varphi_{ix_0}(x_0)\int_{x_0}^{b}\left(\prod_{\substack{j=1\\ j\neq i}}^{r}\varphi_{jx_0}(x)\right)dx\right] =$$

$$\sum_{i=1}^{r}\left[\int_{x_0}^{b}\left(\prod_{\substack{j=1\\ j\neq i}}^{r}\varphi_{jx_0}(x)\right)\left(I_{x_0+}^{\mu;\psi\, H}\mathbb{D}_{x_0+}^{\alpha,\beta;\psi} f_i(x)\right)dx\right], \tag{4.75}$$

and

$$r\int_{a}^{x_0}\left(\prod_{\lambda=1}^{r}\varphi_{\lambda x_0}(x)\right)dx - \sum_{i=1}^{r}\left[\varphi_{ix_0}(x_0)\int_{a}^{x_0}\left(\prod_{\substack{j=1\\ j\neq i}}^{r}\varphi_{jx_0}(x)\right)dx\right] =$$

$$\sum_{i=1}^{r}\left[\int_{a}^{x_0}\left(\prod_{\substack{j=1\\ j\neq i}}^{r}\varphi_{jx_0}(x)\right)\left(I_{x_0-}^{\mu;\psi\, H}\mathbb{D}_{x_0-}^{\alpha,\beta;\psi} f_i(x)\right)dx\right]. \tag{4.76}$$

Finally adding (4.75) and (4.76) we obtain the useful and nice identity (4.64).

Identity (4.64) implies

$$\left|\theta^{\psi}\left(f_1,\ldots,f_r\right)(x_0)\right| \leq \sum_{i=1}^{r}\left[\left[\int_a^{x_0}\left(\prod_{\substack{j=1\\j\neq i}}^{r}\left|\varphi_{jx_0}(x)\right|\right)\left(I_{x_0-}^{\mu;\psi}\left|{}^{H}\mathbb{D}_{x_0-}^{\alpha,\beta;\psi}f_i\right|(x)\right)dx\right]+\right.$$

$$\left.\left[\int_{x_0}^{b}\left(\prod_{\substack{j=1\\j\neq i}}^{r}\left|\varphi_{jx_0}(x)\right|\right)\left(I_{x_0+}^{\mu;\psi}\left|{}^{H}\mathbb{D}_{x_0+}^{\alpha,\beta;\psi}f_i\right|(x)\right)dx\right]\right]= \tag{4.77}$$

$$\sum_{i=1}^{r}\left[\left[\int_a^{x_0}\left(\prod_{\substack{j=1\\j\neq i}}^{r}\left|\varphi_{jx_0}(x)\right|\right)\frac{1}{\Gamma(\mu)}\right.\right.$$

$$\left.\left(\int_x^{x_0}\psi'(t)\left(\psi(t)-\psi(x)\right)^{\mu-1}\left|\left({}^{H}\mathbb{D}_{x_0-}^{\alpha,\beta;\psi}f_i\right)(t)\right|dt\right)dx\right]+$$

$$\left[\int_{x_0}^{b}\left(\prod_{\substack{j=1\\j\neq i}}^{r}\left|\varphi_{jx_0}(x)\right|\right)\frac{1}{\Gamma(\mu)}\right.$$

$$\left.\left.\left(\int_{x_0}^{x}\psi'(t)\left(\psi(x)-\psi(t)\right)^{\mu-1}\left|\left({}^{H}\mathbb{D}_{x_0+}^{\alpha,\beta;\psi}f_i\right)(t)\right|dt\right)dx\right]\right]\leq$$

$$\frac{1}{\Gamma(\mu+1)}\sum_{i=1}^{r}\left[\left[\left\|{}^{H}\mathbb{D}_{x_0-}^{\alpha,\beta;\psi}f_i\right\|_{\infty,[a,x_0]}\int_a^{x_0}\left(\psi(x_0)-\psi(x)\right)^{\mu}\left(\prod_{\substack{j=1\\j\neq i}}^{r}\left|\varphi_{jx_0}(x)\right|\right)dx\right]\right.$$

$$\left.+\left[\left\|{}^{H}\mathbb{D}_{x_0+}^{\alpha,\beta;\psi}f_i\right\|_{\infty,[x_0,b]}\int_{x_0}^{b}\left(\psi(x)-\psi(x_0)\right)^{\mu}\left(\prod_{\substack{j=1\\j\neq i}}^{r}\left|\varphi_{jx_0}(x)\right|\right)dx\right]\right]\leq \tag{4.78}$$

$$\frac{1}{\Gamma(\mu+1)}\sum_{i=1}^{r}\left[\left[\left\|{}^{H}\mathbb{D}_{x_0-}^{\alpha,\beta;\psi}f_i\right\|_{\infty,[a,x_0]}\left(\psi(x_0)-\psi(a)\right)^{\mu}\left\|\prod_{\substack{j=1\\j\neq i}}^{r}\varphi_{jx_0}\right\|_{1,[a,x_0]}\right]+\right.$$

$$\left[\left\| {}^{H}\mathbb{D}_{x_0+}^{\alpha,\beta;\psi} f_i \right\|_{\infty,[x_0,b]} (\psi(b) - \psi(x_0))^{\mu} \left\| \prod_{\substack{j=1 \\ j\neq i}}^{r} \varphi_{jx_0} \right\|_{1,[x_0,b]} \right]\right] = \tag{4.79}$$

$$\frac{1}{\Gamma(\mu+1)} \left\{ \left(\sum_{i=1}^{r} \left\| {}^{H}\mathbb{D}_{x_0-}^{\alpha,\beta;\psi} f_i \right\|_{\infty,[a,x_0]} \left\| \prod_{\substack{j=1 \\ j\neq i}}^{r} \varphi_{jx_0} \right\|_{1,[a,x_0]} \right) (\psi(x_0) - \psi(a))^{\mu} + \right.$$

$$\left. \left(\sum_{i=1}^{r} \left\| {}^{H}\mathbb{D}_{x_0+}^{\alpha,\beta;\psi} f_i \right\|_{\infty,[x_0,b]} \left\| \prod_{\substack{j=1 \\ j\neq i}}^{r} \varphi_{jx_0} \right\|_{1,[x_0,b]} \right) (\psi(b) - \psi(x_0))^{\mu} \right\}, \tag{4.80}$$

proving (4.66).

If $\alpha \notin \mathbb{N}$ and $\alpha > 1$, then $n = \lceil \alpha \rceil > 1$, and $n - 1 \geq 1 > \beta(n - \alpha)$. Hence $n - \beta(n - \alpha) > 1$ and $\mu > 1$. So we have

$$\left|\theta^{\psi}(f_1, ..., f_r)(x_0)\right| \leq \sum_{i=1}^{r} \left[\left[\int_a^{x_0} \left(\prod_{\substack{j=1 \\ j\neq i}}^{r} \left|\varphi_{jx_0}(x)\right| \right) \frac{1}{\Gamma(\mu)} \right.\right.$$

$$\left. \left(\int_x^{x_0} \psi'(t) (\psi(t) - \psi(x))^{\mu-1} \left| \left({}^{H}\mathbb{D}_{x_0-}^{\alpha,\beta;\psi} f_i \right)(t) \right| dt \right) dx \right] +$$

$$\left[\int_{x_0}^{b} \left(\prod_{\substack{j=1 \\ j\neq i}}^{r} \left|\varphi_{jx_0}(x)\right| \right) \frac{1}{\Gamma(\mu)} \right.$$

$$\left.\left.\left(\int_{x_0}^{x} \psi'(t) (\psi(x) - \psi(t))^{\mu-1} \left| \left({}^{H}\mathbb{D}_{x_0+}^{\alpha,\beta;\psi} f_i \right)(t) \right| dt \right) dx \right]\right] \leq \tag{4.81}$$

$$\frac{1}{\Gamma(\mu)} \sum_{i=1}^{r} \left[\left[\left\| {}^{H}\mathbb{D}_{x_0-}^{\alpha,\beta;\psi} f_i \right\|_{L_1([a,x_0],\psi)} \int_a^{x_0} (\psi(x_0) - \psi(x))^{\mu-1} \left(\prod_{\substack{j=1 \\ j\neq i}}^{r} \left|\varphi_{jx_0}(x)\right| \right) dx \right] \right.$$

$$\left. + \left[\left\| {}^{H}\mathbb{D}_{x_0+}^{\alpha,\beta;\psi} f_i \right\|_{L_1([x_0,b],\psi)} \int_{x_0}^{b} (\psi(x) - \psi(x_0))^{\mu-1} \left(\prod_{\substack{j=1 \\ j\neq i}}^{r} \left|\varphi_{jx_0}(x)\right| \right) dx \right]\right] \leq$$

$$\frac{1}{\Gamma(\mu)}\sum_{i=1}^{r}\left[\left[\left\|{}^{H}\mathbb{D}_{x_0-}^{\alpha,\beta;\psi}f_i\right\|_{L_1([a,x_0],\psi)}(\psi(x_0)-\psi(a))^{\mu-1}\left\|\prod_{\substack{j=1\\ j\neq i}}^{r}\varphi_{jx_0}\right\|_{1,[a,x_0]}\right]+\right.$$
$$\left.\left[\left\|{}^{H}\mathbb{D}_{x_0+}^{\alpha,\beta;\psi}f_i\right\|_{L_1([x_0,b],\psi)}(\psi(b)-\psi(x_0))^{\mu-1}\left\|\prod_{\substack{j=1\\ j\neq i}}^{r}\varphi_{jx_0}\right\|_{1,[x_0,b]}\right]\right]= \tag{4.82}$$

$$\frac{1}{\Gamma(\mu)}\left\{\left(\sum_{i=1}^{r}\left\|{}^{H}\mathbb{D}_{x_0-}^{\alpha,\beta;\psi}f_i\right\|_{L_1([a,x_0],\psi)}\left\|\prod_{\substack{j=1\\ j\neq i}}^{r}\varphi_{jx_0}\right\|_{1,[a,x_0]}\right)(\psi(x_0)-\psi(a))^{\mu-1}+\right.$$
$$\left.\left(\sum_{i=1}^{r}\left\|{}^{H}\mathbb{D}_{x_0+}^{\alpha,\beta;\psi}f_i\right\|_{L_1([x_0,b],\psi)}\left\|\prod_{\substack{j=1\\ j\neq i}}^{r}\varphi_{jx_0}\right\|_{1,[x_0,b]}\right)(\psi(b)-\psi(x_0))^{\mu-1}\right\}, \tag{4.83}$$

proving (4.67).

Let $\alpha > 0$ with $\lceil\alpha\rceil = n \in \mathbb{N}$, and let $p, q > 1 : \frac{1}{p}+\frac{1}{q} = 1$, with $\alpha > \frac{1}{q}$. Clearly $n > \frac{1}{q}$. Let $0 < \beta \le 1$, then $\alpha\beta > \frac{\beta}{q}$, furthermore $\mu = n(1-\beta)+\beta\alpha > \frac{\beta}{q} + n(1-\beta) \ge \frac{\beta}{q}+1-\beta = \frac{\beta}{q}+\frac{1}{p}+\frac{1}{q}-\frac{\beta}{p}-\frac{\beta}{q} = \frac{1}{q}+\frac{1}{p}(1-\beta) \ge \frac{1}{q}$. That is $\mu > \frac{1}{q}$.

From (4.81), by using Hölder's inequality twice, we have

$$\left|\theta^{\psi}(f_1,\dots,f_r)(x_0)\right| \le \frac{1}{\Gamma(\mu)}$$

$$\sum_{i=1}^{r}\left[\left[\int_a^{x_0}\left(\prod_{\substack{j=1\\ j\neq i}}^{r}\left|\varphi_{jx_0}(x)\right|\right)\frac{(\psi(x_0)-\psi(x))^{\frac{p(\mu-1)+1}{p}}}{(p(\mu-1)+1)^{\frac{1}{p}}}\left\|{}^{H}\mathbb{D}_{x_0-}^{\alpha,\beta;\psi}f_i\right\|_{L_q([a,x_0],\psi)}dx\right]+\right.$$
$$\left.\left[\int_{x_0}^{b}\left(\prod_{\substack{j=1\\ j\neq i}}^{r}\left|\varphi_{jx_0}(x)\right|\right)\frac{(\psi(x)-\psi(x_0))^{\frac{p(\mu-1)+1}{p}}}{(p(\mu-1)+1)^{\frac{1}{p}}}\left\|{}^{H}\mathbb{D}_{x_0+}^{\alpha,\beta;\psi}f_i\right\|_{L_q([x_0,b],\psi)}dx\right]\right]= \tag{4.84}$$

$$\frac{1}{\Gamma(\mu)(p(\mu-1)+1)^{\frac{1}{p}}}$$

$$\sum_{i=1}^{r}\left[\left[\int_{a}^{x_0}\left(\prod_{\substack{j=1\\j\neq i}}^{r}\left|\varphi_{jx_0}(x)\right|\right)(\psi(x_0)-\psi(x))^{\mu-\frac{1}{q}}\left\|{}^{H}\mathbb{D}_{x_0-}^{\alpha,\beta;\psi}f_i\right\|_{L_q([a,x_0],\psi)}dx\right]+\right.$$

$$\left.\left[\int_{x_0}^{b}\left(\prod_{\substack{j=1\\j\neq i}}^{r}\left|\varphi_{jx_0}(x)\right|\right)(\psi(x)-\psi(x_0))^{\mu-\frac{1}{q}}\left\|{}^{H}\mathbb{D}_{x_0+}^{\alpha,\beta;\psi}f_i\right\|_{L_q([x_0,b],\psi)}dx\right]\right]\leq$$

$$\frac{1}{\Gamma(\mu)(p(\mu-1)+1)^{\frac{1}{p}}}$$

$$\sum_{i=1}^{r}\left[\left[\left\|{}^{H}\mathbb{D}_{x_0-}^{\alpha,\beta;\psi}f_i\right\|_{L_q([a,x_0],\psi)}(\psi(x_0)-\psi(a))^{\mu-\frac{1}{q}}\left\|\prod_{\substack{j=1\\j\neq i}}^{r}\varphi_{jx_0}\right\|_{1,[a,x_0]}\right]+\right. \tag{4.85}$$

$$\left.\left[\left\|{}^{H}\mathbb{D}_{x_0+}^{\alpha,\beta;\psi}f_i\right\|_{L_q([x_0,b],\psi)}(\psi(b)-\psi(x_0))^{\mu-\frac{1}{q}}\left\|\prod_{\substack{j=1\\j\neq i}}^{r}\varphi_{jx_0}\right\|_{1,[x_0,b]}\right]\right]=$$

$$\frac{1}{\Gamma(\mu)(p(\mu-1)+1)^{\frac{1}{p}}}$$

$$\left\{\left(\sum_{i=1}^{r}\left\|{}^{H}\mathbb{D}_{x_0-}^{\alpha,\beta;\psi}f_i\right\|_{L_q([a,x_0],\psi)}\left\|\prod_{\substack{j=1\\j\neq i}}^{r}\varphi_{jx_0}\right\|_{1,[a,x_0]}\right)(\psi(x_0)-\psi(a))^{\mu-\frac{1}{q}}+\right. \tag{4.86}$$

$$\left.\left(\sum_{i=1}^{r}\left\|{}^{H}\mathbb{D}_{x_0+}^{\alpha,\beta;\psi}f_i\right\|_{L_q([x_0,b],\psi)}\left\|\prod_{\substack{j=1\\j\neq i}}^{r}\varphi_{jx_0}\right\|_{1,[x_0,b]}\right)(\psi(b)-\psi(x_0))^{\mu-\frac{1}{q}}\right\},$$

proving (4.68).

■

Next we present a ψ-Hilfer-Hilbert-Pachpatte left fractional inequality:

Theorem 4.15 *Let* $i=1,2$; $\psi_i, f_i\in C^{n_i}([a_i,b_i])$, *with* ψ_i *being strictly increasing over* $[a_i,b_i]$, *where* $n_i-1<\alpha_i<n_i$, $0\leq\beta_i\leq1$, *and* $\gamma_i=\alpha_i+\beta_i(n_i-\alpha_i)$, $x_i\in[a_i,b_i]$. *Assume that* $f_{i\psi_i}^{[n_i-k_i]}\left(I_{a_i+}^{(1-\beta_i)(n_i-\alpha_i);\psi_i}f_i\right)(a_i)=0$, *for* $k_i=1,\dots,n_i-1$. *Let*

also $p, q > 1 : \frac{1}{p} + \frac{1}{q} = 1$, *such that* $\alpha_1 > \frac{1}{q}$ *and* $\alpha_2 > \frac{1}{p}$. *Then*

$$\int_{a_1}^{b_1} \int_{a_2}^{b_2} \frac{|f_1(x_1)| \, |f_2(x_2)| \, dx_1 dx_2}{\left(\frac{(\psi_1(x_1)-\psi_1(a_1))^{p(\alpha_1-1)+1}}{p(p(\alpha_1-1)|+1)} + \frac{(\psi_2(x_2)-\psi_2(a_2))^{q(\alpha_2-1)+1}}{q(q(\alpha_2-1)+1)} \right)} \leq$$

$$\frac{(b_1 - a_1)(b_2 - a_2)}{\Gamma(\alpha_1) \Gamma(\alpha_2)} \left\| {}^H\mathbb{D}_{a_1+}^{\alpha_1,\beta_1;\psi_1} f_1 \right\|_{L_q([a_1,b_1],\psi_1)} \left\| {}^H\mathbb{D}_{a_2+}^{\alpha_2,\beta_2;\psi_2} f_2 \right\|_{L_p([a_2,b_2],\psi_2)}. \quad (4.87)$$

Proof By Theorem 4.4 we have

$$f_i(x_i) = \frac{1}{\Gamma(\alpha_i)} \int_{a_i}^{x_i} \psi_i'(t_i) (\psi_i(x_i) - \psi_i(t_i))^{\alpha_i - 1} \, {}^H\mathbb{D}_{a_i+}^{\alpha_i,\beta_i;\psi_i} f_i(t_i) \, dt_i, \quad (4.88)$$

$\forall \, x_i \in [a_i, b_i]$, $i = 1, 2$.

Then

$$|f_i(x_i)| \leq \frac{1}{\Gamma(\alpha_i)} \int_{a_i}^{x_i} \psi_i'(t_i) (\psi_i(x_i) - \psi_i(t_i))^{\alpha_i - 1} \left| {}^H\mathbb{D}_{a_i+}^{\alpha_i,\beta_i;\psi_i} f_i(t_i) \right| dt_i, \quad (4.89)$$

$i = 1, 2$, $\forall \, x_i \in [a_i, b_i]$.

By Hölder's inequality we obtain

$$|f_1(x_1)| \leq \frac{1}{\Gamma(\alpha_1)} \frac{(\psi_1(x_1) - \psi_1(a_1))^{\frac{p(\alpha_1-1)+1}{p}}}{(p(\alpha_1 - 1) + 1)^{\frac{1}{p}}} \left\| {}^H\mathbb{D}_{a_1+}^{\alpha_1,\beta_1;\psi_1} f_1 \right\|_{L_q([a_1,b_1],\psi_1)}, \quad (4.90)$$

$\forall \, x_1 \in [a_1, b_1]$,

and

$$|f_2(x_2)| \leq \frac{1}{\Gamma(\alpha_2)} \frac{(\psi_2(x_2) - \psi_2(a_2))^{\frac{q(\alpha_2-1)+1}{q}}}{(q(\alpha_2 - 1) + 1)^{\frac{1}{q}}} \left\| {}^H\mathbb{D}_{a_2+}^{\alpha_2,\beta_2;\psi_2} f_2 \right\|_{L_p([a_2,b_2],\psi_2)}, \quad (4.91)$$

$\forall \, x_2 \in [a_2, b_2]$.

Hence we have

$$|f_1(x_1)| \, |f_2(x_2)| \leq \frac{1}{\Gamma(\alpha_1) \Gamma(\alpha_2) (p(\alpha_1 - 1) + 1)^{\frac{1}{p}} (q(\alpha_2 - 1) + 1)^{\frac{1}{q}}}$$

$$(\psi_1(x_1) - \psi_1(a_1))^{\frac{p(\alpha_1-1)+1}{p}} (\psi_2(x_2) - \psi_2(a_2))^{\frac{q(\alpha_2-1)+1}{q}} \quad (4.92)$$

$$\left\| {}^H\mathbb{D}_{a_1+}^{\alpha_1,\beta_1;\psi_1} f_1 \right\|_{L_q([a_1,b_1],\psi_1)} \left\| {}^H\mathbb{D}_{a_2+}^{\alpha_2,\beta_2;\psi_2} f_2 \right\|_{L_p([a_2,b_2],\psi_2)} \leq$$

(using Young's inequality for $a, b \geq 0$, $a^{\frac{1}{p}} b^{\frac{1}{q}} \leq \frac{a}{p} + \frac{b}{q}$)

$$\frac{1}{\Gamma(\alpha_1)\Gamma(\alpha_2)}\left(\frac{(\psi_1(x_1)-\psi_1(a_1))^{p(\alpha_1-1)+1}}{p(p(\alpha_1-1)+1)}+\frac{(\psi_2(x_2)-\psi_2(a_2))^{q(\alpha_2-1)+1}}{q(q(\alpha_2-1)+1)}\right) \tag{4.93}$$

$$\left\| {}^{H}\mathbb{D}_{a_1+}^{\alpha_1,\beta_1;\psi_1} f_1 \right\|_{L_q([a_1,b_1],\psi_1)} \left\| {}^{H}\mathbb{D}_{a_2+}^{\alpha_2,\beta_2;\psi_2} f_2 \right\|_{L_p([a_2,b_2],\psi_2)},$$

$\forall\, x_i \in [a_i, b_i];\ i = 1, 2.$

So far we have

$$\frac{|f_1(x_1)|\,|f_2(x_2)|}{\left(\frac{(\psi_1(x_1)-\psi_1(a_1))^{p(\alpha_1-1)+1}}{p(p(\alpha_1-1)+1)}+\frac{(\psi_2(x_2)-\psi_2(a_2))^{q(\alpha_2-1)+1}}{q(q(\alpha_2-1)+1)}\right)} \leq \tag{4.94}$$

$$\frac{\left\| {}^{H}\mathbb{D}_{a_1+}^{\alpha_1,\beta_1;\psi_1} f_1 \right\|_{L_q([a_1,b_1],\psi_1)} \left\| {}^{H}\mathbb{D}_{a_2+}^{\alpha_2,\beta_2;\psi_2} f_2 \right\|_{L_p([a_2,b_2],\psi_2)}}{\Gamma(\alpha_1)\Gamma(\alpha_2)},$$

$\forall\, x_i \in [a_i, b_i];\ i = 1, 2.$

The denominator in (4.94) can be zero only when $x_1 = a_1$ and $x_2 = a_2$.

Therefore we obtain (4.87), by integrating (4.94) over $[a_1, b_1] \times [a_2, b_2]$. ■

It follows the right side analog of last theorem.

Theorem 4.16 *Let $i = 1, 2$; $\psi_i, f_i \in C^{n_i}([a_i, b_i])$, with ψ_i being strictly increasing over $[a_i, b_i]$, where $n_i - 1 < \alpha_i < n_i$, $0 \leq \beta_i \leq 1$, and $\gamma_i = \alpha_i + \beta_i(n_i - \alpha_i)$, $x_i \in [a_i, b_i]$. Let also $p, q > 1 : \frac{1}{p} + \frac{1}{q} = 1$, such that $\alpha_1 > \frac{1}{q}$ and $\alpha_2 > \frac{1}{p}$. Assume that $f_{i\psi_i}^{[n_i-k_i]}\left(I_{b_i-}^{(1-\beta_i)(n_i-\alpha_i);\psi_i} f_i\right)(b_i) = 0$, for $k_i = 1, \ldots, n_i - 1$. Then*

$$\int_{a_1}^{b_1}\int_{a_2}^{b_2} \frac{|f_1(x_1)|\,|f_2(x_2)|\,dx_1dx_2}{\left(\frac{(\psi_1(b_1)-\psi_1(x_1))^{p(\alpha_1-1)+1}}{p(p(\alpha_1-1)|+1)}+\frac{(\psi_2(b_2)-\psi_2(x_2))^{q(\alpha_2-1)+1}}{q(q(\alpha_2-1)+1)}\right)} \leq$$

$$\frac{(b_1-a_1)(b_2-a_2)}{\Gamma(\alpha_1)\Gamma(\alpha_2)} \left\| {}^{H}\mathbb{D}_{b_1-}^{\alpha_1,\beta_1;\psi_1} f_1 \right\|_{L_q([a_1,b_1],\psi_1)} \left\| {}^{H}\mathbb{D}_{b_2-}^{\alpha_2,\beta_2;\psi_2} f_2 \right\|_{L_p([a_2,b_2],\psi_2)}. \tag{4.95}$$

Proof Similar to Theorem 4.15, by the use of (4.30). ■

We continue with other Hilfer-Hilbert-Pachpatte fractional inequalities.

Theorem 4.17 *Let $i = 1, 2$; $\alpha_i > 0$, $\alpha_i \notin \mathbb{N}$, $\lceil \alpha_i \rceil = n_i$, $0 < \beta_i < 1$, $f_i \in C^{n_i}([a_i, b_i])$, $[a_i, b_i] \subset \mathbb{R}$ and set $\gamma_i = \alpha_i + \beta_i(n_i - \alpha_i)$. Assume further that $\Delta_{a_i+}^{\gamma_i} f_i \in C([a_i, b_i]) : \Delta_{a_i+}^{\gamma_i - j_i} f_i(a_i) = 0$, for $j_i = 1, \ldots, n_i$. Let also $\overline{\alpha}_i > 0 : \lceil \overline{\alpha}_i \rceil = \overline{n}_i$, with $\overline{\gamma}_i = \overline{\alpha}_i + \beta_i(\overline{n}_i - \overline{\alpha}_i)$, and assume that $\alpha_i > \overline{\alpha}_i$ and $\gamma_i > \overline{\gamma}_i$. Furthermore let $p, q > 1 : \frac{1}{p} + \frac{1}{q} = 1$, such that $\alpha_1 > \frac{1}{q}$ and $\alpha_2 > \frac{1}{p}$. Then*

$$\int_{a_1}^{b_1}\int_{a_2}^{b_2}\frac{\left|{}^{H}\mathbb{D}_{a_1+}^{\overline{\alpha}_1,\beta_1}f_1(x_1)\right|\left|{}^{H}\mathbb{D}_{a_2+}^{\overline{\alpha}_2,\beta_2}f_2(x_2)\right|dx_1dx_2}{\left(\frac{(x_1-a_1)^{p(\alpha_1-\overline{\alpha}_1-1)+1}}{p(p(\alpha_1-\overline{\alpha}_1-1)|+1)}+\frac{(x_2-a_2)^{q(\alpha_2-\overline{\alpha}_2-1)+1}}{q(q(\alpha_2-\overline{\alpha}_2-1)+1)}\right)}\leq$$

$$\frac{(b_1-a_1)(b_2-a_2)}{\Gamma(\alpha_1-\overline{\alpha}_1)\Gamma(\alpha_2-\overline{\alpha}_2)}\left\|{}^{H}\mathbb{D}_{a_1+}^{\alpha_1,\beta_1}f_1\right\|_{L_q([a_1,b_1])}\left\|{}^{H}\mathbb{D}_{a_2+}^{\alpha_2,\beta_2}f_2\right\|_{L_p([a_2,b_2])}. \tag{4.96}$$

Proof Similar to Theorem 4.15, by the use of Theorem 4.6. ∎

It follows

Theorem 4.18 *Let $i=1,2$; $\alpha_i>0$, $\alpha_i\notin\mathbb{N}$, $\lceil\alpha_i\rceil=n_i$, $0<\beta_i<1$, $f_i\in C^{n_i}([a_i,b_i])$, $[a_i,b_i]\subset\mathbb{R}$ and set $\gamma_i=\alpha_i+\beta_i(n_i-\alpha_i)$. Assume further that $\Delta_{b_i-}^{\gamma_i}f_i\in C([a_i,b_i]):\Delta_{b_i-}^{\gamma_i-j_i}f_i(b_i)=0$, for $j_i=1,\dots,n_i$. Let also $\overline{\alpha}_i>0:\lceil\overline{\alpha}_i\rceil=\overline{n}_i$, with $\overline{\gamma}_i=\overline{\alpha}_i+\beta_i(\overline{n}_i-\overline{\alpha}_i)$, and assume that $\alpha_i>\overline{\alpha}_i$ and $\gamma_i>\overline{\gamma}_i$. Furthermore let $p,q>1:\frac{1}{p}+\frac{1}{q}=1$, such that $\alpha_1>\frac{1}{q}$ and $\alpha_2>\frac{1}{p}$. Then*

$$\int_{a_1}^{b_1}\int_{a_2}^{b_2}\frac{\left|{}^{H}\mathbb{D}_{b_1-}^{\overline{\alpha}_1,\beta_1}f_1(x_1)\right|\left|{}^{H}\mathbb{D}_{b_2-}^{\overline{\alpha}_2,\beta_2}f_2(x_2)\right|dx_1dx_2}{\left(\frac{(b_1-x_1)^{p(\alpha_1-\overline{\alpha}_1-1)+1}}{p(p(\alpha_1-\overline{\alpha}_1-1)|+1)}+\frac{(b_2-x_2)^{q(\alpha_2-\overline{\alpha}_2-1)+1}}{q(q(\alpha_2-\overline{\alpha}_2-1)+1)}\right)}\leq$$

$$\frac{(b_1-a_1)(b_2-a_2)}{\Gamma(\alpha_1-\overline{\alpha}_1)\Gamma(\alpha_2-\overline{\alpha}_2)}\left\|{}^{H}\mathbb{D}_{b_1-}^{\alpha_1,\beta_1}f_1\right\|_{L_q([a_1,b_1])}\left\|{}^{H}\mathbb{D}_{b_2-}^{\alpha_2,\beta_2}f_2\right\|_{L_p([a_2,b_2])}. \tag{4.97}$$

Proof Similar to Theorem 4.15, by the use of Theorem 4.7. ∎

We finish with two applications:

Corollary 4.19 *All as in Theorem 4.15, with $\psi_1(x_1)=e^{x_1}$, $\psi_2(x_2)=e^{x_2}$. Then*

$$\int_{a_1}^{b_1}\int_{a_2}^{b_2}\frac{|f_1(x_1)||f_2(x_2)|dx_1dx_2}{\left(\frac{(e^{x_1}-e^{a_1})^{p(\alpha_1-1)+1}}{p(p(\alpha_1-1)|+1)}+\frac{(e^{x_2}-e^{a_2})^{q(\alpha_2-1)+1}}{q(q(\alpha_2-1)+1)}\right)}\leq$$

$$\frac{(b_1-a_1)(b_2-a_2)}{\Gamma(\alpha_1)\Gamma(\alpha_2)}\left\|{}^{H}\mathbb{D}_{a_1+}^{\alpha_1,\beta_1;e^{x_1}}f_1\right\|_{L_q([a_1,b_1],e^{x_1})}\left\|{}^{H}\mathbb{D}_{a_2+}^{\alpha_2,\beta_2;e^{x_2}}f_2\right\|_{L_p([a_2,b_2],e^{x_2})}. \tag{4.98}$$

Proof By Theorem 4.15. ∎

Corollary 4.20 *All as in Theorem 4.16, with $[a_i,b_i]\subset(0,+\infty)$, $i=1,2$; and $\psi_1(x_1)=\ln x_1$, $\psi_2(x_2)=\ln x_2$. Then*

$$\int_{a_1}^{b_1}\int_{a_2}^{b_2}\frac{|f_1(x_1)||f_2(x_2)|dx_1dx_2}{\left(\frac{\left(\ln\frac{b_1}{x_1}\right)^{p(\alpha_1-1)+1}}{p(p(\alpha_1-1)|+1)}+\frac{\left(\ln\frac{b_2}{x_2}\right)^{q(\alpha_2-1)+1}}{q(q(\alpha_2-1)+1)}\right)}\leq$$

$$\frac{(b_1 - a_1)(b_2 - a_2)}{\Gamma(\alpha_1)\Gamma(\alpha_2)} \left\| {}^H\mathbb{D}_{b_1-}^{\alpha_1,\beta_1;\ln x_1} f_1 \right\|_{L_q([a_1,b_1],\ln x_1)} \left\| {}^H\mathbb{D}_{b_2-}^{\alpha_2,\beta_2;\ln x_2} f_2 \right\|_{L_p([a_2,b_2],\ln x_2)}. \tag{4.99}$$

Proof By Theorem 4.16.

■

References

1. Almeida, R.: A Caputo fractional derivative of a function with respect to another function. Commun. Nonlinear Sci. Numer. Simulat. **44**, 460–481 (2017)
2. Anastassiou, G.A.: Intelligent Comparisons: Analytic Inequalities. Springer, Heidelberg (2016)
3. Anastassiou, G.: Advancements on ψ -Hilfer fractional calculus and fractional integral inequalities. Nonlinearity and Complexity, accepted, Discontinuity (2021)
4. Anastassiou, G.: Hilfer-Polya, ψ-Hilfer Ostrowski and ψ-Hilfer-Hilbert-Pachpatte fractional inequalities. Symmetry **13**, 463. http://doi.org./10.3390.sym13030463 (2021)
5. Kilbas, A.A., Srivastava, H.M., Trujillo, J.J.: Theory and Applications of Fractional Differentiation Equations. North Holland, Amsterdam (2006)
6. Polya, G.: Ein mittelwertsatz für Funktionen mehrerer Veründerlichen. Tohoku Math. J. **19**, 1–3 (1921)
7. Polya, G., Szegö, G.: Aufgaben und Lehrsätze aus der Analysis, vol. I. Springer, Berlin (1925).(German)
8. Polya, G., Szegö, G.: Problems and Theorems in Analysis, Volume I, Classics in Mathematics. Spriger, Berlin (1972)
9. Polya, G., Szegö, G.: Problems and Theorems in Analysis, Volume I, Chinese Edition (1984)
10. Samko, S.G., Kilbas, A.A., Marichev, O.I.: Fractional Integrals and Derivatives: Theory and Applications. Gordon and Breach Science Publishers, Switzerland (1993)
11. Tomovski, Ž, Hilfer, R., Srivastava, H.M.: Fractional and operational calculus with generalized fractional derivative operators and Mittag-Leffler type functions. Integral Transforms Spec. Funct. **21**(11), 797–814 (2010)
12. Vanterler da, J., Sousa, C., Capelas de Oliveira, E.: On the ψ-Hilfer fractional derivative. Commun. Nonlinear Sci. Numer. Simul. **60**, 72–91 (2018)

Chapter 5
Generalized Hilfer Fractional Approximation of Csiszar's f-Divergence

Here are given tight probabilistic inequalities that provide nearly best estimates for the Csiszar's f-divergence. These use the right and left ψ-Hilfer fractional derivatives of the directing function f. Csiszar's f-divergence or the so called Csiszar's discrimination is used as a measure of dependence between two random variables which is a very essential aspect of stochastics, we apply our results there. The Csiszar's discrimination is the most important and general measure for the comparison between two probability measures. We give also other applications. It follows [3].

5.1 Background-I

Throughout this chapter we use the following.

Let f be a convex function from $(0, +\infty)$ into $\mathbb{R}$ that is strictly convex at 1 with $f(1) = 0$. Let $(X, \mathcal{A}, \lambda)$ be a measure space, where λ is a finite or a σ-finite measure on $(X, \mathcal{A})$. And let μ_1, μ_2 be two probability measures on $(X, \mathcal{A})$ such that $\mu_1 \ll \lambda$, $\mu_2 \ll \lambda$ (absolutely continuous); for example, $\lambda = \mu_1 + \mu_2$. Denote by $p = \frac{d\mu_1}{d\lambda}$, $q = \frac{d\mu_2}{d\lambda}$ (densities) Radon-Nikodym derivatives of μ_1, μ_2 with respect to λ. Here we assume that

$$0 < a \leq \frac{p}{q} \leq b, \text{ a.e. on } X$$

and $a \leq 1 \leq b$.

The quantity

$$\Gamma_f(\mu_1, \mu_2) = \int_X q(x) f\left(\frac{p(x)}{q(x)}\right) d\lambda(x), \tag{5.1}$$

was introduced by I. Csiszar in 1967 (see [6]), and is called f-divergence of the probability measures μ_1 and μ_2. By Lemma 1.1 of [6], the integral (5.1) is well

G. A. Anastassiou, *Unification of Fractional Calculi with Applications*,
Studies in Systems, Decision and Control 398,
https://doi.org/10.1007/978-3-030-86920-5_5

defined and $\Gamma_f(\mu_1,\mu_2)\geq 0$ with equality only when $\mu_1=\mu_2$. In [6] the author without proof mentions that $\Gamma_f(\mu_1,\mu_2)$ does not depend on the choice of λ.

For a proof of the last see [2], Lemma 1.1.

The concept of f-divergence was introduced first in [5] as a generalization of Kullback's "information for discrimination" or I-divergence (generalized entropy) [9, 10] and of Rényi's "information gain" (I-divergence of order α) [12]. In fact the I-divergence of order 1 equals

$$\Gamma_{u\log_2 u}(\mu_1,\mu_2).$$

The choice $f(u)=(u-1)^2$ again produces a known measure of difference of distributions called χ^2-divergence; of course the total variation distance $|\mu_1-\mu_2|=\int_X|p(x)-q(x)|\,d\lambda(x)$ equals $\Gamma_{|u-1|}(\mu_1,\mu_2)$.

Here by assuming $f(1)=0$ we can consider $\Gamma_f(\mu_1,\mu_2)$ as a measure of the difference between the probability measures μ_1,μ_2. The f-divergence is in general asymmetric in μ_1 and μ_2. But because f is convex and strictly convex at 1 (see Lemma 2, [2]) so is

$$f^*(u)=uf\left(\frac{1}{u}\right)$$

and as in [6] we get

$$\Gamma_f(\mu_2,\mu_1)=\Gamma_{f^*}(\mu_1,\mu_2).$$

In information theory and statistics many other concrete divergences are used that are special cases of the above general Csiszar f-divergence, such as Hellinger distance D_H, α-divergence D_α, Bhattacharyya distance D_B, harmonic distance D_{H_α}, Jeffrey's distance D_J, and triangular discrimination D_Δ; for all these, see, for example [4, 7]. The problem of finding and estimating the proper distance (or difference or discrimination) of two probability distributions is one of the major questions in probability theory.

The above f-divergence measures in their various forms have also been applied to anthropology, genetics, finance, economics, political science, biology, approximation of probability distributions, signal processing, and pattern recognition. A great inspiration for this chapter has been the very important monograph on the topic by Dragomir [7].

5.2 Background-II

Let $-\infty<a<b<\infty$, the left and right Riemann-Liouville fractional integrals of order $\alpha\in\mathbb{C}$ ($\mathcal{R}(\alpha)>0$) are defined by

$$\left(I_{a+}^{\alpha}f\right)(x)=\frac{1}{\Gamma(\alpha)}\int_a^x(x-t)^{\alpha-1}f(t)\,dt, \tag{5.2}$$

$x > a$; where Γ stands for the gamma function,
and

$$\left(I_{b-}^{\alpha} f\right)(x) = \frac{1}{\Gamma(\alpha)} \int_{x}^{b} (t - x)^{\alpha - 1} f(t)\, dt, \tag{5.3}$$

$x < b$.

The Riemann-Liouville left and right fractional derivatives of order $\alpha \in \mathbb{C}$ ($\mathcal{R}(\alpha) \geq 0$) are defined by

$$\left(\Delta_{a+}^{\alpha} y\right)(x) = \left(\frac{d}{dx}\right)^{n} \left(I_{a+}^{n-\alpha} y\right)(x) = \frac{1}{\Gamma(n-\alpha)} \left(\frac{d}{dx}\right)^{n} \int_{a}^{x} (x - t)^{n-\alpha-1} y(t)\, dt \tag{5.4}$$

($n = \lceil \mathcal{R}(\alpha) \rceil$, $\lceil \cdot \rceil$ means ceiling of the number; $x > a$)

$$\left(\Delta_{b-}^{\alpha} y\right)(x) = (-1)^{n} \left(\frac{d}{dx}\right)^{n} \left(I_{b-}^{n-\alpha} y\right)(x)$$

$$= \frac{(-1)^{n}}{\Gamma(n-\alpha)} \left(\frac{d}{dx}\right)^{n} \int_{x}^{b} (t - x)^{n-\alpha-1} y(t)\, dt \tag{5.5}$$

($n = \lceil \mathcal{R}(\alpha) \rceil$; $x < b$), respectively, where $\mathcal{R}(\alpha)$ is the real part of α.

In particular, when $\alpha = n \in \mathbb{Z}_{+}$, then

$$\left(\Delta_{a+}^{0} y\right)(x) = \left(\Delta_{b-}^{0} y\right)(x) = y(x);$$

$$\left(\Delta_{a+}^{n} y\right)(x) = y^{(n)}(x), \text{ and } \left(\Delta_{b-}^{n} y\right)(x) = (-1)^{n} y^{(n)}(x), \; n \in \mathbb{N},$$

see [13].

Let $\alpha > 0$, $I = [a, b] \subset \mathbb{R}$, f an integrable function defined on I and $\psi \in C^{1}(I)$ an increasing function such that $\psi'(x) \neq 0$, for all $x \in I$. Left fractional integrals and left Riemann-Liouville fractional derivatives of a function f with respect to another function ψ are defined as [8, 13]

$$I_{a+}^{\alpha,\psi} f(x) = \frac{1}{\Gamma(\alpha)} \int_{a}^{x} \psi'(t) (\psi(x) - \psi(t))^{\alpha-1} f(t)\, dt, \tag{5.6}$$

and

$$\Delta_{a+}^{\alpha,\psi} f(x) = \left(\frac{1}{\psi'(x)} \frac{d}{dx}\right)^{n} I_{a+}^{n-\alpha,\psi} f(x) = \tag{5.7}$$

$$\frac{1}{\Gamma(n-\alpha)} \left(\frac{1}{\psi'(x)} \frac{d}{dx}\right)^{n} \int_{a}^{x} \psi'(t) (\psi(x) - \psi(t))^{n-\alpha-1} f(t)\, dt,$$

respectively, where $n = \lceil \alpha \rceil$.

Similarly, we define the right ones:

$$I_{b-}^{\alpha,\psi} f(x) = \frac{1}{\Gamma(\alpha)} \int_x^b \psi'(t) (\psi(t) - \psi(x))^{\alpha-1} f(t) \, dt, \tag{5.8}$$

and

$$\Delta_{b-}^{\alpha,\psi} f(x) = \left(-\frac{1}{\psi'(x)} \frac{d}{dx} \right)^n I_{b-}^{n-\alpha,\psi} f(x) =$$

$$\frac{1}{\Gamma(n-\alpha)} \left(-\frac{1}{\psi'(x)} \frac{d}{dx} \right)^n \int_x^b \psi'(t) (\psi(t) - \psi(x))^{n-\alpha-1} f(t) \, dt. \tag{5.9}$$

The following semigroup property holds; if $\alpha, \beta > 0$, $f \in C(I)$, then

$$I_{a+}^{\alpha,\psi} I_{a+}^{\beta,\psi} f = I_{a+}^{\alpha+\beta,\psi} f \text{ and } I_{b-}^{\alpha,\psi} I_{b-}^{\beta,\psi} f = I_{b-}^{\alpha+\beta,\psi} f.$$

Next let again $\alpha > 0$, $n = \lceil \alpha \rceil$, $I = [a, b]$, $f, \psi \in C^n(I) : \psi$ is increasing and $\psi'(x) \neq 0$, for all $x \in I$. The left ψ-Caputo fractional derivative of f of order α is given by Almeida [1]

$${}^C D_{a+}^{\alpha,\psi} f(x) = I_{a+}^{n-\alpha,\psi} \left(\frac{1}{\psi'(x)} \frac{d}{dx} \right)^n f(x), \tag{5.10}$$

and the right ψ-Caputo fractional derivative [1]

$${}^C D_{b-}^{\alpha,\psi} f(x) = I_{b-}^{n-\alpha,\psi} \left(-\frac{1}{\psi'(x)} \frac{d}{dx} \right)^n f(x). \tag{5.11}$$

We set

$$f_\psi^{[n]}(x) := f_\psi^{(n)} f(x) := \left(\frac{1}{\psi'(x)} \frac{d}{dx} \right)^n f(x). \tag{5.12}$$

Clearly, when $\alpha = m \in \mathbb{N}$ we have

$${}^C D_{a+}^{\alpha,\psi} f(x) = f_\psi^{[m]}(x) \text{ and } {}^C D_{b-}^{\alpha,\psi} f(x) = (-1)^m f_\psi^{[m]}(x),$$

and if $\alpha \notin \mathbb{N}$, then

$${}^C D_{a+}^{\alpha,\psi} f(x) = \frac{1}{\Gamma(n-\alpha)} \int_a^x \psi'(t) (\psi(x) - \psi(t))^{n-\alpha-1} f_\psi^{[n]}(t) \, dt, \tag{5.13}$$

and

$${}^C D_{b-}^{\alpha,\psi} f(x) = \frac{(-1)^n}{\Gamma(n-\alpha)} \int_x^b \psi'(t) (\psi(t) - \psi(x))^{n-\alpha-1} f_\psi^{[n]}(t) \, dt. \tag{5.14}$$

If $\psi(x) = x$, then we get the usual left and right Caputo fractional derivatives

$$ {}^{C}D_{a+}^{m} f(x) = f^{(m)}(x), \quad {}^{C}D_{b-}^{m} f(x) = (-1)^{m} f^{(m)}(x), $$

for $m \in \mathbb{N}$, and ($\alpha \notin \mathbb{N}$)

$$ D_{*a}^{\alpha} f(x) = {}^{C}D_{a+}^{\alpha} f(x) = \frac{1}{\Gamma(n-\alpha)} \int_{a}^{x} (x-t)^{n-\alpha-1} f^{(n)}(t)\, dt, \tag{5.15} $$

$$ D_{b-}^{\alpha}(x) = {}^{C}D_{b-}^{\alpha} f(x) = \frac{(-1)^{n}}{\Gamma(n-\alpha)} \int_{x}^{b} (t-x)^{n-\alpha-1} f^{(n)}(t)\, dt. \tag{5.16} $$

Also we set

$$ {}^{C}D_{a+}^{0,\psi} f(x) = {}^{C}D_{b-}^{0,\psi} f(x) = f(x). $$

Next we will deal with the ψ-Hilfer fractional derivative.

Definition 5.1 ([15]) Let $n-1 < \alpha < n$, $n \in \mathbb{N}$, $I = [a, b] \subset \mathbb{R}$ and $f, \psi \in C^{n}([a, b])$, ψ is increasing and $\psi'(x) \neq 0$, for all $x \in I$. The ψ-Hilfer fractional derivative (left-sided and right-sided) ${}^{H}\mathbb{D}_{a+(b-)}^{\alpha,\beta;\psi} f$ of order α and type $0 \leq \beta \leq 1$, respectively, are defined by

$$ {}^{H}\mathbb{D}_{a+}^{\alpha,\beta;\psi} f(x) = I_{a+}^{\beta(n-\alpha);\psi} \left(\frac{1}{\psi'(x)} \frac{d}{dx} \right)^{n} I_{a+}^{(1-\beta)(n-\alpha);\psi} f(x), \tag{5.17} $$

and

$$ {}^{H}\mathbb{D}_{b-}^{\alpha,\beta;\psi} f(x) = I_{b-}^{\beta(n-\alpha);\psi} \left(-\frac{1}{\psi'(x)} \frac{d}{dx} \right)^{n} I_{b-}^{(1-\beta)(n-\alpha);\psi} f(x), \; x \in [a, b]. \tag{5.18} $$

The original Hilfer fractional derivatives [14] come from $\psi(x) = x$, and are denoted by ${}^{H}\mathbb{D}_{a+}^{\alpha,\beta} f(x)$ and ${}^{H}\mathbb{D}_{b-}^{\alpha,\beta} f(x)$.

When $\beta = 0$, we get Riemann-Liouville fractional derivatives, while when $\beta = 1$ we have Caputo type fractional derivatives.

We define $\gamma = \alpha + \beta(n-\alpha)$. We notice that $n-1 < \alpha \leq \alpha + \beta(n-\alpha) \leq \alpha + n - \alpha = n$, hence $\lceil \gamma \rceil = n$. We can easily write that [15]

$$ {}^{H}\mathbb{D}_{a+}^{\alpha,\beta;\psi} f(x) = I_{a+}^{\gamma-\alpha;\psi} \Delta_{a+}^{\gamma;\psi} f(x), \tag{5.19} $$

and

$$ {}^{H}\mathbb{D}_{b-}^{\alpha,\beta;\psi} f(x) = I_{b-}^{\gamma-\alpha;\psi} \Delta_{b-}^{\gamma;\psi} f(x), \; x \in [a, b]. \tag{5.20} $$

We have that [15]

$$\Delta_{a+}^{\gamma,\psi} f(x) = \left(\frac{1}{\psi'(x)} \frac{d}{dx} \right)^n I_{a+}^{(1-\beta)(n-\alpha);\psi} f(x), \tag{5.21}$$

and

$$\Delta_{b-}^{\gamma,\psi} f(x) = \left(-\frac{1}{\psi'(x)} \frac{d}{dx} \right)^n I_{b-}^{(1-\beta)(n-\alpha);\psi} f(x). \tag{5.22}$$

In particular, when $0 < \alpha < 1$ and $0 \leq \beta \leq 1$; $\gamma = \alpha + \beta(1-\alpha)$, we have that

$$^H\mathbb{D}_{a+}^{\alpha,\beta;\psi} f(x) = \frac{1}{\Gamma(\gamma-\alpha)} \int_a^x \psi'(t)(\psi(x)-\psi(t))^{\gamma-\alpha-1} \Delta_{a+}^{\gamma;\psi} f(t)\, dt, \tag{5.23}$$

and

$$^H\mathbb{D}_{b-}^{\alpha,\beta;\psi} f(x) = \frac{1}{\Gamma(\gamma-\alpha)} \int_x^b \psi'(t)(\psi(t)-\psi(x))^{\gamma-\alpha-1} \Delta_{b-}^{\gamma;\psi} f(t)\, dt, \tag{5.24}$$

$x \in [a,b]$.

Remark 5.2 ([15]) Let $\mu = n(1-\beta) + \beta\alpha$, then $\lceil \mu \rceil = n$.
Assume that $g(x) = I_{a+}^{(1-\beta)(n-\alpha);\psi} f(x) \in C^n([a,b])$, we have that

$$^H\mathbb{D}_{a+}^{\alpha,\beta;\psi} f(x) = I_{a+}^{n-\mu;\psi} \left(\frac{1}{\psi'(x)} \frac{d}{dx} \right)^n g(x). \tag{5.25}$$

Thus

$$^H\mathbb{D}_{a+}^{\alpha,\beta;\psi} f = {}^C D_{a+}^{\mu;\psi} g(x) = {}^C D_{a+}^{\mu;\psi} \left[I_{a+}^{(1-\beta)(n-\alpha);\psi} f(x) \right]. \tag{5.26}$$

Assume that $w(x) = I_{b-}^{(1-\beta)(n-\alpha);\psi} f(x) \in C^n([a,b])$. Hence

$$^H\mathbb{D}_{b-}^{\alpha,\beta;\psi} f(x) = I_{b-}^{\beta(n-\alpha);\psi} \left(-\frac{1}{\psi'(x)} \frac{d}{dx} \right)^n w(x) = I_{b-}^{n-\mu;\psi} \left(-\frac{1}{\psi'(x)} \frac{d}{dx} \right)^n w(x). \tag{5.27}$$

Thus

$$^H\mathbb{D}_{b-}^{\alpha,\beta;\psi} f = {}^C D_{b-}^{\mu;\psi} w(x) = {}^C D_{b-}^{\mu;\psi} \left(I_{b-}^{(1-\beta)(n-\alpha);\psi} f(x) \right). \tag{5.28}$$

We mention the simplified ψ-Hilfer fractional Taylor formulae:

Theorem 5.3 *(see also [15]) Let $\psi, f \in C^n([a,b])$, with ψ being increasing such that $\psi'(x) \neq 0$ over $[a,b]$, where $n-1 < \alpha < n$, $0 \leq \beta \leq 1$, and $\gamma = \alpha + \beta(n-\alpha)$, $x \in [a,b]$. Then*

$$f(x) - \sum_{k=1}^{n-1} \frac{(\psi(x) - \psi(a))^{\gamma-k}}{\Gamma(\gamma-k+1)} f_{\psi}^{[n-k]} \left(I_{a+}^{(1-\beta)(n-\alpha);\psi} f \right)(a) =$$

$$\frac{1}{\Gamma(\alpha)} \int_a^x \psi'(t) (\psi(x) - \psi(t))^{\alpha-1} \, {}^H\mathbb{D}_{a+}^{\alpha,\beta;\psi} f(t) \, dt, \tag{5.29}$$

and

$$f(x) - \sum_{k=1}^{n-1} \frac{(-1)^k (\psi(b) - \psi(x))^{\gamma-k}}{\Gamma(\gamma-k+1)} f_{\psi}^{[n-k]} \left(I_{b-}^{(1-\beta)(n-\alpha);\psi} f \right)(b) =$$

$$\frac{1}{\Gamma(\alpha)} \int_x^b \psi'(t) (\psi(t) - \psi(x))^{\alpha-1} \, {}^H\mathbb{D}_{b-}^{\alpha,\beta;\psi} f(t) \, dt. \tag{5.30}$$

Here notice that $\left(I_{a+}^{(1-\beta)(n-\alpha);\psi} f \right)(a) = \left(I_{b-}^{(1-\beta)(n-\alpha);\psi} f \right)(b) = 0.$

5.3 Main Results-I

Here f and the whole setting is as in Sect. 5.1 the Background-I. Additionally, we assume that $\psi, f \in C^n([a,b])$, with ψ being increasing such that $\psi'(x) \neq 0$ over $[a,b]$, where $n-1 < \alpha < n$, $0 \leq \beta \leq 1$, and $\gamma = \alpha + \beta(n-\alpha)$. Furthermore we stress that $0 < a \leq \frac{p(x)}{q(x)} \leq b$, a.e. on X.

Next we present estimations for $\Gamma_f(\mu_1, \mu_2)$. We give first left side results:

Theorem 5.4 *Additionally assume that* $f_{\psi}^{[n-k]} \left(I_{a+}^{(1-\beta)(n-\alpha);\psi} f \right)(a) = 0$, $k = 1, ..., n-1$. *Then*

$$\Gamma_f(\mu_1, \mu_2) \leq \frac{\left\| {}^H\mathbb{D}_{a+}^{\alpha,\beta;\psi} f \right\|_{\infty,[a,b]}}{\Gamma(\alpha+1)} \int_X q(x) \left(\psi\left(\frac{p(x)}{q(x)} \right) - \psi(a) \right)^{\alpha} d\lambda. \tag{5.31}$$

Proof We have that (by 5.29)

$$f(x) = \frac{1}{\Gamma(\alpha)} \int_a^x \psi'(t) (\psi(x) - \psi(t))^{\alpha-1} \, {}^H\mathbb{D}_{a+}^{\alpha,\beta;\psi} f(t) \, dt, \tag{5.32}$$

$\forall\, x \in [a,b]$.

Hence it holds

$$|f(x)| \leq \frac{1}{\Gamma(\alpha)} \int_a^x \psi'(t) (\psi(x) - \psi(t))^{\alpha-1} \left| {}^H\mathbb{D}_{a+}^{\alpha,\beta;\psi} f(t) \right| dt$$

$$\frac{\left\| {}^{H}\mathbb{D}_{a+}^{\alpha,\beta;\psi} f \right\|_{\infty,[a,b]}}{\Gamma(\alpha+1)} (\psi(x) - \psi(a))^{\alpha}, \forall x \in [a,b]. \tag{5.33}$$

Consequently we obtain

$$\Gamma_f(\mu_1,\mu_2) = \int_X q(x) f\left(\frac{p(x)}{q(x)}\right) d\lambda \le \int_X q(x) \left| f\left(\frac{p(x)}{q(x)}\right) \right| d\lambda \overset{(5.33)}{\le}$$

$$\frac{\left\| {}^{H}\mathbb{D}_{a+}^{\alpha,\beta;\psi} f \right\|_{\infty,[a,b]}}{\Gamma(\alpha+1)} \int_X q(x) \left(\psi\left(\frac{p(x)}{q(x)}\right) - \psi(a) \right)^{\alpha} d\lambda, \tag{5.34}$$

proving (5.31). ■

We continue with the L_1-analog:

Theorem 5.5 *All as in Theorem 5.4, with* $\alpha > 1$. *Then*

$$\Gamma_f(\mu_1,\mu_2) \le \frac{\left\| {}^{H}\mathbb{D}_{a+}^{\alpha,\beta;\psi} f \right\|_{L_1([a,b],\psi)}}{\Gamma(\alpha)} \int_X q(x) \left(\psi\left(\frac{p(x)}{q(x)}\right) - \psi(a) \right)^{\alpha-1} d\lambda. \tag{5.35}$$

Proof Since $\alpha > 1$, by (5.33), we have

$$|f(x)| \le \frac{(\psi(x) - \psi(a))^{\alpha-1}}{\Gamma(\alpha)} \int_a^x \left| {}^{H}\mathbb{D}_{a+}^{\alpha,\beta;\psi} f(t) \right| d\psi(t) \le$$

$$\frac{\left\| {}^{H}\mathbb{D}_{a+}^{\alpha,\beta;\psi} f \right\|_{L_1([a,b],\psi)}}{\Gamma(\alpha)} (\psi(x) - \psi(a))^{\alpha-1}, \forall x \in [a,b]. \tag{5.36}$$

Hence it holds

$$\Gamma_f(\mu_1,\mu_2) \le \int_X q(x) \left| f\left(\frac{p(x)}{q(x)}\right) \right| d\lambda \overset{(5.36)}{\le}$$

$$\frac{\left\| {}^{H}\mathbb{D}_{a+}^{\alpha,\beta;\psi} f \right\|_{L_1([a,b],\psi)}}{\Gamma(\alpha)} \int_X q(x) \left(\psi\left(\frac{p(x)}{q(x)}\right) - \psi(a) \right)^{\alpha-1} d\lambda, \tag{5.37}$$

proving (5.35). ■

Next comes to L_q-analog:

Theorem 5.6 *All as in Theorem 5.4. Let also $p, q > 1 : \frac{1}{p} + \frac{1}{q} = 1$, with $\alpha > \frac{1}{q}$. Then*

$$\Gamma_f(\mu_1, \mu_2) \leq \frac{\left\| {}^H\mathbb{D}_{a+}^{\alpha,\beta;\psi} f \right\|_{L_q([a,b],\psi)}}{\Gamma(\alpha)(p(\alpha-1)+1)^{\frac{1}{p}}} \int_X q(x) \left(\psi\left(\frac{p(x)}{q(x)} \right) - \psi(a) \right)^{\alpha - \frac{1}{q}} d\lambda. \tag{5.38}$$

Proof By Hölder's inequality we have

$$|f(x)| \overset{(5.33.)}{\leq} \frac{1}{\Gamma(\alpha)} \int_a^x \psi'(t)(\psi(x) - \psi(t))^{\alpha-1} \left| {}^H\mathbb{D}_{a+}^{\alpha,\beta;\psi} f(t) \right| dt =$$

$$\frac{1}{\Gamma(\alpha)} \int_a^x (\psi(x) - \psi(t))^{\alpha-1} \left| {}^H\mathbb{D}_{a+}^{\alpha,\beta;\psi} f(t) \right| d\psi(t) \leq$$

$$\frac{1}{\Gamma(\alpha)} \frac{(\psi(x) - \psi(a))^{\frac{p(\alpha-1)+1}{p}}}{(p(\alpha-1)+1)^{\frac{1}{p}}} \left\| {}^H\mathbb{D}_{a+}^{\alpha,\beta;\psi} f \right\|_{L_q([a,b],\psi)} =$$

$$\frac{\left\| {}^H\mathbb{D}_{a+}^{\alpha,\beta;\psi} f \right\|_{L_q([a,b],\psi)}}{\Gamma(\alpha)(p(\alpha-1)+1)^{\frac{1}{p}}} (\psi(x) - \psi(a))^{\alpha-\frac{1}{q}}, \ \forall x \in [a, b], \tag{5.39}$$

and $\alpha > \frac{1}{q}$.

Consequently we derive

$$\Gamma_f(\mu_1, \mu_2) \leq \int_X q(x) \left| f\left(\frac{p(x)}{q(x)} \right) \right| d\lambda \overset{(5.39)}{\leq}$$

$$\frac{\left\| {}^H\mathbb{D}_{a+}^{\alpha,\beta;\psi} f \right\|_{L_q([a,b],\psi)}}{\Gamma(\alpha)(p(\alpha-1)+1)^{\frac{1}{p}}} \int_X q(x) \left(\psi\left(\frac{p(x)}{q(x)} \right) - \psi(a) \right)^{\alpha-\frac{1}{q}} d\lambda, \tag{5.40}$$

where $\alpha > \frac{1}{q}$, proving (5.38). ∎

We continue with right side results:

Theorem 5.7 *Additionally assume that $f_\psi^{[n-k]}\left(I_{b-}^{(1-\beta)(n-\alpha);\psi} f \right)(b) = 0$, $k = 1, ..., n-1$. Then*

$$\Gamma_f(\mu_1, \mu_2) \leq \frac{\left\| {}^H\mathbb{D}_{b-}^{\alpha,\beta;\psi} f \right\|_{\infty,[a,b]}}{\Gamma(\alpha+1)} \int_X q(x) \left(\psi(b) - \psi\left(\frac{p(x)}{q(x)} \right) \right)^{\alpha} d\lambda. \tag{5.41}$$

Proof We have that (by 5.30)

$$f(x) = \frac{1}{\Gamma(\alpha)} \int_x^b \psi'(t) (\psi(t) - \psi(x))^{\alpha-1} \, {}^H\mathbb{D}_{b-}^{\alpha,\beta;\psi} f(t) \, dt, \tag{5.42}$$

$\forall \, x \in [a, b]$.

Hence it holds

$$|f(x)| \leq \frac{1}{\Gamma(\alpha)} \int_x^b \psi'(t) (\psi(t) - \psi(x))^{\alpha-1} \left| {}^H\mathbb{D}_{b-}^{\alpha,\beta;\psi} f(t) \right| dt \leq$$

$$\frac{\left\| {}^H\mathbb{D}_{b-}^{\alpha,\beta;\psi} f \right\|_{\infty,[a,b]}}{\Gamma(\alpha+1)} (\psi(b) - \psi(x))^{\alpha}, \forall x \in [a, b]. \tag{5.43}$$

Therefore we derive

$$\Gamma_f(\mu_1, \mu_2) \leq \int_X q(x) \left| f\left(\frac{p(x)}{q(x)}\right) \right| d\lambda \overset{(5.43)}{\leq}$$

$$\frac{\left\| {}^H\mathbb{D}_{b-}^{\alpha,\beta;\psi} f \right\|_{\infty,[a,b]}}{\Gamma(\alpha+1)} \int_X q(x) \left(\psi(b) - \psi\left(\frac{p(x)}{q(x)}\right) \right)^{\alpha} d\lambda, \tag{5.44}$$

proving (5.41). ■

It follows the L_1-analog:

Theorem 5.8 *All as in Theorem 5.7,* $\alpha > 1$. *Then*

$$\Gamma_f(\mu_1, \mu_2) \leq \frac{\left\| {}^H\mathbb{D}_{b-}^{\alpha,\beta;\psi} f \right\|_{L_1([a,b],\psi)}}{\Gamma(\alpha)} \int_X q(x) \left(\psi(b) - \psi\left(\frac{p(x)}{q(x)}\right) \right)^{\alpha-1} d\lambda. \tag{5.45}$$

Proof Since $\alpha > 1$, by (5.43), we have

$$|f(x)| \leq \frac{(\psi(b) - \psi(x))^{\alpha-1}}{\Gamma(\alpha)} \int_x^b \left| {}^H\mathbb{D}_{b-}^{\alpha,\beta;\psi} f(t) \right| d\psi(t) \leq$$

$$\frac{\left\| {}^H\mathbb{D}_{b-}^{\alpha,\beta;\psi} f \right\|_{L_1([a,b],\psi)}}{\Gamma(\alpha)} (\psi(b) - \psi(x))^{\alpha-1}, \forall x \in [a, b]. \tag{5.46}$$

Hence it holds

$$\Gamma_f(\mu_1, \mu_2) \leq \int_X q(x) \left| f\left(\frac{p(x)}{q(x)}\right) \right| d\lambda \overset{(5.46)}{\leq}$$

$$\frac{\left\| {}^{H}\mathbb{D}_{b-}^{\alpha,\beta;\psi} f \right\|_{L_1([a,b],\psi)}}{\Gamma(\alpha)} \int_X q(x) \left(\psi(b) - \psi\left(\frac{p(x)}{q(x)}\right)\right)^{\alpha-1} d\lambda, \tag{5.47}$$

proving (5.45). ■

Next comes the L_q-analog from the right side.

Theorem 5.9 *All as in Theorem 5.7. Let also* $p, q > 1 : \frac{1}{p} + \frac{1}{q} = 1$, *with* $\alpha > \frac{1}{q}$. *Then*

$$\Gamma_f(\mu_1, \mu_2) \le \frac{\left\| {}^{H}\mathbb{D}_{b-}^{\alpha,\beta;\psi} f \right\|_{L_q([a,b],\psi)}}{\Gamma(\alpha)(p(\alpha-1)+1)^{\frac{1}{p}}} \int_X q(x) \left(\psi(b) - \psi\left(\frac{p(x)}{q(x)}\right)\right)^{\alpha-\frac{1}{q}} d\lambda. \tag{5.48}$$

Proof We apply Hölder's inequality to (5.43) to get

$$|f(x)| \le \frac{1}{\Gamma(\alpha)} \int_x^b (\psi(t) - \psi(x))^{\alpha-1} \left| {}^{H}\mathbb{D}_{b-}^{\alpha,\beta;\psi} f(t) \right| d\psi(t) \le$$

$$\frac{1}{\Gamma(\alpha)} \frac{(\psi(b) - \psi(x))^{\frac{p(\alpha-1)+1}{p}}}{(p(\alpha-1)+1)^{\frac{1}{p}}} \left\| {}^{H}\mathbb{D}_{b-}^{\alpha,\beta;\psi} f \right\|_{L_q([a,b],\psi)} =$$

$$\frac{\left\| {}^{H}\mathbb{D}_{b-}^{\alpha,\beta;\psi} f \right\|_{L_q([a,b],\psi)}}{\Gamma(\alpha)(p(\alpha-1)+1)^{\frac{1}{p}}} (\psi(b) - \psi(x))^{\alpha-\frac{1}{q}}, \ \forall x \in [a,b], \alpha > \frac{1}{q}. \tag{5.49}$$

Consequently we derive

$$\Gamma_f(\mu_1, \mu_2) \le \int_X q(x) \left| f\left(\frac{p(x)}{q(x)}\right)\right| d\lambda \overset{(5.49)}{\le}$$

$$\frac{\left\| {}^{H}\mathbb{D}_{b-}^{\alpha,\beta;\psi} f \right\|_{L_q([a,b],\psi)}}{\Gamma(\alpha)(p(\alpha-1)+1)^{\frac{1}{p}}} \int_X q(x) \left(\psi(b) - \psi\left(\frac{p(x)}{q(x)}\right)\right)^{\alpha-\frac{1}{q}} d\lambda, \tag{5.50}$$

where $\alpha > \frac{1}{q}$, proving (5.48). ■

We continue with

Remark 5.10 (I) Assume that ${}^{H}\mathbb{D}_{a+}^{\alpha,\beta;\psi} f \underset{(\le 0)}{\ge} 0$ over $[a,b]$, then

$$f(x) \underset{(\le)}{\ge} \sum_{k=1}^{n-1} \frac{(\psi(x) - \psi(a))^{\gamma-k}}{\Gamma(\gamma-k+1)} f_{\psi}^{[n-k]} \left(I_{a+}^{(1-\beta)(n-\alpha);\psi} f \right)(a), \tag{5.51}$$

$\forall\, x \in [a, b]$.

Hence

$$qf\left(\frac{p}{q}\right) \underset{(\leq)}{\geq} \sum_{k=1}^{n-1} q \frac{\left(\psi\left(\frac{p}{q}\right) - \psi(a)\right)^{\gamma-k}}{\Gamma(\gamma-k+1)} f_{\psi}^{[n-k]}\left(I_{a+}^{(1-\beta)(n-\alpha);\psi} f\right)(a). \tag{5.52}$$

Consequently we derive

$$\Gamma_f(\mu_1, \mu_2) \underset{(\leq)}{\overset{(5.52.)}{\geq}} \sum_{k=1}^{n-1} \frac{f_{\psi}^{[n-k]}\left(I_{a+}^{(1-\beta)(n-\alpha);\psi} f\right)(a)}{\Gamma(\gamma-k+1)} \int_X q\left(\psi\left(\frac{p}{q}\right) - \psi(a)\right)^{\gamma-k} d\lambda. \tag{5.53}$$

(II) Assume that ${}^{H}\mathbb{D}_{b-}^{\alpha,\beta;\psi} f \underset{(\leq 0)}{\geq} 0$ over $[a, b]$, then

$$f(x) \underset{(\leq)}{\overset{(5.30.)}{\geq}} \sum_{k=1}^{n-1} \frac{(-1)^k (\psi(b) - \psi(x))^{\gamma-k}}{\Gamma(\gamma-k+1)} f_{\psi}^{[n-k]}\left(I_{b-}^{(1-\beta)(n-\alpha);\psi} f\right)(b), \tag{5.54}$$

$\forall\, x \in [a, b]$.

Hence

$$qf\left(\frac{p}{q}\right) \underset{(\leq)}{\geq} \sum_{k=1}^{n-1} \frac{(-1)^k q\left(\psi(b) - \psi\left(\frac{p}{q}\right)\right)^{\gamma-k}}{\Gamma(\gamma-k+1)} f_{\psi}^{[n-k]}\left(I_{b-}^{(1-\beta)(n-\alpha);\psi} f\right)(b). \tag{5.55}$$

Consequently we derive

$$\Gamma_f(\mu_1, \mu_2) \underset{(\leq)}{\overset{(5.55.)}{\geq}} \sum_{k=1}^{n-1} \frac{(-1)^k f_{\psi}^{[n-k]}\left(I_{b-}^{(1-\beta)(n-\alpha);\psi} f\right)(b)}{\Gamma(\gamma-k+1)} \int_X q\left(\psi(b) - \psi\left(\frac{p}{q}\right)\right)^{\gamma-k} d\lambda. \tag{5.56}$$

We state

Corollary 5.11 (to Theorem 5.4) *Case of $\psi(x) = x$. So, additionally assume that $\left(I_{a+}^{(1-\beta)(n-\alpha)} f\right)^{(n-k)}(a) = 0$, $k = 1, ..., n-1$. Then*

$$\Gamma_f(\mu_1, \mu_2) \leq \frac{\left\| {}^{H}\mathbb{D}_{a+}^{\alpha,\beta} f \right\|_{\infty,[a,b]}}{\Gamma(\alpha+1)} \int_X q(x)^{1-\alpha} (p(x) - aq(x))^{\alpha} d\lambda. \tag{5.57}$$

It follows

Corollary 5.12 (to Theorem 5.4) *Case of* $\psi(x) = e^x$. *So, additionally assume that* $f_{e^x}^{[n-k]}\left(I_{a+}^{(1-\beta)(n-\alpha);e^x} f\right)(a) = 0$, $k = 1, ..., n-1$. *Then*

$$\Gamma_f(\mu_1, \mu_2) \leq \frac{\left\| {}^H\mathbb{D}_{a+}^{\alpha,\beta;e^x} f \right\|_{\infty,[a,b]}}{\Gamma(\alpha+1)} \int_X q(x) \left(e^{\frac{p(x)}{q(x)}} - e^a\right)^\alpha d\lambda. \tag{5.58}$$

5.4 Background-III

Next we use the following. Let f by a convex function from $(0, +\infty)$ into $\mathbb{R}$ which is strictly convex at 1 with $f(1) = 0$. Let $\left(\mathbb{R}^2, \mathcal{B}^2, \lambda\right)$ be the measure space, where λ is the product Lebesgue measure on $\left(\mathbb{R}^2, \mathcal{B}^2\right)$ with $\mathcal{B}$ being the Borel σ-field. And let $X, Y : \Omega \to \mathbb{R}$ be random variables on the probability space (Ω, P). Consider the probability distributions μ_{XY} and $\mu_X \times \mu_Y$ on $\mathbb{R}^2$, where $\mu_{XY}, \mu_X \times \mu_Y$ stand for the joint distribution of X and Y and their marginal distributions, respectively.

Here we assume as existing the following probability density functions, the joint pdf of μ_{XY} to be $t(x, y)$, $x, y \in \mathbb{R}$, the pdf of μ_X to be $p(x)$ and pdf of μ_Y to be $q(y)$. Clearly $\mu_X \times \mu_Y$ has pdf $p(x) q(y)$. Here we further assume that $0 < a \leq \frac{t}{pq} \leq b$, a.e. on $\mathbb{R}^2$ and $a \leq 1 \leq b$.

The quantity

$$\Gamma_f(\mu_{XY}, \mu_X \times \mu_Y) = \int_{\mathbb{R}^2} p(x) q(y) f\left(\frac{t(x,y)}{p(x) q(y)}\right) d\lambda(x,y), \tag{5.59}$$

is the Csiszar's distance or f-divergence between μ_{XY} and $\mu_X \times \mu_Y$.

Here X, Y are less dependent the closer the distributions μ_{XY} and $\mu_X \times \mu_Y$ are, thus $\Gamma_f(\mu_{XY}, \mu_X \times \mu_Y)$ can be considered as a measure of dependence of X and Y. For $f(u) = u \log_2 u$ we obtain the mutual information of X, and Y,

$$I(X, Y) = I(\mu_{XY} \| \mu_X \times \mu_Y) = \Gamma_{u \log_2 u}(\mu_{XY}, \mu_X \times \mu_Y),$$

see [6]. For $f(u) = (u-1)^2$ we get the mean square contingency:

$$\varphi^2(X, Y) = \Gamma_{(u-1)^2}(\mu_{XY}, \mu_X \times \mu_Y),$$

see [11]. In the last we need $\mu_{XY} \ll \mu_X \times \mu_Y$, where $\ll$ denotes absolute continuity, but to cover the case of $\mu_{XY} \not\ll \mu_X \times \mu_Y$ we set $\varphi^2(X, Y) = +\infty$, then the last formula is always valid.

Clearly here $\mu_{XY}, \mu_X \times \mu_Y \ll \lambda$, also $\Gamma_f(\mu_{XY}, \mu_X \times \mu_Y) > 0$ with equality only when $\mu_{XY} = \mu_X \times \mu_Y$, i.e. when X, Y are independent r.v.'s.

5.5 Main Results-II

Here f and the whole setting is as in Sect. 5.4, the Background-III. Additionally, we assume that $\psi, f \in C^n([a,b])$, with ψ being increasing such that $\psi'(x) \neq 0$ over $[a,b]$, where $n-1<\alpha<n$, $0 \le \beta \le 1$, and $\gamma = \alpha + \beta(n-\alpha)$. Furthermore we stress that $0 < a \le \frac{t}{pq} \le b$, a.e. on $\mathbb{R}^2$.

Next we present estimations for $\Gamma_f(\mu_{XY}, \mu_X \times \mu_Y)$. We apply our results of Sect. 5.3, the Main Results-I.

We give first left side results:

Theorem 5.13 *Additionally assume that* $f_\psi^{[n-k]}\left(I_{a+}^{(1-\beta)(n-\alpha);\psi} f\right)(a) = 0$, $k = 1, ..., n-1$. *Then*

$$\Gamma_f(\mu_{XY}, \mu_X \times \mu_Y) \le \frac{\left\| {}^H\mathbb{D}_{a+}^{\alpha,\beta;\psi} f \right\|_{\infty,[a,b]}}{\Gamma(\alpha+1)}$$

$$\int_{\mathbb{R}^2} (p(x)q(y)) \left(\psi\left(\frac{t(x,y)}{p(x)q(y)}\right) - \psi(a)\right)^{\alpha} d\lambda(x,y). \tag{5.60}$$

Proof By Theorem 5.4. ■

We give the L_1-analog:

Theorem 5.14 *All as in Theorem 5.13, with* $\alpha > 1$. *Then*

$$\Gamma_f(\mu_{XY}, \mu_X \times \mu_Y) \le \frac{\left\| {}^H\mathbb{D}_{a+}^{\alpha,\beta;\psi} f \right\|_{L_1([a,b],\psi)}}{\Gamma(\alpha)}$$

$$\int_{\mathbb{R}^2} (p(x)q(y)) \left(\psi\left(\frac{t(x,y)}{p(x)q(y)}\right) - \psi(a)\right)^{\alpha-1} d\lambda(x,y). \tag{5.61}$$

Proof By Theorem 5.5. ■

The L_q-analog follows:

Theorem 5.15 *All as in Theorem 5.13. Let also* $p, q > 1 : \frac{1}{p} + \frac{1}{q} = 1$, *with* $\alpha > \frac{1}{q}$. *Then*

$$\Gamma_f(\mu_{XY}, \mu_X \times \mu_Y) \le \frac{\left\| {}^H\mathbb{D}_{a+}^{\alpha,\beta;\psi} f \right\|_{L_q([a,b],\psi)}}{\Gamma(\alpha)(p(\alpha-1)+1)^{\frac{1}{p}}}$$

$$\int_{\mathbb{R}^2} (p(x)q(y)) \left(\psi\left(\frac{t(x,y)}{p(x)q(y)}\right) - \psi(a)\right)^{\alpha-\frac{1}{q}} d\lambda(x,y). \tag{5.62}$$

Proof By Theorem 5.6. ■

We continue with right side results:

Theorem 5.16 *Additionally assume that* $f_{\psi}^{[n-k]}\left(I_{b-}^{(1-\beta)(n-\alpha);\psi}f\right)(b)=0$, $k=1,...,n-1$. *Then*

$$\Gamma_f\left(\mu_{XY},\mu_X\times\mu_Y\right)\leq\frac{\left\|{}^H\mathbb{D}_{b-}^{\alpha,\beta;\psi}f\right\|_{\infty,[a,b]}}{\Gamma\left(\alpha+1\right)}$$

$$\int_{\mathbb{R}^2}\left(p\left(x\right)q\left(y\right)\right)\left(\psi\left(b\right)-\psi\left(\frac{t\left(x,y\right)}{p\left(x\right)q\left(y\right)}\right)\right)^{\alpha}d\lambda\left(x,y\right). \tag{5.63}$$

Proof By Theorem 5.7. ■

It follows the L_1-analog:

Theorem 5.17 *All as in Theorem 5.16,* $\alpha>1$. *Then*

$$\Gamma_f\left(\mu_{XY},\mu_X\times\mu_Y\right)\leq\frac{\left\|{}^H\mathbb{D}_{b-}^{\alpha,\beta;\psi}f\right\|_{L_1([a,b],\psi)}}{\Gamma\left(\alpha\right)}$$

$$\int_{\mathbb{R}^2}\left(p\left(x\right)q\left(y\right)\right)\left(\psi\left(b\right)-\psi\left(\frac{t\left(x,y\right)}{p\left(x\right)q\left(y\right)}\right)\right)^{\alpha-1}d\lambda\left(x,y\right). \tag{5.64}$$

Proof By Theorem 5.8. ■

The L_q-analog comes next from the right side.

Theorem 5.18 *All as in Theorem 5.16. Let also* $p,q>1:\frac{1}{p}+\frac{1}{q}=1$, *with* $\alpha>\frac{1}{q}$. *Then*

$$\Gamma_f\left(\mu_{XY},\mu_X\times\mu_Y\right)\leq\frac{\left\|{}^H\mathbb{D}_{b-}^{\alpha,\beta;\psi}f\right\|_{L_q([a,b],\psi)}}{\Gamma\left(\alpha\right)\left(p\left(\alpha-1\right)+1\right)^{\frac{1}{p}}}$$

$$\int_{\mathbb{R}^2}\left(p\left(x\right)q\left(y\right)\right)\left(\psi\left(b\right)-\psi\left(\frac{t\left(x,y\right)}{p\left(x\right)q\left(y\right)}\right)\right)^{\alpha-\frac{1}{q}}d\lambda\left(x,y\right). \tag{5.65}$$

Proof By Theorem 5.9. ■

We make

Remark 5.19 (see Remark 5.10) (I) Assume that ${}^H\mathbb{D}_{a+}^{\alpha,\beta;\psi}f\underset{(\leq 0)}{\geq}0$ over $[a,b]$, then

$$\Gamma_f(\mu_{XY}, \mu_X \times \mu_Y) \underset{(\leq)}{\geq} \sum_{k=1}^{n-1} \frac{f_{\psi}^{[n-k]}\left(I_{a+}^{(1-\beta)(n-\alpha);\psi} f\right)(a)}{\Gamma(\gamma - k + 1)} \tag{5.66}$$

$$\int_{\mathbb{R}^2} (p(x) q(y)) \left(\psi\left(\frac{t(x,y)}{p(x) q(y)}\right) - \psi(a)\right)^{\gamma-k} d\lambda(x,y).$$

(II) Assume that ${}^H\mathbb{D}_{b-}^{\alpha,\beta;\psi} f \underset{(\leq 0)}{\geq} 0$ over $[a,b]$, then

$$\Gamma_f(\mu_{XY}, \mu_X \times \mu_Y) \underset{(\leq)}{\geq} \sum_{k=1}^{n-1} \frac{(-1)^k f_{\psi}^{[n-k]}\left(I_{b-}^{(1-\beta)(n-\alpha);\psi} f\right)(b)}{\Gamma(\gamma - k + 1)} \tag{5.67}$$

$$\int_{\mathbb{R}^2} (p(x) q(y)) \left(\psi(b) - \psi\left(\frac{t(x,y)}{p(x) q(y)}\right)\right)^{\gamma-k} d\lambda(x,y).$$

We state

Corollary 5.20 (to Theorem 5.13) *Case of* $\psi(x) = x$. *So, additionally assume that* $\left(I_{a+}^{(1-\beta)(n-\alpha)} f\right)^{(n-k)}(a) = 0$, $k = 1, ..., n-1$. *Then*

$$\Gamma_f(\mu_{XY}, \mu_X \times \mu_Y) \leq \frac{\left\| {}^H\mathbb{D}_{a+}^{\alpha,\beta} f \right\|_{\infty,[a,b]}}{\Gamma(\alpha + 1)} \tag{5.68}$$

$$\int_{\mathbb{R}^2} (p(x) q(y))^{1-\alpha} (t(x,y) - a p(x) q(y))^{\alpha} d\lambda(x,y).$$

It follows

Corollary 5.21 (to Theorem 5.13) *Case of* $\psi(x) = e^x$. *So, additionally assume that* $f_{e^x}^{[n-k]}\left(I_{a+}^{(1-\beta)(n-\alpha);e^x} f\right)(a) = 0$, $k = 1, ..., n-1$. *Then*

$$\Gamma_f(\mu_{XY}, \mu_X \times \mu_Y) \leq \frac{\left\| {}^H\mathbb{D}_{a+}^{\alpha,\beta;e^x} f \right\|_{\infty,[a,b]}}{\Gamma(\alpha + 1)} \tag{5.69}$$

$$\int_{\mathbb{R}^2} (p(x) q(y)) \left(e^{\frac{t(x,y)}{p(x) q(y)}} - e^a\right)^{\alpha} d\lambda(x,y).$$

References

1. Almeida, R.: A Caputo fractional derivative of a function with respect to another function. Commun. Nonlinear Sci. Numer. Simul. **44**, 460–481 (2017)
2. Anastassiou, G.A.: Fractional and other approximation of Csiszar's f-divergence. In: Altomare, F. (ed.) Proceedings of FAAT 04. Rendiconti Del Circolo Matematica Di Palermo, Serie II, Suppl. 76, pp. 197–212 (2005)
3. Anastassiou, G.: ψ-Hilfer fractional approximations of Csiszar's f-divergence. Adv. Nonlinear Var. Inequal. **24**(2), 51–66 (2021)
4. Barnett, N.S., Gerone, P., Dragomir, S.S., Sofo, A.: Approximating Csiszar's f-divergence by the use of Taylor's formula with integral remainder (paper #10, pp. 16). In: Dragomir, S.S. (ed.) Inequalities for Csiszar f-Divergence in Information Theory, Victoria University, Melbourne, Australia (2000) On-line: http://rgmia.vu.edu.au
5. Csiszar, I.: Eine Informationstheoretische Ungleichung und ihre Anwendung auf den Beweis der Ergodizität von Markoffschen Ketten, Magyar Trud. Akad. Mat. Kutato Int. Közl. **8**, 85–108 (1963)
6. Csiszar, I.: Information-type measures of difference of probability distributions and indirect observations. Studia Math. Hungarica **2**, 299–318 (1967)
7. Dragomir, S.S.: (ed) Inequalities for Csiszar f-Divergence in Information Theory. Victoria University, Melbourne, Australia (2000). On line: http://rgmia.vu.edu.au
8. Kilbas, A.A., Srivastava, H.M., Trujillo, J.J.: Theory and Applications of Fractional Differentiation Equations. North Holland, Amsterdam, New York (2006)
9. Kullback, S.: Information Theory and Statistics. Wiley, New York (1959)
10. Kullback, S., Leibler, R.: On information and sufficiency. Ann. Math. Statist. **22**, 79–86 (1951)
11. Rényi, A.: On measures of dependence. Acta Math. Acad. Sci. Hungar. **10**, 441–451 (1959)
12. Rényi, A..: On measures of entropy and information. In: Proceedings of the 4rth Berkeley Symposium on Mathematical Statistic and Probability, I, Berkeley, pp. 547–561 (1960)
13. Samko, S.G., Kilbas, A.A., Marichev, O.I.: Fractional Integrals and Derivatives: Theory and Applications. Gordon and Breach Science Publishers, Switzerland (1993)
14. Tomovski, Ž, Hilfer, R., Srivastava, H.M.: Fractional and operational calculus with generalized fractional derivative operators and Mittag-Leffler type functions. Integ. Transforms Spec. Funct. **21**(11), 797–814 (2010)
15. Vanterler da, J., Sousa, J., Capelas de Oliveira, E.: On the ψ-Hilfer fractional derivative, Commun. Nonlinear Sci. Numer. Simul. **60**, 72–91 (2018)

Chapter 6
Generalized Hilfer Fractional Self Adjoint Operator Inequalities

We give here ψ-Hilfer and Hilfer fractional self adjoint operator inner product comparison, Poincaré, Sobolev and Opial type inequalities. At first we give right and left ψ-Hilfer fractional representation formulae in the self adjoint operator sense. Operator inequalities are based in the self adjoint operator order over a Hilbert space. It follows [3].

6.1 Background—I

Let A be a selfadjoint linear operator on a complex Hilbert space $(H; \langle \cdot, \cdot \rangle)$. The Gelfand map establishes a $*-$isometrically isomorphism Φ between the set $C(Sp(A))$ of all continuous functions defined on the spectrum of A, denoted $Sp(A)$, and the C^*-algebra $C^*(A)$ generated by A and the identity operator 1_H on H as follows (see e.g. [7, p. 3]):

For any $f, g \in C(Sp(A))$ and any $\alpha, \beta \in \mathbb{C}$ we have

(i) $\Phi(\alpha f + \beta g) = \alpha \Phi(f) + \beta \Phi(g)$;

(ii) $\Phi(fg) = \Phi(f)\Phi(g)$ (the operation composition is on the right) and $\Phi\left(\overline{f}\right) = (\Phi(f))^*$;

(iii) $\|\Phi(f)\| = \|f\| := \sup_{t \in Sp(A)} |f(t)|$;

(iv) $\Phi(f_0) = 1_H$ and $\Phi(f_1) = A$, where $f_0(t) = 1$ and $f_1(t) = t$, for $t \in Sp(A)$.

With this notation we define

$$f(A) := \Phi(f), \text{ for all } f \in C(Sp(A)),$$

and we call it the continuous functional calculus for a selfadjoint operator A.

G. A. Anastassiou, *Unification of Fractional Calculi with Applications*, Studies in Systems, Decision and Control 398,
https://doi.org/10.1007/978-3-030-86920-5_6

If A is a selfadjoint operator and f is a real valued continuous function on $Sp(A)$ then $f(t) \geq 0$ for any $t \in Sp(A)$ implies that $f(A) \geq 0$, i.e. $f(A)$ is a positive operator on H. Moreover, if both f and g are real valued continuous functions on $Sp(A)$ then the following important property holds:

(P) $f(t) \geq g(t)$ for any $t \in Sp(A)$, implies that $f(A) \geq g(A)$ in the operator order of $B(H)$ (the Banach algebra of all bounded linear operators from H into itself).

Equivalently, we use (see [6], pp. 7–8):

Let U be a selfadjoint operator on the complex Hilbert space $(H, \langle \cdot, \cdot \rangle)$ with the spectrum $Sp(U)$ included in the interval $[m, M]$ for some real numbers $m < M$ and $\{E_\lambda\}_\lambda$ be its spectral family.

Then for any continuous function $f : [m, M] \to \mathbb{C}$, it is well known that we have the following spectral representation in terms of the Riemann-Stieljes integral:

$$\langle f(U)x, y\rangle = \int_{m-0}^{M} f(\lambda)\, d\left(\langle E_\lambda x, y\rangle\right), \tag{6.1}$$

for any $x, y \in H$. The function $g_{x,y}(\lambda) := \langle E_\lambda x, y\rangle$ is of bounded variation on the interval $[m, M]$, and

$$g_{x,y}(m-0) = 0 \text{ and } g_{x,y}(M) = \langle x, y\rangle, \tag{6.2}$$

for any $x, y \in H$. Furthermore, it is known that $g_x(\lambda) := \langle E_\lambda x, x\rangle$ is increasing and right continuous on $[m, M]$.

We have also the formula

$$\langle f(U)x, x\rangle = \int_{m-0}^{M} f(\lambda)\, d\left(\langle E_\lambda x, x\rangle\right), \ \forall x \in H. \tag{6.3}$$

As a symbol we can write

$$f(U) = \int_{m-0}^{M} f(\lambda)\, dE_\lambda. \tag{6.4}$$

Above, $m = \min\{\lambda | \lambda \in Sp(U)\} := \min Sp(U)$, $M = \max\{\lambda | \lambda \in Sp(U)\} := \max Sp(U)$. The projections $\{E_\lambda\}_{\lambda \in \mathbb{R}}$, are called the spectral family of A, with the properties:

(a) $E_\lambda \leq E_{\lambda'}$ for $\lambda \leq \lambda'$;

(b) $E_{m-0} = 0_H$ (zero operator), $E_M = 1_H$ (identity operator) and $E_{\lambda+0} = E_\lambda$ for all $\lambda \in \mathbb{R}$.

Furthermore

$$E_\lambda := \varphi_\lambda(U), \ \forall \lambda \in \mathbb{R},$$

is a projection which reduces U, with

$$\varphi_{\lambda}(s) := \begin{cases} 1, \text{ for } -\infty < s \leq \lambda, \\ 0, \text{ for } \lambda < s < +\infty. \end{cases} \tag{6.5}$$

The spectral family $\{E_{\lambda}\}_{\lambda \in \mathbb{R}}$ determines uniquely the self-adjoint operator U and vice versa.

For more on the topic see [8], pp. 256–266, and for more details see there pp. 157–266. See also [5].

Some more basics are given (we follow [6], pp. 1–5):

Let $(H; \langle \cdot, \cdot \rangle)$ be a Hilbert space over $\mathbb{C}$. A bounded linear operator A defined on H is selfjoint, i.e., $A = A^*$, iff $\langle Ax, x \rangle \in \mathbb{R}$, $\forall\, x \in H$, and if A is selfadjoint, then

$$\|A\| = \sup_{x \in H : \|x\| = 1} |\langle Ax, x \rangle| . \tag{6.6}$$

Let A, B be selfadjoint operators on H. Then $A \leq B$ iff $\langle Ax, x \rangle \leq \langle Bx, x \rangle$, $\forall\, x \in H$.

In particular, A is called positive if $A \geq 0$.

Denote by

$$\mathcal{P} := \left\{ \varphi(s) := \sum_{k=0}^{n} \alpha_k s^k | n \geq 0, \alpha_k \in \mathbb{C}, 0 \leq k \leq n \right\}. \tag{6.7}$$

If $A \in \mathcal{B}(H)$ is selfadjoint, and $\varphi(s) \in \mathcal{P}$ has real coefficients, then $\varphi(A)$ is selfadjoint, and

$$\|\varphi(A)\| = \max \{ |\varphi(\lambda)|, \lambda \in Sp(A) \} . \tag{6.8}$$

If φ is any function defined on $\mathbb{R}$ we define

$$\|\varphi\|_A := \sup \{ |\varphi(\lambda)|, \lambda \in Sp(A) \} . \tag{6.9}$$

If A is selfadjoint operator on Hilbert space H and φ is continuous and given that $\varphi(A)$ is selfadjoint, then $\|\varphi(A)\| = \|\varphi\|_A$. And if φ is a continuous real valued function so it is $|\varphi|$, then $\varphi(A)$ and $|\varphi|(A) = |\varphi(A)|$ are selfadjoint operators (by [6], p. 4, Theorem 7).

Hence it holds

$$\| |\varphi(A)| \| = \| |\varphi| \|_A = \sup \{ ||\varphi(\lambda)||, \lambda \in Sp(A) \} \tag{6.10}$$

$$= \sup \{ |\varphi(\lambda)|, \lambda \in Sp(A) \} = \|\varphi\|_A = \|\varphi(A)\| ,$$

that is

$$\| |\varphi(A)| \| = \|\varphi(A)\| . \tag{6.11}$$

For a selfadjoint operator $A \in \mathcal{B}(H)$ which is positive, there exists a unique positive selfadjoint operator $B := \sqrt{A} \in \mathcal{B}(H)$ such that $B^2 = A$, that is $\left(\sqrt{A}\right)^2 = A$. We call B the square root of A.

Let $A \in \mathcal{B}(H)$, then A^*A is selfadjoint and positive. Define the "operator absolute value" $|A| := \sqrt{A^*A}$. If $A = A^*$, then $|A| = \sqrt{A^2}$.

For a continuous real valued function φ we observe the following:

$$|\varphi(A)| \text{ (the functional absolute value) } = \int_{m-0}^{M} |\varphi(\lambda)|\, dE_\lambda =$$

$$\int_{m-0}^{M} \sqrt{(\varphi(\lambda))^2} dE_\lambda = \sqrt{(\varphi(A))^2} = |\varphi(A)| \text{ (operator absolute value),}$$

where A is a selfadjoint operator.

That is we have

$$|\varphi(A)| \text{ (functional absolute value) } = |\varphi(A)| \text{ (operator absolute value).} \quad (6.12)$$

6.2 Background—II

Let $-\infty < a < b < \infty$, the left and right Riemann-Liouville fractional integrals of order $\alpha \in \mathbb{C}$ ($\mathcal{R}(\alpha) > 0$) are defined by

$$\left(I_{a+}^{\alpha} f\right)(x) = \frac{1}{\Gamma(\alpha)} \int_a^x (x-t)^{\alpha-1} f(t)\, dt, \quad (6.13)$$

$x > a$; where Γ stands for the gamma function,

and

$$\left(I_{b-}^{\alpha} f\right)(x) = \frac{1}{\Gamma(\alpha)} \int_x^b (t-x)^{\alpha-1} f(t)\, dt, \quad (6.14)$$

$x < b$.

The Riemann-Liouville left and right fractional derivatives of order $\alpha \in \mathbb{C}$ ($\mathcal{R}(\alpha) \geq 0$) are defined by

$$\left(\Delta_{a+}^{\alpha} y\right)(x) = \left(\frac{d}{dx}\right)^n \left(I_{a+}^{n-\alpha} y\right)(x) = \frac{1}{\Gamma(n-\alpha)} \left(\frac{d}{dx}\right)^n \int_a^x (x-t)^{n-\alpha-1} y(t)\, dt \quad (6.15)$$

($n = \lceil \mathcal{R}(\alpha) \rceil$, $\lceil \cdot \rceil$ means ceiling of the number; $x > a$)

$$\left(\Delta_{b-}^{\alpha} y\right)(x) = (-1)^n \left(\frac{d}{dx}\right)^n \left(I_{b-}^{n-\alpha} y\right)(x) =$$

$$\frac{(-1)^n}{\Gamma(n-\alpha)}\left(\frac{d}{dx}\right)^n \int_x^b (t-x)^{n-\alpha-1}\, y(t)\, dt \tag{6.16}$$

($n = \lceil \mathcal{R}(\alpha) \rceil$; $x < b$), respectively, where $\mathcal{R}(\alpha)$ is the real part of α.

In particular, when $\alpha = n \in \mathbb{Z}_+$, then

$$\left(\Delta_{a+}^{0} y\right)(x) = \left(\Delta_{b-}^{0} y\right)(x) = y(x);$$

$$\left(\Delta_{a+}^{n} y\right)(x) = y^{(n)}(x), \text{ and } \left(\Delta_{b-}^{n} y\right)(x) = (-1)^n y^{(n)}(x), \ n \in \mathbb{N},$$

see [11].

Let $\alpha > 0$, $I = [a, b] \subset \mathbb{R}$, f an integrable function defined on I and $\psi \in C^1(I)$ an increasing function such that $\psi'(x) \neq 0$, for all $x \in I$. Left fractional integrals and left Riemann-Liouville fractional derivatives of a function f with respect to another function ψ are defined as ([9, 11])

$$I_{a+}^{\alpha,\psi} f(x) = \frac{1}{\Gamma(\alpha)} \int_a^x \psi'(t)\,(\psi(x) - \psi(t))^{\alpha-1} f(t)\, dt, \tag{6.17}$$

and

$$\Delta_{a+}^{\alpha,\psi} f(x) = \left(\frac{1}{\psi'(x)}\frac{d}{dx}\right)^n I_{a+}^{n-\alpha,\psi} f(x) = \tag{6.18}$$

$$\frac{1}{\Gamma(n-\alpha)}\left(\frac{1}{\psi'(x)}\frac{d}{dx}\right)^n \int_a^x \psi'(t)\,(\psi(x) - \psi(t))^{n-\alpha-1} f(t)\, dt,$$

respectively, where $n = \lceil \alpha \rceil$.

Similarly, we define the right ones:

$$I_{b}^{\alpha,\psi} f(x) = \frac{1}{\Gamma(\alpha)} \int_x^b \psi'(t)\,(\psi(t) - \psi(x))^{\alpha-1} f(t)\, dt, \tag{6.19}$$

and

$$\Delta_{b-}^{\alpha,\psi} f(x) = \left(-\frac{1}{\psi'(x)}\frac{d}{dx}\right)^n I_{b-}^{n-\alpha,\psi} f(x) =$$

$$\frac{1}{\Gamma(n-\alpha)}\left(-\frac{1}{\psi'(x)}\frac{d}{dx}\right)^n \int_x^b \psi'(t)\,(\psi(t) - \psi(x))^{n-\alpha-1} f(t)\, dt. \tag{6.20}$$

The following semigroup property holds; if $\alpha, \beta > 0$, $f \in C(I)$, then

$$I_{a+}^{\alpha,\psi} I_{a+}^{\beta,\psi} f = I_{a+}^{\alpha+\beta,\psi} f \text{ and } I_{b-}^{\alpha,\psi} I_{b-}^{\beta,\psi} f = I_{b-}^{\alpha+\beta,\psi} f.$$

Next let again $\alpha > 0$, $n = \lceil \alpha \rceil$, $I = [a, b]$, $f, \psi \in C^n(I)$: ψ is increasing and $\psi'(x) \neq 0$, for all $x \in I$. The left ψ-Caputo fractional derivative of f of order α is given by ([1])

$$^{C}D_{a+}^{\alpha,\psi} f(x) = I_{a+}^{n-\alpha,\psi} \left(\frac{1}{\psi'(x)} \frac{d}{dx} \right)^n f(x), \tag{6.21}$$

and the right ψ-Caputo fractional derivative ([1])

$$^{C}D_{b-}^{\alpha,\psi} f(x) = I_{b-}^{n-\alpha,\psi} \left(-\frac{1}{\psi'(x)} \frac{d}{dx} \right)^n f(x). \tag{6.22}$$

We set

$$f_{\psi}^{[n]}(x) := f_{\psi}^{(n)} f(x) := \left(\frac{1}{\psi'(x)} \frac{d}{dx} \right)^n f(x). \tag{6.23}$$

Clearly, when $\alpha = m \in \mathbb{N}$ we have

$$^{C}D_{a+}^{\alpha,\psi} f(x) = f_{\psi}^{[m]}(x) \text{ and } {}^{C}D_{b-}^{\alpha,\psi} f(x) = (-1)^m f_{\psi}^{[m]}(x),$$

and if $\alpha \notin \mathbb{N}$, then

$$^{C}D_{a+}^{\alpha,\psi} f(x) = \frac{1}{\Gamma(n-\alpha)} \int_a^x \psi'(t) (\psi(x) - \psi(t))^{n-\alpha-1} f_{\psi}^{[n]}(t)\, dt, \tag{6.24}$$

and

$$^{C}D_{b-}^{\alpha,\psi} f(x) = \frac{(-1)^n}{\Gamma(n-\alpha)} \int_x^b \psi'(t) (\psi(t) - \psi(x))^{n-\alpha-1} f_{\psi}^{[n]}(t)\, dt. \tag{6.25}$$

If $\psi(x) = x$, then we get the usual left and right Caputo fractional derivatives

$$^{C}D_{a+}^{m} f(x) = f^{(m)}(x), \quad {}^{C}D_{b-}^{m} f(x) = (-1)^m f^{(m)}(x),$$

for $m \in \mathbb{N}$, and ($\alpha \notin \mathbb{N}$)

$$D_{*a}^{\alpha} f(x) = {}^{C} D_{a+}^{\alpha} f(x) = \frac{1}{\Gamma(n-\alpha)} \int_a^x (x-t)^{n-\alpha-1} f^{(n)}(t)\, dt, \tag{6.26}$$

$$D_{b-}^{\alpha}(x) = {}^{C} D_{b-}^{\alpha} f(x) = \frac{(-1)^n}{\Gamma(n-\alpha)} \int_x^b (t-x)^{n-\alpha-1} f^{(n)}(t)\, dt. \tag{6.27}$$

Also we set

$$^{C}D_{a+}^{0,\psi} f(x) = {}^{C} D_{b-}^{0,\psi} f(x) = f(x).$$

Next we will deal with the ψ-Hilfer fractional derivative.

Definition 6.1 ([4]) Let $n-1<\alpha<n$, $n\in\mathbb{N}$, $I=[a,b]\subset\mathbb{R}$ and $f,\psi\in C^n([a,b])$, ψ is increasing and $\psi'(x)\neq 0$, for all $x\in I$. The ψ-Hilfer fractional derivative (left-sided and right-sided) ${}^{H}\mathbb{D}_{a+(b-)}^{\alpha,\beta;\psi}f$ of order α and type $0\leq\beta\leq 1$, respectively, are defined by

$$ {}^{H}\mathbb{D}_{a+}^{\alpha,\beta;\psi}f(x)=I_{a+}^{\beta(n-\alpha);\psi}\left(\frac{1}{\psi'(x)}\frac{d}{dx}\right)^{n}I_{a+}^{(1-\beta)(n-\alpha);\psi}f(x), \tag{6.28}$$

and

$$ {}^{H}\mathbb{D}_{b-}^{\alpha,\beta;\psi}f(x)=I_{b-}^{\beta(n-\alpha);\psi}\left(-\frac{1}{\psi'(x)}\frac{d}{dx}\right)^{n}I_{b-}^{(1-\beta)(n-\alpha);\psi}f(x),\ x\in[a,b]. \tag{6.29}$$

The original Hilfer fractional derivatives ([12]) come from $\psi(x)=x$, and are denoted by ${}^{H}\mathbb{D}_{a+}^{\alpha,\beta}f(x)$ and ${}^{H}\mathbb{D}_{b-}^{\alpha,\beta}f(x)$.

When $\beta=0$, we get Riemann-Liouville fractional derivatives, while when $\beta=1$ we have Caputo type fractional derivatives.

We define $\gamma=\alpha+\beta(n-\alpha)$. We notice that $n-1<\alpha\leq\alpha+\beta(n-\alpha)\leq\alpha+n-\alpha=n$, hence $\lceil\gamma\rceil=n$. We can easily write that ([4])

$$ {}^{H}\mathbb{D}_{a+}^{\alpha,\beta;\psi}f(x)=I_{a+}^{\gamma-\alpha;\psi}\Delta_{a+}^{\gamma;\psi}f(x), \tag{6.30}$$

and

$$ {}^{H}\mathbb{D}_{b-}^{\alpha,\beta;\psi}f(x)=I_{b-}^{\gamma-\alpha;\psi}\Delta_{b-}^{\gamma;\psi}f(x),\ x\in[a,b]. \tag{6.31}$$

We have that ([4])

$$\Delta_{a+}^{\gamma,\psi}f(x)=\left(\frac{1}{\psi'(x)}\frac{d}{dx}\right)^{n}I_{a+}^{(1-\beta)(n-\alpha);\psi}f(x), \tag{6.32}$$

and

$$\Delta_{b-}^{\gamma,\psi}f(x)=\left(-\frac{1}{\psi'(x)}\frac{d}{dx}\right)^{n}I_{b-}^{(1-\beta)(n-\alpha);\psi}f(x). \tag{6.33}$$

In particular, when $0<\alpha<1$ and $0\leq\beta\leq 1$; $\gamma=\alpha+\beta(1-\alpha)$, we have that

$$ {}^{H}\mathbb{D}_{a+}^{\alpha,\beta;\psi}f(x)=\frac{1}{\Gamma(\gamma-\alpha)}\int_{a}^{x}\psi'(t)(\psi(x)-\psi(t))^{\gamma-\alpha-1}\Delta_{a+}^{\gamma;\psi}f(t)\,dt, \tag{6.34}$$

and

$$ {}^{H}\mathbb{D}_{b-}^{\alpha,\beta;\psi}f(x)=\frac{1}{\Gamma(\gamma-\alpha)}\int_{x}^{b}\psi'(t)(\psi(t)-\psi(x))^{\gamma-\alpha-1}\Delta_{b-}^{\gamma;\psi}f(t)\,dt, \tag{6.35}$$

$x\in[a,b]$.

Remark 6.2 ([4]) Let $\mu = n(1-\beta) + \beta\alpha$, then $\lceil \mu \rceil = n$.
Assume that $g(x) = I_{a+}^{(1-\beta)(n-\alpha);\psi} f(x) \in C^n([a,b])$, we have that

$$^H\mathbb{D}_{a+}^{\alpha,\beta;\psi} f(x) = I_{a+}^{n-\mu;\psi} \left(\frac{1}{\psi'(x)} \frac{d}{dx} \right)^n g(x). \tag{6.36}$$

Thus

$$^H\mathbb{D}_{a+}^{\alpha,\beta;\psi} f = {}^C D_{a+}^{\mu;\psi} g(x) = {}^C D_{a+}^{\mu;\psi} \left[I_{a+}^{(1-\beta)(n-\alpha);\psi} f(x) \right]. \tag{6.37}$$

Assume that $w(x) = I_{b-}^{(1-\beta)(n-\alpha);\psi} f(x) \in C^n([a,b])$. Hence

$$^H\mathbb{D}_{b-}^{\alpha,\beta;\psi} f(x) = I_{b-}^{\beta(n-\alpha);\psi} \left(-\frac{1}{\psi'(x)} \frac{d}{dx} \right)^n w(x) = I_{b-}^{n-\mu;\psi} \left(-\frac{1}{\psi'(x)} \frac{d}{dx} \right)^n w(x). \tag{6.38}$$

Thus

$$^H\mathbb{D}_{b-}^{\alpha,\beta;\psi} f = {}^C D_{b-}^{\mu;\psi} w(x) = {}^C D_{b-}^{\mu;\psi} \left(I_{b-}^{(1-\beta)(n-\alpha);\psi} f(x) \right). \tag{6.39}$$

We mention the simplified ψ-Hilfer fractional Taylor formulae:

Theorem 6.3 *(see also [4]) Let $\psi, f \in C^n([a,b])$, with ψ being increasing such that $\psi'(x) \neq 0$ over $[a,b]$, where $n-1 < \alpha < n$, $0 \leq \beta \leq 1$, and $\gamma = \alpha + \beta(n-\alpha)$, $x \in [a,b]$. Then*

$$f(x) - \sum_{k=1}^{n-1} \frac{(\psi(x) - \psi(a))^{\gamma-k}}{\Gamma(\gamma-k+1)} f_\psi^{[n-k]} \left(I_{a+}^{(1-\beta)(n-\alpha);\psi} f \right)(a) =$$

$$\frac{1}{\Gamma(\alpha)} \int_a^x \psi'(t) (\psi(x) - \psi(t))^{\alpha-1}\, {}^H\mathbb{D}_{a+}^{\alpha,\beta;\psi} f(t)\, dt, \tag{6.40}$$

and

$$f(x) - \sum_{k=1}^{n-1} \frac{(-1)^k (\psi(b) - \psi(x))^{\gamma-k}}{\Gamma(\gamma-k+1)} f_\psi^{[n-k]} \left(I_{b-}^{(1-\beta)(n-\alpha);\psi} f \right)(b) =$$

$$\frac{1}{\Gamma(\alpha)} \int_x^b \psi'(t) (\psi(t) - \psi(x))^{\alpha-1}\, \mathbb{D}_{b-}^{\alpha,\beta;\psi} f(t)\, dt. \tag{6.41}$$

Here notice that $\left(I_{a+}^{(1-\beta)(n-\alpha);\psi} f \right)(a) = \left(I_{b-}^{(1-\beta)(n-\alpha);\psi} f \right)(b) = 0$.

6.3 Main Results

Let A be a selfadjoint operator on the Hilbert space H with the spectrum $Sp(A) \subseteq [m, M]$, $m < M$; $m, M \in \mathbb{R}$.

In the next we obtain ψ-Hilfer and Hilfer fractional comparison, Poincaré, Sobolev and Opial type inequalities in the operator order of $B(H)$ (the Banach algebra of all bounded linear operators from H into itself). All of our functions from now on in this chapter are real valued.

We give at first the following ψ-Hilfer fractional operator representation formulae:

Theorem 6.4 *Let A be a selfadjoint operator on the Hilbert space H with the spectrum $Sp(A) \subseteq [m, M]$ for some real numbers $m < M$, $\{E_\lambda\}_\lambda$ be its spectral family and I be a closed subinterval on $\mathbb{R}$ with $[m, M] \subset \overset{\circ}{I}$ (the interior of I). Let $\psi, f : I \to \mathbb{R}$ such that $\psi, f \in C^n([m, M])$, with ψ being increasing and $\psi'(\lambda) \neq 0$ over $[m, M]$, where $n - 1 < \alpha < n$, $0 \leq \beta \leq 1$, and $\gamma = \alpha + \beta(n - \alpha)$, $\lambda \in [m, M]$.*

Then

(i)

$$f(A) = \sum_{k=1}^{n-1} \frac{f_\psi^{[n-k]}\left(I_{m+}^{(1-\beta)(n-\alpha);\psi} f\right)(m)}{\Gamma(\gamma - k + 1)} (\psi(A) - \psi(m) 1_H)^{\gamma - k} \tag{6.42}$$

$$+R_n^{(1)}(f, \psi, m, M),$$

where

$$R_n^{(1)}(f, \psi, m, M) :=$$

$$\frac{1}{\Gamma(\alpha)} \int_{m-0}^{M} \left(\int_m^\lambda \psi'(t)(\psi(\lambda) - \psi(t))^{\alpha-1} {}^H\mathbb{D}_{m+}^{\alpha,\beta;\psi} f(t)\, dt \right) dE_\lambda, \tag{6.43}$$

and

(ii)

$$f(A) = \sum_{k=1}^{n-1} \frac{(-1)^k f_\psi^{[n-k]}\left(I_{M-}^{(1-\beta)(n-\alpha);\psi} f\right)(M)}{\Gamma(\gamma - k + 1)} (\psi(M) 1_H - \psi(A))^{\gamma - k} \tag{6.44}$$

$$+R_n^{(2)}(f, \psi, m, M),$$

where

$$R_n^{(2)}(f, \psi.m, M) :=$$

$$\frac{1}{\Gamma(\alpha)} \int_{m-0}^{M} \left(\int_\lambda^M \psi'(t)(\psi(t) - \psi(\lambda))^{\alpha-1} {}^H\mathbb{D}_{M-}^{\alpha,\beta;\psi} f(t)\, dt \right) dE_\lambda. \tag{6.45}$$

Proof By Theorem 6.3 we have

$$f(\lambda) = \sum_{k=1}^{n-1} \frac{(\psi(\lambda) - \psi(m))^{\gamma-k}}{\Gamma(\gamma-k+1)} f_{\psi}^{[n-k]} \left(I_{m+}^{(1-\beta)(n-\alpha);\psi} f \right)(m) + \tag{6.46}$$

$$\frac{1}{\Gamma(\alpha)} \int_{m}^{\lambda} \psi'(t) (\psi(\lambda) - \psi(t))^{\alpha-1\,H} \mathbb{D}_{m+}^{\alpha,\beta;\psi} f(t)\, dt,$$

and

$$f(\lambda) = \sum_{k=1}^{n-1} \frac{(-1)^k (\psi(M) - \psi(\lambda))^{\gamma-k}}{\Gamma(\gamma-k+1)} f_{\psi}^{[n-k]} \left(I_{M-}^{(1-\beta)(n-\alpha);\psi} f \right)(M) + \tag{6.47}$$

$$\frac{1}{\Gamma(\alpha)} \int_{\lambda}^{M} \psi'(t) (\psi(t) - \psi(\lambda))^{\alpha-1\,H} \mathbb{D}_{M-}^{\alpha,\beta;\psi} f(t)\, dt,$$

$\forall\, \lambda \in [m, M]$.

Then we integrate (6.46), (6.47) against E_λ to get

$$\int_{m-0}^{M} f(\lambda)\, dE_\lambda = \sum_{k=1}^{n-1} \frac{f_{\psi}^{[n-k]} \left(I_{m+}^{(1-\beta)(n-\alpha);\psi} f \right)(m)}{\Gamma(\gamma-k+1)} \int_{m-0}^{M} (\psi(\lambda) - \psi(m))^{\gamma-k}\, dE_\lambda +$$

$$\frac{1}{\Gamma(\alpha)} \int_{m-0}^{M} \left(\int_{m}^{\lambda} \psi'(t) (\psi(\lambda) - \psi(t))^{\alpha-1\,H} \mathbb{D}_{m+}^{\alpha,\beta;\psi} f(t)\, dt \right) dE_\lambda, \tag{6.48}$$

and

$$\int_{m-0}^{M} f(\lambda)\, dE_\lambda = \sum_{k=1}^{n-1} \frac{(-1)^k f_{\psi}^{[n-k]} \left(I_{M-}^{(1-\beta)(n-\alpha);\psi} f \right)(M)}{\Gamma(\gamma-k+1)} \int_{m-0}^{M} (\psi(M) - \psi(\lambda))^{\gamma-k}\, dE_\lambda +$$

$$\frac{1}{\Gamma(\alpha)} \int_{m-0}^{M} \left(\int_{\lambda}^{M} \psi'(t) (\psi(t) - \psi(\lambda))^{\alpha-1\,H} \mathbb{D}_{M-}^{\alpha,\beta;\psi} f(t)\, dt \right) dE_\lambda. \tag{6.49}$$

By the spectral representation theorem we obtain

$$f(A) = \sum_{k=1}^{n-1} \frac{f_{\psi}^{[n-k]} \left(I_{m+}^{(1-\beta)(n-\alpha);\psi} f \right)(m)}{\Gamma(\gamma-k+1)} (\psi(A) - \psi(m)\, 1_H)^{\gamma-k} + \tag{6.50}$$

$$\frac{1}{\Gamma(\alpha)} \int_{m-0}^{M} \left(\int_{m}^{\lambda} \psi'(t) (\psi(\lambda) - \psi(t))^{\alpha-1\,H} \mathbb{D}_{m+}^{\alpha,\beta;\psi} f(t)\, dt \right) dE_\lambda,$$

and

$$f(A) = \sum_{k=1}^{n-1} \frac{(-1)^k f_{\psi}^{[n-k]} \left(I_{M-}^{(1-\beta)(n-\alpha);\psi} f \right)(M)}{\Gamma(\gamma - k + 1)} (\psi(M) 1_H - \psi(A))^{\gamma-k} + \tag{6.51}$$

$$\frac{1}{\Gamma(\alpha)} \int_{m-0}^{M} \left(\int_{\lambda}^{M} \psi'(t) (\psi(t) - \psi(\lambda))^{\alpha-1} {}^{H}\mathbb{D}_{M-}^{\alpha,\beta;\psi} f(t)\, dt \right) dE_{\lambda},$$

proving the claim. ■

We make

Remark 6.5 (to Theorem 6.4) Assume that $f_{\psi}^{[n-k]} \left(I_{m+}^{(1-\beta)(n-\alpha);\psi} f \right)(m) = 0$, $k = 1, ..., n-1$, then (see (6.40))

$$f(\lambda) = \frac{1}{\Gamma(\alpha)} \int_{m}^{\lambda} \psi'(t) (\psi(\lambda) - \psi(t))^{\alpha-1} {}^{H}\mathbb{D}_{m+}^{\alpha,\beta;\psi} f(t)\, dt, \tag{6.52}$$

$\forall \lambda \in [m, M]$.

Hence it holds

$$|f(\lambda)| \leq \frac{\left\| {}^{H}\mathbb{D}_{m+}^{\alpha,\beta;\psi} f \right\|_{\infty,[m,M]}}{\Gamma(\alpha+1)} (\psi(\lambda) - \psi(m))^{\alpha}, \tag{6.53}$$

$\forall \lambda \in [m, M]$.

If $\alpha > 1$, by (6.52) we get

$$|f(\lambda)| \leq \frac{(\psi(\lambda) - \psi(m))^{\alpha-1}}{\Gamma(\alpha)} \int_{m}^{\lambda} \left| {}^{H}\mathbb{D}_{m+}^{\alpha,\beta;\psi} f(t) \right| d\psi(t) \leq$$

$$\frac{(\psi(\lambda) - \psi(m))^{\alpha-1}}{\Gamma(\alpha)} \int_{m}^{M} \left| {}^{H}\mathbb{D}_{m+}^{\alpha,\beta;\psi} f(t) \right| d\psi(t). \tag{6.54}$$

That is, when $\alpha > 1$, we obtained

$$|f(\lambda)| \leq \frac{\left\| {}^{H}\mathbb{D}_{m+}^{\alpha,\beta;\psi} f \right\|_{L_1([m,M],\psi)}}{\Gamma(\alpha)} (\psi(\lambda) - \psi(m))^{\alpha-1}, \tag{6.55}$$

$\forall \lambda \in [m, M]$.

Next, let $p, q > 1 : \frac{1}{p} + \frac{1}{q} = 1$, with $\alpha > \frac{1}{q}$. Then, by (6.52) and Hölder's inequality, we derive

$$|f(\lambda)| \leq \frac{\left\| {}^{H}\mathbb{D}_{m+}^{\alpha,\beta;\psi} f \right\|_{L_q([m,M],\psi)}}{\Gamma(\alpha)(p(\alpha-1)+1)^{\frac{1}{p}}} (\psi(\lambda) - \psi(m))^{\alpha-\frac{1}{q}}, \tag{6.56}$$

$\forall \lambda \in [m, M]$.

By (6.42), (6.43) and $f_{\psi}^{[n-k]}\left(I_{m+}^{(1-\beta)(n-\alpha);\psi} f\right)(m) = 0$, $k = 1, \ldots, n-1$, we have

$$f(A) = \frac{1}{\Gamma(\alpha)} \int_{m-0}^{M} \left(\int_{m}^{\lambda} \psi'(t)(\psi(\lambda) - \psi(t))^{\alpha-1H} \mathbb{D}_{m+}^{\alpha,\beta;\psi} f(t)\, dt \right) dE_{\lambda}. \tag{6.57}$$

Therefore it holds

$$\langle f(A)x, y \rangle = \frac{1}{\Gamma(\alpha)} \int_{m-0}^{M} \left(\int_{m}^{\lambda} \psi'(t)(\psi(\lambda) - \psi(t))^{\alpha-1H} \mathbb{D}_{m+}^{\alpha,\beta;\psi} f(t)\, dt \right) d\langle E_{\lambda}x, y \rangle, \tag{6.58}$$

$\forall x, y \in H$.

The function $g_{x,y}(\lambda) := \langle E_{\lambda}x, y \rangle$ is of bounded variation on $[m, M]$, $g_{x,y}(m-0) = 0$ and $g_{x,y}(M) = \langle x, y \rangle$, $\forall x, y \in H$.

It is also well known that $g_x(\lambda) := \langle E_{\lambda}x, x \rangle$ is nondecreasing and right continuous on $[m, M]$. One has

$$\langle f(A)x, x \rangle = \frac{1}{\Gamma(\alpha)} \int_{m-0}^{M} \left(\int_{m}^{\lambda} \psi'(t)(\psi(\lambda) - \psi(t))^{\alpha-1H} \mathbb{D}_{m+}^{\alpha,\beta;\psi} f(t)\, dt \right) d\langle E_{\lambda}x, x \rangle, \tag{6.59}$$

$\forall x \in H$.

Therefore it holds

$$|\langle f(A)x, x \rangle| \overset{(6.53.)}{\leq} \frac{\left\| {}^{H}\mathbb{D}_{m+}^{\alpha,\beta;\psi} f \right\|_{\infty,[m,M]}}{\Gamma(\alpha+1)} \int_{m-0}^{M} (\psi(\lambda) - \psi(m))^{\alpha} d\langle E_{\lambda}x, x \rangle = \tag{6.60}$$

$$\frac{\left\| {}^{H}\mathbb{D}_{m+}^{\alpha,\beta;\psi} f \right\|_{\infty,[m,M]}}{\Gamma(\alpha+1)} \langle (\psi(A) - \psi(m) 1_H)^{\alpha} x, x \rangle, \ \forall x \in H.$$

Similarly, when $\alpha > 1$, by (6.55), we get

$$|\langle f(A)x, x \rangle| \leq \frac{\left\| {}^{H}\mathbb{D}_{m+}^{\alpha,\beta;\psi} f \right\|_{L_1([m,M],\psi)}}{\Gamma(\alpha)} \int_{m-0}^{M} (\psi(\lambda) - \psi(m))^{\alpha-1} d\langle E_{\lambda}x, x \rangle = \tag{6.61}$$

$$\frac{\left\| {}^{H}\mathbb{D}_{m+}^{\alpha,\beta;\psi} f \right\|_{L_1([m,M],\psi)}}{\Gamma(\alpha)} \left\langle (\psi(A) - \psi(m) 1_H)^{\alpha-1} x, x \right\rangle, \ \forall x \in H.$$

Let now $p, q > 1 : \frac{1}{p} + \frac{1}{q} = 1$, with $\alpha > \frac{1}{q}$. Then, by (6.56) we derive

$$|\langle f(A)x,x\rangle| \leq \frac{\left\| {}^{H}\mathbb{D}_{m+}^{\alpha,\beta;\psi} f \right\|_{L_q([m,M],\psi)}}{\Gamma(\alpha)(p(\alpha-1)+1)^{\frac{1}{p}}} \left\langle (\psi(A) - \psi(m) 1_H)^{\alpha-\frac{1}{q}} x, x \right\rangle, \ \forall x \in H. \tag{6.62}$$

We have proved the following ψ-Hilfer fractional left side comparison results.

Theorem 6.6 *All as in Theorem 6.4, plus* $f_{\psi}^{[n-k]}\left(I_{m+}^{(1-\beta)(n-\alpha);\psi} f\right)(m) = 0,\ k = 1, ..., n-1$. *Then*

i)

$$|\langle f(A)x,x\rangle| \leq \frac{\left\| {}^{H}\mathbb{D}_{m+}^{\alpha,\beta;\psi} f \right\|_{\infty,[m,M]}}{\Gamma(\alpha+1)} \langle (\psi(A) - \psi(m) 1_H)^{\alpha} x, x \rangle, \ \forall x \in H. \tag{6.63}$$

Inequality (6.63) means that

$$\|f(A)\| \leq \frac{\left\| {}^{H}\mathbb{D}_{m+}^{\alpha,\beta;\psi} f \right\|_{\infty,[m,M]}}{\Gamma(\alpha+1)} \|(\psi(A) - \psi(m) 1_H)^{\alpha}\|, \tag{6.64}$$

and in particular

$$f(A) \leq \frac{\left\| {}^{H}\mathbb{D}_{m+}^{\alpha,\beta;\psi} f \right\|_{\infty,[m,M]}}{\Gamma(\alpha+1)} (\psi(A) - \psi(m) 1_H)^{\alpha}. \tag{6.65}$$

ii) when $\alpha > 1$ *we have*

$$|\langle f(A)x,x\rangle| \leq \frac{\left\| {}^{H}\mathbb{D}_{m+}^{\alpha,\beta;\psi} f \right\|_{L_1([m,M],\psi)}}{\Gamma(\alpha)} \left\langle (\psi(A) - \psi(m) 1_H)^{\alpha-1} x, x \right\rangle, \ \forall x \in H. \tag{6.66}$$

Inequality (6.66) means that

$$\|f(A)\| \leq \frac{\left\| {}^{H}\mathbb{D}_{m+}^{\alpha,\beta;\psi} f \right\|_{L_1([m,M],\psi)}}{\Gamma(\alpha)} \left\|(\psi(A) - \psi(m) 1_H)^{\alpha-1}\right\|, \tag{6.67}$$

and in particular

$$f(A) \leq \frac{\left\| {}^{H}\mathbb{D}_{m+}^{\alpha,\beta;\psi} f \right\|_{L_1([m,M],\psi)}}{\Gamma(\alpha)} (\psi(A) - \psi(m) 1_H)^{\alpha-1}. \tag{6.68}$$

iii) when $p, q > 1 : \frac{1}{p} + \frac{1}{q} = 1$, *with* $\alpha > \frac{1}{q}$, *we have*

$$|\langle f(A)x,x\rangle| \leq \frac{\left\| {}^{H}\mathbb{D}_{m+}^{\alpha,\beta;\psi} f \right\|_{L_q([m,M],\psi)}}{\Gamma(\alpha)(p(\alpha-1)+1)^{\frac{1}{p}}} \left\langle (\psi(A) - \psi(m)1_H)^{\alpha-\frac{1}{q}} x, x \right\rangle, \ \forall x \in H. \tag{6.69}$$

Inequality (6.69) means that

$$\|f(A)\| \leq \frac{\left\| {}^{H}\mathbb{D}_{m+}^{\alpha,\beta;\psi} f \right\|_{L_q([m,M],\psi)}}{\Gamma(\alpha)(p(\alpha-1)+1)^{\frac{1}{p}}} \left\| (\psi(A) - \psi(m)1_H)^{\alpha-\frac{1}{q}} \right\|, \tag{6.70}$$

and in particular

$$f(A) \leq \frac{\left\| {}^{H}\mathbb{D}_{m+}^{\alpha,\beta;\psi} f \right\|_{L_q([m,M],\psi)}}{\Gamma(\alpha)(p(\alpha-1)+1)^{\frac{1}{p}}} (\psi(A) - \psi(m)1_H)^{\alpha-\frac{1}{q}}. \tag{6.71}$$

We make

Remark 6.7 (to Theorem 6.4) Here we assume that $f_{\psi}^{[n-k]}\left(I_{M-}^{(1-\beta)(n-\alpha);\psi} f\right)(M) = 0$, $k = 1, ..., n-1$, then (see (6.41))

$$f(\lambda) = \frac{1}{\Gamma(\alpha)} \int_{\lambda}^{M} \psi'(t)(\psi(t) - \psi(\lambda))^{\alpha-1} \, {}^{H}\mathbb{D}_{M-}^{\alpha,\beta;\psi} f(t)\, dt, \tag{6.72}$$

$\forall \lambda \in [m, M]$.

Hence it holds

$$|f(\lambda)| \leq \frac{\left\| {}^{H}\mathbb{D}_{M-}^{\alpha,\beta;\psi} f \right\|_{\infty,[m,M]}}{\Gamma(\alpha+1)} (\psi(M) - \psi(\lambda))^{\alpha}, \tag{6.73}$$

$\forall \lambda \in [m, M]$.

If $\alpha > 1$, we get

$$|f(\lambda)| \leq \frac{\left\| {}^{H}\mathbb{D}_{M-}^{\alpha,\beta;\psi} f \right\|_{L_1([m,M],\psi)}}{\Gamma(\alpha)} (\psi(M) - \psi(\lambda))^{\alpha-1}, \tag{6.74}$$

$\forall \lambda \in [m, M]$.

Next, let $p, q > 1 : \frac{1}{p} + \frac{1}{q} = 1$, with $\alpha > \frac{1}{q}$. Then

$$|f(\lambda)| \leq \frac{\left\| {}^{H}\mathbb{D}_{M-}^{\alpha,\beta;\psi} f \right\|_{L_q([m,M],\psi)}}{\Gamma(\alpha)(p(\alpha-1)+1)^{\frac{1}{p}}} (\psi(M) - \psi(\lambda))^{\alpha-\frac{1}{q}}, \tag{6.75}$$

$\forall \lambda \in [m, M]$.

Here also it holds (by (6.44), (6.45))

$$f(A) = \frac{1}{\Gamma(\alpha)} \int_{m-0}^{M} \left(\int_{\lambda}^{M} \psi'(t) (\psi(t) - \psi(\lambda))^{\alpha-1} {}^{H}\mathbb{D}_{M-}^{\alpha,\beta;\psi} f(t)\, dt \right) dE_{\lambda}. \quad (6.76)$$

Consequently we have

$$\langle f(A)x, y \rangle = \frac{1}{\Gamma(\alpha)} \int_{m-0}^{M} \left(\int_{\lambda}^{M} \psi'(t) (\psi(t) - \psi(\lambda))^{\alpha-1} {}^{H}\mathbb{D}_{M-}^{\alpha,\beta;\psi} f(t)\, dt \right) d\langle E_{\lambda}x, y \rangle, \quad (6.77)$$

$\forall\, x, y \in H$.

In particular we have

$$\langle f(A)x, x \rangle = \frac{1}{\Gamma(\alpha)} \int_{m-0}^{M} \left(\int_{\lambda}^{M} \psi'(t) (\psi(t) - \psi(\lambda))^{\alpha-1} {}^{H}\mathbb{D}_{M-}^{\alpha,\beta;\psi} f(t)\, dt \right) d\langle E_{\lambda}x, x \rangle, \quad (6.78)$$

$\forall\, x \in H$.

We state the following ψ-Hilfer fractional right side comparison results.

Theorem 6.8 *All as in Theorem 6.4, plus* $f_{\psi}^{[n-k]}\left(I_{M-}^{(1-\beta)(n-\alpha);\psi} f\right)(M) = 0$, $k = 1, ..., n-1$. *Then*

i)

$$|\langle f(A)x, x \rangle| \leq \frac{\left\| {}^{H}\mathbb{D}_{M-}^{\alpha,\beta;\psi} f \right\|_{\infty,[m,M]}}{\Gamma(\alpha+1)} \langle (\psi(M) 1_H - \psi(A))^{\alpha} x, x \rangle, \ \forall x \in H. \quad (6.79)$$

Inequality (6.79) means that

$$\|f(A)\| \leq \frac{\left\| {}^{H}\mathbb{D}_{M-}^{\alpha,\beta;\psi} f \right\|_{\infty,[m,M]}}{\Gamma(\alpha+1)} \|(\psi(M) 1_H - \psi(A))^{\alpha}\|, \quad (6.80)$$

and in particular

$$f(A) \leq \frac{\left\| {}^{H}\mathbb{D}_{M-}^{\alpha,\beta;\psi} f \right\|_{\infty,[m,M]}}{\Gamma(\alpha+1)} (\psi(M) 1_H - \psi(A))^{\alpha}. \quad (6.81)$$

ii) when $\alpha > 1$ *we have*

$$|\langle f(A)x, x\rangle| \leq \frac{\left\| {}^H\mathbb{D}_{M-}^{\alpha,\beta;\psi} f \right\|_{L_1([m,M],\psi)}}{\Gamma(\alpha)} \left\langle (\psi(M)1_H - \psi(A))^{\alpha-1} x, x \right\rangle, \ \forall x \in H. \tag{6.82}$$

Inequality (6.82) means that

$$\|f(A)\| \leq \frac{\left\| {}^H\mathbb{D}_{M-}^{\alpha,\beta;\psi} f \right\|_{L_1([m,M],\psi)}}{\Gamma(\alpha)} \left\| (\psi(M)1_H - \psi(A))^{\alpha-1} \right\|, \tag{6.83}$$

and in particular

$$f(A) \leq \frac{\left\| {}^H\mathbb{D}_{M-}^{\alpha,\beta;\psi} f \right\|_{L_1([m,M],\psi)}}{\Gamma(\alpha)} (\psi(M)1_H - \psi(A))^{\alpha-1}. \tag{6.84}$$

(iii) when $p, q > 1 : \frac{1}{p} + \frac{1}{q} = 1$*, with* $\alpha > \frac{1}{q}$*, we have*

$$|\langle f(A)x, x\rangle| \leq \frac{\left\| {}^H\mathbb{D}_{M-}^{\alpha,\beta;\psi} f \right\|_{L_q([m,M],\psi)}}{\Gamma(\alpha)(p(\alpha-1)+1)^{\frac{1}{p}}} \left\langle (\psi(M)1_H - \psi(A))^{\alpha-\frac{1}{q}} x, x \right\rangle, \ \forall x \in H. \tag{6.85}$$

Inequality (6.85) means that

$$\|f(A)\| \leq \frac{\left\| {}^H\mathbb{D}_{M-}^{\alpha,\beta;\psi} f \right\|_{L_q([m,M],\psi)}}{\Gamma(\alpha)(p(\alpha-1)+1)^{\frac{1}{p}}} \left\| (\psi(M)1_H - \psi(A))^{\alpha-\frac{1}{q}} \right\|, \tag{6.86}$$

and in particular

$$f(A) \leq \frac{\left\| {}^H\mathbb{D}_{M-}^{\alpha,\beta;\psi} f \right\|_{L_q([m,M],\psi)}}{\Gamma(\alpha)(p(\alpha-1)+1)^{\frac{1}{p}}} (\psi(M)1_H - \psi(A))^{\alpha-\frac{1}{q}}. \tag{6.87}$$

Proof Similar to the proof of Theorem 6.6, by the use of Remark 6.7. ■

We need

Definition 6.9 Let the real valued function $f \in C([m, M])$, and we consider

$$g(t) = \int_m^t f(z)\,dz, \ \forall t \in [m, M], \tag{6.88}$$

then $g \in C([m, M])$.

We denote by

$$\int_{m1_H}^{A} f := \Phi(g) = g(A). \tag{6.89}$$

We understand and write that $(r > 0)$

$$g^r(A) = \Phi(g^r) =: \left(\int_{m1_H}^{A} f\right)^r. \tag{6.90}$$

Clearly $\left(\int_{m1_H}^{A} f\right)^r$ is a self adjoint operator on H, for any $r > 0$.

Definition 6.10 Let $f : [m, M] \to \mathbb{R}$ be continuous. We consider

$$g(t) = \int_t^M f(z)\,dz, \ \forall t \in [m, M], \tag{6.91}$$

then $g \in C([m, M])$.

We denote by

$$\int_A^{M1_H} f := \Phi(g) = g(A). \tag{6.92}$$

We denote also

$$g^r(A) = \Phi(g^r) =: \left(\int_A^{M1_H} f\right)^r, \ r > 0. \tag{6.93}$$

Clearly $\left(\int_A^{M1_H} f\right)^r$ is a self adjoint operator on H, for any $r > 0$.

We mention the following Hilfer fractional derivatives representation formulae:

Theorem 6.11 *([2]) Let $\alpha > 0$, $\alpha \notin \mathbb{N}$, $\lceil \alpha \rceil = n$, $0 < \beta < 1$; $f \in C^n([m, M])$, $[m, M] \subset \mathbb{R}$; and set $\gamma = \alpha + \beta(n - \alpha)$. Assume further that $\Delta_{m+}^{\gamma} f \in C([a, b])$: $\Delta_{m+}^{\gamma - j} f(m) = 0$, for $j = 1, ..., n$. Let also $\overline{\alpha} > 0 : \lceil \overline{\alpha} \rceil = \overline{n}$, with $\overline{\gamma} = \overline{\alpha} + \beta(\overline{n} - \overline{\alpha})$, and assume that $\alpha > \overline{\alpha}$ and $\gamma > \overline{\gamma}$. Then*

$${}^{H}\mathbb{D}_{m+}^{\overline{\alpha},\beta} f(\lambda) = \frac{1}{\Gamma(\alpha - \overline{\alpha})} \int_m^{\lambda} (\lambda - t)^{\alpha - \overline{\alpha} - 1} \, {}^{H}\mathbb{D}_{m+}^{\alpha,\beta} f(t)\,dt, \tag{6.94}$$

$\forall \lambda \in [m, M]$,

furthermore ${}^{H}\mathbb{D}_{m+}^{\overline{\alpha},\beta} f \in AC([m, M])$ (absolutely continuous functions) if $\alpha - \overline{\alpha} \geq 1$ and ${}^{H}\mathbb{D}_{m+}^{\overline{\alpha},\beta} f \in C([m, M])$ if $\alpha - \overline{\alpha} \in (0, 1)$.

Theorem 6.12 *([2]) Let $\alpha > 0$, $\alpha \notin \mathbb{N}$, $\lceil \alpha \rceil = n$, $0 < \beta < 1$; $f \in C^n([m, M])$, $[m, M] \subset \mathbb{R}$; and set $\gamma = \alpha + \beta(n - \alpha)$. Assume further that $\Delta_{M-}^{\gamma} f \in C([a, b])$: $\Delta_{M-}^{\gamma - j} f(M) = 0$, $j = 1, ..., n$. Let also $\overline{\alpha} > 0 : \lceil \overline{\alpha} \rceil = \overline{n}$, with $\overline{\gamma} = \overline{\alpha} + \beta(\overline{n} - \overline{\alpha})$, and assume that $\alpha > \overline{\alpha}$ and $\gamma > \overline{\gamma}$. Then*

$$ {}^{H}\mathbb{D}_{M-}^{\overline{\alpha},\beta} f(\lambda) = \frac{1}{\Gamma(\alpha-\overline{\alpha})} \int_{\lambda}^{M} (t-\lambda)^{\alpha-\overline{\alpha}-1}\, {}^{H}\mathbb{D}_{M-}^{\alpha,\beta} f(t)\, dt, \tag{6.95} $$

$\forall\, \lambda \in [m, M]$,

furthermore ${}^{H}\mathbb{D}_{M-}^{\overline{\alpha},\beta} f \in AC([m, M])$ *if* $\alpha - \overline{\alpha} \geq 1$ *and* ${}^{H}\mathbb{D}_{M-}^{\overline{\alpha},\beta} f \in C([m, M])$ *if* $\alpha - \overline{\alpha} \in (0, 1)$.

We mention the following left and right Hilfer fractional Poincaré type inequalities:

Theorem 6.13 *([2]) All as in Theorem 6.11,* $0 \leq \beta < 1$*. Let also* $p, q > 1 : \frac{1}{p} + \frac{1}{q} = 1$ *and assume that* $\alpha - \overline{\alpha} > \frac{1}{q}$*. Then*

$$ \left\| {}^{H}\mathbb{D}_{m+}^{\overline{\alpha},\beta} f \right\|_{q,[m,\lambda]} \leq \frac{(\lambda - m)^{\alpha-\overline{\alpha}}}{\Gamma(\alpha-\overline{\alpha}) \left(q(\alpha-\overline{\alpha})\right)^{\frac{1}{q}} \left(p(\alpha-\overline{\alpha}-1)+1\right)^{\frac{1}{p}}} \left\| {}^{H}\mathbb{D}_{m+}^{\alpha,\beta} f \right\|_{q,[m,\lambda]}, \tag{6.96} $$

$\forall\, \lambda \in [m, M]$.

Theorem 6.14 *([2]) All as in Theorem 6.12,* $0 \leq \beta < 1$*. Let also* $p, q > 1 : \frac{1}{p} + \frac{1}{q} = 1$ *and assume that* $\alpha - \overline{\alpha} > \frac{1}{q}$*. Then*

$$ \left\| {}^{H}\mathbb{D}_{M-}^{\overline{\alpha},\beta} f \right\|_{q,[\lambda,M]} \leq \frac{(M - \lambda)^{\alpha-\overline{\alpha}}}{\Gamma(\alpha-\overline{\alpha}) \left(q(\alpha-\overline{\alpha})\right)^{\frac{1}{q}} \left(p(\alpha-\overline{\alpha}-1)+1\right)^{\frac{1}{p}}} \left\| {}^{H}\mathbb{D}_{M-}^{\alpha,\beta} f \right\|_{q,[\lambda,M]}, \tag{6.97} $$

$\forall\, \lambda \in [m, M]$.

We present the following left Hilfer fractional operator Poincaré type inequality.

Theorem 6.15 *Let* A *be a selfadjoint operator on the Hilbert space* H *with the spectrum* $Sp(A) \subseteq [m, M]$ *for some real numbers* $m < M$*. We suppose also the assumptions of Theorem 6.13. Then*

$$ \int_{m1_H}^{A} \left| {}^{H}\mathbb{D}_{m+}^{\overline{\alpha},\beta} f \right|^{q} \leq $$

$$ \frac{(A - m1_H)^{(\alpha-\overline{\alpha})q}}{\left(\Gamma(\alpha-\overline{\alpha})\right)^{q} \left(q(\alpha-\overline{\alpha})\right) \left(p(\alpha-\overline{\alpha}-1)+1\right)^{q-1}} \left(\int_{m1_H}^{A} \left| {}^{H}\mathbb{D}_{m+}^{\alpha,\beta} f \right|^{q} \right). \tag{6.98} $$

Proof By Theorem 6.13 and properties (P) and (ii), see Sect. 6.1. ■

We present also the following right Hilfer fractional operator Poincaré type inequality.

Theorem 6.16 *The setting as in Theorem 6.15, but now under the assumptions of Theorem 6.14. Then*

$$\int_{A}^{M1_H} \left| {}^{H}\mathbb{D}_{M-}^{\overline{\alpha},\beta} f \right|^q \leq$$

$$\frac{(M1_H - A)^{(\alpha-\overline{\alpha})q}}{(\Gamma(\alpha-\overline{\alpha}))^q (q(\alpha-\overline{\alpha}))(p(\alpha-\overline{\alpha}-1)+1)^{q-1}} \left(\int_{A}^{M1_H} \left| {}^{H}\mathbb{D}_{M-}^{\alpha,\beta} f \right|^q \right). \quad (6.99)$$

Proof By Theorem 6.14 and properties (P) and (ii), see Sect. 6.1. ■

Next we mention some Hilfer fractional left and right Opial type inequalities ([10]).

Theorem 6.17 *([2]) Let all as in Theorem 6.11,* $0 \leq \beta < 1$. *Let also* $p, q > 1 :$ $\frac{1}{p} + \frac{1}{q} = 1$, *and assume that* $\alpha - \overline{\alpha} > \frac{1}{q}$. *Then*

$$\int_{m}^{\lambda} \left| {}^{H}\mathbb{D}_{m+}^{\overline{\alpha},\beta} f(w) \right| \left| {}^{H}\mathbb{D}_{m+}^{\alpha,\beta} f(w) \right| dw \leq$$

$$\frac{2^{-\frac{1}{q}} (\lambda - m)^{(\alpha-\overline{\alpha})+\left(\frac{1}{p}-\frac{1}{q}\right)}}{\Gamma(\alpha-\overline{\alpha}) \left[(p(\alpha-\overline{\alpha}-1)+1)(p(\alpha-\overline{\alpha}-1)+2) \right]^{\frac{1}{p}}} \left(\int_{m}^{\lambda} \left| {}^{H}\mathbb{D}_{m+}^{\alpha,\beta} f(w) \right|^q dw \right)^{\frac{2}{q}}, \quad (6.100)$$

$\forall \lambda \in [m, M]$.

Theorem 6.18 *([2]) Let all as in Theorem 6.12,* $0 \leq \beta < 1$. *Let also* $p, q > 1 :$ $\frac{1}{p} + \frac{1}{q} = 1$, *and assume that* $\alpha - \overline{\alpha} > \frac{1}{q}$. *Then*

$$\int_{\lambda}^{M} \left| {}^{H}\mathbb{D}_{M-}^{\overline{\alpha},\beta} f(w) \right| \left| {}^{H}\mathbb{D}_{M-}^{\alpha,\beta} f(w) \right| dw \leq$$

$$\frac{2^{-\frac{1}{q}} (M - \lambda)^{(\alpha-\overline{\alpha})+\left(\frac{1}{p}-\frac{1}{q}\right)}}{\Gamma(\alpha-\overline{\alpha}) \left[(p(\alpha-\overline{\alpha}-1)+1)(p(\alpha-\overline{\alpha}-1)+2) \right]^{\frac{1}{p}}} \left(\int_{\lambda}^{M} \left| {}^{H}\mathbb{D}_{M-}^{\alpha,\beta} f(w) \right|^q dw \right)^{\frac{2}{q}}, \quad (6.101)$$

$\forall \lambda \in [m, M]$.

We give the following operator Hilfer fractional left and right Opial type inequalities. These come by the use of Theorems 6.17, 6.18 and property (P) with (ii), see Sect. 6.1.

Theorem 6.19 *All as in Theorem 6.17. Then*

$$\int_{m1_H}^{A} \left| {}^{H}\mathbb{D}_{m+}^{\overline{\alpha},\beta} f \right| \left| {}^{H}\mathbb{D}_{m+}^{\alpha,\beta} f \right| \leq \frac{2^{-\frac{1}{q}}}{\Gamma(\alpha-\overline{\alpha}) \left[(p(\alpha-\overline{\alpha}-1)+1)(p(\alpha-\overline{\alpha}-1)+2) \right]^{\frac{1}{p}}}$$

$$(A - m1_H)^{(\alpha-\overline{\alpha})+\left(\frac{1}{p}-\frac{1}{q}\right)} \left(\int_{m1_H}^{A} \left| {}^{H}\mathbb{D}_{m+}^{\alpha,\beta} f \right|^q \right)^{\frac{2}{q}}. \quad (6.102)$$

Theorem 6.20 *All as in Theorem 6.18. Then*

$$\int_A^{M1_H} \left| {}^H\mathbb{D}_{M-}^{\overline{\alpha},\beta} f \right| \left| {}^H\mathbb{D}_{M-}^{\alpha,\beta} f \right| \leq \frac{2^{-\frac{1}{q}}}{\Gamma(\alpha-\overline{\alpha}) \left[(p(\alpha-\overline{\alpha}-1)+1)(p(\alpha-\overline{\alpha}-1)+2) \right]^{\frac{1}{p}}}$$

$$(M1_H - A)^{(\alpha-\overline{\alpha})+\left(\frac{1}{p}-\frac{1}{q}\right)} \left(\int_A^{M1_H} \left| {}^H\mathbb{D}_{M-}^{\alpha,\beta} f \right|^q \right)^{\frac{2}{q}}. \quad (6.103)$$

We give the following Hilfer left fractional Poincaré type inequality.

Theorem 6.21 *Let $f \in C^n([m, M])$, $n-1 < \alpha < n$, $0 \leq \beta \leq 1$. Assume that $\left(I_{m+}^{(1-\beta)(n-\alpha)} f \right)^{(n-k)}(m) = 0$, $k = 1, ..., n-1$. Let $p, q > 1 : \frac{1}{p} + \frac{1}{q} = 1$, with $\alpha > \frac{1}{q}$. Then*

$$\int_m^\lambda |f(\lambda_1)|^q d\lambda_1 \leq \frac{(\lambda-m)^{\alpha q}}{(\alpha q)(\Gamma(\alpha))^q (p(\alpha-1)+1)^{(q-1)}} \int_m^\lambda \left| {}^H\mathbb{D}_{m+}^{\alpha,\beta} f(t) \right|^q dt, \quad (6.104)$$

$\forall \lambda \in [m, M]$.

Proof Assume $\left(I_{m+}^{(1-\beta)(n-\alpha)} f \right)^{(n-k)}(m) = 0$, $k = 1, ..., n-1$, then (see (6.40))

$$f(\lambda) = \frac{1}{\Gamma(\alpha)} \int_m^\lambda (\lambda - t)^{\alpha-1} {}^H\mathbb{D}_{m+}^{\alpha,\beta} f(t) dt, \quad (6.105)$$

$\forall \lambda \in [m, M]$.

Let $p, q > 1 : \frac{1}{p} + \frac{1}{q} = 1$, with $\alpha > \frac{1}{q}$. By Hölder's inequality we obtain

$$|f(\lambda)| \leq \frac{1}{\Gamma(\alpha)} \int_m^\lambda (\lambda - t)^{\alpha-1} \left| {}^H\mathbb{D}_{m+}^{\alpha,\beta} f(t) \right| dt \leq$$

$$\frac{1}{\Gamma(\alpha)} \frac{(\lambda-m)^{\frac{(p(\alpha-1)+1)}{p}}}{(p(\alpha-1)+1)^{\frac{1}{p}}} \left(\int_m^\lambda \left| {}^H\mathbb{D}_{m+}^{\alpha,\beta} f(t) \right|^q dt \right)^{\frac{1}{q}} =$$

$$\frac{(\lambda-m)^{\alpha-\frac{1}{q}}}{\Gamma(\alpha)(p(\alpha-1)+1)^{\frac{1}{p}}} \left(\int_m^\lambda \left| {}^H\mathbb{D}_{m+}^{\alpha,\beta} f(t) \right|^q dt \right)^{\frac{1}{q}}, \ \forall \lambda \in [m, M]. \quad (6.106)$$

One can write

$$|f(\lambda_1)| \leq \frac{(\lambda_1-m)^{\alpha-\frac{1}{q}}}{\Gamma(\alpha)(p(\alpha-1)+1)^{\frac{1}{p}}} \left(\int_m^{\lambda_1} \left| {}^H\mathbb{D}_{m+}^{\alpha,\beta} f(t) \right|^q dt \right)^{\frac{1}{q}}, \quad (6.107)$$

$\forall \lambda_1 \in [m, \lambda]$, where $\lambda \in [m, M]$.

That is

$$|f(\lambda_1)|^q \leq \frac{(\lambda_1 - m)^{\alpha q - 1}}{(\Gamma(\alpha))^q (p(\alpha-1)+1)^{\frac{q}{p}}} \left(\int_m^{\lambda} \left| {}^H\mathbb{D}_{m+}^{\alpha,\beta} f(t) \right|^q dt \right), \tag{6.108}$$

$\forall\, \lambda_1 \in [m, \lambda]$.

Consequently, we derive

$$\int_m^{\lambda} |f(\lambda_1)|^q d\lambda_1 \leq \frac{\left(\int_m^{\lambda} (\lambda_1 - m)^{\alpha q - 1} d\lambda_1 \right)}{(\Gamma(\alpha))^q (p(\alpha-1)+1)^{\frac{q}{p}}} \left(\int_m^{\lambda} \left| {}^H\mathbb{D}_{m+}^{\alpha,\beta} f(t) \right|^q dt \right) = \tag{6.109}$$

$$\frac{(\lambda - m)^{\alpha q}}{(\alpha q)(\Gamma(\alpha))^q (p(\alpha-1)+1)^{q-1}} \left(\int_m^{\lambda} \left| {}^H\mathbb{D}_{m+}^{\alpha,\beta} f(t) \right|^q dt \right),$$

$\forall\, \lambda \in [m, M]$, proving the claim. ■

We also give the following Hilfer right fractional Poincaré type inequality.

Theorem 6.22 *Let* $f \in C^n([m, M])$, $n-1 < \alpha < n$, $0 \leq \beta \leq 1$. *Assume that* $\left(I_{M-}^{(1-\beta)(n-\alpha)} f \right)^{(n-k)} (M) = 0$, $k = 1, ..., n-1$. *Let* $p, q > 1 : \frac{1}{p} + \frac{1}{q} = 1$, *with* $\alpha > \frac{1}{q}$. *Then*

$$\int_{\lambda}^{M} |f(\lambda_1)|^q d\lambda_1 \leq \frac{(M - \lambda)^{\alpha q}}{(\alpha q)(\Gamma(\alpha))^q (p(\alpha-1)+1)^{(q-1)}} \int_{\lambda}^{M} \left| {}^H\mathbb{D}_{M-}^{\alpha,\beta} f(t) \right|^q dt, \tag{6.110}$$

$\forall\, \lambda \in [m, M]$.

Proof As similar to the proof of Theorem 6.21 is omitted. We use (6.41), etc. ■

Next comes a Hilfer left fractional Sobolev type inequality.

Theorem 6.23 *All as in Theorem 6.21 and* $r \geq 1$. *Then*

$$\int_m^{\lambda} |f(\lambda_1)|^r d\lambda_1 \leq$$

$$\frac{(\lambda - m)^{r\left(\alpha - \frac{1}{q}\right)+1}}{(\Gamma(\alpha))^r (p(\alpha-1)+1)^{\frac{r}{p}} \left(r\left(\alpha - \frac{1}{q}\right) + 1 \right)} \left(\int_m^{\lambda} \left| {}^H\mathbb{D}_{m+}^{\alpha,\beta} f(t) \right|^q dt \right)^{\frac{r}{q}}, \tag{6.111}$$

$\forall\, \lambda \in [m, M]$.

Proof By (6.107) and $r \geq 1$, we find

$$|f(\lambda_1)|^r \leq \frac{(\lambda_1 - m)^{r\left(\alpha - \frac{1}{q}\right)}}{(\Gamma(\alpha))^r (p(\alpha - 1) + 1)^{\frac{r}{p}}} \left(\int_m^{\lambda} \left| {}^H\mathbb{D}_{m+}^{\alpha,\beta} f(t) \right|^q dt \right)^{\frac{r}{q}}, \tag{6.112}$$

$\forall \lambda_1 \in [m, \lambda]$, where $\lambda \in [m, M]$.

Consequently, it holds

$$\int_m^{\lambda} |f(\lambda_1)|^r d\lambda_1 \leq \frac{\int_m^{\lambda} (\lambda_1 - m)^{r\left(\alpha - \frac{1}{q}\right)} d\lambda_1}{(\Gamma(\alpha))^r (p(\alpha - 1) + 1)^{\frac{r}{p}}} \left(\int_m^{\lambda} \left| {}^H\mathbb{D}_{m+}^{\alpha,\beta} f(t) \right|^q dt \right)^{\frac{r}{q}} = \tag{6.113}$$

$$\frac{(\lambda - m)^{r\left(\alpha - \frac{1}{q}\right)+1}}{(\Gamma(\alpha))^r (p(\alpha - 1) + 1)^{\frac{r}{p}} \left(r\left(\alpha - \frac{1}{q}\right) + 1\right)} \left(\int_m^{\lambda} \left| {}^H\mathbb{D}_{m+}^{\alpha,\beta} f(t) \right|^q dt \right)^{\frac{r}{q}},$$

$\forall \lambda \in [m, M]$. The claim is proved. ■

It follows the Hilfer right fractional Sobolev type inequality.

Theorem 6.24 *All as in Theorem 6.22 and* $r \geq 1$. *Then*

$$\int_{\lambda}^{M} |f(\lambda_1)|^r d\lambda_1 \leq \frac{(M - \lambda)^{r\left(\alpha - \frac{1}{q}\right)+1}}{(\Gamma(\alpha))^r (p(\alpha - 1) + 1)^{\frac{r}{p}} \left(r\left(\alpha - \frac{1}{q} + 1\right)\right)}$$

$$\left(\int_{\lambda}^{M} \left| {}^H\mathbb{D}_{M-}^{\alpha,\beta} f(t) \right|^q dt \right)^{\frac{r}{q}}, \tag{6.114}$$

$\forall \lambda \in [m, M]$.

Proof As similar to the proof of Theorem 6.23 is omitted. ■

Next come left and right operator Hilfer fractional Poincaré type inequalities.

Theorem 6.25 *Let A be a selfadjoint operator on the Hilbert space H with* $Sp(A) \subseteq [m, M]$, $m < M$. *We suppose all as in Theorem 6.21. Then*

$$\int_{m1_H}^{A} |f|^q \leq \frac{(A - m1_H)^{\alpha q}}{(\alpha q)(\Gamma(\alpha))^q (p(\alpha - 1) + 1)^{q-1}} \left(\int_{m1_H}^{A} \left| {}^H\mathbb{D}_{m+}^{\alpha,\beta} f \right|^q \right). \tag{6.115}$$

Proof By Theorem 6.21 and properties (P) and (ii), see Sect. 6.1. ■

Theorem 6.26 *Let A be a selfadjoint operator on H with* $Sp(A) \subseteq [m, M]$, $m < M$. *We suppose all as in Theorem 6.22. Then*

$$\int_{A}^{M1_H} |f|^q \leq \frac{(M1_H - A)^{\alpha q}}{(\alpha q)(\Gamma(\alpha))^q (p(\alpha - 1) + 1)^{q-1}} \left(\int_{A}^{M1_H} \left| {}^H\mathbb{D}_{M-}^{\alpha,\beta} f \right|^q \right). \tag{6.116}$$

Proof By Theorem 6.22 and properties (P) and (ii), see Sect. 6.1. ■

We finish with operator Hilfer left and right fractional Sobolev type inequalities.

Theorem 6.27 *Let A be a selfadjoint operator on H with Sp (A) ⊆ [m, M], m < M. We suppose all as in Theorem 6.23. Then*

$$\int_{m1_H}^{A} |f|^r \leq \frac{(A - m1_H)^{r\left(\alpha-\frac{1}{q}\right)+1}}{(\Gamma(\alpha))^r (p(\alpha-1)+1)^{\frac{r}{p}} \left(r\left(\alpha - \frac{1}{q} + 1\right)\right)} \left(\int_{m1_H}^{A} \left| {}^{H}\mathbb{D}_{m+}^{\alpha,\beta} f \right|^q \right)^{\frac{r}{q}}. \tag{6.117}$$

Proof By Theorem 6.23 and properties (P) and (ii), see Sect. 6.1. ■

Theorem 6.28 *Let A be a selfadjoint operator on H with Sp (A) ⊆ [m, M], m < M. We suppose all as in Theorem 6.24. Then*

$$\int_{A}^{M1_H} |f|^r \leq \frac{(M1_H - A)^{r\left(\alpha-\frac{1}{q}\right)+1}}{(\Gamma(\alpha))^r (p(\alpha-1)+1)^{\frac{r}{p}} \left(r\left(\alpha - \frac{1}{q} + 1\right)\right)} \left(\int_{A}^{M1_H} \left| {}^{H}\mathbb{D}_{M-}^{\alpha,\beta} f \right|^q \right)^{\frac{r}{q}}. \tag{6.118}$$

Proof By Theorem 6.24 and properties (P) and (ii), see Sect. 6.1. ■

References

1. Almeida, R.: A Caputo fractional derivative of a function with respect to another function. Commun. Nonlinear Sci. Numer. Simulat. **44**, 460–481 (2017)
2. Anastassiou, G.: Advancements on ψ-Hilfer fractional calculus and fractional integral inequalities. Nonlinearity and Complexity (2021). [Accepted, Discontinuity]
3. Anastassiou, G.: ψ-Hilfer and Hilfer fractional self adjoint operator analytic inequalities. Asia Mathematika **5**(1), 83–102 (2021)
4. da Sousa, J.V.C., de Oliveira, C.: On the ψ-Hilfer fractional derivative. Commun. Nonlinear Sci. Numer. Simul. **60**, 72–91 (2018)
5. Dragomir, S.S.: Inequalities for functions of selfadjoint operators on Hilbert spaces (2011), https://ajmaa.org/RGMIA/monographs/InFuncOp.pdf
6. Dragomir, S.: Operator Inequalities of Ostrowski and Trapezoidal Type. Springer, New York (2012)
7. Furuta, T., Mićić Hot, J., Pečaric, J., Seo, Y.: Mond-Pečaric Method in Operator Inequalities
8. Helmberg, G.: Introduction to Spectral Theory in Hilbert Space. Wiley, New York (1969)
9. Kilbas, A.A., Srivastava, H.M., Trujillo, J.J.: Theory and Applications of Fractional Differentiation Equations. North Holland, Amsterdam, New York (2006)
10. Opial, Z.: Sur une inégalité. Ann. Polon. Math. **8**, 29–32 (1960)
11. Samko, S.G., Kilbas, A.A., Marichev, O.I.: Fractional Integrals and Derivatives: Theory and Applications. Gordon and Breach Science Publishers, Switzerland (1993)
12. Tomovski, Ž, Hilfer, R., Srivastava, H.M.: Fractional and operational calculus with generalized fractional derivative operators and Mittag-Leffler type functions. Integral Transforms Spec Funct. **21**(11), 797–814 (2010)

Chapter 7
Essential Forward and Reverse Generalized Hilfer-Hardy Fractional Inequalities

Here we present forward and reverse left and right L_p fractional integral inequalities of Hardy type. These engage fractional derivatives and fractional integrals from the ψ-Hilfer and Hilfer fractional calculi. It follows [4].

7.1 Background

Let $-\infty < a < b < \infty$, the left and right Riemann-Liouville fractional integrals of order $\alpha \in \mathbb{C}$ $(\mathcal{R}(\alpha) > 0)$ are defined by

$$\left(I_{a+}^{\alpha} f\right)(x) = \frac{1}{\Gamma(\alpha)} \int_a^x (x-t)^{\alpha-1} f(t)\, dt, \tag{7.1}$$

$x > a$; where Γ stands for the gamma function,

and

$$\left(I_{b-}^{\alpha} f\right)(x) = \frac{1}{\Gamma(\alpha)} \int_x^b (t-x)^{\alpha-1} f(t)\, dt, \tag{7.2}$$

$x < b$.

The Riemann-Liouville left and right fractional derivatives of order $\alpha \in \mathbb{C}$ $(\mathcal{R}(\alpha) \geq 0)$ are defined by

$$\left(\Delta_{a+}^{\alpha} y\right)(x) = \left(\frac{d}{dx}\right)^n \left(I_{a+}^{n-\alpha} y\right)(x) = \frac{1}{\Gamma(n-\alpha)} \left(\frac{d}{dx}\right)^n \int_a^x (x-t)^{n-\alpha-1} y(t)\, dt \tag{7.3}$$

G. A. Anastassiou, *Unification of Fractional Calculi with Applications*,
Studies in Systems, Decision and Control 398,
https://doi.org/10.1007/978-3-030-86920-5_7

($n = \lceil \mathcal{R}(\alpha) \rceil$, $\lceil \cdot \rceil$ means ceiling of the number; $x > a$)

$$\left(\Delta_{b-}^{\alpha} y\right)(x) = (-1)^n \left(\frac{d}{dx}\right)^n \left(I_{b-}^{n-\alpha} y\right)(x) =$$

$$\frac{(-1)^n}{\Gamma(n-\alpha)} \left(\frac{d}{dx}\right)^n \int_x^b (t-x)^{n-\alpha-1} y(t)\, dt \tag{7.4}$$

($n = \lceil \mathcal{R}(\alpha) \rceil$; $x < b$), respectively, where $\mathcal{R}(\alpha)$ is the real part of α.

In particular, when $\alpha = n \in \mathbb{Z}_{+}$, then

$$\begin{gathered}\left(\Delta_{a+}^{0} y\right)(x) = \left(\Delta_{b-}^{0} y\right)(x) = y(x); \\ \left(\Delta_{a+}^{n} y\right)(x) = y^{(n)}(x), \text{ and } \left(\Delta_{b-}^{n} y\right)(x) = (-1)^n y^{(n)}(x), \; n \in \mathbb{N},\end{gathered} \tag{7.5}$$

see [9].

Let $\alpha > 0$, $I = [a, b] \subset \mathbb{R}$, f an integrable function defined on I and $\psi \in C^1(I)$ an increasing function such that $\psi'(x) \neq 0$, for all $x \in I$. Left fractional integrals and left Riemann-Liouville fractional derivatives of a function f with respect to another function ψ are defined as ([8], [9])

$$I_{a+}^{\alpha,\psi} f(x) = \frac{1}{\Gamma(\alpha)} \int_a^x \psi'(t) (\psi(x) - \psi(t))^{\alpha-1} f(t)\, dt, \tag{7.6}$$

and

$$\Delta_{a+}^{\alpha,\psi} f(x) = \left(\frac{1}{\psi'(x)} \frac{d}{dx}\right)^n I_{a+}^{n-\alpha,\psi} f(x) = \tag{7.7}$$

$$\frac{1}{\Gamma(n-\alpha)} \left(\frac{1}{\psi'(x)} \frac{d}{dx}\right)^n \int_a^x \psi'(t) (\psi(x) - \psi(t))^{n-\alpha-1} f(t)\, dt,$$

respectively, where $n = \lceil \alpha \rceil$.

Similarly, we define the right ones:

$$I_{b-}^{\alpha,\psi} f(x) = \frac{1}{\Gamma(\alpha)} \int_x^b \psi'(t) (\psi(t) - \psi(x))^{\alpha-1} f(t)\, dt, \tag{7.8}$$

and

$$\Delta_{b-}^{\alpha,\psi} f(x) = \left(-\frac{1}{\psi'(x)} \frac{d}{dx}\right)^n I_{b-}^{n-\alpha,\psi} f(x) =$$

$$\frac{1}{\Gamma(n-\alpha)} \left(-\frac{1}{\psi'(x)} \frac{d}{dx}\right)^n \int_x^b \psi'(t) (\psi(t) - \psi(x))^{n-\alpha-1} f(t)\, dt. \tag{7.9}$$

The following semigroup property holds; if $\alpha, \beta > 0, f \in C(I)$, then

$$I_{a+}^{\alpha,\psi} I_{a+}^{\beta,\psi} f = I_{a+}^{\alpha+\beta,\psi} f \text{ and } I_{b-}^{\alpha,\psi} I_{b-}^{\beta,\psi} f = I_{b-}^{\alpha+\beta,\psi} f.$$

Next let again $\alpha > 0$, $n = \lceil \alpha \rceil$, $I = [a, b]$, $f, \psi \in C^n(I) : \psi$ is increasing and $\psi'(x) \neq 0$, for all $x \in I$. The left ψ-Caputo fractional derivative of f of order α is given by ([1])

$$^{C}D_{a+}^{\alpha,\psi} f(x) = I_{a+}^{n-\alpha,\psi} \left(\frac{1}{\psi'(x)} \frac{d}{dx} \right)^n f(x), \tag{7.10}$$

and the right ψ-Caputo fractional derivative ([1])

$$^{C}D_{b-}^{\alpha,\psi} f(x) = I_{b-}^{n-\alpha,\psi} \left(-\frac{1}{\psi'(x)} \frac{d}{dx} \right)^n f(x). \tag{7.11}$$

We set

$$f_{\psi}^{[n]}(x) := f_{\psi}^{(n)} f(x) := \left(\frac{1}{\psi'(x)} \frac{d}{dx} \right)^n f(x). \tag{7.12}$$

Clearly, when $\alpha = m \in \mathbb{N}$ we have

$$^{C}D_{a+}^{\alpha,\psi} f(x) = f_{\psi}^{[m]}(x) \text{ and } ^{C}D_{b-}^{\alpha,\psi} f(x) = (-1)^m f_{\psi}^{[m]}(x),$$

and if $\alpha \notin \mathbb{N}$, then

$$^{C}D_{a+}^{\alpha,\psi} f(x) = \frac{1}{\Gamma(n-\alpha)} \int_a^x \psi'(t) (\psi(x) - \psi(t))^{n-\alpha-1} f_{\psi}^{[n]}(t)\, dt, \tag{7.13}$$

and

$$^{C}D_{b-}^{\alpha,\psi} f(x) = \frac{(-1)^n}{\Gamma(n-\alpha)} \int_x^b \psi'(t) (\psi(t) - \psi(x))^{n-\alpha-1} f_{\psi}^{[n]}(t)\, dt. \tag{7.14}$$

If $\psi(x) = x$, then we get the usual left and right Caputo fractional derivatives

$$^{C}D_{a+}^{m} f(x) = f^{(m)}(x), \quad ^{C}D_{b-}^{m} f(x) = (-1)^m f^{(m)}(x),$$

for $m \in \mathbb{N}$, and ($\alpha \notin \mathbb{N}$)

$$D_{*a}^{\alpha} f(x) = {}^{C}D_{a+}^{\alpha} f(x) = \frac{1}{\Gamma(n-\alpha)} \int_a^x (x-t)^{n-\alpha-1} f^{(n)}(t)\, dt, \tag{7.15}$$

$$D_{b-}^{\alpha}(x) = {}^{C}D_{b-}^{\alpha} f(x) = \frac{(-1)^n}{\Gamma(n-\alpha)} \int_x^b (t-x)^{n-\alpha-1} f^{(n)}(t)\, dt. \tag{7.16}$$

Also we set

$$ {}^{C}D_{a+}^{0,\psi} f(x) = {}^{C}D_{b-}^{0,\psi} f(x) = f(x). $$

Next we will deal with the ψ-Hilfer fractional derivative.

Definition 7.1 ([6]) Let $n-1 < \alpha < n$, $n \in \mathbb{N}$, $I = [a,b] \subset \mathbb{R}$ and $f, \psi \in C^n([a,b])$, ψ is increasing and $\psi'(x) \neq 0$, for all $x \in I$. The ψ-Hilfer fractional derivative (left-sided and right-sided) ${}^{H}\mathbb{D}_{a+(b-)}^{\alpha,\beta;\psi} f$ of order α and type $0 \leq \beta \leq 1$, respectively, are defined by

$$ {}^{H}\mathbb{D}_{a+}^{\alpha,\beta;\psi} f(x) = I_{a+}^{\beta(n-\alpha);\psi} \left(\frac{1}{\psi'(x)} \frac{d}{dx} \right)^n I_{a+}^{(1-\beta)(n-\alpha);\psi} f(x), \quad (7.17) $$

and

$$ {}^{H}\mathbb{D}_{b-}^{\alpha,\beta;\psi} f(x) = I_{b-}^{\beta(n-\alpha);\psi} \left(-\frac{1}{\psi'(x)} \frac{d}{dx} \right)^n I_{b-}^{(1-\beta)(n-\alpha);\psi} f(x), \quad x \in [a,b]. \quad (7.18) $$

The original Hilfer fractional derivatives ([10]) come from $\psi(x) = x$, and are denoted by ${}^{H}\mathbb{D}_{a+}^{\alpha,\beta} f(x)$ and ${}^{H}\mathbb{D}_{b-}^{\alpha,\beta} f(x)$.

When $\beta = 0$, we get Riemann-Liouville fractional derivatives, while when $\beta = 1$ we have Caputo type fractional derivatives.

We define $\gamma = \alpha + \beta(n-\alpha)$. We notice that $n-1 < \alpha \leq \alpha + \beta(n-\alpha) \leq \alpha + n - \alpha = n$, hence $\lceil \gamma \rceil = n$. We can easily write that ([6])

$$ {}^{H}\mathbb{D}_{a+}^{\alpha,\beta;\psi} f(x) = I_{a+}^{\gamma-\alpha;\psi} \Delta_{a+}^{\gamma;\psi} f(x), \quad (7.19) $$

and

$$ {}^{H}\mathbb{D}_{b-}^{\alpha,\beta;\psi} f(x) = I_{b-}^{\gamma-\alpha;\psi} \Delta_{b-}^{\gamma;\psi} f(x), \quad x \in [a,b]. \quad (7.20) $$

We have that ([6])

$$ \Delta_{a+}^{\gamma,\psi} f(x) = \left(\frac{1}{\psi'(x)} \frac{d}{dx} \right)^n I_{a+}^{(1-\beta)(n-\alpha);\psi} f(x), \quad (7.21) $$

and

$$ \Delta_{b-}^{\gamma,\psi} f(x) = \left(-\frac{1}{\psi'(x)} \frac{d}{dx} \right)^n I_{b-}^{(1-\beta)(n-\alpha);\psi} f(x). \quad (7.22) $$

In particular, when $0 < \alpha < 1$ and $0 \leq \beta \leq 1$; $\gamma = \alpha + \beta(1-\alpha)$, we have that

$$ {}^{H}\mathbb{D}_{a+}^{\alpha,\beta;\psi} f(x) = \frac{1}{\Gamma(\gamma-\alpha)} \int_a^x \psi'(t) (\psi(x) - \psi(t))^{\gamma-\alpha-1} \Delta_{a+}^{\gamma;\psi} f(t)\, dt, \quad (7.23) $$

and

$$
{}^{H}\mathbb{D}_{b-}^{\alpha,\beta;\psi} f(x) = \frac{1}{\Gamma(\gamma-\alpha)} \int_x^b \psi'(t)(\psi(t)-\psi(x))^{\gamma-\alpha-1} \Delta_{b-}^{\gamma;\psi} f(t)\,dt, \quad (7.24)
$$

$x \in [a, b]$.

Remark 7.2 ([6]) Let $\mu = n(1-\beta) + \beta\alpha$, then $\lceil \mu \rceil = n$.

Assume that $g(x) = I_{a+}^{(1-\beta)(n-\alpha);\psi} f(x) \in C^n([a, b])$, we have that

$$
{}^{H}\mathbb{D}_{a+}^{\alpha,\beta;\psi} f(x) = I_{a+}^{n-\mu;\psi} \left(\frac{1}{\psi'(x)} \frac{d}{dx} \right)^n g(x). \quad (7.25)
$$

Thus

$$
{}^{H}\mathbb{D}_{a+}^{\alpha,\beta;\psi} f = {}^{C}D_{a+}^{\mu;\psi} g(x) = {}^{C}D_{a+}^{\mu;\psi} \left[I_{a+}^{(1-\beta)(n-\alpha);\psi} f(x) \right]. \quad (7.26)
$$

Assume that $w(x) = I_{b-}^{(1-\beta)(n-\alpha);\psi} f(x) \in C^n([a, b])$. Hence

$$
{}^{H}\mathbb{D}_{b-}^{\alpha,\beta;\psi} f(x) = I_{b-}^{\beta(n-\alpha);\psi} \left(-\frac{1}{\psi'(x)} \frac{d}{dx} \right)^n w(x) = I_{b-}^{n-\mu;\psi} \left(-\frac{1}{\psi'(x)} \frac{d}{dx} \right)^n w(x). \quad (7.27)
$$

Thus

$$
{}^{H}\mathbb{D}_{b-}^{\alpha,\beta;\psi} f = {}^{C}D_{b-}^{\mu;\psi} w(x) = {}^{C}D_{b-}^{\mu;\psi} \left(I_{b-}^{(1-\beta)(n-\alpha);\psi} f(x) \right). \quad (7.28)
$$

We mention the simplified ψ-Hilfer fractional Taylor formulae:

Theorem 7.3 *(see also [6]) Let $\psi, f \in C^n([a, b])$, with ψ being increasing such that $\psi'(x) \neq 0$ over $[a, b]$, where $n-1 < \alpha < n$, $0 \leq \beta \leq 1$, and $\gamma = \alpha + \beta(n-\alpha)$, $x \in [a, b]$. Then*

$$
f(x) - \sum_{k=1}^{n-1} \frac{(\psi(x)-\psi(a))^{\gamma-k}}{\Gamma(\gamma-k+1)} f_{\psi}^{[n-k]} \left(I_{a+}^{(1-\beta)(n-\alpha);\psi} f \right)(a) =
$$

$$
\frac{1}{\Gamma(\alpha)} \int_a^x \psi'(t)(\psi(x)-\psi(t))^{\alpha-1}\, {}^{H}\mathbb{D}_{a+}^{\alpha,\beta;\psi} f(t)\,dt, \quad (7.29)
$$

and

$$
f(x) - \sum_{k=1}^{n-1} \frac{(-1)^k (\psi(b)-\psi(x))^{\gamma-k}}{\Gamma(\gamma-k+1)} f_{\psi}^{[n-k]} \left(I_{b-}^{(1-\beta)(n-\alpha);\psi} f \right)(b) =
$$

$$
\frac{1}{\Gamma(\alpha)} \int_x^b \psi'(t)(\psi(t)-\psi(x))^{\alpha-1}\, {}^{H}\mathbb{D}_{b-}^{\alpha,\beta;\psi} f(t)\,dt. \quad (7.30)
$$

Here notice that $\left(I_{a+}^{(1-\beta)(n-\alpha);\psi} f \right)(a) = \left(I_{b-}^{(1-\beta)(n-\alpha);\psi} f \right)(b) = 0$.

We also mention the following alternative ψ-Hilfer fractional Taylor formulae:

Theorem 7.4 *([3]) Let* $f, \psi \in C^n([a,b])$, *with* ψ *being increasing,* $\psi'(x) \neq 0$ *over* $[a,b] \subset \mathbb{R}$, $\alpha > 0 : \lceil \alpha \rceil = n$, $0 \leq \beta \leq 1$, $\mu = n(1-\beta) + \beta\alpha$. *Assume that* $g(x) = I_{a+}^{(1-\beta)(n-\alpha);\psi} f(x)$, $w(x) = I_{b-}^{(1-\beta)(n-\alpha);\psi} f(x) \in C^n([a,b])$.

Then

(1)

$$I_{a+}^{\mu;\psi} {}^H\mathbb{D}_{a+}^{\alpha,\beta;\psi} f(x) = g(x) - \sum_{k=0}^{n-1} \frac{g_\psi^{[k]}(a)}{k!} (\psi(x) - \psi(a))^k, \tag{7.31}$$

where

$$g_\psi^{[k]}(x) = \left(\frac{1}{\psi'(x)} \frac{d}{dx} \right)^k g(x), \ k = 0, 1, ..., n-1,$$

and

(2)

$$I_{b-}^{\mu;\psi} {}^H\mathbb{D}_{b-}^{\alpha,\beta;\psi} f(x) = w(x) - \sum_{k=0}^{n-1} \frac{(-1)^k w_\psi^{[k]}(b)}{k!} (\psi(b) - \psi(x))^k, \tag{7.32}$$

where

$$w_\psi^{[k]}(x) = \left(\frac{1}{\psi'(x)} \frac{d}{dx} \right)^k w(x), \ k = 0, 1, ..., n-1; x \in [a,b].$$

Next we list two Hilfer fractional derivatives representation formulae:

Theorem 7.5 *([3]) Let* $\alpha > 0, \alpha \notin \mathbb{N}, \lceil \alpha \rceil = n, 0 < \beta < 1; f \in C^n([a,b]), [a,b] \subset \mathbb{R}$; *and set* $\gamma = \alpha + \beta(n-\alpha)$. *Assume further that* $\Delta_{a+}^{\gamma} f \in C([a,b]) : \Delta_{a+}^{\gamma-j} f(a) = 0$, *for* $j = 1, ..., n$. *Let also* $\overline{\alpha} > 0 : \lceil \overline{\alpha} \rceil = \overline{n}$, *with* $\overline{\gamma} = \overline{\alpha} + \beta(\overline{n} - \overline{\alpha})$, *and assume that* $\alpha > \overline{\alpha}$ *and* $\gamma > \overline{\gamma}$. *Then*

$${}^H\mathbb{D}_{a+}^{\overline{\alpha},\beta} f(x) = \frac{1}{\Gamma(\alpha - \overline{\alpha})} \int_a^x (x-t)^{\alpha - \overline{\alpha} - 1} \, {}^H\mathbb{D}_{a+}^{\alpha,\beta} f(t) \, dt, \tag{7.33}$$

$\forall x \in [a,b]$,

furthermore ${}^H\mathbb{D}_{a+}^{\overline{\alpha},\beta} f \in AC([a,b])$ *(absolutely continuous functions) if* $\alpha - \overline{\alpha} \geq 1$ *and* ${}^H\mathbb{D}_{a+}^{\overline{\alpha},\beta} f \in C([a,b])$ *if* $\alpha - \overline{\alpha} \in (0,1)$.

Theorem 7.6 *([3]) Let* $\alpha > 0, \alpha \notin \mathbb{N}, \lceil \alpha \rceil = n, 0 < \beta < 1; f \in C^n([a,b]), [a,b] \subset \mathbb{R}$; *and set* $\gamma = \alpha + \beta(n-\alpha)$. *Assume further that* $\Delta_{b-}^{\gamma} f \in C([a,b]) : \Delta_{b-}^{\gamma-j} f(b) = 0$, $j = 1, ..., n$. *Let also* $\overline{\alpha} > 0 : \lceil \overline{\alpha} \rceil = \overline{n}$, *with* $\overline{\gamma} = \overline{\alpha} + \beta(\overline{n} - \overline{\alpha})$, *and assume that* $\alpha > \overline{\alpha}$ *and* $\gamma > \overline{\gamma}$. *Then*

$${}^H\mathbb{D}_{b-}^{\overline{\alpha},\beta} f(x) = \frac{1}{\Gamma(\alpha - \overline{\alpha})} \int_x^b (t-x)^{\alpha - \overline{\alpha} - 1} \, {}^H\mathbb{D}_{b-}^{\alpha,\beta} f(t) \, dt, \tag{7.34}$$

$\forall\, x \in [a, b]$,

furthermore ${}^{H}\mathbb{D}_{b-}^{\overline{\alpha},\beta} f \in AC([a, b])$ *if* $\alpha - \overline{\alpha} \geq 1$ *and* ${}^{H}\mathbb{D}_{b-}^{\overline{\alpha},\beta} f \in C([a, b])$ *if* $\alpha - \overline{\alpha} \in (0, 1)$.

The fractional integral operator $I_{a+}^{\alpha} f$ and $I_{b-}^{\alpha} f$, $\alpha > 0$, are bounded in $L_p(a, b)$, $1 \leq p \leq \infty$, that is

$$\left\| I_{a+}^{\alpha} f \right\|_p \leq K \|f\|_p, \quad \left\| I_{b-}^{\alpha} f \right\|_p \leq K \|f\|_p, \tag{7.35}$$

where

$$K = \frac{(b-a)^{\alpha}}{\Gamma(\alpha+1)}. \tag{7.36}$$

The left inequality (7.35) was proved by H.G. Hardy in one of his first papers, see [7].

In this chapter we prove Hardy type fractional integral inequalities generated by the above background and by applying Chap. 6, see [5], and the results of [2].

7.2 Main Results

First we present forward results.

We need

Theorem 7.7 *([5], p. 102) Let* $p, q > 1 : \frac{1}{p} + \frac{1}{q} = 1$; $\alpha_i > \frac{1}{q}$, $i = 1, ..., m$. *Let* $a, b \in \mathbb{R}$ *and* $f_i : (a, b) \to \mathbb{R}$, *be Lebesgue measurable functions so that* $\|f_i\|_{L_q(\psi)}$ *is finite,* $i = 1, ..., m$. *Here* ψ *is strictly increasing over* $[a, b]$. *Then*

$$\left\| \prod_{i=1}^{m} \left(I_{a+}^{\alpha_i;\psi} f_i \right) \right\|_{L_p(\psi)} \leq$$

$$\frac{(\psi(b) - \psi(a))^{\sum_{i=1}^{m} \alpha_i - \frac{m}{q} + \frac{1}{p}}}{\left[\left(p \sum_{i=1}^{m} \alpha_i + m(1-p) + 1 \right)^{\frac{1}{p}} \left(\prod_{i=1}^{m} \Gamma(\alpha_i) \left(p(\alpha_i - 1) + 1 \right)^{\frac{1}{p}} \right) \right]} \left(\prod_{i=1}^{m} \|f_i\|_{L_q(\psi)} \right). \tag{7.37}$$

By Definition 7.1, (7.19) and (7.37), we conclude:

Theorem 7.8 *Let $p, q > 1 : \frac{1}{p} + \frac{1}{q} = 1$, $i = 1, ..., m$. Here $0 < \alpha_i \notin \mathbb{N} : \lceil \alpha_i \rceil = n_i \in \mathbb{N}$, $[a, b] \subset \mathbb{R}$ and $f_i \in C^{n_i}([a, b])$, $\psi \in C^{\infty}([a, b])$, ψ is strictly increasing, $0 < \beta_i \leq 1$, $\gamma_i := \alpha_i + \beta_i (n_i - \alpha_i)$; $\gamma_i - \alpha_i > \frac{1}{q}$. Assume that $\Delta_{a+}^{\gamma_i;\psi} f_i \in C([a, b])$, $i = 1, ..., m$. Then*

$$\left\| \prod_{i=1}^{m} \left({}^{H}\mathbb{D}_{a+}^{\alpha_i,\beta_i;\psi} f_i \right) \right\|_{L_p(\psi)} = \left\| \prod_{i=1}^{m} \left(I_{a+}^{\gamma_i-\alpha_i;\psi} \Delta_{a+}^{\gamma_i;\psi} f_i \right) \right\|_{L_p(\psi)} \leq$$

$$\frac{(\psi(b) - \psi(a))^{\sum_{i=1}^{m}(\gamma_i-\alpha_i) - \frac{m}{q} + \frac{1}{p}}}{\left[\left(p \sum_{i=1}^{m} (\gamma_i - \alpha_i) + m(1-p) + 1 \right)^{\frac{1}{p}} \left(\prod_{i=1}^{m} \Gamma(\gamma_i - \alpha_i) (p(\gamma_i - \alpha_i - 1) + 1)^{\frac{1}{p}} \right) \right]}$$

$$\left(\prod_{i=1}^{m} \left\| \Delta_{a+}^{\gamma_i;\psi} f_i \right\|_{L_q(\psi)} \right). \tag{7.38}$$

We mention

Theorem 7.9 *([5], p. 104) Let $p, q > 1 : \frac{1}{p} + \frac{1}{q} = 1$; $\alpha_i > \frac{1}{q}$, $i = 1, ..., m$; $r > 0$. Here $a, b \in \mathbb{R}$, $a < b$ and strictly increasing ψ on $[a, b]$. Let $f_i : (a, b) \to \mathbb{R}$, be Lebesgue measurable functions and $\|f_i\|_{L_q(\psi)}$ is finite, $i = 1, ..., m$. Then*

$$\left\| \prod_{i=1}^{m} \left(I_{a+}^{\alpha_i;\psi} f_i \right) \right\|_{L_r(\psi)} \leq$$

$$\frac{(\psi(b) - \psi(a))^{\sum_{i=1}^{m} \alpha_i - m + \frac{m}{p} + \frac{1}{r}}}{\left[\left(r \left(\sum_{i=1}^{m} \alpha_i - m + \frac{m}{p} \right) + 1 \right)^{\frac{1}{r}} \left(\prod_{i=1}^{m} \Gamma(\alpha_i) (p(\alpha_i - 1) + 1)^{\frac{1}{p}} \right) \right]} \left(\prod_{i=1}^{m} \|f_i\|_{L_q(\psi)} \right). \tag{7.39}$$

By Definition 7.1, (7.19) and (7.39), we conclude:

Theorem 7.10 *All as in Theorem 7.8, $r > 0$. Then*

$$\left\| \prod_{i=1}^{m} \left({}^{H}\mathbb{D}_{a+}^{\alpha_i,\beta_i;\psi} f_i \right) \right\|_{L_r(\psi)} = \left\| \prod_{i=1}^{m} \left(I_{a+}^{\gamma_i-\alpha_i;\psi} \Delta_{a+}^{\gamma_i;\psi} f_i \right) \right\|_{L_r(\psi)} \leq$$

$$\frac{(\psi(b) - \psi(a))^{\sum_{i=1}^{m}(\gamma_i-\alpha_i)-m+\frac{m}{p}+\frac{1}{r}}}{\left[\left(r \left(\sum_{i=1}^{m} (\gamma_i - \alpha_i) - m + \frac{m}{p} \right) + 1 \right)^{\frac{1}{r}} \left(\prod_{i=1}^{m} \Gamma(\gamma_i - \alpha_i) \left(p(\gamma_i - \alpha_i - 1) + 1 \right)^{\frac{1}{p}} \right) \right]}$$

$$\left(\prod_{i=1}^{m} \left\| \Delta_{a+}^{\gamma_i;\psi} f_i \right\|_{L_q(\psi)} \right). \tag{7.40}$$

We need

Theorem 7.11 *([5], p. 105) Let $p, q > 1 : \frac{1}{p} + \frac{1}{q} = 1$; $\alpha_i > \frac{1}{q}$, $i = 1, ..., m$. Let $a, b \in \mathbb{R}$ and $f_i : (a, b) \to \mathbb{R}$, be Lebesgue measurable functions so that $\|f_i\|_{L_q(\psi)}$ is finite, $i = 1, ..., m$. Here ψ is strictly increasing over $[a, b]$. Then*

$$\left\| \prod_{i=1}^{m} \left(I_{b-}^{\alpha_i;\psi} f_i \right) \right\|_{L_p(\psi)} \leq$$

$$\frac{(\psi(b) - \psi(a))^{\sum_{i=1}^{m}\alpha_i-\frac{m}{q}+\frac{1}{p}}}{\left[\left(p \sum_{i=1}^{m} \alpha_i + m(1-p) + 1 \right)^{\frac{1}{p}} \left(\prod_{i=1}^{m} \Gamma(\alpha_i) \left(p(\alpha_i - 1) + 1 \right)^{\frac{1}{p}} \right) \right]} \left(\prod_{i=1}^{m} \|f_i\|_{L_q(\psi)} \right). \tag{7.41}$$

By Definition 7.1, (7.20) and (7.41), we conclude:

Theorem 7.12 *Let $p, q > 1 : \frac{1}{p} + \frac{1}{q} = 1$, $i = 1, ..., m$. Here $0 < \alpha_i \notin \mathbb{N} : \lceil \alpha_i \rceil = n_i \in \mathbb{N}$, $[a, b] \subset \mathbb{R}$ and $f_i \in C^{n_i}([a, b])$, $\psi \in C^{\infty}([a, b])$, ψ is strictly increasing, $0 < \beta_i \leq 1$, $\gamma_i := \alpha_i + \beta_i (n_i - \alpha_i)$; $\gamma_i - \alpha_i > \frac{1}{q}$. Assume that $\Delta_{b-}^{\gamma_i;\psi} f_i \in C([a, b])$, $i = 1, ..., m$. Then*

$$\left\| \prod_{i=1}^{m} \left({}^{H}\mathbb{D}_{b-}^{\alpha_i,\beta_i;\psi} f_i \right) \right\|_{L_p(\psi)} = \left\| \prod_{i=1}^{m} \left(I_{b-}^{\gamma_i-\alpha_i;\psi} \Delta_{b-}^{\gamma_i;\psi} f_i \right) \right\|_{L_p(\psi)} \leq$$

$$\frac{(\psi(b) - \psi(a))^{\sum_{i=1}^{m}(\gamma_i-\alpha_i)-\frac{m}{q}+\frac{1}{p}}}{\left[\left(p \sum_{i=1}^{m} (\gamma_i - \alpha_i) + m(1-p) + 1 \right)^{\frac{1}{p}} \left(\prod_{i=1}^{m} \Gamma(\gamma_i - \alpha_i) \left(p(\gamma_i - \alpha_i - 1) + 1 \right)^{\frac{1}{p}} \right) \right]}$$

$$\left(\prod_{i=1}^{m} \left\| \Delta_{b-}^{\gamma_i;\psi} f_i \right\|_{L_q(\psi)} \right). \tag{7.42}$$

We mention

Theorem 7.13 *([5], p. 107) Let* $p, q > 1 : \frac{1}{p} + \frac{1}{q} = 1$; $\alpha_i > \frac{1}{q}$, $i = 1, ..., m$; $r > 0$. *Here* $a, b \in \mathbb{R}$, $a < b$ *and strictly increasing* ψ *on* $[a, b]$. *Let* $f_i : (a, b) \to \mathbb{R}$, *be Lebesgue measurable functions and* $\|f_i\|_{L_q(\psi)}$ *is finite,* $i = 1, ..., m$. *Then*

$$\left\| \prod_{i=1}^{m} \left(I_{b-}^{\alpha_i;\psi} f_i \right) \right\|_{L_r(\psi)} \leq$$

$$\frac{(\psi(b) - \psi(a))^{\sum_{i=1}^{m} \alpha_i - m + \frac{m}{p} + \frac{1}{r}}}{\left[\left(r \left(\sum_{i=1}^{m} \alpha_i - m + \frac{m}{p} \right) + 1 \right)^{\frac{1}{r}} \left(\prod_{i=1}^{m} \Gamma(\alpha_i) (p(\alpha_i - 1) + 1)^{\frac{1}{p}} \right) \right]} \left(\prod_{i=1}^{m} \|f_i\|_{L_q(\psi)} \right). \tag{7.43}$$

By Definition 7.1, (7.20) and (7.43), we conclude:

Theorem 7.14 *All as in Theorem 7.12,* $r > 0$. *Then*

$$\left\| \prod_{i=1}^{m} \left({}^{H}\mathbb{D}_{b-}^{\alpha_i,\beta_i;\psi} f_i \right) \right\|_{L_r(\psi)} = \left\| \prod_{i=1}^{m} \left(I_{b-}^{\gamma_i-\alpha_i;\psi} \Delta_{b-}^{\gamma_i;\psi} f_i \right) \right\|_{L_r(\psi)} \leq$$

$$\frac{(\psi(b) - \psi(a))^{\sum_{i=1}^{m} (\gamma_i - \alpha_i) - m + \frac{m}{p} + \frac{1}{r}}}{\left[\left(r \left(\sum_{i=1}^{m} (\gamma_i - \alpha_i) - m + \frac{m}{p} \right) + 1 \right)^{\frac{1}{r}} \left(\prod_{i=1}^{m} \Gamma(\gamma_i - \alpha_i) (p(\gamma_i - \alpha_i - 1) + 1)^{\frac{1}{p}} \right) \right]}$$

$$\left(\prod_{i=1}^{m} \left\| \Delta_{b-}^{\gamma_i;\psi} f_i \right\|_{L_q(\psi)} \right). \tag{7.44}$$

Theorems 7.7–7.14 have great applications for various ψ, expecially when $\psi(x) = x$.

Notice that (7.33) can be written as

$${}^{H}\mathbb{D}_{a+}^{\overline{\alpha},\beta} f(x) = \left(I_{a+}^{\alpha-\overline{\alpha}} {}^{H}\mathbb{D}_{a+}^{\alpha,\beta} f \right)(x), \tag{7.45}$$

$\forall x \in [a, b]$, $0 < \beta < 1$,

and that (7.34) can be written as

$$ {}^{H}\mathbb{D}_{b-}^{\overline{\alpha},\beta} f(x) = \left(I_{b-}^{\alpha-\overline{\alpha}\,H}\mathbb{D}_{b-}^{\alpha,\beta} f \right)(x), \tag{7.46} $$

$\forall\, x \in [a,b]$, $0 < \beta < 1$.

By Theorems 7.5, 7.7, $\psi(x) = x$, and (7.45) we derive

Theorem 7.15 *Here* $i = 1, ..., m$. *Let* $\alpha_i > 0$, $\alpha_i \notin \mathbb{N} : \lceil \alpha_i \rceil = n_i$, $0 < \beta_i < 1$; $f_i \in C^{n_i}([a,b])$, $[a,b] \subset \mathbb{R}$; *and set* $\gamma_i := \alpha_i + \beta_i (n_i - \alpha_i)$. *Assume further that* $\Delta_{a+}^{\gamma_i} f_i \in C([a,b]) : \Delta_{a+}^{\gamma_i - j_i} f_i(a) = 0$, *for* $j_i = 1, ..., n_i$. *Let also* $\overline{\alpha}_i > 0 : \lceil \overline{\alpha}_i \rceil = \overline{n}_i$, *with* $\overline{\gamma}_i := \overline{\alpha}_i + \beta_i (\overline{n}_i - \overline{\alpha}_i)$, *and assume that* $\alpha_i > \overline{\alpha}_i$ *and* $\gamma_i > \overline{\gamma}_i$. *Finally let* $p, q > 1 : \frac{1}{p} + \frac{1}{q} = 1$ *with* $\alpha_i - \overline{\alpha}_i > \frac{1}{q}$. *Then*

$$ \left\| \prod_{i=1}^{m} \left({}^{H}\mathbb{D}_{a+}^{\overline{\alpha}_i,\beta_i} f_i \right) \right\|_{L_p([a,b])} \le $$

$$ \frac{(b-a)^{\sum_{i=1}^{m}(\alpha_i - \overline{\alpha}_i) - \frac{m}{q} + \frac{1}{p}}}{\left[\left(p \sum_{i=1}^{m} (\alpha_i - \overline{\alpha}_i) + m(1-p) + 1 \right)^{\frac{1}{p}} \left(\prod_{i=1}^{m} \Gamma(\alpha_i - \overline{\alpha}_i) \left(p(\alpha_i - \overline{\alpha}_i - 1) + 1 \right)^{\frac{1}{p}} \right) \right]} $$

$$ \left(\prod_{i=1}^{m} \left\| {}^{H}\mathbb{D}_{a+}^{\alpha_i,\beta_i} f_i \right\|_{L_q([a,b])} \right). \tag{7.47} $$

We continue with

Theorem 7.16 *All as in Theorem 7.15, plus* $r > 0$. *Then*

$$ \left\| \prod_{i=1}^{m} \left({}^{H}\mathbb{D}_{a+}^{\overline{\alpha}_i,\beta_i} f_i \right) \right\|_{L_r([a,b])} \le $$

$$ \frac{(b-a)^{\sum_{i=1}^{m}(\alpha_i - \overline{\alpha}_i) - m + \frac{m}{p} + \frac{1}{r}}}{\left[\left(r \left(\sum_{i=1}^{m} (\alpha_i - \overline{\alpha}_i) - m + \frac{m}{p} \right) + 1 \right)^{\frac{1}{r}} \left(\prod_{i=1}^{m} \Gamma(\alpha_i - \overline{\alpha}_i) \left(p(\alpha_i - \overline{\alpha}_i - 1) + 1 \right)^{\frac{1}{p}} \right) \right]} $$

$$ \left(\prod_{i=1}^{m} \left\| {}^{H}\mathbb{D}_{a+}^{\alpha_i,\beta_i} f_i \right\|_{L_q([a,b])} \right). \tag{7.48} $$

Proof By Theorems 7.5, 7.9, $\psi(x) = x$, and (7.45). ■

By Theorems 7.6, 7.11, $\psi(x) = x$, and (7.46) we obtain

Theorem 7.17 *Here $i = 1, ..., m$. Let $\alpha_i > 0$, $\alpha_i \notin \mathbb{N} : \lceil \alpha_i \rceil = n_i$, $0 < \beta_i < 1$; $f_i \in C^{n_i}([a,b])$, $[a,b] \subset \mathbb{R}$; and set $\gamma_i := \alpha_i + \beta_i(n_i - \alpha_i)$. Assume further that $\Delta_{b-}^{\gamma_i} f_i \in C([a,b]) : \Delta_{b-}^{\gamma_i - j_i} f_i(b) = 0$, for $j_i = 1, ..., n_i$. Let also $\overline{\alpha}_i > 0 : \lceil \overline{\alpha}_i \rceil = \overline{n}_i$, with $\overline{\gamma}_i := \overline{\alpha}_i + \beta_i(\overline{n}_i - \overline{\alpha}_i)$, and assume that $\alpha_i > \overline{\alpha}_i$ and $\gamma_i > \overline{\gamma}_i$. Finally let $p, q > 1 : \frac{1}{p} + \frac{1}{q} = 1$ with $\alpha_i - \overline{\alpha}_i > \frac{1}{q}$. Then*

$$\left\| \prod_{i=1}^{m} \left({}^{H}\mathbb{D}_{b-}^{\overline{\alpha}_i,\beta_i} f_i \right) \right\|_{L_p([a,b])} \leq$$

$$\frac{(b-a)^{\sum_{i=1}^{m}(\alpha_i - \overline{\alpha}_i) - \frac{m}{q} + \frac{1}{p}}}{\left[\left(p \sum_{i=1}^{m} (\alpha_i - \overline{\alpha}_i) + m(1-p) + 1 \right)^{\frac{1}{p}} \left(\prod_{i=1}^{m} \Gamma(\alpha_i - \overline{\alpha}_i) \left(p(\alpha_i - \overline{\alpha}_i - 1) + 1 \right)^{\frac{1}{p}} \right) \right]}$$

$$\left(\prod_{i=1}^{m} \left\| {}^{H}\mathbb{D}_{b-}^{\alpha_i,\beta_i} f_i \right\|_{L_q([a,b])} \right). \tag{7.49}$$

We continue with

Theorem 7.18 *All as in Theorem 7.17, plus $r > 0$. Then*

$$\left\| \prod_{i=1}^{m} \left({}^{H}\mathbb{D}_{b-}^{\overline{\alpha}_i,\beta_i} f_i \right) \right\|_{L_r([a,b])} \leq$$

$$\frac{(b-a)^{\sum_{i=1}^{m}(\alpha_i - \overline{\alpha}_i) - m + \frac{m}{p} + \frac{1}{r}}}{\left[\left(r \left(\sum_{i=1}^{m} (\alpha_i - \overline{\alpha}_i) - m + \frac{m}{p} \right) + 1 \right)^{\frac{1}{r}} \left(\prod_{i=1}^{m} \Gamma(\alpha_i - \overline{\alpha}_i) \left(p(\alpha_i - \overline{\alpha}_i - 1) + 1 \right)^{\frac{1}{p}} \right) \right]}$$

$$\left(\prod_{i=1}^{m} \left\| {}^{H}\mathbb{D}_{b-}^{\alpha_i,\beta_i} f_i \right\|_{L_q([a,b])} \right). \tag{7.50}$$

Proof By Theorems 7.6, 7.13, $\psi(x) = x$, and (7.46). ■

On Theorem 7.3:
Given that $f_{\psi}^{[n-k]} \left(I_{a+}^{(1-\beta)(n-\alpha);\psi} f \right)(a) = 0$, $k = 1, ..., n-1$, we obtain that

$$f(x) = \left(I_{a+}^{\alpha;\psi} \, {}^{H}\mathbb{D}_{a+}^{\alpha,\beta;\psi} f \right)(x), \quad \forall x \in [a,b]. \tag{7.51}$$

And, given that $f_{\psi}^{[n-k]}\left(I_{b-}^{(1-\beta)(n-\alpha);\psi}f\right)(b)=0$, $k=1,...,n-1$, we obtain that

$$f(x)=\left(I_{b-}^{\alpha;\psi H}\mathbb{D}_{b-}^{\alpha,\beta;\psi}f\right)(x), \ \forall x\in[a,b]. \tag{7.52}$$

Based on Theorem 7.3 (7.29), (7.51) and Theorem 7.7 we present

Theorem 7.19 *Here $i=1,...,m$. Let $f_i\in C^{n_i}([a,b])$, with $\psi\in C^{\infty}([a,b])$ and strictly increasing, where $0<\alpha_i\notin\mathbb{N}$, $\lceil\alpha_i\rceil=n_i\in\mathbb{N}$, $0\leq\beta_i\leq 1$. Let also $p,q>1$: $\frac{1}{p}+\frac{1}{q}=1$; $\alpha_i>\frac{1}{q}$. Assume also $f_{i\psi}^{[n_i-k_i]}\left(I_{a+}^{(1-\beta_i)(n_i-\alpha_i);\psi}f_i\right)(a)=0$, $k_i=1,...,n_i-1$. Then*

$$\left\|\prod_{i=1}^{m}f_i\right\|_{L_p(\psi)}\leq\frac{(\psi(b)-\psi(a))^{\sum_{i=1}^{m}\alpha_i-\frac{m}{q}+\frac{1}{p}}}{\left[\left(p\sum_{i=1}^{m}\alpha_i+m(1-p)+1\right)^{\frac{1}{p}}\left(\prod_{i=1}^{m}\Gamma(\alpha_i)(p(\alpha_i-1)+1)^{\frac{1}{p}}\right)\right]}$$

$$\left(\prod_{i=1}^{m}\left\|{}^{H}\mathbb{D}_{a+}^{\alpha_i,\beta_i;\psi}f_i\right\|_{L_q(\psi)}\right). \tag{7.53}$$

Based on Theorem 7.3 (7.30), (7.52) and Theorem 7.11 we give

Theorem 7.20 *Here $i=1,...,m$. Let $f_i\in C^{n_i}([a,b])$, with $\psi\in C^{\infty}([a,b])$ and strictly increasing, where $0<\alpha_i\notin\mathbb{N}$, $\lceil\alpha_i\rceil=n_i\in\mathbb{N}$, $0\leq\beta_i\leq 1$. Let also $p,q>1$: $\frac{1}{p}+\frac{1}{q}=1$; $\alpha_i>\frac{1}{q}$. Assume also $f_{i\psi}^{[n_i-k_i]}\left(I_{b-}^{(1-\beta_i)(n_i-\alpha_i);\psi}f_i\right)(b)=0$, $k_i=1,...,n_i-1$. Then*

$$\left\|\prod_{i-1}^{m}f_i\right\|_{L_p(\psi)}\leq\frac{(\psi(b)-\psi(a))^{\sum_{i=1}^{m}\alpha_i-\frac{m}{q}+\frac{1}{p}}}{\left[\left(p\sum_{i=1}^{m}\alpha_i+m(1-p)+1\right)^{\frac{1}{p}}\left(\prod_{i=1}^{m}\Gamma(\alpha_i)(p(\alpha_i-1)+1)^{\frac{1}{p}}\right)\right]}$$

$$\left(\prod_{i=1}^{m}\left\|{}^{H}\mathbb{D}_{b-}^{\alpha_i,\beta_i;\psi}f_i\right\|_{L_q(\psi)}\right). \tag{7.54}$$

On Theorem 7.4:
Assume $g_{\psi}^{[k]}(a)=0$, $k=0,1,...,n-1$, then

$$g(x)=I_{a+}^{\mu;\psi H}\mathbb{D}_{a+}^{\alpha,\beta;\psi}f(x), \ \forall x\in[a,b]. \tag{7.55}$$

And, if $w_{\psi}^{[k]}(b)=0$, $k=0,1,...,n-1$, then

$$w(x)=I_{b-}^{\mu;\psi H}\mathbb{D}_{b-}^{\alpha,\beta;\psi}f(x), \ \forall x\in[a,b]. \tag{7.56}$$

Based on Theorem 7.4 (7.31), (7.55) and Theorem 7.7 we get:

Theorem 7.21 *Here $i = 1, ..., m$. Let $f_i \in C^{n_i}([a, b])$, with $\psi \in C^{\infty}([a, b])$ and strictly increasing, where $0 < \alpha_i \notin \mathbb{N}$, $\lceil \alpha_i \rceil = n_i \in \mathbb{N}$, $0 \leq \beta_i \leq 1$, $\mu_i = n_i(1 - \beta_i) + \beta_i \alpha_i$. Assume that $g_i(x) = I_{a+}^{(1-\beta_i)(n_i-\alpha_i);\psi} f_i(x) \in C^{n_i}([a, b])$. Assume that $g_{i\psi}^{[k_i]}(a) = 0$, where $g_{i\psi}^{[k_i]}(x) = \left(\frac{1}{\psi'(x)}\frac{d}{dx}\right)^{k_i} g_i(x)$, $k_i = 0, 1, ..., n_i - 1$, $x \in [a, b]$. Let $p, q > 1$: $\frac{1}{p} + \frac{1}{q} = 1$; $\mu_i > \frac{1}{q}$. Then*

$$\left\| \prod_{i=1}^{m} g_i \right\|_{L_p(\psi)} \leq \frac{(\psi(b) - \psi(a))^{\sum_{i=1}^{m} \mu_i - \frac{m}{q} + \frac{1}{p}}}{\left[\left(p \sum_{i=1}^{m} \mu_i + m(1-p) + 1 \right)^{\frac{1}{p}} \left(\prod_{i=1}^{m} \Gamma(\mu_i)(p(\mu_i - 1) + 1)^{\frac{1}{p}} \right) \right]}$$

$$\left(\prod_{i=1}^{m} \left\| {}^{H}\mathbb{D}_{a+}^{\alpha_i, \beta_i; \psi} f_i \right\|_{L_q(\psi)} \right). \tag{7.57}$$

By Theorem 7.4 (7.32), (7.56) and Theorem 7.11 we get

Theorem 7.22 *Here $i = 1, ..., m$. Let $f_i \in C^{n_i}([a, b])$, with $\psi \in C^{\infty}([a, b])$ and strictly increasing, where $0 < \alpha_i \notin \mathbb{N}$, $\lceil \alpha_i \rceil = n_i \in \mathbb{N}$, $0 \leq \beta_i \leq 1$, $\mu_i = n_i(1 - \beta_i) + \beta_i \alpha_i$. Assume that $w_i(x) = I_{b-}^{(1-\beta_i)(n_i-\alpha_i);\psi} f_i(x) \in C^{n_i}([a, b])$. Assume that $w_{i\psi}^{[k_i]}(b) = 0$, where $w_{i\psi}^{[k_i]}(x) = \left(\frac{1}{\psi'(x)}\frac{d}{dx}\right)^{k_i} w_i(x)$, $k_i = 0, 1, ..., n_i - 1$, $x \in [a, b]$. Let $p, q > 1$: $\frac{1}{p} + \frac{1}{q} = 1$; $\mu_i > \frac{1}{q}$. Then*

$$\left\| \prod_{i=1}^{m} w_i \right\|_{L_p(\psi)} \leq \frac{(\psi(b) - \psi(a))^{\sum_{i=1}^{m} \mu_i - \frac{m}{q} + \frac{1}{p}}}{\left[\left(p \sum_{i=1}^{m} \mu_i + m(1-p) + 1 \right)^{\frac{1}{p}} \left(\prod_{i=1}^{m} \Gamma(\mu_i)(p(\mu_i - 1) + 1)^{\frac{1}{p}} \right) \right]}$$

$$\left(\prod_{i=1}^{m} \left\| {}^{H}\mathbb{D}_{b-}^{\alpha_i, \beta_i; \psi} f_i \right\|_{L_q(\psi)} \right). \tag{7.58}$$

Next come reverse results.

We will use the following results:

Theorem 7.23 *([2], p. 447) Let $0 < p < 1$, $q < 0$ such that $\frac{1}{p} + \frac{1}{q} = 1$; $\alpha_i > 0$, $i = 1, ..., m$. Here $a, b \in \mathbb{R}$ and strictly increasing ψ. Let $f_i : (a, b) \to \mathbb{R}$, $i = 1, ..., m$, of fixed strict sign a.e., which are Lebesgue measurable functions, so that $\left\| \prod_{i=1}^{m} \left(I_{a+}^{\alpha_i;\psi} f_i \right) \right\|_{L_p(\psi)}$, $\prod_{i=1}^{m} \|f_i\|_{L_q(\psi)}$ are finite. Then*

$$\left\| \prod_{i=1}^{m} \left(I_{a+}^{\alpha_i;\psi} f_i \right) \right\|_{L_p(\psi)} \geq$$

$$\frac{(\psi(b) - \psi(a))^{\sum_{i=1}^{m} \alpha_i + m\left(\frac{1}{p} - 1\right) + \frac{1}{p}}}{\left[\left(p \sum_{i=1}^{m} \alpha_i + m(1-p) + 1 \right)^{\frac{1}{p}} \left(\prod_{i=1}^{m} \Gamma(\alpha_i) (p(\alpha_i - 1) + 1)^{\frac{1}{p}} \right) \right]} \left(\prod_{i=1}^{m} \|f_i\|_{L_q(\psi)} \right). \tag{7.59}$$

Theorem 7.24 *([2], p. 449) Let $0 < p < 1$, $q < 0$ such that $\frac{1}{p} + \frac{1}{q} = 1$; $\alpha_i > 0$, $i = 1, ..., m$; $r > 0$. Here $a, b \in \mathbb{R}$ and strictly increasing ψ. Let $f_i : (a, b) \to \mathbb{R}$, $i = 1, ..., m$, of fixed strict sign a.e., which are Lebesgue measurable functions, and $\left\| \prod_{i=1}^{m} \left(I_{a+}^{\alpha_i;\psi} f_i \right) \right\|_{L_r(\psi)}$, $\prod_{i=1}^{m} \|f_i\|_{L_q(\psi)}$ are finite. Then*

$$\left\| \prod_{i=1}^{m} \left(I_{a+}^{\alpha_i;\psi} f_i \right) \right\|_{L_r(\psi)} \geq$$

$$\frac{(\psi(b) - \psi(a))^{\sum_{i=1}^{m} \alpha_i - m + \frac{m}{p} + \frac{1}{r}}}{\left[\left(r \left(\sum_{i=1}^{m} \alpha_i - m + \frac{m}{p} \right) + 1 \right)^{\frac{1}{r}} \left(\prod_{i=1}^{m} \Gamma(\alpha_i) (p(\alpha_i - 1) + 1)^{\frac{1}{p}} \right) \right]} \left(\prod_{i=1}^{m} \|f_i\|_{L_q(\psi)} \right). \tag{7.60}$$

Theorem 7.25 *([2], p. 450) Let $0 < p < 1$, $q < 0$ such that $\frac{1}{p} + \frac{1}{q} = 1$; $\alpha_i > 0$, $i = 1, ..., m$. Here $a, b \in \mathbb{R}$ and strictly increasing ψ. Let $f_i : (a, b) \to \mathbb{R}$, $i = 1, ..., m$, of fixed strict sign a.e., which are Lebesgue measurable functions, so that $\left\| \prod_{i=1}^{m} \left(I_{b-}^{\alpha_i;\psi} f_i \right) \right\|_{L_p(\psi)}$, $\prod_{i=1}^{m} \|f_i\|_{L_q(\psi)}$ are finite. Then*

$$\left\| \prod_{i=1}^{m} \left(I_{b-}^{\alpha_i;\psi} f_i \right) \right\|_{L_p(\psi)} \geq$$

$$\frac{(\psi(b) - \psi(a))^{\sum_{i=1}^{m} \alpha_i + m\left(\frac{1}{p} - 1\right) + \frac{1}{p}}}{\left[\left(p \sum_{i=1}^{m} \alpha_i + m(1-p) + 1 \right)^{\frac{1}{p}} \left(\prod_{i=1}^{m} \Gamma(\alpha_i) (p(\alpha_i - 1) + 1)^{\frac{1}{p}} \right) \right]} \left(\prod_{i=1}^{m} \|f_i\|_{L_q(\psi)} \right). \tag{7.61}$$

Theorem 7.26 *([2], p. 451) Let $0 < p < 1$, $q < 0$ such that $\frac{1}{p} + \frac{1}{q} = 1$; $\alpha_i > 0$, $i = 1, ..., m$; $r > 0$. Here $a, b \in \mathbb{R}$ and strictly increasing ψ. Let $f_i : (a, b) \to \mathbb{R}$,*

$i = 1, ..., m$, of fixed strict sign a.e., which are Lebesgue measurable functions, and $\left\| \prod_{i=1}^{m} \left(I_{b-}^{\alpha_i;\psi} f_i \right) \right\|_{L_r(\psi)}$, $\prod_{i=1}^{m} \|f_i\|_{L_q(\psi)}$ are finite. Then

$$\left\| \prod_{i=1}^{m} \left(I_{b-}^{\alpha_i;\psi} f_i \right) \right\|_{L_r(\psi)} \geq$$

$$\frac{(\psi(b) - \psi(a))^{\sum_{i=1}^{m} \alpha_i - m + \frac{m}{p} + \frac{1}{r}}}{\left[\left(r \left(\sum_{i=1}^{m} \alpha_i - m + \frac{m}{p} \right) + 1 \right)^{\frac{1}{r}} \left(\prod_{i=1}^{m} \Gamma(\alpha_i) (p(\alpha_i - 1) + 1)^{\frac{1}{p}} \right) \right]} \left(\prod_{i=1}^{m} \|f_i\|_{L_q(\psi)} \right). \tag{7.62}$$

We give the following reverse fractional results:
By Definition 7.1, (7.19) and (7.59) we conclude:

Theorem 7.27 *Let $0 < p < 1$, $q < 0 : \frac{1}{p} + \frac{1}{q} = 1$, $i = 1, ..., m$. Here $0 < \alpha_i \notin \mathbb{N}$: $\lceil \alpha_i \rceil = n_i \in \mathbb{N}$, $[a, b] \subset \mathbb{R}$ and $f_i \in C^{n_i}([a, b])$, $\psi \in C^{\infty}([a, b])$, ψ is strictly increasing, $0 < \beta_i \leq 1$, $\gamma_i := \alpha_i + \beta_i (n_i - \alpha_i)$. Assume that $\Delta_{a+}^{\gamma_i;\psi} f_i \in C([a, b])$, and of fixed strict sign a.e., $i = 1, ..., m$, so that $\left(\prod_{i=1}^{m} \left\| \Delta_{a+}^{\gamma_i;\psi} f_i \right\|_{L_q(\psi)} \right)$ is finite. Then*

$$\left\| \prod_{i=1}^{m} \left({}^{H}\mathbb{D}_{a+}^{\alpha_i,\beta_i;\psi} f_i \right) \right\|_{L_p(\psi)} = \left\| \prod_{i=1}^{m} \left(I_{a+}^{\gamma_i - \alpha_i;\psi} \Delta_{a+}^{\gamma_i;\psi} f_i \right) \right\|_{L_p(\psi)} \geq$$

$$\frac{(\psi(b) - \psi(a))^{\sum_{i=1}^{m} (\gamma_i - \alpha_i) - \frac{m}{q} + \frac{1}{p}}}{\left[\left(p \sum_{i=1}^{m} (\gamma_i - \alpha_i) + m(1 - p) + 1 \right)^{\frac{1}{p}} \left(\prod_{i=1}^{m} \Gamma(\gamma_i - \alpha_i) (p(\gamma_i - \alpha_i - 1) + 1)^{\frac{1}{p}} \right) \right]}$$

$$\left(\prod_{i=1}^{m} \left\| \Delta_{a+}^{\gamma_i;\psi} f_i \right\|_{L_q(\psi)} \right). \tag{7.63}$$

We give

Theorem 7.28 *All as in Theorem 7.27, $r > 0$. Then*

$$\left\| \prod_{i=1}^{m} \left({}^{H}\mathbb{D}_{a+}^{\alpha_i,\beta_i;\psi} f_i \right) \right\|_{L_r(\psi)} = \left\| \prod_{i=1}^{m} \left(I_{a+}^{\gamma_i - \alpha_i;\psi} \Delta_{a+}^{\gamma_i;\psi} f_i \right) \right\|_{L_r(\psi)} \geq$$

$$\frac{(\psi(b)-\psi(a))^{\sum_{i=1}^{m}(\gamma_i-\alpha_i)-m+\frac{m}{p}+\frac{1}{r}}}{\left[\left(r\left(\sum_{i=1}^{m}(\gamma_i-\alpha_i)-m+\frac{m}{p}\right)+1\right)^{\frac{1}{r}}\left(\prod_{i=1}^{m}\Gamma(\gamma_i-\alpha_i)(p(\gamma_i-\alpha_i-1)+1)^{\frac{1}{p}}\right)\right]}$$

$$\left(\prod_{i=1}^{m}\left\|\Delta_{a+}^{\gamma_i;\psi}f_i\right\|_{L_q(\psi)}\right). \tag{7.64}$$

Proof See Definition 7.1, (7.19) and apply Theorem 7.24. ■

By Definition 7.1, (7.20) and (7.61), we conclude:

Theorem 7.29 *Let* $0<p<1$, $q<0$: $\frac{1}{p}+\frac{1}{q}=1$, $i=1,...,m$. *Here* $0<\alpha_i\notin\mathbb{N}$: $\lceil\alpha_i\rceil=n_i\in\mathbb{N}$, $[a,b]\subset\mathbb{R}$ *and* $f_i\in C^{n_i}([a,b])$, $\psi\in C^{\infty}([a,b])$, ψ *is strictly increasing,* $0<\beta_i\leq 1$, $\gamma_i:=\alpha_i+\beta_i(n_i-\alpha_i)$. *Assume that* $\Delta_{b-}^{\gamma_i;\psi}f_i\in C([a,b])$, *and of fixed strict sign a.e.,* $i=1,...,m$, *so that* $\left(\prod_{i=1}^{m}\left\|\Delta_{b-}^{\gamma_i;\psi}f_i\right\|_{L_q(\psi)}\right)$ *is finite. Then*

$$\left\|\prod_{i=1}^{m}\left({}^{H}\mathbb{D}_{b-}^{\alpha_i,\beta_i;\psi}f_i\right)\right\|_{L_p(\psi)}=\left\|\prod_{i=1}^{m}\left(I_{b-}^{\gamma_i-\alpha_i;\psi}\Delta_{b-}^{\gamma_i;\psi}f_i\right)\right\|_{L_p(\psi)}\geq$$

$$\frac{(\psi(b)-\psi(a))^{\sum_{i=1}^{m}(\gamma_i-\alpha_i)-\frac{m}{q}+\frac{1}{p}}}{\left[\left(p\sum_{i=1}^{m}(\gamma_i-\alpha_i)+m(1-p)+1\right)^{\frac{1}{p}}\left(\prod_{i=1}^{m}\Gamma(\gamma_i-\alpha_i)(p(\gamma_i-\alpha_i-1)+1)^{\frac{1}{p}}\right)\right]}$$

$$\left(\prod_{i=1}^{m}\left\|\Delta_{b-}^{\gamma_i;\psi}f_i\right\|_{L_q(\psi)}\right). \tag{7.65}$$

We give

Theorem 7.30 *All as in Theorem 7.29,* $r>0$. *Then*

$$\left\|\prod_{i=1}^{m}\left({}^{H}\mathbb{D}_{b-}^{\alpha_i,\beta_i;\psi}f_i\right)\right\|_{L_r(\psi)}=\left\|\prod_{i=1}^{m}\left(I_{b-}^{\gamma_i-\alpha_i;\psi}\Delta_{b-}^{\gamma_i;\psi}f_i\right)\right\|_{L_r(\psi)}\geq$$

$$\frac{(\psi(b)-\psi(a))^{\sum_{i=1}^{m}(\gamma_i-\alpha_i)-m+\frac{m}{p}+\frac{1}{r}}}{\left[\left(r\left(\sum_{i=1}^{m}(\gamma_i-\alpha_i)-m+\frac{m}{p}\right)+1\right)^{\frac{1}{r}}\left(\prod_{i=1}^{m}\Gamma(\gamma_i-\alpha_i)(p(\gamma_i-\alpha_i-1)+1)^{\frac{1}{p}}\right)\right]}$$

$$\left(\prod_{i=1}^{m}\left\|\Delta_{b-}^{\gamma_i;\psi}f_i\right\|_{L_q(\psi)}\right). \tag{7.66}$$

Proof See Definition 7.1, (7.20) and apply Theorem 7.26. ■

We present

Theorem 7.31 *Here* $i=1,...,m$. *Let* $f_i \in C^{n_i}([a,b])$, *with* $\psi \in C^{\infty}([a,b])$ *and strictly increasing, where* $0<\alpha_i \notin \mathbb{N}$, $\lceil \alpha_i \rceil = n_i \in \mathbb{N}$, $0 \le \beta_i \le 1$. *Let also* $0<p<1$, $q<0: \frac{1}{p}+\frac{1}{q}=1$; *and* ${}^{H}\mathbb{D}_{a+}^{\alpha_i,\beta_i;\psi}f_i$ *is of fixed strict sign a.e., so that* $f_{i\psi}^{[n_i-k_i]}\left(I_{a+}^{(1-\beta_i)(n_i-\alpha_i);\psi}f_i\right)(a)=0$, $k_i=1,...,n_i-1$. *Assume that* $\left(\prod_{i=1}^{m}\left\|{}^{H}\mathbb{D}_{a+}^{\alpha_i,\beta_i;\psi}f_i\right\|_{L_q(\psi)}\right)$ *is finite. Then*

$$\left\|\prod_{i=1}^{m}f_i\right\|_{L_p(\psi)} \ge \frac{(\psi(b)-\psi(a))^{\sum_{i=1}^{m}\alpha_i-\frac{m}{q}+\frac{1}{p}}}{\left[\left(p\sum_{i=1}^{m}\alpha_i+m(1-p)+1\right)^{\frac{1}{p}}\left(\prod_{i=1}^{m}\Gamma(\alpha_i)(p(\alpha_i-1)+1)^{\frac{1}{p}}\right)\right]}$$

$$\left(\prod_{i=1}^{m}\left\|{}^{H}\mathbb{D}_{a+}^{\alpha_i,\beta_i;\psi}f_i\right\|_{L_q(\psi)}\right). \tag{7.67}$$

Proof By using Theorems 7.3, 7.23 and (7.51). ■

We also give

Theorem 7.32 *Here* $i=1,...,m$. *Let* $f_i \in C^{n_i}([a,b])$, *with* $\psi \in C^{\infty}([a,b])$ *and strictly increasing, where* $0<\alpha_i \notin \mathbb{N}$, $\lceil \alpha_i \rceil = n_i \in \mathbb{N}$, $0 \le \beta_i \le 1$. *Let also* $0<p<1$, $q<0: \frac{1}{p}+\frac{1}{q}=1$; *and* ${}^{H}\mathbb{D}_{b-}^{\alpha_i,\beta_i;\psi}f_i$ *is of fixed strict sign a.e., so that* $f_{i\psi}^{[n_i-k_i]}\left(I_{b-}^{(1-\beta_i)(n_i-\alpha_i);\psi}f_i\right)(b)=0$, $k_i=1,...,n_i-1$. *Assume that* $\left(\prod_{i=1}^{m}\left\|{}^{H}\mathbb{D}_{b-}^{\alpha_i,\beta_i;\psi}f_i\right\|_{L_q(\psi)}\right)$ *is finite. Then*

$$\left\|\prod_{i=1}^{m}f_i\right\|_{L_p(\psi)} \ge \frac{(\psi(b)-\psi(a))^{\sum_{i=1}^{m}\alpha_i-\frac{m}{q}+\frac{1}{p}}}{\left[\left(p\sum_{i=1}^{m}\alpha_i+m(1-p)+1\right)^{\frac{1}{p}}\left(\prod_{i=1}^{m}\Gamma(\alpha_i)(p(\alpha_i-1)+1)^{\frac{1}{p}}\right)\right]}$$

$$\left(\prod_{i=1}^{m}\left\|{}^{H}\mathbb{D}_{b-}^{\alpha_i,\beta_i;\psi}f_i\right\|_{L_q(\psi)}\right). \tag{7.68}$$

Proof By using Theorems 7.3, 7.25 and (7.52). ■

We continue with

Theorem 7.33 *Here* $i = 1, ..., m$. *Let* $f_i \in C^{n_i}([a,b])$, *with* $\psi \in C^{\infty}([a,b])$ *and strictly increasing, where* $0 < \alpha_i \notin \mathbb{N}$, $\lceil \alpha_i \rceil = n_i \in \mathbb{N}$, $0 \leq \beta_i \leq 1$, $\mu_i = n_i(1-\beta_i) + \beta_i \alpha_i$. *Assume that* $g_i(x) = I_{a+}^{(1-\beta_i)(n_i-\alpha_i);\psi} f_i(x) \in C^{n_i}([a,b])$. *And assume that* $g_{i\psi}^{[k_i]}(a) = 0$, *where* $g_{i\psi}^{[k_i]}(x) = \left(\frac{1}{\psi'(x)}\frac{d}{dx}\right)^{k_i} g_i(x)$, $k_i = 0, 1, ..., n_i - 1$, $x \in [a,b]$. *Let* $0 < p < 1$, $q < 0 : \frac{1}{p} + \frac{1}{q} = 1$. *Suppose also that* ${}^{H}\mathbb{D}_{a+}^{\alpha_i,\beta_i;\psi} f_i$ *is of fixed strict sign a.e., and* $\left(\prod_{i=1}^{m} \left\| {}^{H}\mathbb{D}_{a+}^{\alpha_i,\beta_i;\psi} f_i \right\|_{L_q(\psi)}\right)$ *is finite. Then*

$$\left\| \prod_{i=1}^{m} g_i \right\|_{L_p(\psi)} \geq \frac{(\psi(b) - \psi(a))^{\sum_{i=1}^{m} \mu_i - \frac{m}{q} + \frac{1}{p}}}{\left[\left(p \sum_{i=1}^{m} \mu_i + m(1-p) + 1\right)^{\frac{1}{p}} \left(\prod_{i=1}^{m} \Gamma(\mu_i)\left(p(\mu_i - 1) + 1\right)^{\frac{1}{p}}\right)\right]} \left(\prod_{i=1}^{m} \left\| {}^{H}\mathbb{D}_{a+}^{\alpha_i,\beta_i;\psi} f_i \right\|_{L_q(\psi)}\right). \quad (7.69)$$

Proof By Theorem 7.4 (7.31), (7.55) and Theorem 7.23. ∎

We finish with

Theorem 7.34 *Here* $i = 1, ..., m$. *Let* $f_i \in C^{n_i}([a,b])$, *with* $\psi \in C^{\infty}([a,b])$ *and strictly increasing, where* $0 < \alpha_i \notin \mathbb{N}$, $\lceil \alpha_i \rceil = n_i \in \mathbb{N}$, $0 \leq \beta_i \leq 1$, $\mu_i = n_i(1-\beta_i) + \beta_i \alpha_i$. *Assume that* $w_i(x) = I_{b-}^{(1-\beta_i)(n_i-\alpha_i);\psi} f_i(x) \in C^{n_i}([a,b])$. *And assume that* $w_{i\psi}^{[k_i]}(b) = 0$, *where* $w_{i\psi}^{[k_i]}(x) = \left(\frac{1}{\psi'(x)}\frac{d}{dx}\right)^{k_i} w_i(x)$, $k_i = 0, 1, ..., n_i - 1$, $x \in [a,b]$. *Let* $0 < p < 1$, $q < 0 : \frac{1}{p} + \frac{1}{q} = 1$. *Suppose also that* ${}^{H}\mathbb{D}_{b-}^{\alpha_i,\beta_i;\psi} f_i$ *is of fixed strict sign a.e., and* $\left(\prod_{i=1}^{m} \left\| {}^{H}\mathbb{D}_{b-}^{\alpha_i,\beta_i;\psi} f_i \right\|_{L_q(\psi)}\right)$ *is finite. Then*

$$\left\| \prod_{i=1}^{m} w_i \right\|_{L_p(\psi)} \geq \frac{(\psi(b) - \psi(a))^{\sum_{i=1}^{m} \mu_i - \frac{m}{q} + \frac{1}{p}}}{\left[\left(p \sum_{i=1}^{m} \mu_i + m(1-p) + 1\right)^{\frac{1}{p}} \left(\prod_{i=1}^{m} \Gamma(\mu_i)\left(p(\mu_i - 1) + 1\right)^{\frac{1}{p}}\right)\right]} \left(\prod_{i=1}^{m} \left\| {}^{H}\mathbb{D}_{b-}^{\alpha_i,\beta_i;\psi} f_i \right\|_{L_q(\psi)}\right). \quad (7.70)$$

Proof By Theorem 7.4 (7.32), (7.56) and Theorem 7.25. ∎

References

1. Almeida, R.: A Caputo fractional derivative of a function with respect to another function. Commun. Nonlinear Sci. Numer. Simulat. **44**, 460–481 (2017)
2. Anastassiou, G.A., Mezei, R.A.: Reverse and forward fractional integral inequalities. In: Anastassiou, G., Duman, O. (eds.) Advances in Applied Mathematics and Approximation Theory, Contributions from AMAT 2012, pp. 441–478. Springer, New York (2013)
3. Anastassiou, G.: Advancements on ψ-Hilfer fractional calculus and fractional integral inequalities. Nonlinear Complex. (2021). [Accepted, Discontinuity]
4. Anastassiou, G.: Basic forward and reverse ψ-Hilfer-Hardy fractional inequalities. Commun. Appl. Nonlinear. Anal. **28**(2), 59–78 (2021)
5. Anastassiou, G.A.: Intelligent Comparisons: Analytic Inequalities. Springer, Heidelberg, New York (2016)
6. da Sousa, J.V.C., de Oliveira, E.C.: On the ψ-Hilfer fractional derivative. Commun. Nonlinear Sci. Numer. Simul. **60**, 72–91 (2018)
7. Hardy, H.G.: Notes on some points in the integral calculus. Messenger Math. **47**(10), 145–150 (1918)
8. Kilbas, A.A., Srivastava, H.M., Trujillo, J.J.: Theory and Applications of Fractional Differentiation Equations. North Holland, Amsterdam, New York (2006)
9. Samko, S.G., Kilbas, A.A., Marichev, O.I.: Fractional Integrals and Derivatives: Theory and Applications. Gordon and Breach Science Publishers, Switzerland (1993)
10. Tomovski, Ž, Hilfer, R., Srivastava, H.M.: Fractional and operational calculus with generalized fractional derivative operators and Mittag-Leffler type functions. Integr. Transforms Spec. Funct. **21**(11), 797–814 (2010)

Chapter 8
Principles of Generalized Prabhakar-Hilfer Fractional Calculus and Applications

Here we introduce the generalized Prabhakar fractional calculus and we also combine it with the generalized Hilfer calculus. We prove that the generalized left and right side Prabhakar fractional integrals preserve continuity and we find tight upper bounds for them. We present several left and right side generalized Prabhakar fractional inequalities of Hardy, Opial and Hilbert-Pachpatte types. We apply these in the setting of generalized Hilfer-calculus. It follows [1].

8.1 Background

We are inspired by [2, 8, 9]. We start by introducing our own generalized ψ-Prabhakar type of fractional calculus, then mixing it with the ψ-Hilfer fractional calculus. Then, we prove a variety of generalized Hardy, Opial and Hilbert-Pachpatte type left and right fractional integral inequalities related to ψ-Hilfer ([4]) and ψ-Prabhakar fractional calculi. We involve several functions.

We consider the Prabhakar function (also known as the three parameter Mittag-Laffler function), (see [6], p. 97; [5])

$$E_{\alpha,\beta}^{\gamma}(z)=\sum_{k=0}^{\infty}\frac{(\gamma)_k}{k!\Gamma(\alpha k+\beta)}z^k, \tag{8.1}$$

where Γ is the gamma function; $\alpha,\beta,\gamma\in\mathbb{R}:\alpha,\beta>0,\ z\in\mathbb{R}$, and $(\gamma)_k=\gamma(\gamma+1)\ldots(\gamma+k-1)$. It is $E_{\alpha,\beta}^{0}(z)=\frac{1}{\Gamma(\beta)}$.

G. A. Anastassiou, *Unification of Fractional Calculi with Applications*,
Studies in Systems, Decision and Control 398,
https://doi.org/10.1007/978-3-030-86920-5_8

Let $a, b \in \mathbb{R}$, $a < b$ and $x \in [a, b]$; $f \in C([a, b])$. Let also $\psi \in C^1([a, b])$ which is increasing. The left and right Prabhakar fractional integrals with respect to ψ are defined as follows:

$$\left(e^{\gamma;\psi}_{\rho,\mu,\omega,a+}f\right)(x) = \int_a^x \psi'(t)(\psi(x) - \psi(t))^{\mu-1} E^{\gamma}_{\rho,\mu}\left[\omega(\psi(x) - \psi(t))^{\rho}\right] f(t)\,dt, \quad (8.2)$$

and

$$\left(e^{\gamma;\psi}_{\rho,\mu,\omega,b-}f\right)(x) = \int_x^b \psi'(t)(\psi(t) - \psi(x))^{\mu-1} E^{\gamma}_{\rho,\mu}\left[\omega(\psi(t) - \psi(x))^{\rho}\right] f(t)\,dt, \quad (8.3)$$

where $\rho, \mu > 0$; $\gamma, \omega \in \mathbb{R}$.

Functions (8.2) and (8.3) are continuous, see Theorem 8.15.

Next, additionally, assume that $\psi'(x) \neq 0$ over $[a, b]$.

Let $\psi, f \in C^N([a, b])$, where $N = \lceil \mu \rceil$, ($\lceil \cdot \rceil$ is the ceiling of the number), $0 < \mu \notin \mathbb{N}$. We define the ψ-Prabhakar-Caputo left and right fractional derivatives of order μ as follows ($x \in [a, b]$):

$$\left({}^C D^{\gamma;\psi}_{\rho,\mu,\omega,a+}f\right)(x) = \int_a^x \psi'(t)(\psi(x) - \psi(t))^{N-\mu-1}$$

$$E^{-\gamma}_{\rho,N-\mu}\left[\omega(\psi(x) - \psi(t))^{\rho}\right]\left(\frac{1}{\psi'(t)}\frac{d}{dt}\right)^N f(t)\,dt, \quad (8.4)$$

and

$$\left({}^C D^{\gamma;\psi}_{\rho,\mu,\omega,b-}f\right)(x) = (-1)^N \int_x^b \psi'(t)(\psi(t) - \psi(x))^{N-\mu-1}$$

$$E^{-\gamma}_{\rho,N-\mu}\left[\omega(\psi(t) - \psi(x))^{\rho}\right]\left(\frac{1}{\psi'(t)}\frac{d}{dt}\right)^N f(t)\,dt. \quad (8.5)$$

One can write (see (8.4), (8.5))

$$\left({}^C D^{\gamma;\psi}_{\rho,\mu,\omega,a+}f\right)(x) = \left(e^{-\gamma;\psi}_{\rho,N-\mu,\omega,a+}f^{[N]}_{\psi}\right)(x), \quad (8.6)$$

and

$$\left({}^C D^{\gamma;\psi}_{\rho,\mu,\omega,b-}f\right)(x) = (-1)^N \left(e^{-\gamma;\psi}_{\rho,N-\mu,\omega,b-}f^{[N]}_{\psi}\right)(x), \quad (8.7)$$

where

$$f^{[N]}_{\psi}(x) = f^{(N)}_{\psi}f(x) := \left(\frac{1}{\psi'(x)}\frac{d}{dx}\right)^N f(x), \quad (8.8)$$

$\forall\, x \in [a, b]$.

Functions (8.6) and (8.7) are continuous on $[a, b]$.

Next we define the ψ-Prabhakar-Riemann Liouville left and right fractional derivatives of order μ as follows ($x \in [a, b]$):

$$\left({}^{RL}D^{\gamma;\psi}_{\rho,\mu,\omega,a+} f \right)(x) = \left(\frac{1}{\psi'(x)} \frac{d}{dx} \right)^N \int_a^x \psi'(t) \left(\psi(x) - \psi(t) \right)^{N-\mu-1}$$

$$E^{-\gamma}_{\rho,N-\mu} \left[\omega \left(\psi(x) - \psi(t) \right)^{\rho} \right] f(t)\, dt, \tag{8.9}$$

and

$$\left({}^{RL}D^{\gamma;\psi}_{\rho,\mu,\omega,b-} f \right)(x) = \left(-\frac{1}{\psi'(x)} \frac{d}{dx} \right)^N \int_x^b \psi'(t) \left(\psi(t) - \psi(x) \right)^{N-\mu-1}$$

$$E^{-\gamma}_{\rho,N-\mu} \left[\omega \left(\psi(t) - \psi(x) \right)^{\rho} \right] f(t)\, dt. \tag{8.10}$$

That is we have

$$\left({}^{RL}D^{\gamma;\psi}_{\rho,\mu,\omega,a+} f \right)(x) = \left(\frac{1}{\psi'(x)} \frac{d}{dx} \right)^N \left(e^{-\gamma;\psi}_{\rho,N-\mu,\omega,a+} f \right)(x), \tag{8.11}$$

and

$$\left({}^{RL}D^{\gamma;\psi}_{\rho,\mu,\omega,b-} f \right)(x) = \left(-\frac{1}{\psi'(x)} \frac{d}{dx} \right)^N \left(e^{-\gamma;\psi}_{\rho,N-\mu,\omega,b-} f \right)(x), \tag{8.12}$$

$\forall\, x \in [a, b]$.

We define also the ψ-Hilfer-Prabhakar left and right fractional derivatives of order μ and type $0 \leq \beta \leq 1$, as follows

$$\left({}^{H}\mathbb{D}^{\gamma,\beta;\psi}_{\rho,\mu,\omega,a+} f \right)(x) = e^{-\gamma\beta;\psi}_{\rho,\beta(N-\mu),\omega,a+} \left(\frac{1}{\psi'(x)} \frac{d}{dx} \right)^N e^{-\gamma(1-\beta);\psi}_{\rho,(1-\beta)(N-\mu),\omega,a+} f(x), \tag{8.13}$$

and

$$\left({}^{H}\mathbb{D}^{\gamma,\beta;\psi}_{\rho,\mu,\omega,b-} f \right)(x) = e^{-\gamma\beta;\psi}_{\rho,\beta(N-\mu),\omega,b-} \left(-\frac{1}{\psi'(x)} \frac{d}{dx} \right)^N e^{-\gamma(1-\beta);\psi}_{\rho,(1-\beta)(N-\mu),\omega,b-} f(x), \tag{8.14}$$

$\forall\, x \in [a, b]$.

When $\beta = 0$, we get the Riemann-Liouville version, and when $\beta = 1$, we get the Caputo version.

We call $\xi = \mu + \beta (N - \mu)$, we have that $N - 1 < \mu \leq \mu + \beta (N - \mu) \leq \mu + N - \mu = N$, hence $\lceil \xi \rceil = N$.

We can easily write that

$$\left({}^{H}\mathbb{D}^{\gamma,\beta;\psi}_{\rho,\mu,\omega,a+} f \right)(x) = e^{-\gamma\beta;\psi}_{\rho,\xi-\mu,\omega,a+} \, {}^{RL}D^{\gamma(1-\beta);\psi}_{\rho,\xi,\omega,a+} f(x), \tag{8.15}$$

and

$$\left({}^{H}\mathbb{D}^{\gamma,\beta;\psi}_{\rho,\mu,\omega,b-} f \right)(x) = e^{-\gamma\beta;\psi}_{\rho,\xi-\mu,\omega,b-} \, {}^{RL}D^{\gamma(1-\beta);\psi}_{\rho,\xi,\omega,b-} f(x), \tag{8.16}$$

$\forall\, x \in [a,b]$.

8.2 Main Results

We start with a left ψ-Prabhakar fractional Hardy type integral inequality involving several functions.

Theorem 8.1 *Here $i = 1, ..., m$; $f_i \in C([a,b])$, $\psi \in C^1([a,b])$ and ψ is increasing. Let $\rho_i, \mu_i > 0$, $\gamma_i, \omega_i \in \mathbb{R}$. Also let $r_1, r_2, r_3 > 1 : \frac{1}{r_1} + \frac{1}{r_2} + \frac{1}{r_3} = 1$, and assume that $\mu_i > \frac{1}{r_2} + \frac{1}{r_3}$, for all $i = 1, ..., m$.*

Then

$$\left\| \prod_{i=1}^{m} e^{\gamma_i;\psi}_{\rho_i,\mu_i,\omega_i,a+} f_i \right\|_{L_{r_1}([a,b],\psi)} \leq$$

$$\frac{(\psi(b) - \psi(a))^{\left[\sum_{i=1}^{m} \mu_i - m + \frac{m}{r_1} + \frac{1}{r_1} - \frac{1}{r_1 r_2}\right]}}{\left(r_1 r_3 \left(\sum_{i=1}^{m} \mu_i - m\right) + m r_3 + 1\right)^{\frac{1}{r_1 r_3}} \left(\prod_{i=1}^{m} \left(r_1(\mu_i - 1) + 1\right)\right)^{\frac{1}{r_1}}}$$

$$\left\{ \int_a^b \left[\prod_{i=1}^{m} \left(\int_a^x \left| E^{\gamma_i}_{\rho_i,\mu_i} \left[\omega_i (\psi(x) - \psi(t))^{\rho_i} \right] \right|^{r_2} d\psi(t) \right) \right]^{r_1} d\psi(x) \right\}^{\frac{1}{r_1 r_2}}$$

$$\left(\prod_{i=1}^{m} \| f_i \|_{L_{r_3}([a,b],\psi)} \right). \tag{8.17}$$

Proof By (8.2) we have

$$\left(e^{\gamma_i;\psi}_{\rho_i,\mu_i,\omega_i,a+} f_i \right)(x) =$$

$$\int_a^x \psi'(t) (\psi(x) - \psi(t))^{\mu_i - 1} E^{\gamma_i}_{\rho_i,\mu_i} \left[\omega_i (\psi(x) - \psi(t))^{\rho_i} \right] f_i(t)\, dt, \tag{8.18}$$

$i = 1, ..., m$; $\forall\, x \in [a,b]$.

By Hölder's inequality and (8.18) we obtain

$$\left|\left(e^{\gamma_i;\psi}_{\rho_i,\mu_i,\omega_i,a+}f_i\right)(x)\right| \leq$$

$$\int_a^x \psi'(t)(\psi(x)-\psi(t))^{\mu_i-1}\left|E^{\gamma_i}_{\rho_i,\mu_i}\left[\omega_i(\psi(x)-\psi(t))^{\rho_i}\right]\right||f_i(t)|\,dt \leq$$

$$\left(\int_a^x (\psi(x)-\psi(t))^{r_1(\mu_i-1)}\,d\psi(t)\right)^{\frac{1}{r_1}}$$

$$\left(\int_a^x \left|E^{\gamma_i}_{\rho_i,\mu_i}\left[\omega_i(\psi(x)-\psi(t))^{\rho_i}\right]\right|^{r_2}d\psi(t)\right)^{\frac{1}{r_2}}\left(\int_a^x |f_i(t)|^{r_3}\,d\psi(t)\right)^{\frac{1}{r_3}} \leq \tag{8.19}$$

$$\frac{(\psi(x)-\psi(a))^{\mu_i-1+\frac{1}{r_1}}}{(r_1(\mu_i-1)+1)^{\frac{1}{r_1}}}$$

$$\left(\int_a^x \left|E^{\gamma_i}_{\rho_i,\mu_i}\left[\omega_i(\psi(x)-\psi(t))^{\rho_i}\right]\right|^{r_2}d\psi(t)\right)^{\frac{1}{r_2}}\|f_i\|_{L_{r_3}([a,b],\psi)}\,.$$

So far we have

$$\left|\left(e^{\gamma_i;\psi}_{\rho_i,\mu_i,\omega_i,a+}f_i\right)(x)\right| \leq \frac{(\psi(x)-\psi(a))^{\mu_i-1+\frac{1}{r_1}}}{(r_1(\mu_i-1)+1)^{\frac{1}{r_1}}}$$

$$\left(\int_a^x \left|E^{\gamma_i}_{\rho_i,\mu_i}\left[\omega_i(\psi(x)-\psi(t))^{\rho_i}\right]\right|^{r_2}d\psi(t)\right)^{\frac{1}{r_2}}\|f_i\|_{L_{r_3}([a,b],\psi)}\,, \tag{8.20}$$

$\forall\, x \in [a,b]$, with $\mu_i > \frac{1}{r_2} + \frac{1}{r_3}$, for any $i = 1, ..., m$.

Hence it holds

$$\left(\prod_{i=1}^m \left|\left(e^{\gamma_i;\psi}_{\rho_i,\mu_i,\omega_i,a+}f_i\right)(x)\right|\right)^{r_1} \leq \frac{(\psi(x)-\psi(a))^{r_1\sum_{i=1}^m \mu_i - mr_1+m}}{\left(\prod_{i=1}^m (r_1(\mu_i-1)+1)\right)}$$

$$\left[\prod_{i=1}^m \left(\int_a^x \left|E^{\gamma_i}_{\rho_i,\mu_i}\left[\omega_i(\psi(x)-\psi(t))^{\rho_i}\right]\right|^{r_2}d\psi(t)\right)\right]^{\frac{r_1}{r_2}}\left(\prod_{i=1}^m \|f_i\|_{L_{r_3}([a,b],\psi)}\right)^{r_1}, \tag{8.21}$$

$\forall\, x \in [a,b]$.

Therefore we obtain

$$\int_a^b \left(\prod_{i=1}^m \left| \left(e^{\gamma_i;\psi}_{\rho_i,\mu_i,\omega_i,a+} f_i \right)(x) \right| \right)^{r_1} d\psi(x) \leq \frac{\left(\prod_{i=1}^m \| f_i \|_{L_{r_3}([a,b],\psi)} \right)^{r_1}}{\left(\prod_{i=1}^m (r_1(\mu_i - 1) + 1) \right)} \tag{8.22}$$

$$\left[\int_a^b (\psi(x) - \psi(a))^{r_1 \sum\limits_{i=1}^m \mu_i - m r_1 + m} \right.$$

$$\left. \left[\prod_{i=1}^m \left(\int_a^x \left| E^{\gamma_i}_{\rho_i,\mu_i} \left[\omega_i (\psi(x) - \psi(t))^{\rho_i} \right] \right|^{r_2} d\psi(t) \right) \right]^{\frac{r_1}{r_2}} d\psi(x) \right]$$

(again by Hölder's inequality)

$$\leq \frac{\left(\prod_{i=1}^m \| f_i \|_{L_{r_3}([a,b],\psi)} \right)^{r_1}}{\left(\prod_{i=1}^m (r_1(\mu_i - 1) + 1) \right)} \left(\int_a^b (\psi(x) - \psi(a))^{r_1 r_3 \sum\limits_{i=1}^m \mu_i - m r_1 r_3 + m r_3} d\psi(x) \right)^{\frac{1}{r_3}}$$

$$\left\{ \int_a^b \left[\prod_{i=1}^m \left(\int_a^x \left| E^{\gamma_i}_{\rho_i,\mu_i} \left[\omega_i (\psi(x) - \psi(t))^{\rho_i} \right] \right|^{r_2} d\psi(t) \right) \right]^{r_1} d\psi(x) \right\}^{\frac{1}{r_2}} (\psi(b) - \psi(a))^{\frac{1}{r_1}} =$$

$$\frac{\left(\prod_{i=1}^m \| f_i \|_{L_{r_3}([a,b],\psi)} \right)^{r_1} (\psi(b) - \psi(a))^{r_1 \sum_{i=1}^m \mu_i - m r_1 + m + 1 - \frac{1}{r_2}}}{\left(\prod_{i=1}^m (r_1(\mu_i - 1) + 1) \right) \left(r_1 r_3 \sum_{i=1}^m \mu_i - m r_1 r_3 + m r_3 + 1 \right)^{\frac{1}{r_3}}} \tag{8.23}$$

$$\left\{ \int_a^b \left[\prod_{i=1}^m \left(\int_a^x \left| E^{\gamma_i}_{\rho_i,\mu_i} \left[\omega_i (\psi(x) - \psi(t))^{\rho_i} \right] \right|^{r_2} d\psi(t) \right) \right]^{r_1} d\psi(x) \right\}^{\frac{1}{r_2}},$$

where $\mu_i > \frac{1}{r_2} + \frac{1}{r_3}$, $i = 1, ..., m$.

The claim is proved. ■

We continue with a right ψ-Prabhakar fractional Hardy type integral inequality involving several functions.

Theorem 8.2 *All as in Theorem 8.1. It holds*

$$\left\| \prod_{i=1}^{m} e_{\rho_i,\mu_i,\omega_i,b-}^{\gamma_i;\psi} f_i \right\|_{L_{r_1}([a,b],\psi)} \leq$$

$$\frac{(\psi(b)-\psi(a))^{\left[\sum_{i=1}^{m}\mu_i - m + \frac{m}{r_1} + \frac{1}{r_1} - \frac{1}{r_1 r_2}\right]}}{\left(r_1 r_3 \left(\sum_{i=1}^{m}\mu_i - m\right) + m r_3 + 1\right)^{\frac{1}{r_1 r_3}} \left(\prod_{i=1}^{m}\left(r_1(\mu_i - 1) + 1\right)\right)^{\frac{1}{r_1}}}$$

$$\left\{\int_a^b \left[\prod_{i=1}^{m}\left(\int_x^b \left|E_{\rho_i,\mu_i}^{\gamma_i}\left[\omega_i(\psi(t)-\psi(x))^{\rho_i}\right]\right|^{r_2} d\psi(t)\right)\right]^{r_1} d\psi(x)\right\}^{\frac{1}{r_1 r_2}}$$

$$\left(\prod_{i=1}^{m} \|f_i\|_{L_{r_3}([a,b],\psi)}\right). \tag{8.24}$$

Proof Similar to the proof of Theorem 8.1 and omitted. ■

Next we apply Theorems 8.1, 8.2.
We give the related Hardy type inequalities:

Theorem 8.3 *Here* $i = 1, ..., m$; $f_i \in C^{N_i}([a,b])$, *where* $N_i = \lceil \mu_i \rceil$, $0 < \mu_i \notin \mathbb{N}$; $\theta := \max\{N_1, ..., N_m\}$, $\psi \in C^{\theta}([a,b])$ *with* $\psi' \neq 0$ *and increasing. Let* $\rho_i > 0$, $\gamma_i, \omega_i \in \mathbb{R}$. *Also let* $r_1, r_2, r_3 > 1 : \frac{1}{r_1} + \frac{1}{r_2} + \frac{1}{r_3} = 1$, *and assume that* $N_i - \mu_i > \frac{1}{r_2} + \frac{1}{r_3}$, *for all* $i = 1, ..., m$.
Then
(i)

$$\left\| \prod_{i=1}^{m} {}^{C}D_{\rho_i,\mu_i,\omega_i,a+}^{\gamma_i;\psi} f_i \right\|_{L_{r_1}([a,b],\psi)} \leq$$

$$\frac{(\psi(b)-\psi(a))^{\left[\sum_{i=1}^{m}(N_i-\mu_i) - m + \frac{m}{r_1} + \frac{1}{r_1} - \frac{1}{r_1 r_2}\right]}}{\left(r_1 r_3 \left(\sum_{i=1}^{m}(N_i-\mu_i) - m\right) + m r_3 + 1\right)^{\frac{1}{r_1 r_3}} \left(\prod_{i=1}^{m}\left(r_1(N_i-\mu_i - 1) + 1\right)\right)^{\frac{1}{r_1}}}$$

$$\left\{\int_a^b \left[\prod_{i=1}^{m}\left(\int_a^x \left|E_{\rho_i,N_i-\mu_i}^{-\gamma_i}\left[\omega_i(\psi(x)-\psi(t))^{\rho_i}\right]\right|^{r_2} d\psi(t)\right)\right]^{r_1} d\psi(x)\right\}^{\frac{1}{r_1 r_2}}$$

$$\left(\prod_{i=1}^{m} \left\|f_{i\psi}^{[N_i]}\right\|_{L_{r_3}([a,b],\psi)}\right), \tag{8.25}$$

and

(ii)

$$\left\| \prod_{i=1}^{m} {}^{C}D^{\gamma_i;\psi}_{\rho_i,\mu_i,\omega_i,b-} f_i \right\|_{L_{r_1}([a,b],\psi)} \leq$$

$$\frac{(\psi(b)-\psi(a))^{\left[\sum_{i=1}^{m}(N_i-\mu_i)-m+\frac{m}{r_1}+\frac{1}{r_1}-\frac{1}{r_1 r_2}\right]}}{\left(r_1 r_3\left(\sum_{i=1}^{m}(N_i-\mu_i)-m\right)+m r_3+1\right)^{\frac{1}{r_1 r_3}}\left(\prod_{i=1}^{m}(r_1(N_i-\mu_i-1)+1)\right)^{\frac{1}{r_1}}}$$

$$\left\{\int_a^b\left[\prod_{i=1}^{m}\left(\int_x^b\left|E^{-\gamma_i}_{\rho_i,N_i-\mu_i}\left[\omega_i(\psi(t)-\psi(x))^{\rho_i}\right]\right|^{r_2}d\psi(t)\right)\right]^{r_1}d\psi(x)\right\}^{\frac{1}{r_1 r_2}}$$

$$\left(\prod_{i=1}^{m}\left\|f_{i\psi}^{[N_i]}\right\|_{L_{r_3}([a,b],\psi)}\right), \tag{8.26}$$

Proof By (8.6), (8.7) and Theorems 8.1, 8.2. ∎

We also present other Hardy type related inequalities:

Theorem 8.4 *Here $i=1,...,m$; $f_i\in C^{N_i}([a,b])$, where $N_i=\lceil\mu_i\rceil$, $0<\mu_i\notin\mathbb{N}$; $\theta:=\max\{N_1,...,N_m\}$, $\psi\in C^{\theta}([a,b])$, $\psi'\neq 0$, and ψ is increasing. Let $\rho_i>0$, $\gamma_i,\omega_i\in\mathbb{R}$, $0\leq\beta_i\leq 1$, $\xi_i=\mu_i+\beta_i(N_i-\mu_i)$. Also let $r_1,r_2,r_3>1:\frac{1}{r_1}+\frac{1}{r_2}+\frac{1}{r_3}=1$, and assume that $\xi_i-\mu_i>\frac{1}{r_2}+\frac{1}{r_3}$, for all $i=1,...,m$.*
Also assume that ${}^{RL}D^{\gamma_i(1-\beta_i);\psi}_{\rho_i,\xi_i,\omega_i,a+}f_i$, ${}^{RL}D^{\gamma_i(1-\beta_i);\psi}_{\rho_i,\xi_i,\omega_i,b-}f_i\in C([a,b])$, $i=1,...,m$.
Then
(i)

$$\left\| \prod_{i=1}^{m} {}^{H}\mathbb{D}^{\gamma_i,\beta_i;\psi}_{\rho_i,\mu_i,\omega_i,a+} f_i \right\|_{L_{r_1}([a,b],\psi)} \leq$$

$$\frac{(\psi(b)-\psi(a))^{\left[\sum_{i=1}^{m}(\xi_i-\mu_i)-m+\frac{m}{r_1}+\frac{1}{r_1}-\frac{1}{r_1 r_2}\right]}}{\left(r_1 r_3\left(\sum_{i=1}^{m}(\xi_i-\mu_i)-m\right)+m r_3+1\right)^{\frac{1}{r_1 r_3}}\left(\prod_{i=1}^{m}(r_1(\xi_i-\mu_i-1)+1)\right)^{\frac{1}{r_1}}} \tag{8.27}$$

$$\left\{\int_a^b\left[\prod_{i=1}^{m}\left(\int_a^x\left|E^{-\gamma_i\beta_i}_{\rho_i,\xi_i-\mu_i}\left[\omega_i(\psi(x)-\psi(t))^{\rho_i}\right]\right|^{r_2}d\psi(t)\right)\right]^{r_1}d\psi(x)\right\}^{\frac{1}{r_1 r_2}}$$

$$\left(\prod_{i=1}^{m}\left\|{}^{RL}D^{\gamma_i(1-\beta_i);\psi}_{\rho_i,\xi_i,\omega_i,a+}f_i\right\|_{L_{r_3}([a,b],\psi)}\right),$$

and

(ii)

$$\left\| \prod_{i=1}^{m} {}^{H}\mathbb{D}_{\rho_i,\mu_i,\omega_i,b-}^{\gamma_i,\beta_i;\psi} f_i \right\|_{L_{r_1}([a,b],\psi)} \leq$$

$$\frac{(\psi(b)-\psi(a))^{\left[\sum_{i=1}^{m}(\xi_i-\mu_i)-m+\frac{m}{r_1}+\frac{1}{r_1}-\frac{1}{r_1 r_2}\right]}}{\left(r_1 r_3\left(\sum_{i=1}^{m}(\xi_i-\mu_i)-m\right)+m r_3+1\right)^{\frac{1}{r_1 r_3}}\left(\prod_{i=1}^{m}\left(r_1(\xi_i-\mu_i-1)+1\right)\right)^{\frac{1}{r_1}}} \tag{8.28}$$

$$\left\{\int_a^b \left[\prod_{i=1}^{m}\left(\int_x^b \left|E_{\rho_i,\xi_i-\mu_i}^{-\gamma_i\beta_i}\left[\omega_i(\psi(t)-\psi(x))^{\rho_i}\right]\right|^{r_2} d\psi(t)\right)\right]^{r_1} d\psi(x)\right\}^{\frac{1}{r_1 r_2}}$$

$$\left(\prod_{i=1}^{m}\left\| {}^{RL}D_{\rho_i,\xi_i,\omega_i,b-}^{\gamma_i(1-\beta_i);\psi} f_i \right\|_{L_{r_3}([a,b],\psi)}\right).$$

Proof By (8.15), (8.16) and Theorems 8.1, 8.2. ■

From now on all entities are according and respectively to 8.1. Background. Next we give Opial type inequalities related to Prabhakar fractional calculus. A left side one follows:

Theorem 8.5 *Let* $p, q > 1 : \frac{1}{p} + \frac{1}{q} = 1$. *Then*

$$\int_a^x \left|\left(e_{\rho,\mu,\omega,a+}^{\gamma;\psi} f\right)(w)\right| |f(w)| \, \psi'(w) \, dw \leq 2^{-\frac{1}{q}}$$

$$\left[\int_a^x \left\{\int_a^w (\psi(w)-\psi(t))^{p(\mu-1)} \left|E_{p,\mu}^{\gamma}\left[\omega(\psi(w)-\psi(t))^{\rho}\right]\right|^p dt\right\} dw\right]^{\frac{1}{p}}$$

$$\left(\int_a^x |f(w)|^q \left(\psi'(w)\right)^q dw\right)^{\frac{2}{q}}, \tag{8.29}$$

$\forall\, x \in [a,b]$.

Proof By (8.2), using Hölder's inequality, we have

$$\left|\left(e_{\rho,\mu,\omega,a+}^{\gamma;\psi} f\right)(x)\right| \leq \int_a^x \psi'(t)(\psi(x)-\psi(t))^{\mu-1} \left|E_{\rho,\mu}^{\gamma}\left[\omega(\psi(x)-\psi(t))^{\rho}\right]\right| |f(t)| \, dt$$

$$\leq \left(\int_a^x (\psi(x)-\psi(t))^{p(\mu-1)} \left|E_{\rho,\mu}^{\gamma}\left[\omega(\psi(x)-\psi(t))^{\rho}\right]\right|^p dt\right)^{\frac{1}{p}}$$

$$\left(\int_a^x \left(\psi'(t)\,|f(t)|\right)^q dt\right)^{\frac{1}{q}}. \tag{8.30}$$

Call

$$\phi(x) = \int_a^x (\psi'(t)|f(t)|)^q \, dt, \quad \phi(a) = 0. \tag{8.31}$$

Thus

$$\phi'(x) = (\psi'(x)|f(x)|)^q \geq 0, \tag{8.32}$$

and

$$(\phi'(x))^{\frac{1}{q}} = \psi'(x)|f(x)| \geq 0, \quad \forall\, x \in [a, b].$$

Consequently, we get

$$\left|\left(e^{\gamma;\psi}_{\rho,\mu,\omega,a+} f\right)(w)\right| \psi'(w)|f(w)| \leq$$

$$\left(\int_a^w (\psi(w) - \psi(t))^{p(\mu-1)} \left|E^{\gamma}_{\rho,\mu}\left[\omega(\psi(w) - \psi(t))^{\rho}\right]\right|^p dt\right)^{\frac{1}{p}}$$

$$(\phi(w)\phi'(w))^{\frac{1}{q}}, \quad \forall\, w \in [a, b]. \tag{8.33}$$

Then, by applying again Hölder's inequality:

$$\int_a^x \left|\left(e^{\gamma;\psi}_{\rho,\mu,\omega,a+} f\right)(w)\right| |f(w)|\psi'(w)\, dw \leq \tag{8.34}$$

$$\int_a^x \left\{\int_a^w (\psi(w) - \psi(t))^{p(\mu-1)} \left|E^{\gamma}_{\rho,\mu}\left[\omega(\psi(w) - \psi(t))^{\rho}\right]\right|^p dt\right\}^{\frac{1}{p}}$$

$$(\phi(w)\phi'(w))^{\frac{1}{q}}\, dw \leq$$

$$\left[\int_a^x \left\{\int_a^w (\psi(w) - \psi(t))^{p(\mu-1)} \left|E^{\gamma}_{\rho,\mu}\left[\omega(\psi(w) - \psi(t))^{\rho}\right]\right|^p dt\right\} dw\right]^{\frac{1}{p}}$$

$$\left(\int_a^x \phi(w)\, d\phi(w)\right)^{\frac{1}{q}} =$$

$$\left[\int_a^x \left\{\int_a^w (\psi(w) - \psi(t))^{p(\mu-1)} \left|E^{\gamma}_{\rho,\mu}\left[\omega(\psi(w) - \psi(t))^{\rho}\right]\right|^p dt\right\} dw\right]^{\frac{1}{p}}$$

$$\left(\frac{\phi^2(x)}{2}\right)^{\frac{1}{q}} = 2^{-\frac{1}{q}}$$

$$\left[\int_a^x \left\{\int_a^w (\psi(w)-\psi(t))^{p(\mu-1)} \left|E_{\rho,\mu}^{\gamma}\left[\omega(\psi(w)-\psi(t))^{\rho}\right]\right|^p dt\right\} dw\right]^{\frac{1}{p}}$$

$$\left(\int_a^x (\psi'(w)\,|f(w)|)^q\, dw\right)^{\frac{2}{q}}. \tag{8.35}$$

The theorem is proved. ■

The right side Opial inequality follows:

Theorem 8.6 *Let* $p, q > 1 : \frac{1}{p} + \frac{1}{q} = 1$. *Then*

$$\int_x^b \left|\left(e_{\rho,\mu,\omega,b-}^{\gamma;\psi} f\right)(w)\right| |f(w)|\, \psi'(w)\, dw \leq 2^{-\frac{1}{q}}$$

$$\left[\int_x^b \left\{\int_w^b (\psi(t)-\psi(w))^{p(\mu-1)} \left|E_{\rho,\mu}^{\gamma}\left[\omega(\psi(t)-\psi(w))^{\rho}\right]\right|^p dt\right\} dw\right]^{\frac{1}{p}}$$

$$\left(\int_x^b |f(w)|^q \left(\psi'(w)\right)^q dw\right)^{\frac{2}{q}}, \tag{8.36}$$

$\forall\, x \in [a, b]$.

Proof As similar to the proof of Theorem 8.5 is omitted. ■

We continue with more interesting Opial type Prabhakar-Caputo fractional inequalities:

Theorem 8.7 *Let* $p, q > 1 : \frac{1}{p} + \frac{1}{q} = 1$. *Then*

(i)

$$\int_a^x \left|\left({}^{C}D_{\rho,\mu,\omega,a+}^{\gamma;\psi} f\right)(w)\right| \left|f_{\psi}^{[N]}(w)\right| \psi'(w)\, dw \leq 2^{-\frac{1}{q}}$$

$$\left[\int_a^x \left\{\int_a^w (\psi(w)-\psi(t))^{p(N-\mu-1)} \left|E_{\rho,N-\mu}^{-\gamma}\left[\omega(\psi(w)-\psi(t))^{\rho}\right]\right|^p dt\right\} dw\right]^{\frac{1}{p}}$$

$$\left(\int_a^x \left|f_{\psi}^{[N]}(w)\right|^q \left(\psi'(w)\right)^q dw\right)^{\frac{2}{q}}, \tag{8.37}$$

and

(ii)

$$\int_x^b \left|\left({}^{C}D_{\rho,\mu,\omega,b-}^{\gamma;\psi} f\right)(w)\right| \left|f_{\psi}^{[N]}(w)\right| \psi'(w)\, dw \leq 2^{-\frac{1}{q}}$$

$$\left[\int_x^b \left\{\int_w^b (\psi(t)-\psi(w))^{p(N-\mu-1)} \left|E_{\rho,N-\mu}^{-\gamma}\left[\omega(\psi(t)-\psi(w))^{\rho}\right]\right|^p dt\right\} dw\right]^{\frac{1}{p}}$$

$$\left(\int_x^b \left|f_{\psi}^{[N]}(w)\right|^q (\psi'(w))^q\, dw\right)^{\frac{2}{q}}, \tag{8.38}$$

$\forall\, x \in [a,b]$.

Proof By Theorems 8.5, 8.6 and (8.6)-(8.8). ∎

Next come ψ-Hilfer-Prabhakar left and right Opial type fractional inequalities:

Theorem 8.8 *Let* $p, q>1 : \frac{1}{p}+\frac{1}{q}=1$. *Additionally here assume that* ${}^{RL}D_{\rho,\xi,\omega,a+}^{\gamma(1-\beta);\psi} f$, ${}^{RL}D_{\rho,\xi,\omega,b-}^{\gamma(1-\beta);\psi} f \in C([a,b])$. *Then*

(i)

$$\int_a^x \left|\left({}^{H}\mathbb{D}_{\rho,\mu,\omega,a+}^{\gamma,\beta;\psi} f\right)(w)\right| \left|\left({}^{RL}D_{\rho,\xi,\omega,a+}^{\gamma(1-\beta);\psi} f\right)(w)\right| \psi'(w)\, dw \le 2^{-\frac{1}{q}}$$

$$\left[\int_a^x \left\{\int_a^w (\psi(w)-\psi(t))^{p(\xi-\mu-1)} \left|E_{\rho,\xi-\mu}^{-\gamma\beta}\left[\omega(\psi(w)-\psi(t))^{\rho}\right]\right|^p dt\right\} dw\right]^{\frac{1}{p}}$$

$$\left(\int_a^x \left|\left({}^{RL}D_{\rho,\xi,\omega,a+}^{\gamma(1-\beta);\psi} f\right)(w)\right|^q (\psi'(w))^q\, dw\right)^{\frac{2}{q}}, \tag{8.39}$$

and

(ii)

$$\int_x^b \left|\left({}^{H}\mathbb{D}_{\rho,\mu,\omega,b-}^{\gamma,\beta;\psi} f\right)(w)\right| \left|\left({}^{RL}D_{\rho,\xi,\omega,b-}^{\gamma(1-\beta);\psi} f\right)(w)\right| \psi'(w)\, dw \le 2^{-\frac{1}{q}}$$

$$\left[\int_x^b \left\{\int_w^b (\psi(t)-\psi(w))^{p(\xi-\mu-1)} \left|E_{\rho,\xi-\mu}^{-\gamma\beta}\left[\omega(\psi(t)-\psi(w))^{\rho}\right]\right|^p dt\right\} dw\right]^{\frac{1}{p}}$$

$$\left(\int_x^b \left|\left({}^{RL}D_{\rho,\xi,\omega,b-}^{\gamma(1-\beta);\psi} f\right)(w)\right|^q (\psi'(w))^q\, dw\right)^{\frac{2}{q}}, \tag{8.40}$$

$\forall\, x \in [a,b]$.

Proof By Theorems 8.5, 8.6 and (8.15), (8.16). ∎

Next we give several Prabhakar Hilbert-Pachpatte fractional inequalities. We start with a left side one.

Theorem 8.9 *Let $p, q>1 : \frac{1}{p}+\frac{1}{q} = 1$; $i = 1, 2$. Let $[a_i, b_i] \subset \mathbb{R}$, $\psi_i \in C^1([a_i, b_i])$ and strictly increasing, $f_i \in C([a_i, b_i])$; $\rho_i, \mu_i > 0$, $\gamma_i, \omega_i \in \mathbb{R}$. Then*

$$\int_{a_1}^{b_1}\int_{a_2}^{b_2} \frac{\left|\left(e^{\gamma_1;\psi_1}_{\rho_1,\mu_1,\omega_1,a_1+}f_1\right)(x_1)\right|\left|\left(e^{\gamma_2;\psi_2}_{\rho_2,\mu_2,\omega_2,a_2+}f_2\right)(x_2)\right| dx_1dx_2}{\left\{\frac{\left[\int_{a_1}^{x_1}\{(\psi_1(x_1)-\psi_1(t_1))^{\mu_1-1}\left|E^{\gamma_1}_{\rho_1,\mu_1}\left[\omega_1(\psi_1(x_1)-\psi_1(t_1))^{\rho_1}\right]\right|\}^p dt_1\right]}{p} + \frac{\left[\int_{a_2}^{x_2}\{(\psi_2(x_2)-\psi_2(t_2))^{\mu_2-1}\left|E^{\gamma_2}_{\rho_2,\mu_2}\left[\omega_2(\psi_2(x_2)-\psi_2(t_2))^{\rho_2}\right]\right|\}^q dt_2\right]}{q}\right\}}$$

$$\leq (b_1 - a_1)(b_2 - a_2)\left\|\psi_1' f_1\right\|_q \left\|\psi_2' f_2\right\|_p . \tag{8.41}$$

Proof We have that ($i = 1, 2$)

$$\left(e^{\gamma_i;\psi_i}_{\rho_i,\mu_i,\omega_i,a_i+}f_i\right)(x_i) \overset{(8.2)}{=}$$

$$\int_{a_i}^{x_i} \psi_i'(t_i)(\psi_i(x_i) - \psi_i(t_i))^{\mu_i-1} E^{\gamma_i}_{\rho_i,\mu_i}\left[\omega_i(\psi_i(x_i) - \psi_i(t_i))^{\rho_i}\right] f_i(t_i)\, dt_i, \tag{8.42}$$

$\forall x_i \in [a_i, b_i]$, where $\rho_i, \mu_i > 0$; $\gamma_i, \omega_i \in \mathbb{R}$.

Then

$$\left|\left(e^{\gamma_i;\psi_i}_{\rho_i,\mu_i,\omega_i,a_i+}f_i\right)(x_i)\right| \leq$$

$$\int_{a_i}^{x_i} \psi_i'(t_i)(\psi_i(x_i) - \psi_i(t_i))^{\mu_i-1}\left|E^{\gamma_i}_{\rho_i,\mu_i}\left[\omega_i(\psi_i(x_i) - \psi_i(t_i))^{\rho_i}\right]\right| |f_i(t_i)|\, dt_i, \tag{8.43}$$

$i = 1, 2$, $\forall x_i \in [a_i, b_i]$.

By applying Hölder's inequality twice we get:

$$\left|\left(e^{\gamma_1;\psi_1}_{\rho_1,\mu_1,\omega_1,a_1+}f_1\right)(x_1)\right| \leq$$

$$\left[\int_{a_1}^{x_1}\left\{(\psi_1(x_1) - \psi_1(t_1))^{\mu_1-1}\left|E^{\gamma_1}_{\rho_1,\mu_1}\left[\omega_1(\psi_1(x_1) - \psi_1(t_1))^{\rho_1}\right]\right|\right\}^p dt_1\right]^{\frac{1}{p}}$$

$$\left(\int_{a_1}^{x_1}\left(\psi_1'(t_1)|f_1(t_1)|\right)^q dt_1\right)^{\frac{1}{q}}, \tag{8.44}$$

$\forall x_1 \in [a_1, b_1]$,

and

$$\left|\left(e^{\gamma_2;\psi_2}_{\rho_2,\mu_2,\omega_2,a_2+}f_2\right)(x_2)\right| \leq$$

$$\left[\int_{a_2}^{x_2}\left\{(\psi_2(x_2) - \psi_2(t_2))^{\mu_2-1}\left|E^{\gamma_2}_{\rho_2,\mu_2}\left[\omega_2(\psi_2(x_2) - \psi_2(t_2))^{\rho_2}\right]\right|\right\}^q dt_2\right]^{\frac{1}{q}}$$

$$\left(\int_{a_2}^{x_2} \left(\psi_2'(t_2) |f_2(t_2)|\right)^p dt_2\right)^{\frac{1}{p}}, \tag{8.45}$$

$\forall\, x_2 \in [a_2, b_2]$.

Hence we have (by (8.44), (8.45))

$$\left|\left(e^{\gamma_1;\psi_1}_{\rho_1,\mu_1,\omega_1,a_1+} f_1\right)(x_1)\right| \left|\left(e^{\gamma_2;\psi_2}_{\rho_2,\mu_2,\omega_2,a_2+} f_2\right)(x_2)\right| \leq$$

$$\left[\int_{a_1}^{x_1} \left\{(\psi_1(x_1) - \psi_1(t_1))^{\mu_1 - 1} \left|E^{\gamma_1}_{\rho_1,\mu_1}\left[\omega_1 (\psi_1(x_1) - \psi_1(t_1))^{\rho_1}\right]\right|\right\}^p dt_1\right]^{\frac{1}{p}}$$

$$\left[\int_{a_2}^{x_2} \left\{(\psi_2(x_2) - \psi_2(t_2))^{\mu_2 - 1} \left|E^{\gamma_2}_{\rho_2,\mu_2}\left[\omega_2 (\psi_2(x_2) - \psi_2(t_2))^{\rho_2}\right]\right|\right\}^q dt_2\right]^{\frac{1}{q}}$$

$$\left\|\psi_1' f_1\right\|_q \left\|\psi_2' f_2\right\|_p \leq \tag{8.46}$$

(using Yang's inequality for $a, b \geq 0$, $a^{\frac{1}{p}} b^{\frac{1}{q}} \leq \frac{a}{p} + \frac{b}{q}$)

$$\left\{\frac{\left[\int_{a_1}^{x_1} \left\{(\psi_1(x_1) - \psi_1(t_1))^{\mu_1 - 1} \left|E^{\gamma_1}_{\rho_1,\mu_1}\left[\omega_1 (\psi_1(x_1) - \psi_1(t_1))^{\rho_1}\right]\right|\right\}^p dt_1\right]}{p} + \right.$$

$$\left.\frac{\left[\int_{a_2}^{x_2} \left\{(\psi_2(x_2) - \psi_2(t_2))^{\mu_2 - 1} \left|E^{\gamma_2}_{\rho_2,\mu_2}\left[\omega_2 (\psi_2(x_2) - \psi_2(t_2))^{\rho_2}\right]\right|\right\}^q dt_2\right]}{q}\right\}$$

$$\left\|\psi_1' f_1\right\|_q \left\|\psi_2' f_2\right\|_p,$$

$\forall\, x_i \in [a_i, b_i]$, $i = 1, 2$.

So far we have

$$\frac{\left|\left(e^{\gamma_1;\psi_1}_{\rho_1,\mu_1,\omega_1,a_1+} f_1\right)(x_1)\right| \left|\left(e^{\gamma_2;\psi_2}_{\rho_2,\mu_2,\omega_2,a_2+} f_2\right)(x_2)\right|}{\left\{\frac{\left[\int_{a_1}^{x_1} \left\{(\psi_1(x_1)-\psi_1(t_1))^{\mu_1-1} \left|E^{\gamma_1}_{\rho_1,\mu_1}\left[\omega_1(\psi_1(x_1)-\psi_1(t_1))^{\rho_1}\right]\right|\right\}^p dt_1\right]}{p} + \frac{\left[\int_{a_2}^{x_2} \left\{(\psi_2(x_2)-\psi_2(t_2))^{\mu_2-1} \left|E^{\gamma_2}_{\rho_2,\mu_2}\left[\omega_2(\psi_2(x_2)-\psi_2(t_2))^{\rho_2}\right]\right|\right\}^q dt_2\right]}{q}\right\}}$$

$$\leq \left\|\psi_1' f_1\right\|_q \left\|\psi_2' f_2\right\|_p, \tag{8.47}$$

$\forall\, x_i \in [a_i, b_i]$, $i = 1, 2$.

The denominator in (8.47) can be zero only when $x_1 = a_1$ and $x_2 = a_2$. Therefore we obtain (8.41) by integrating (8.47) over $[a_1, b_1] \times [a_2, b_2]$. ■

It follows the corresponding to (8.41) right side inequality.

Theorem 8.10 *All as in Theorem 8.9. Then*

$$\int_{a_1}^{b_1}\int_{a_2}^{b_2} \frac{\left|\left(e^{\gamma_1;\psi_1}_{\rho_1,\mu_1,\omega_1,b_1-}f_1\right)(x_1)\right|\left|\left(e^{\gamma_2;\psi_2}_{\rho_2,\mu_2,\omega_2,b_2-}f_2\right)(x_2)\right|dx_1dx_2}{\left\{\frac{\left[\int_{x_1}^{b_1}\{(\psi_1(t_1)-\psi_1(x_1))^{\mu_1-1}|E^{\gamma_1}_{\rho_1,\mu_1}[\omega_1(\psi_1(t_1)-\psi_1(x_1))^{\rho_1}]|\}^p dt_1\right]}{p}+\frac{\left[\int_{x_2}^{b_2}\{(\psi_2(t_2)-\psi_2(x_2))^{\mu_2-1}|E^{\gamma_2}_{\rho_2,\mu_2}[\omega_2(\psi_2(t_2)-\psi_2(x_2))^{\rho_2}]|\}^q dt_2\right]}{q}\right\}}$$

$$\leq (b_1-a_1)(b_2-a_2)\left\|\psi_1' f_1\right\|_q \left\|\psi_2' f_2\right\|_p. \quad (8.48)$$

Proof As similar to the proof of Theorem 8.9 is omitted. ∎

We continue with applications of Theorems 8.9, 8.10.

Theorem 8.11 *Let* $p, q > 1 : \frac{1}{p}+\frac{1}{q}=1; i=1,2.$ *Let* $[a_i, b_i] \subset \mathbb{R}$, $\psi_i \in C^{\max(N_1,N_2)}([a_i, b_i])$, $\psi_i' \neq 0$, *and strictly increasing;* $f_i \in C^{N_i}([a_i, b_i])$, *where* $N_i = \lceil \mu_i \rceil$, $0 < \mu_i \notin \mathbb{N}$. *Here* $\rho_i > 0$; $\gamma_i, \omega_i \in \mathbb{R}$. *Then*

$$\int_{a_1}^{b_1}\int_{a_2}^{b_2} \frac{\left|\left({}^C D^{\gamma_1;\psi_1}_{\rho_1,\mu_1,\omega_1,a_1+}f_1\right)(x_1)\right|\left|\left({}^C D^{\gamma_2;\psi_2}_{\rho_2,\mu_2,\omega_2,a_2+}f_2\right)(x_2)\right|dx_1dx_2}{\left\{\frac{\left[\int_{a_1}^{x_1}\left\{(\psi_1(x_1)-\psi_1(t_1))^{N_1-\mu_1-1}\left|E^{-\gamma_1}_{\rho_1,N_1-\mu_1}[\omega_1(\psi_1(x_1)-\psi_1(t_1))^{\rho_1}]\right|\right\}^p dt_1\right]}{p}+\frac{\left[\int_{a_2}^{x_2}\left\{(\psi_2(x_2)-\psi_2(t_2))^{N_2-\mu_2-1}\left|E^{-\gamma_2}_{\rho_2,N_2-\mu_2}[\omega_2(\psi_2(x_2)-\psi_2(t_2))^{\rho_2}]\right|\right\}^q dt_2\right]}{q}\right\}}$$

$$\leq (b_1-a_1)(b_2-a_2)\left\|\psi_1' f^{[N_1]}_{1\psi_1}\right\|_q \left\|\psi_2' f^{[N_2]}_{2\psi_2}\right\|_p. \quad (8.49)$$

Proof By Theorem 8.9 and (8.2), (8.6). ∎

We also give

Theorem 8.12 *All as in Theorem 8.11. Then*

$$\int_{a_1}^{b_1}\int_{a_2}^{b_2} \frac{\left|\left({}^C D^{\gamma_1;\psi_1}_{\rho_1,\mu_1,\omega_1,b_1-}f_1\right)(x_1)\right|\left|\left({}^C D^{\gamma_2;\psi_2}_{\rho_2,\mu_2,\omega_2,b_2-}f_2\right)(x_2)\right|dx_1dx_2}{\left\{\frac{\left[\int_{x_1}^{b_1}\left\{(\psi_1(t_1)-\psi_1(x_1))^{N_1-\mu_1-1}\left|E^{-\gamma_1}_{\rho_1,N_1-\mu_1}[\omega_1(\psi_1(t_1)-\psi_1(x_1))^{\rho_1}]\right|\right\}^p dt_1\right]}{p}+\frac{\left[\int_{x_2}^{b_2}\left\{(\psi_2(t_2)-\psi_2(x_2))^{N_2-\mu_2-1}\left|E^{-\gamma_2}_{\rho_2,N_2-\mu_2}[\omega_2(\psi_2(t_2)-\psi_2(x_2))^{\rho_2}]\right|\right\}^q dt_2\right]}{q}\right\}}$$

$$\leq (b_1-a_1)(b_2-a_2)\left\|\psi_1' f^{[N_1]}_{1\psi_1}\right\|_q \left\|\psi_2' f^{[N_2]}_{2\psi_2}\right\|_p. \quad (8.50)$$

Proof By Theorem 8.10 and (8.3), (8.7). ∎

We present

Theorem 8.13 *Let* $p, q > 1 : \frac{1}{p}+\frac{1}{q} = 1; i = 1, 2.$ *Let* $[a_i, b_i] \subset \mathbb{R}$, $\psi_i \in C^{\max(N_1,N_2)}([a_i, b_i])$, $\psi_i' \neq 0$, *and strictly increasing;* $f_i \in C^{N_i}([a_i, b_i])$, *where* $N_i = \lceil \mu_i \rceil$, $0 < \mu_i \notin \mathbb{N}$. *Here* $\rho_i > 0$; $\gamma_i, \omega_i \in \mathbb{R}$ *and* $\xi_i = \mu_i + \beta_i (N_i - \mu_i)$, $i = 1, 2$, *where* $0 \leq \beta_i \leq 1$. *Then*

$$\int_{a_1}^{b_1}\int_{a_2}^{b_2} \frac{\left|\left({}^{H}\mathbb{D}^{\gamma_1,\beta_1;\psi_1}_{\rho_1,\mu_1,\omega_1,a_1+}f_1\right)(x_1)\right|\left|\left({}^{H}\mathbb{D}^{\gamma_2,\beta_2;\psi_2}_{\rho_2,\mu_2,\omega_2,a_2+}f_2\right)(x_2)\right| dx_1 dx_2}{\left\{\frac{\left[\int_{a_1}^{x_1}\left\{(\psi_1(x_1)-\psi_1(t_1))^{\xi_1-\mu_1-1}\left|E^{-\gamma_1\beta_1}_{\rho_1,\xi_1-\mu_1}\left[\omega_1(\psi_1(x_1)-\psi_1(t_1))^{\rho_1}\right]\right|\right\}^p dt_1\right]}{p}+\frac{\left[\int_{a_2}^{x_2}\left\{(\psi_2(x_2)-\psi_2(t_2))^{\xi_2-\mu_2-1}\left|E^{-\gamma_2\beta_2}_{\rho_2,\xi_2-\mu_2}\left[\omega_2(\psi_2(x_2)-\psi_2(t_2))^{\rho_2}\right]\right|\right\}^q dt_2\right]}{q}\right\}}$$

$$\leq (b_1 - a_1)(b_2 - a_2)\left\|\psi_1' \, {}^{RL}D^{\gamma_1(1-\beta_1);\psi_1}_{\rho_1,\xi_1,\omega_1,a_1+}f_1\right\|_q \left\|\psi_2' \, {}^{RL}D^{\gamma_2(1-\beta_2);\psi_2}_{\rho_2,\xi_2,\omega_2,a_2+}f_2\right\|_p. \quad (8.51)$$

Proof By Theorem 8.9 and (8.15). ∎

We also give

Theorem 8.14 *All as in Theorem 8.13. Then*

$$\int_{a_1}^{b_1}\int_{a_2}^{b_2} \frac{\left|\left({}^{H}\mathbb{D}^{\gamma_1,\beta_1;\psi_1}_{\rho_1,\mu_1,\omega_1,b_1-}f_1\right)(x_1)\right|\left|\left({}^{H}\mathbb{D}^{\gamma_2,\beta_2;\psi_2}_{\rho_2,\mu_2,\omega_2,b_2-}f_2\right)(x_2)\right| dx_1 dx_2}{\left\{\frac{\left[\int_{x_1}^{b_1}\left\{(\psi_1(t_1)-\psi_1(x_1))^{\xi_1-\mu_1-1}\left|E^{-\gamma_1\beta_1}_{\rho_1,\xi_1-\mu_1}\left[\omega_1(\psi_1(t_1)-\psi_1(x_1))^{\rho_1}\right]\right|\right\}^p dt_1\right]}{p}+\frac{\left[\int_{x_2}^{b_2}\left\{(\psi_2(t_2)-\psi_2(x_2))^{\xi_2-\mu_2-1}\left|E^{-\gamma_2\beta_2}_{\rho_2,\xi_2-\mu_2}\left[\omega_2(\psi_2(t_2)-\psi_2(x_2))^{\rho_2}\right]\right|\right\}^q dt_2\right]}{q}\right\}}$$

$$\leq (b_1 - a_1)(b_2 - a_2)\left\|\psi_1' \, {}^{RL}D^{\gamma_1(1-\beta_1);\psi_1}_{\rho_1,\xi_1,\omega_1,b_1-}f_1\right\|_q \left\|\psi_2' \, {}^{RL}D^{\gamma_2(1-\beta_2);\psi_2}_{\rho_2,\xi_2,\omega_2,b_2-}f_2\right\|_p. \quad (8.52)$$

Proof By Theorem 8.10 and (8.16). ∎

8.3 Appendix

We give the following important fundamental results:

Theorem 8.15 *Let* $\rho, \mu > 0$; $\gamma, \omega \in \mathbb{R}$; *and* $\psi \in C^1([a, b])$ *increasing,* $f \in C([a, b])$. *Then* $\left(e^{\gamma;\psi}_{\rho,\mu,\omega,a+}f\right), \left(e^{\gamma;\psi}_{\rho,\mu,\omega,b-}f\right) \in C([a, b])$.

Proof We only prove $\left(e^{\gamma;\psi}_{\rho,\mu,\omega,a+}f\right) \in C([a, b])$. We skip the proof for the other as similar.

We consider the power series

$$\overline{E}^{\gamma}_{\rho,\mu}(z) = \sum_{k=0}^{\infty} \frac{|(\gamma)_k|}{k!\Gamma(\rho k + \mu)(\rho k + \mu)} z^k, \quad z \in \mathbb{R}. \quad (8.53)$$

We form

$$\overline{R}^{-1} := \lim_{k\to\infty} \frac{\frac{|(\gamma)_{k+1}|}{(k+1)!\Gamma(\rho(k+1)+\mu)(\rho(k+1)+\mu)}}{\frac{|(\gamma)_k|}{k!\Gamma(\rho k+\mu)(\rho k+\mu)}} = \lim_{k\to\infty} \frac{\frac{|\gamma+k|}{(k+1)\Gamma(\rho(k+1)+\mu)(\rho(k+1)+\mu)}}{\frac{1}{\Gamma(\rho k+\mu)(\rho k+\mu)}} = \tag{8.54}$$

$$\lim_{k\to\infty} \frac{|\gamma+k|\,\Gamma(\rho k+\mu)(\rho k+\mu)}{(k+1)\,\Gamma(\rho(k+1)+\mu)(\rho(k+1)+\mu)} =$$

$$\lim_{k\to\infty}\left(\frac{|\gamma+k|\,\Gamma(\rho k+\mu)}{(k+1)\,\Gamma(\rho(k+1)+\mu)}\right)\lim_{k\to\infty}\left(\frac{\rho k+\mu}{(\rho k+\mu)+\rho}\right) =: (\Xi)\,. \tag{8.55}$$

Notice that

$$\lim_{k\to\infty}\left(\frac{\rho k+\mu}{(\rho k+\mu)+\rho}\right) = 1. \tag{8.56}$$

From (8.1) we have that its radius of convergence is

$$R = \lim_{k\to\infty} \frac{\frac{|(\gamma)_k|}{k!\Gamma(\rho k+\mu)}}{\frac{|(\gamma)_{k+1}|}{(k+1)!\Gamma(\rho(k+1)+\mu)}} = \lim_{k\to\infty} \frac{\frac{1}{\Gamma(\rho k+\mu)}}{\frac{|\gamma+k|}{(k+1)\Gamma(\rho(k+1)+\mu)}} =$$

$$\lim_{k\to\infty} \frac{(k+1)\,\Gamma(\rho(k+1)+\mu)}{|\gamma+k|\,\Gamma(\rho k+\mu)} = \infty,$$

because (8.1) is an entire function.

Therefore, we have that

$$\lim_{k\to\infty} \frac{|\gamma+k|\,\Gamma(\rho k+\mu)}{(k+1)\,\Gamma(\rho(k+1)+\mu)} = 0.$$

Consequently by (8.55), (8.56), we get that $(\Xi) = 0$. Thus $\overline{R}^{-1} = 0$ and the radius of convergence of $\overline{E}^{\gamma}_{\rho,\mu}(z)$, see (8.53), is $\overline{R} = \infty$, hence (8.53) is convergent everywhere.

Consequently it holds

$$\sum_{k=0}^{\infty} \frac{|(\gamma)_k|\,(|\omega|\,(\psi(x)-\psi(a))^{\rho})^k}{k!\Gamma(\rho k+\mu)(\rho k+\mu)} < \infty, \tag{8.57}$$

$\forall\, x \in [a,b]$.

We notice that

$$\sum_{k=0}^{\infty} \frac{|(\gamma)_k|\,|\omega|^k}{k!\Gamma(\rho k+\mu)} \int_a^x \psi'(t)\,(\psi(x)-\psi(t))^{(\rho k+\mu)-1}\,|f(t)|\,dt \leq$$

$$\|f\|_{\infty}\sum_{k=0}^{\infty}\frac{\left|(\gamma)_k\right||\omega|^k}{k!\Gamma(\rho k+\mu)}\frac{(\psi(x)-\psi(a))^{\rho k+\mu}}{\rho k+\mu}\leq \tag{8.58}$$

$$\|f\|_{\infty}(\psi(b)-\psi(a))^{\mu}\sum_{k=0}^{\infty}\frac{\left|(\gamma)_k\right|\left(|\omega|(\psi(x)-\psi(a))^{\rho}\right)^k}{k!\Gamma(\rho k+\mu)(\rho k+\mu)}\overset{(8.57)}{<}\infty.$$

Consequently, by [7], p. 175, we derive

$$\left(e^{\gamma;\psi}_{\rho,\mu,\omega,a+}f\right)(x)\overset{(8.2)}{=}\int_a^x\psi'(t)(\psi(x)-\psi(t))^{\mu-1}$$

$$\left(\sum_{k=0}^{\infty}\frac{(\gamma)_k}{k!\Gamma(\rho k+\mu)}\left(\omega(\psi(x)-\psi(t))^{\rho}\right)^k\right)f(t)\,dt=$$

$$\sum_{k=0}^{\infty}\frac{(\gamma)_k\,\omega^k}{k!\Gamma(\rho k+\mu)}\int_a^x\psi'(t)(\psi(x)-\psi(t))^{(\rho k+\mu)-1}f(t)\,dt, \tag{8.59}$$

$\forall\, x\in[a,b]$.

By [3], p. 98, we obtain that the function

$$\lambda^{(k)}_{\rho,\mu}(f,x)=\int_a^x\psi'(t)(\psi(x)-\psi(t))^{(\rho k+\mu)-1}f(t)\,dt,$$

$x\in[a,b]$, is absolutely continuous for $\rho k+\mu\geq 1$ and continuous for $\rho k+\mu\in(0,1)$; $\psi\in C^1([a,b])$ and increasing.

That is always $\lambda^{(k)}_{\rho,\mu}(|f|,x)\in C([a,b])$, for all $k=0,1,\ldots$

By (8.57), one can derive that

$$\sum_{k=0}^{\infty}\frac{\left|(\gamma)_k\right||\omega|^k}{k!\Gamma(\rho k+\mu)}\lambda^{(k)}_{\rho,\mu}(|f|,x)\leq$$

$$\|f\|_{\infty}(\psi(b)-\psi(a))^{\mu}\sum_{k=0}^{\infty}\frac{\left|(\gamma)_k\right|\left(|\omega|(\psi(b)-\psi(a))^{\rho}\right)^k}{k!\Gamma(\rho k+\mu)(\rho k+\mu)}<\infty. \tag{8.60}$$

Notice that

$$\left|\lambda^{(k)}_{\rho,\mu}(f,x)\right|\leq\lambda^{(k)}_{\rho,\mu}(|f|,x)=\int_a^x\psi'(t)(\psi(x)-\psi(t))^{(\rho k+\mu)-1}|f(t)|\,dt\leq \tag{8.61}$$

$$\|f\|_{\infty}\frac{(\psi(b)-\psi(a))^{(\rho k+\mu)}}{(\rho k+\mu)},\quad k=0,1,\ldots$$

And even more we get:

$$\frac{|(\gamma)_k| \, |\omega|^k}{k!\Gamma(\rho k+\mu)} \left|\lambda_{\rho,\mu}^{(k)}(f,x)\right| \leq \frac{|(\gamma)_k| \, |\omega|^k}{k!\Gamma(\rho k+\mu)} \lambda_{\rho,\mu}^{(k)}(|f|,x) \leq$$

$$\left(\frac{|(\gamma)_k| \, |\omega|^k}{k!\Gamma(\rho k+\mu)}\right) \frac{\|f\|_\infty (\psi(b)-\psi(a))^{(\rho k+\mu)}}{(\rho k+\mu)} =: M_k, \quad k=0,1,...; \tag{8.62}$$

and by (8.60) that $\sum_{k=0}^{\infty} M_k < \infty$, converges.

By Weierstrass M-test we get that $\sum_{k=0}^{\infty} \frac{(\gamma)_k \omega^k}{k!\Gamma(\rho k+\mu)} \lambda_{\rho,\mu}^{(k)}(f,x)$ is uniformly and absolutely convergent for $x \in [a,b]$.

Consequently by (8.59) we derive that $\left(e_{\rho,\mu,\omega,a+}^{\gamma;\psi} f\right) \in C([a,b])$. The proof is completed. ∎

We finish with

Corollary 8.16 *All as in Theorem 8.15. We have that*

$$\left\| e_{\rho,\mu,\omega,a+(b-)}^{\gamma;\psi} f \right\|_\infty \leq \left(\sum_{k=0}^{\infty} \frac{|(\gamma)_k| \, |\omega|^k (\psi(b)-\psi(a))^{\rho k+\mu}}{k!\Gamma(\rho k+\mu+1)} \right) \|f\|_\infty < +\infty. \tag{8.63}$$

That is $e_{\rho,\mu,\omega,a+(b-)}^{\gamma;\psi}$ *are bounded linear operators and positive operators if* $\gamma, \omega > 0$.

Proof By (8.59), (8.60). ∎

References

1. Anastassiou, G.A.: Foundations of Generalized Prabhakar-Hilfer fractional Calculus with Applications (2021). [Submitted]
2. Anastassiou, G.A.: Fractional Differentiation Inequalities. Springer, Heidelberg, New York (2009)
3. Anastassiou, G.A.: Intelligent computations: abstract fractional calculus inequalities. Approximations. Springer, Heidelberg, New York (2018)
4. da Sousa, J.V.C., de Oliveira, E.C.: On the ψ-Hilfer fractional derivative. Commun. Nonlinear Sci. Numer. Simul. **60**, 72–91 (2018)
5. Giusti, A., et al.: A practical guide to Prabhakar fractional calculus. Fractional Calc. Appl. Anal. **23**(1), 9–54 (2020)
6. Gorenflo, R., Kilbas, A., Mainardi, F., Rogosin, S.: Mittag-Leffler functions. Related Topics and Application. Springer, Heidelberg, New York (2014)

7. Hewith, E., Stromberg, K.: Real and Abstract Analysis. Springer, New York, Heidelberg, Berlin (1965)
8. Polito, F., Tomovski, Z.: Some properties of Prabhakar-type fractional calculus operators. Fractional Differ. Calc. **6**(1), 73–94 (2016)
9. Prabhakar, T.R.: A singular integral equation with a generalized Mittag Leffler function in the kernel. Yokohama Math. J. **19**, 7–15 (1971)

Chapter 9
Advanced and General Hilfer-Prabhakar-Hardy Fractional Inequalities

We present a series of left and right side Hardy type fractional inequalities under convexity in the setting of generalized Hilfer and Prabhakar fractional Calculi. It follows [2].

9.1 Background

Let $-\infty < a < b < \infty$, the left and right Riemann-Liouville fractional integrals of order $\alpha \in \mathbb{C}$ ($\mathcal{R}(\alpha) > 0$) are defined by

$$\left(I_{a+}^{\alpha} f\right)(x) = \frac{1}{\Gamma(\alpha)} \int_{a}^{x} (x-t)^{\alpha-1} f(t)\, dt, \tag{9.1}$$

$x > a$; where Γ stands for the gamma function,

and

$$\left(I_{b-}^{\alpha} f\right)(x) = \frac{1}{\Gamma(\alpha)} \int_{x}^{b} (t-x)^{\alpha-1} f(t)\, dt, \tag{9.2}$$

$x < b$.

The Riemann-Liouville left and right fractional derivatives of order $\alpha \in \mathbb{C}$ ($\mathcal{R}(\alpha) \geq 0$) are defined by

$$\left(\Delta_{a+}^{\alpha} y\right)(x) = \left(\frac{d}{dx}\right)^{n} \left(I_{a+}^{n-\alpha} y\right)(x) = \frac{1}{\Gamma(n-\alpha)} \left(\frac{d}{dx}\right)^{n} \int_{a}^{x} (x-t)^{n-\alpha-1} y(t)\, dt \tag{9.3}$$

($n = \lceil \mathcal{R}(\alpha) \rceil$, $\lceil \cdot \rceil$ means ceiling of the number; $x > a$)

G. A. Anastassiou, *Unification of Fractional Calculi with Applications*,
Studies in Systems, Decision and Control 398,
https://doi.org/10.1007/978-3-030-86920-5_9

$$\left(\Delta_{b-}^{\alpha} y\right)(x) = (-1)^n \left(\frac{d}{dx}\right)^n \left(I_{b-}^{n-\alpha} y\right)(x) =$$

$$\frac{(-1)^n}{\Gamma(n-\alpha)} \left(\frac{d}{dx}\right)^n \int_x^b (t-x)^{n-\alpha-1} y(t)\, dt \tag{9.4}$$

($n = \lceil \mathcal{R}(\alpha) \rceil$; $x < b$), respectively, where $\mathcal{R}(\alpha)$ is the real part of α.

In particular, when $\alpha = n \in \mathbb{Z}_+$, then

$$\begin{gathered}\left(\Delta_{a+}^{0} y\right)(x) = \left(\Delta_{b-}^{0} y\right)(x) = y(x); \\ \left(\Delta_{a+}^{n} y\right)(x) = y^{(n)}(x), \text{ and } \left(\Delta_{b-}^{n} y\right)(x) = (-1)^n y^{(n)}(x), \ n \in \mathbb{N},\end{gathered} \tag{9.5}$$

see [14].

Let $\alpha > 0$, $I = [a, b] \subset \mathbb{R}$, f an integrable function defined on I and $\psi \in C^1(I)$ an increasing function such that $\psi'(x) \neq 0$, for all $x \in I$. Left fractional integrals and left Riemann-Liouville fractional derivatives of a function f with respect to another function ψ are defined as ([11, 14])

$$I_{a+}^{\alpha,\psi} f(x) = \frac{1}{\Gamma(\alpha)} \int_a^x \psi'(t) (\psi(x) - \psi(t))^{\alpha-1} f(t)\, dt, \tag{9.6}$$

and

$$\Delta_{a+}^{\alpha,\psi} f(x) = \left(\frac{1}{\psi'(x)} \frac{d}{dx}\right)^n I_{a+}^{n-\alpha,\psi} f(x) = \tag{9.7}$$

$$\frac{1}{\Gamma(n-\alpha)} \left(\frac{1}{\psi'(x)} \frac{d}{dx}\right)^n \int_a^x \psi'(t) (\psi(x) - \psi(t))^{n-\alpha-1} f(t)\, dt,$$

respectively, where $n = \lceil \alpha \rceil$.

Similarly, we define the right ones:

$$I_{b-}^{\alpha,\psi} f(x) = \frac{1}{\Gamma(\alpha)} \int_x^b \psi'(t) (\psi(t) - \psi(x))^{\alpha-1} f(t)\, dt, \tag{9.8}$$

and

$$\Delta_{b-}^{\alpha,\psi} f(x) = \left(-\frac{1}{\psi'(x)} \frac{d}{dx}\right)^n I_{b-}^{n-\alpha,\psi} f(x) =$$

$$\frac{1}{\Gamma(n-\alpha)} \left(-\frac{1}{\psi'(x)} \frac{d}{dx}\right)^n \int_x^b \psi'(t) (\psi(t) - \psi(x))^{n-\alpha-1} f(t)\, dt. \tag{9.9}$$

The following semigroup property holds; if $\alpha, \beta > 0$, $f \in C(I)$, then

$$I_{a+}^{\alpha,\psi} I_{a+}^{\beta,\psi} f = I_{a+}^{\alpha+\beta,\psi} f \ \text{ and } \ I_{b-}^{\alpha,\psi} I_{b-}^{\beta,\psi} f = I_{b-}^{\alpha+\beta,\psi} f.$$

Next let again $\alpha > 0$, $n = \lceil \alpha \rceil$, $I = [a, b]$, $f, \psi \in C^n(I) : \psi$ is increasing and $\psi'(x) \neq 0$, for all $x \in I$. The left ψ-Caputo fractional derivative of f of order α is given by ([1])

$$ {}^C D_{a+}^{\alpha,\psi} f(x) = I_{a+}^{n-\alpha,\psi} \left(\frac{1}{\psi'(x)} \frac{d}{dx} \right)^n f(x), \tag{9.10} $$

and the right ψ-Caputo fractional derivative ([1])

$$ {}^C D_{b-}^{\alpha,\psi} f(x) = I_{b-}^{n-\alpha,\psi} \left(-\frac{1}{\psi'(x)} \frac{d}{dx} \right)^n f(x). \tag{9.11} $$

We set

$$ f_{\psi}^{[n]}(x) := f_{\psi}^{(n)} f(x) := \left(\frac{1}{\psi'(x)} \frac{d}{dx} \right)^n f(x). \tag{9.12} $$

Clearly, when $\alpha = m \in \mathbb{N}$ we have

$$ {}^C D_{a+}^{\alpha,\psi} f(x) = f_{\psi}^{[m]}(x) \quad \text{and} \quad {}^C D_{b-}^{\alpha,\psi} f(x) = (-1)^m f_{\psi}^{[m]}(x), $$

and if $\alpha \notin \mathbb{N}$, then

$$ {}^C D_{a+}^{\alpha,\psi} f(x) = \frac{1}{\Gamma(n-\alpha)} \int_a^x \psi'(t) (\psi(x) - \psi(t))^{n-\alpha-1} f_{\psi}^{[n]}(t)\, dt, \tag{9.13} $$

and

$$ {}^C D_{b-}^{\alpha,\psi} f(x) = \frac{(-1)^n}{\Gamma(n-\alpha)} \int_x^b \psi'(t) (\psi(t) - \psi(x))^{n-\alpha-1} f_{\psi}^{[n]}(t)\, dt. \tag{9.14} $$

If $\psi(x) = x$, then we get the usual left and right Caputo fractional derivatives

$$ {}^C D_{a+}^{m} f(x) = f^{(m)}(x), \; {}^C D_{b-}^{m} f(x) = (-1)^m f^{(m)}(x), $$

for $m \in \mathbb{N}$, and ($\alpha \notin \mathbb{N}$)

$$ D_{*a}^{\alpha} f(x) = {}^C D_{a+}^{\alpha} f(x) = \frac{1}{\Gamma(n-\alpha)} \int_a^x (x-t)^{n-\alpha-1} f^{(n)}(t)\, dt, \tag{9.15} $$

$$ D_{b-}^{\alpha}(x) = {}^C D_{b-}^{\alpha} f(x) = \frac{(-1)^n}{\Gamma(n-\alpha)} \int_x^b (t-x)^{n-\alpha-1} f^{(n)}(t)\, dt. \tag{9.16} $$

Also we set

$$ {}^C D_{a+}^{0,\psi} f(x) = {}^C D_{b-}^{0,\psi} f(x) = f(x). $$

Next we talk about the ψ-Hilfer fractional derivative.

Definition 9.1 ([7]) Let $n-1<\alpha<n$, $n\in\mathbb{N}$, $I=[a,b]\subset\mathbb{R}$ and $f,\psi\in C^n([a,b])$, ψ is increasing and $\psi'(x)\neq 0$, for all $x\in I$. The ψ-Hilfer fractional derivative (left-sided and right-sided) ${}^H\mathbb{D}_{a+(b-)}^{\alpha,\beta;\psi}f$ of order α and type $0\leq\beta\leq 1$, respectively, are defined by

$$ {}^H\mathbb{D}_{a+}^{\alpha,\beta;\psi}f(x)=I_{a+}^{\beta(n-\alpha);\psi}\left(\frac{1}{\psi'(x)}\frac{d}{dx}\right)^n I_{a+}^{(1-\beta)(n-\alpha);\psi}f(x), \tag{9.17} $$

and

$$ {}^H\mathbb{D}_{b-}^{\alpha,\beta;\psi}f(x)=I_{b-}^{\beta(n-\alpha);\psi}\left(-\frac{1}{\psi'(x)}\frac{d}{dx}\right)^n I_{b-}^{(1-\beta)(n-\alpha);\psi}f(x),\quad x\in[a,b]. \tag{9.18} $$

The original Hilfer fractional derivatives ([15]) come from $\psi(x)=x$, and are denoted by ${}^H\mathbb{D}_{a+}^{\alpha,\beta}f(x)$ and ${}^H\mathbb{D}_{b-}^{\alpha,\beta}f(x)$.

When $\beta=0$, we get Riemann-Liouville fractional derivatives, while when $\beta=1$ we have Caputo type fractional derivatives.

We define $\gamma=\alpha+\beta(n-\alpha)$. We notice that $n-1<\alpha\leq\alpha+\beta(n-\alpha)\leq\alpha+n-\alpha=n$, hence $\lceil\gamma\rceil=n$. We can easily write that ([7])

$$ {}^H\mathbb{D}_{a+}^{\alpha,\beta;\psi}f(x)=I_{a+}^{\gamma-\alpha;\psi}\Delta_{a+}^{\gamma;\psi}f(x), \tag{9.19} $$

and

$$ {}^H\mathbb{D}_{b-}^{\alpha,\beta;\psi}f(x)=I_{b-}^{\gamma-\alpha;\psi}\Delta_{b-}^{\gamma;\psi}f(x),\quad x\in[a,b]. \tag{9.20} $$

We have that ([7])

$$ \Delta_{a+}^{\gamma,\psi}f(x)=\left(\frac{1}{\psi'(x)}\frac{d}{dx}\right)^n I_{a+}^{(1-\beta)(n-\alpha);\psi}f(x), \tag{9.21} $$

and

$$ \Delta_{b-}^{\gamma,\psi}f(x)=\left(-\frac{1}{\psi'(x)}\frac{d}{dx}\right)^n I_{b-}^{(1-\beta)(n-\alpha);\psi}f(x). \tag{9.22} $$

In particular, when $0<\alpha<1$ and $0\leq\beta\leq 1$; $\gamma=\alpha+\beta(1-\alpha)$, we have that

$$ {}^H\mathbb{D}_{a+}^{\alpha,\beta;\psi}f(x)=\frac{1}{\Gamma(\gamma-\alpha)}\int_a^x\psi'(t)(\psi(x)-\psi(t))^{\gamma-\alpha-1}\Delta_{a+}^{\gamma;\psi}f(t)\,dt, \tag{9.23} $$

and

$$ {}^H\mathbb{D}_{b-}^{\alpha,\beta;\psi}f(x)=\frac{1}{\Gamma(\gamma-\alpha)}\int_x^b\psi'(t)(\psi(t)-\psi(x))^{\gamma-\alpha-1}\Delta_{b-}^{\gamma;\psi}f(t)\,dt, \tag{9.24} $$

$x \in [a, b]$.

Remark 9.2 ([7]) Let $\mu = n(1-\beta) + \beta\alpha$, then $\lceil \mu \rceil = n$.
Assume that $g(x) = I_{a+}^{(1-\beta)(n-\alpha);\psi} f(x) \in C^n([a, b])$, we have that

$$ {}^{H}\mathbb{D}_{a+}^{\alpha,\beta;\psi} f(x) = I_{a+}^{n-\mu;\psi} \left(\frac{1}{\psi'(x)} \frac{d}{dx} \right)^n g(x). \tag{9.25}$$

Thus

$$ {}^{H}\mathbb{D}_{a+}^{\alpha,\beta;\psi} f = {}^{C}D_{a+}^{\mu;\psi} g(x) = {}^{C}D_{a+}^{\mu;\psi} \left[I_{a+}^{(1-\beta)(n-\alpha);\psi} f(x) \right]. \tag{9.26}$$

Assume that $w(x) = I_{b-}^{(1-\beta)(n-\alpha);\psi} f(x) \in C^n([a, b])$. Hence

$$ {}^{H}\mathbb{D}_{b-}^{\alpha,\beta;\psi} f(x) = I_{b-}^{\beta(n-\alpha);\psi} \left(-\frac{1}{\psi'(x)} \frac{d}{dx} \right)^n w(x) = I_{b-}^{n-\mu;\psi} \left(-\frac{1}{\psi'(x)} \frac{d}{dx} \right)^n w(x). \tag{9.27}$$

Thus

$$ {}^{H}\mathbb{D}_{b-}^{\alpha,\beta;\psi} f = {}^{C}D_{b-}^{\mu;\psi} w(x) = {}^{C}D_{b-}^{\mu;\psi} \left(I_{b-}^{(1-\beta)(n-\alpha);\psi} f(x) \right). \tag{9.28}$$

We mention the simplified ψ-Hilfer fractional Taylor formulae:

Theorem 9.3 *(see also [7]) Let $\psi, f \in C^n([a, b])$, with ψ being increasing such that $\psi'(x) \neq 0$ over $[a, b]$, where $n-1 < \alpha < n$, $0 \leq \beta \leq 1$, and $\gamma = \alpha + \beta(n-\alpha)$, $x \in [a, b]$. Then*

$$ f(x) - \sum_{k=1}^{n-1} \frac{(\psi(x) - \psi(a))^{\gamma-k}}{\Gamma(\gamma-k+1)} f_{\psi}^{[n-k]} \left(I_{a+}^{(1-\beta)(n-\alpha);\psi} f \right)(a) =$$

$$ \frac{1}{\Gamma(\alpha)} \int_a^x \psi'(t) (\psi(x) - \psi(t))^{\alpha-1} \, {}^{H}\mathbb{D}_{a+}^{\alpha,\beta;\psi} f(t) \, dt, \tag{9.29}$$

and

$$ f(x) - \sum_{k=1}^{n-1} \frac{(-1)^k (\psi(b) - \psi(x))^{\gamma-k}}{\Gamma(\gamma-k+1)} f_{\psi}^{[n-k]} \left(I_{b-}^{(1-\beta)(n-\alpha);\psi} f \right)(b) =$$

$$ \frac{1}{\Gamma(\alpha)} \int_x^b \psi'(t) (\psi(t) - \psi(x))^{\alpha-1} \, {}^{H}\mathbb{D}_{b-}^{\alpha,\beta;\psi} f(t) \, dt. \tag{9.30}$$

Here notice that $\left(I_{a+}^{(1-\beta)(n-\alpha);\psi} f \right)(a) = \left(I_{b-}^{(1-\beta)(n-\alpha);\psi} f \right)(b) = 0$.

We also mention the following alternative ψ-Hilfer fractional Taylor formulae:

Theorem 9.4 *([3]) Let $f, \psi \in C^n([a,b])$, with ψ being increasing, $\psi'(x) \neq 0$ over $[a,b] \subset \mathbb{R}$, $\alpha > 0 : \lceil \alpha \rceil = n$, $0 \leq \beta \leq 1$, $\mu = n(1-\beta) + \beta\alpha$. Assume that $g(x) = I_{a+}^{(1-\beta)(n-\alpha);\psi} f(x)$, $w(x) = I_{b-}^{(1-\beta)(n-\alpha);\psi} f(x) \in C^n([a,b])$.*

Then

1)

$$I_{a+}^{\mu;\psi}\ {}^{H}\mathbb{D}_{a+}^{\alpha,\beta;\psi} f(x) = g(x) - \sum_{k=0}^{n-1} \frac{g_{\psi}^{[k]}(a)}{k!} (\psi(x) - \psi(a))^k, \tag{9.31}$$

where

$$g_{\psi}^{[k]}(x) = \left(\frac{1}{\psi'(x)} \frac{d}{dx} \right)^k g(x), \quad k = 0, 1, ..., n-1,$$

and

2)

$$I_{b-}^{\mu;\psi}\ {}^{H}\mathbb{D}_{b-}^{\alpha,\beta;\psi} f(x) = w(x) - \sum_{k=0}^{n-1} \frac{(-1)^k w_{\psi}^{[k]}(b)}{k!} (\psi(b) - \psi(x))^k, \tag{9.32}$$

where

$$w_{\psi}^{[k]}(x) = \left(\frac{1}{\psi'(x)} \frac{d}{dx} \right)^k w(x), \quad k = 0, 1, ..., n-1; \ x \in [a,b].$$

Next we list two Hilfer fractional derivatives representation formulae:

Theorem 9.5 *([3]) Let $\alpha > 0$, $\alpha \notin \mathbb{N}$, $\lceil \alpha \rceil = n$, $0 < \beta < 1$; $f \in C^n([a,b])$, $[a,b] \subset \mathbb{R}$; and set $\gamma = \alpha + \beta(n-\alpha)$. Assume further that $\Delta_{a+}^{\gamma} f \in C([a,b]) : \Delta_{a+}^{\gamma-j} f(a) = 0$, for $j = 1, ..., n$. Let also $\overline{\alpha} > 0 : \lceil \overline{\alpha} \rceil = \overline{n}$, with $\overline{\gamma} = \overline{\alpha} + \beta(\overline{n} - \overline{\alpha})$, and assume that $\alpha > \overline{\alpha}$ and $\gamma > \overline{\gamma}$. Then*

$$^{H}\mathbb{D}_{a+}^{\overline{\alpha},\beta} f(x) = \frac{1}{\Gamma(\alpha - \overline{\alpha})} \int_a^x (x-t)^{\alpha - \overline{\alpha} - 1}\ {}^{H}\mathbb{D}_{a+}^{\alpha,\beta} f(t)\, dt, \tag{9.33}$$

$\forall x \in [a,b]$,

furthermore $^{H}\mathbb{D}_{a+}^{\overline{\alpha},\beta} f \in AC([a,b])$ (absolutely continuous functions) if $\alpha - \overline{\alpha} \geq 1$ and $^{H}\mathbb{D}_{a+}^{\overline{\alpha},\beta} f \in C([a,b])$ if $\alpha - \overline{\alpha} \in (0,1)$.

Theorem 9.6 *([3]) Let $\alpha > 0$, $\alpha \notin \mathbb{N}$, $\lceil \alpha \rceil = n$, $0 < \beta < 1$; $f \in C^n([a,b])$, $[a,b] \subset \mathbb{R}$; and set $\gamma = \alpha + \beta(n-\alpha)$. Assume further that $\Delta_{b-}^{\gamma} f \in C([a,b]) : \Delta_{b-}^{\gamma-j} f(b) = 0$, $j = 1, ..., n$. Let also $\overline{\alpha} > 0 : \lceil \overline{\alpha} \rceil = \overline{n}$, with $\overline{\gamma} = \overline{\alpha} + \beta(\overline{n} - \overline{\alpha})$, and assume that $\alpha > \overline{\alpha}$ and $\gamma > \overline{\gamma}$. Then*

$$ {}^{H}\mathbb{D}_{b-}^{\overline{\alpha},\beta} f(x) = \frac{1}{\Gamma(\alpha-\overline{\alpha})} \int_x^b (t-x)^{\alpha-\overline{\alpha}-1} \, {}^{H}\mathbb{D}_{b-}^{\alpha,\beta} f(t)\, dt, \tag{9.34} $$

$\forall\, x \in [a,b]$,

furthermore ${}^{H}\mathbb{D}_{b-}^{\overline{\alpha},\beta} f \in AC([a,b])$ *if* $\alpha - \overline{\alpha} \geq 1$ *and* ${}^{H}\mathbb{D}_{b-}^{\overline{\alpha},\beta} f \in C([a,b])$ *if* $\alpha - \overline{\alpha} \in (0,1)$.

The fractional integral operator $I_{a+}^{\alpha} f$ and $I_{b-}^{\alpha} f$, $\alpha > 0$, are bounded in $L_p(a,b)$, $1 \leq p \leq \infty$, that is

$$ \left\| I_{a+}^{\alpha} f \right\|_p \leq K \left\| f \right\|_p, \quad \left\| I_{b-}^{\alpha} f \right\|_p \leq K \left\| f \right\|_p, \tag{9.35} $$

where

$$ K = \frac{(b-a)^{\alpha}}{\Gamma(\alpha+1)}. \tag{9.36} $$

The left inequality (9.35) was proved by H.G. Hardy in one of his first papers, see [10].

We continue this Background section with the following material from [4], where the author introduced the genralized ψ-Prabhakar type of fractional calculus and mixed it with the ψ-Hilfer fractional calculus.

So we consider the Prabhakar function (also known as the three parameter Mittag-Laffler function), (see [9], p. 97; [8])

$$ E_{\alpha,\beta}^{\gamma}(z) = \sum_{k=0}^{\infty} \frac{(\gamma)_k}{k!\Gamma(\alpha k+\beta)} z^k, \tag{9.37} $$

where Γ is the gamma function; $\alpha, \beta, \gamma \in \mathbb{R} : \alpha, \beta > 0$, $z \in \mathbb{R}$, and $(\gamma)_k = \gamma(\gamma+1) \ldots (\gamma+k-1)$. It is $E_{\alpha,\beta}^{0}(z) = \frac{1}{\Gamma(\beta)}$.

Let $a, b \in \mathbb{R}$, $a < b$ and $x \in [a,b]$; $f \in C([a,b])$. Let also $\psi \in C^1([a,b])$ which is increasing. The left and right Prabhakar fractional integrals with respect to ψ are defined as follows:

$$ \left(e_{\rho,\mu,\omega,a+}^{\gamma;\psi} f\right)(x) = \int_a^x \psi'(t)\,(\psi(x)-\psi(t))^{\mu-1} E_{\rho,\mu}^{\gamma}\left[\omega(\psi(x)-\psi(t))^{\rho}\right] f(t)\, dt, \tag{9.38} $$

and

$$ \left(e_{\rho,\mu,\omega,b-}^{\gamma;\psi} f\right)(x) = \int_x^b \psi'(t)\,(\psi(t)-\psi(x))^{\mu-1} E_{\rho,\mu}^{\gamma}\left[\omega(\psi(t)-\psi(x))^{\rho}\right] f(t)\, dt, \tag{9.39} $$

where $\rho, \mu > 0$; $\gamma, \omega \in \mathbb{R}$.

Functions (9.38) and (9.39) are continuous ([4]).

Next, additionally, assume that $\psi'(x) \neq 0$ over $[a,b]$.

Let $\psi, f \in C^N([a,b])$, where $N = \lceil \mu \rceil$, ($\lceil \cdot \rceil$ is the ceiling of the number), $0 < \mu \notin \mathbb{N}$. We define the ψ-Prabhakar-Caputo left and right fractional derivatives of order μ as follows ($x \in [a, b]$):

$$\left({}^{C}D^{\gamma;\psi}_{\rho,\mu,\omega,a+}f\right)(x) = \int_a^x \psi'(t)(\psi(x) - \psi(t))^{N-\mu-1}$$

$$E^{-\gamma}_{\rho,N-\mu}\left[\omega(\psi(x) - \psi(t))^{\rho}\right]\left(\frac{1}{\psi'(t)}\frac{d}{dt}\right)^{N} f(t)\,dt, \tag{9.40}$$

and

$$\left({}^{C}D^{\gamma;\psi}_{\rho,\mu,\omega,b-}f\right)(x) = (-1)^N \int_x^b \psi'(t)(\psi(t) - \psi(x))^{N-\mu-1}$$

$$E^{-\gamma}_{\rho,N-\mu}\left[\omega(\psi(t) - \psi(x))^{\rho}\right]\left(\frac{1}{\psi'(t)}\frac{d}{dt}\right)^{N} f(t)\,dt. \tag{9.41}$$

One can write these (see (9.40), (9.41)) as

$$\left({}^{C}D^{\gamma;\psi}_{\rho,\mu,\omega,a+}f\right)(x) = \left(e^{-\gamma;\psi}_{\rho,N-\mu,\omega,a+}f^{[N]}_{\psi}\right)(x), \tag{9.42}$$

and

$$\left({}^{C}D^{\gamma;\psi}_{\rho,\mu,\omega,b-}f\right)(x) = (-1)^N \left(e^{-\gamma;\psi}_{\rho,N-\mu,\omega,b-}f^{[N]}_{\psi}\right)(x), \tag{9.43}$$

where

$$f^{[N]}_{\psi}(x) = f^{(N)}_{\psi} f(x) := \left(\frac{1}{\psi'(x)}\frac{d}{dx}\right)^{N} f(x), \tag{9.44}$$

$\forall\, x \in [a, b]$.

Functions (9.42) and (9.43) are continuous on $[a, b]$.

Next we define the ψ-Prabhakar-Riemann Liouville left and right fractional derivatives of order μ as follows ($x \in [a, b]$):

$$\left({}^{RL}D^{\gamma;\psi}_{\rho,\mu,\omega,a+}f\right)(x) = \left(\frac{1}{\psi'(x)}\frac{d}{dx}\right)^{N} \int_a^x \psi'(t)(\psi(x) - \psi(t))^{N-\mu-1}$$

$$E^{-\gamma}_{\rho,N-\mu}\left[\omega(\psi(x) - \psi(t))^{\rho}\right] f(t)\,dt, \tag{9.45}$$

and

$$\left({}^{RL}D^{\gamma;\psi}_{\rho,\mu,\omega,b-}f\right)(x) = \left(-\frac{1}{\psi'(x)}\frac{d}{dx}\right)^{N} \int_x^b \psi'(t)(\psi(t) - \psi(x))^{N-\mu-1}$$

$$E^{-\gamma}_{\rho,N-\mu}\left[\omega(\psi(t) - \psi(x))^{\rho}\right] f(t)\,dt. \tag{9.46}$$

That is we have

$$\left({}^{RL}D^{\gamma;\psi}_{\rho,\mu,\omega,a+}f\right)(x) = \left(\frac{1}{\psi'(x)}\frac{d}{dx}\right)^N \left(e^{-\gamma;\psi}_{\rho,N-\mu,\omega,a+}f\right)(x), \tag{9.47}$$

and

$$\left({}^{RL}D^{\gamma;\psi}_{\rho,\mu,\omega,b-}f\right)(x) = \left(-\frac{1}{\psi'(x)}\frac{d}{dx}\right)^N \left(e^{-\gamma;\psi}_{\rho,N-\mu,\omega,b-}f\right)(x), \tag{9.48}$$

$\forall\, x \in [a, b]$.

We define also the ψ-Hilfer-Prabhakar left and right fractional derivatives of order μ and type $0 \leq \beta \leq 1$, as follows

$$\left({}^{H}\mathbb{D}^{\gamma,\beta;\psi}_{\rho,\mu,\omega,a+}f\right)(x) = e^{-\gamma\beta;\psi}_{\rho,\beta(N-\mu),\omega,a+}\left(\frac{1}{\psi'(x)}\frac{d}{dx}\right)^N e^{-\gamma(1-\beta);\psi}_{\rho,(1-\beta)(N-\mu),\omega,a+}f(x), \tag{9.49}$$

and

$$\left({}^{H}\mathbb{D}^{\gamma,\beta;\psi}_{\rho,\mu,\omega,b-}f\right)(x) = e^{-\gamma\beta;\psi}_{\rho,\beta(N-\mu),\omega,b-}\left(-\frac{1}{\psi'(x)}\frac{d}{dx}\right)^N e^{-\gamma(1-\beta);\psi}_{\rho,(1-\beta)(N-\mu),\omega,b-}f(x), \tag{9.50}$$

$\forall\, x \in [a, b]$.

When $\beta = 0$, we get the Riemann-Liouville version, and when $\beta = 1$, we get the Caputo version.

We call $\xi = \mu + \beta(N - \mu)$, we have that $N - 1 < \mu \leq \mu + \beta(N - \mu) \leq \mu + N - \mu = N$, hence $\lceil \xi \rceil = N$.

We can easily write that

$$\left({}^{H}\mathbb{D}^{\gamma,\beta;\psi}_{\rho,\mu,\omega,a+}f\right)(x) = e^{-\gamma\beta;\psi}_{\rho,\xi-\mu,\omega,a+}\ {}^{RL}D^{\gamma(1-\beta);\psi}_{\rho,\xi,\omega,a+}f(x), \tag{9.51}$$

and

$$\left({}^{H}\mathbb{D}^{\gamma,\beta;\psi}_{\rho,\mu,\omega,b-}f\right)(x) = e^{-\gamma\beta;\psi}_{\rho,\xi-\mu,\omega,b-}\ {}^{RL}D^{\gamma(1-\beta);\psi}_{\rho,\xi,\omega,b-}f(x), \tag{9.52}$$

$\forall\, x \in [a, b]$.

In this chapter we prove advanced Hardy type inequalities based on the above mentioned fractional background and convexity of functions. Our chapter is inspired by [6, 10, 12, 13].

9.2 Main Results

Our setting follows:

Let $(\Omega_1, \Sigma_1, \mu_1)$ and $(\Omega_2, \Sigma_2, \mu_2)$ be measure spaces with positive σ-finite measures, and let $k_i : \Omega_1 \times \Omega_2 \to \mathbb{R}$ be nonnegative measurable functions, $k_i(x, \cdot)$ measurable on Ω_2, and

$$K_i(x) = \int_{\Omega_2} k_i(x, y)\, d\mu_2(y), \text{ for any } x \in \Omega_1, \tag{9.53}$$

$i = 1, ..., m \in \mathbb{N}$. We assume that $K_i(x) > 0$ a.e. on Ω_1 and the weight functions are nonnegative measurable functions on the related set.

We consider measurable functions $g_i : \Omega_1 \to \mathbb{R}$ with the representation

$$g_i(x) = \int_{\Omega_2} k_i(x, y)\, f_i(y)\, d\mu_2(y), \tag{9.54}$$

where $f_i : \Omega_2 \to \mathbb{R}$ are measurable functions, $i = 1, ..., m$. Here u stands for a weight function on Ω_1 (nonnegative measurable function).

We mention

Theorem 9.7 *([5], p. 29) Let $j \in \{1, ..., m\}$ be fixed. Assume that the function $x \to \left(\frac{u(x) \prod_{i=1}^{m} k_i(x,y)}{\prod_{i=1}^{m} K_i(x)} \right)$ is integrable on Ω_1, for each $y \in \Omega_2$. Define λ_m on Ω_2 by*

$$\lambda_m(y) := \int_{\Omega_1} \left(\frac{u(x) \prod_{i=1}^{m} k_i(x, y)}{\prod_{i=1}^{m} K_i(x)} \right) d\mu_1(x) < \infty. \tag{9.55}$$

Here $\Phi_i : \mathbb{R}_+ \to \mathbb{R}_+$, $i = 1, ..., m$, are convex and increasing functions.

Then

$$I := \int_{\Omega_1} u(x) \prod_{i=1}^{m} \Phi_i \left(\left| \frac{g_i(x)}{K_i(x)} \right| \right) d\mu_1(x) \leq$$

$$\left(\prod_{\substack{i=1 \\ i \neq j}}^{m} \int_{\Omega_2} \Phi_i(|f_i(y)|)\, d\mu_2(y) \right) \left(\int_{\Omega_2} \Phi_j(|f_j(y)|)\, \lambda_m(y)\, d\mu_2(y) \right) := I_j, \tag{9.56}$$

true for all measurable functions, $i = 1, ..., m$, $f_i : \Omega_2 \to \mathbb{R}$ such that:

(i) f_i, $\Phi_i(|f_i|)$, are both $k_i(x, y)\, d\mu_2(y)$ - integrable, μ_1-a.e. in $x \in \Omega_1$,

(ii) $\lambda_m \Phi_j(|f_j|)$; $\Phi_1(|f_1|)$, $\Phi_2(|f_2|)$, $\Phi_3(|f_3|)$, ..., $\widehat{\Phi_j(|f_j|)}$, ..., $\Phi_m(|f_m|)$, are all μ_2-integrable,

and for all corresponding functions g_i given by (9.54). Above $\widehat{\Phi_j(|f_j|)}$ means missing item.

Remark 9.8 ([5]) Replace above (ii) by

(iii) $\Phi_1(|f_1|), ..., \Phi_m(|f_m|); \lambda_m \Phi_1(|f_1|), ..., \lambda_m \Phi_m(|f_m|)$ are all μ_2-integrable functions.

Then

$$I \leq \frac{\sum_{j=1}^{m} I_j}{m}. \tag{9.57}$$

From now on all entities we are dealing with are according to Sect. 9.1. Background.

We present

Theorem 9.9 *Here $i = 1, ..., m$. Let $\alpha_i > 0$, $\alpha_i \notin \mathbb{N}$, $\lceil \alpha_i \rceil = n_i$, $0 < \beta_i < 1$; $f_i \in C^{n_i}([a,b])$, $[a,b] \subset \mathbb{R}$; and set $\gamma_i = \alpha_i + \beta_i (n_i - \alpha_i)$. Assume further that $\Delta_{a+}^{\gamma_i} f_i \in C([a,b]) : \Delta_{a+}^{\gamma_i - j_i} f_i(a) = 0$, for $j_i = 1, ..., n_i$. Let also $\overline{\alpha}_i > 0 : \lceil \overline{\alpha}_i \rceil = \overline{n}_i$, with $\overline{\gamma}_i = \overline{\alpha}_i + \beta_i (\overline{n}_i - \overline{\alpha}_i)$, and assume that $\alpha_i > \overline{\alpha}_i$ and $\gamma_i > \overline{\gamma}_i$.*

Let also $\Phi_i : \mathbb{R}_+ \to \mathbb{R}_+$, $i = 1, ..., m$, be convex and increasing functions; $u \geq 0$ is a weight measurable function on $[a,b]$. We assume that

$$\lambda_m(y) = \prod_{i=1}^{m} (\alpha_i - \overline{\alpha}_i) \int_y^b u(x) \frac{(x-y)^{\sum_{i=1}^{m}(\alpha_i - \overline{\alpha}_i) - m}}{(x-a)^{\sum_{i=1}^{m}(\alpha_i - \overline{\alpha}_i)}} dx < \infty, \tag{9.58}$$

for all $a < y < b$ and λ_m is Lebesgue integrable.

Then

$$\int_a^b u(x) \prod_{i=1}^{m} \Phi_i \left(\frac{\left| {}^{H}\mathbb{D}_{a+}^{\overline{\alpha}_i, \beta_i} f_i(x) \right|}{(x-a)^{\alpha_i - \overline{\alpha}_i}} \Gamma(\alpha_i - \overline{\alpha}_i + 1) \right) dx \leq$$

$$\left(\prod_{\substack{i=1 \\ i \neq j}}^{m} \int_a^b \Phi_i \left(\left| {}^{H}\mathbb{D}_{a+}^{\alpha_i, \beta_i} f_i(y) \right| \right) dy \right) \left(\int_a^b \Phi_j \left(\left| {}^{H}\mathbb{D}_{a+}^{\alpha_j, \beta_j} f_j(y) \right| \right) \lambda_m(y) \, dy \right), \tag{9.59}$$

with $j \in \{1, ..., m\}$.

Proof By Theorems 9.5, 9.7 and from [5], pp. 31–33, see relations there (2.40)–(2.48). ■

Remark 9.10 (to Theorem 9.9) One can have Φ_i =identity map or e^x, or $\Phi_i(x) = x^{p_i}$, $x \in \mathbb{R}_+$, $p_i > 1$, etc.

Also, one can choose $u(x) = (x-a)^{\sum_{i=1}^{m}(\alpha_i - \overline{\alpha}_i)}$.

To save space in this chapter we skip these interesting applications here and later.

We continue with

Theorem 9.11 *Here* $i = 1, ..., m$. *Let* $\alpha_i > 0$, $\alpha_i \notin \mathbb{N}$, $\lceil \alpha_i \rceil = n_i$, $0 < \beta_i < 1$; $f_i \in C^{n_i}([a, b])$, $[a, b] \subset \mathbb{R}$; *and set* $\gamma_i = \alpha_i + \beta_i (n_i - \alpha_i)$. *Assume further that* $\Delta_{b-}^{\gamma_i} f_i \in C([a, b]) : \Delta_{b-}^{\gamma_i - j_i} f_i(b) = 0$, $j_i = 1, ..., n_i$. *Let also* $\overline{\alpha}_i > 0 : \lceil \overline{\alpha}_i \rceil = \overline{n}_i$, *with* $\overline{\gamma}_i = \overline{\alpha}_i + \beta_i (\overline{n}_i - \overline{\alpha}_i)$, *and assume that* $\alpha_i > \overline{\alpha}_i$ *and* $\gamma_i > \overline{\gamma}_i$. *Let also* $\Phi_i : \mathbb{R}_+ \to \mathbb{R}_+$, $i = 1, ..., m$, *be convex and increasing functions;* $u \geq 0$ *is a weight measurable function on* $[a, b]$. *We assume that*

$$\lambda_m(y) := \prod_{i=1}^{m} (\alpha_i - \overline{\alpha}_i) \int_a^y u(x) \frac{(y - x)^{\sum_{i=1}^{m}(\alpha_i - \overline{\alpha}_i) - m}}{(b - x)^{\sum_{i=1}^{m}(\alpha_i - \overline{\alpha}_i)}} dx < \infty, \tag{9.60}$$

for all $a < y < b$ *and* λ_m *is Lebesgue integrable.*

Then

$$\int_a^b u(x) \prod_{i=1}^{m} \Phi_i \left(\frac{\left| {}^H\mathbb{D}_{b-}^{\overline{\alpha}_i, \beta_i} f_i(x) \right|}{(b - x)^{\alpha_i - \overline{\alpha}_i}} \Gamma(\alpha_i - \overline{\alpha}_i + 1) \right) dx \leq$$

$$\left(\prod_{\substack{i=1 \\ i \neq j}}^{m} \int_a^b \Phi_i \left(\left| {}^H\mathbb{D}_{b-}^{\alpha_i, \beta_i} f_i(y) \right| \right) dy \right) \left(\int_a^b \Phi_j \left(\left| {}^H\mathbb{D}_{b-}^{\alpha_j, \beta_j} f_j(y) \right| \right) \lambda_m(y) \, dy \right), \tag{9.61}$$

with $j \in \{1, ..., m\}$.

Proof By Theorems 9.6, 9.7 and from [5], pp. 35–37, see relations there (2.58)–(2.68). ■

We continue with

Theorem 9.12 *Here* $i = 1, ..., m$. *Let* $f_i \in C^{n_i}([a, b])$, $\theta := \max\{n_1, ..., n_m\}$, $\psi \in C^{\theta}([a, b])$, *with* ψ *being increasing:* $\psi'(x) \neq 0$ *over* $[a, b]$, *where* $n_i - 1 < \alpha_i < n_i$, $0 \leq \beta_i \leq 1$, *and* $\gamma_i = \alpha_i + \beta_i (n_i - \alpha_i)$.

Assume that $f_{i\psi}^{[n_i - k_i]} \left(I_{a+}^{(1-\beta_i)(n_i - \alpha_i); \psi} f_i \right)(a) = 0$, $k_i = 1, ..., n_i - 1$; $i = 1, ..., m$.

Let also $\Phi_i : \mathbb{R}_+ \to \mathbb{R}_+$, $i = 1, ..., m$, *be convex and increasing functions;* $u \geq 0$ *is a weight measurable function on* $[a, b]$. *We assume that*

$$\lambda_m^{\psi}(y) = \left(\prod_{i=1}^{m} \alpha_i \right) (\psi'(y))^m \int_y^b u(x) \frac{(\psi(x) - \psi(y))^{\sum_{i=1}^{m} \alpha_i - m}}{(\psi(x) - \psi(a))^{\sum_{i=1}^{m} \alpha_i}} dx < \infty, \tag{9.62}$$

for all $a < y < b$ *and* λ_m^{ψ} *is Lebesgue integrable.*

Then

$$\int_a^b u(x) \prod_{i=1}^{m} \Phi_i \left(\frac{|f_i(x)|}{(\psi(x) - \psi(a))^{\alpha_i}} \Gamma(\alpha_i + 1) \right) dx \leq$$

$$\left(\prod_{\substack{i=1\\i\neq j}}^{m}\int_a^b \Phi_i\left(\left|\left({}^H\mathbb{D}_{a+}^{\alpha_i,\beta_i;\psi}f_i\right)(y)\right|\right)dy\right)\left(\int_a^b \Phi_j\left(\left|\left({}^H\mathbb{D}_{a+}^{\alpha_j,\beta_j;\psi}f_j\right)(y)\right|\right)\lambda_m^{\psi}(y)\,dy\right), \tag{9.63}$$

with $j \in \{1, ..., m\}$, true for continuous ${}^H\mathbb{D}_{a+}^{\alpha_i,\beta_i;\psi}f_i$, $i = 1, ..., m$.

Proof From (9.29) we get:

$$f_i(x) = \frac{1}{\Gamma(\alpha_i)}\int_a^x \psi'(t)(\psi(x)-\psi(t))^{\alpha_i-1}\ {}^H\mathbb{D}_{a+}^{\alpha_i,\beta_i;\psi}f_i(t)\,dt, \tag{9.64}$$

$\forall\, x \in [a,b]$; $i = 1, ..., m$.

Then we apply Theorem 9.7, along with [5], pp. 47–49, see the relations there (2.107)–(2.120). ∎

We also give

Theorem 9.13 *Here $i = 1, ..., m$. Let $f_i \in C^{n_i}([a,b])$, $\theta := \max\{n_1, ..., n_m\}$, $\psi \in C^{\theta}([a,b])$, with ψ being increasing: $\psi'(x) \neq 0$ over $[a,b]$, where $n_i - 1 < \alpha_i < n_i$, $0 \leq \beta_i \leq 1$, and $\gamma_i = \alpha_i + \beta_i(n_i - \alpha_i)$.*

Assume that $f_{i\psi}^{[n_i-k_i]}\left(I_{b-}^{(1-\beta_i)(n_i-\alpha_i);\psi}f_i\right)(b) = 0$, $k_i = 1, ..., n_i - 1$; $i = 1, ..., m$.

Let also $\Phi_i : \mathbb{R}_+ \to \mathbb{R}_+$, $i = 1, ..., m$, be convex and increasing functions; $u \geq 0$ is a weight measurable function on $[a,b]$. We assume that

$$\lambda_m^{\psi}(y) = \left(\prod_{i=1}^{m}\alpha_i\right)(\psi'(y))^m\int_a^y u(x)\frac{(\psi(y)-\psi(x))^{\sum_{i=1}^{m}\alpha_i-m}}{(\psi(b)-\psi(x))^{\sum_{i=1}^{m}\alpha_i}}dx < \infty, \tag{9.65}$$

for all $a < y < b$ and λ_m^{ψ} is Lebesgue integrable.

Then

$$\int_a^b u(x)\prod_{i=1}^{m}\Phi_i\left(\frac{|f_i(x)|}{(\psi(b)-\psi(x))^{\alpha_i}}\Gamma(\alpha_i+1)\right)dx \leq$$

$$\left(\prod_{\substack{i=1\\i\neq j}}^{m}\int_a^b \Phi_i\left(\left|\left({}^H\mathbb{D}_{b-}^{\alpha_i,\beta_i;\psi}f_i\right)(y)\right|\right)dy\right)\left(\int_a^b \Phi_j\left(\left|\left({}^H\mathbb{D}_{b-}^{\alpha_j,\beta_j;\psi}f_j\right)(y)\right|\right)\lambda_m^{\psi}(y)\,dy\right), \tag{9.66}$$

with $j \in \{1, ..., m\}$, true for continuous ${}^H\mathbb{D}_{b-}^{\alpha_i,\beta_i;\psi}f_i$, $i = 1, ..., m$.

Proof From (9.30) we get:

$$f_i(x) = \frac{1}{\Gamma(\alpha_i)}\int_x^b \psi'(t)(\psi(t)-\psi(x))^{\alpha_i-1}\ {}^H\mathbb{D}_{b-}^{\alpha_i,\beta_i;\psi}f_i(t)\,dt, \tag{9.67}$$

$\forall\, x \in [a,b]\,;\, i = 1, ..., m.$

Then we apply Theorem 9.7, along with [5], pp. 51–53, see the relations there (2.132)–(2.143). ■

We present

Theorem 9.14 *Here* $i = 1, ..., m$. *Let* $f_i \in C^{n_i}([a,b])$, $\theta := \max\{n_1, ..., n_m\}$, $\psi \in C^{\theta}([a,b])$, *with* ψ *being increasing:* $\psi'(x) \neq 0$ *over* $[a,b] \subset \mathbb{R}$, $\alpha_i > 0 : \lceil \alpha_i \rceil = n_i$, $0 \leq \beta_i \leq 1$, $\mu_i = n_i(1-\beta_i) + \beta_i \alpha_i$. *Assume that* $g_i(x) := I_{a+}^{(1-\beta_i)(n_i-\alpha_i);\psi} f_i(x) \in C^{n_i}([a,b])$, *and* $g_{i\psi}^{[k_i]}(a) = 0$, $k_i = 0, ..., n_i - 1$, *where* $g_{i\psi}^{[k_i]}(x) = \left(\frac{1}{\psi'(x)}\frac{d}{dx}\right)^{k_i} g_i(x)$, $k_i = 0, 1, ..., n_i - 1$.

Let also $\Phi_i : \mathbb{R}_+ \to \mathbb{R}_+$, $i = 1, ..., m$, *be convex and increasing functions;* $u \geq 0$ *is a weight measurable function on* $[a,b]$. *We assume that*

$$\lambda_m^{\psi}(y) = \left(\prod_{i=1}^{m}\mu_i\right)(\psi'(y))^m \int_y^b u(x) \frac{(\psi(x)-\psi(y))^{\sum_{i=1}^{m}\mu_i - m}}{(\psi(x)-\psi(a))^{\sum_{i=1}^{m}\mu_i}} dx < \infty, \quad (9.68)$$

for all $a < y < b$ *and* λ_m^{ψ} *is Lebesgue integrable.*

Then

$$\int_a^b u(x) \prod_{i=1}^{m} \Phi_i\left(\frac{|g_i(x)|}{(\psi(x)-\psi(a))^{\mu_i}}\Gamma(\mu_i+1)\right) dx \leq$$

$$\left(\prod_{\substack{i=1\\ i\neq j}}^{m} \int_a^b \Phi_i\left(\left|\left({}^H\mathbb{D}_{a+}^{\alpha_i,\beta_i;\psi} f_i\right)(y)\right|\right) dy\right)\left(\int_a^b \Phi_j\left(\left|\left({}^H\mathbb{D}_{a+}^{\alpha_j,\beta_j;\psi} f_j\right)(y)\right|\right)\lambda_m^{\psi}(y)\, dy\right), \quad (9.69)$$

with $j \in \{1, ..., m\}$, *true for continuous* ${}^H\mathbb{D}_{a+}^{\alpha_i,\beta_i;\psi} f_i$, $i = 1, ..., m$.

Proof From (9.31) we get that

$$g_i(x) = I_{a+}^{\mu_i;\psi}\, {}^H\mathbb{D}_{a+}^{\alpha_i,\beta_i;\psi} f_i(x), \quad (9.70)$$

$\forall\, x \in [a,b]\,;\, i = 1, ..., m.$

Then we apply Theorem 9.7, along with [5], pp. 47–49, see the relations there (2.107)–(2.120). ■

We also give

Theorem 9.15 *Here* $i = 1, ..., m$. *Let* $f_i \in C^{n_i}([a,b])$, $\theta := \max\{n_1, ..., n_m\}$, $\psi \in C^{\theta}([a,b])$, *with* ψ *being increasing,* $\psi'(x) \neq 0$ *over* $[a,b] \subset \mathbb{R}$, $\alpha_i > 0 : \lceil \alpha_i \rceil = n_i$, $0 \leq \beta_i \leq 1$, $\mu_i = n_i(1-\beta_i) + \beta_i \alpha_i$. *Assume that* $w_i(x) := I_{b-}^{(1-\beta_i)(n_i-\alpha_i);\psi} f_i(x) \in C^{n_i}([a,b])$, *and* $w_{i\psi}^{[k_i]}(b) = 0$, $k_i = 0, ..., n_i - 1$, *where* $w_{i\psi}^{[k_i]}(x) = \left(\frac{1}{\psi'(x)}\frac{d}{dx}\right)^{k_i} w_i(x)$, $k_i = 0, 1, ..., n_i - 1$.

Let also $\Phi_i : \mathbb{R}_+ \to \mathbb{R}_+$, $i = 1, ..., m$, be convex and increasing functions; $u \geq 0$ is a weight measurable function on $[a, b]$. We assume that

$$\lambda_m^{\psi}(y) = \left(\prod_{i=1}^{m} \mu_i\right) \left(\psi'(y)\right)^m \int_a^y u(x) \frac{(\psi(y) - \psi(x))^{\sum_{i=1}^{m} \mu_i - m}}{(\psi(b) - \psi(x))^{\sum_{i=1}^{m} \mu_i}} dx < \infty, \quad (9.71)$$

for all $a < y < b$ and λ_m^{ψ} is Lebesgue integrable.

Then

$$\int_a^b u(x) \prod_{i=1}^{m} \Phi_i \left(\frac{|w_i(x)|}{(\psi(b) - \psi(x))^{\mu_i}} \Gamma(\mu_i + 1) \right) dx \leq$$

$$\left(\prod_{\substack{i=1 \\ i \neq j}}^{m} \int_a^b \Phi_i \left(\left| \left({}^H\mathbb{D}_{b-}^{\alpha_i, \beta_i; \psi} f_i \right)(y) \right| \right) dy \right) \left(\int_a^b \Phi_j \left(\left| \left({}^H\mathbb{D}_{b-}^{\alpha_j, \beta_j; \psi} f_j \right)(y) \right| \right) \lambda_m^{\psi}(y) \, dy \right), \quad (9.72)$$

with $j \in \{1, ..., m\}$, true for continuous ${}^H\mathbb{D}_{b-}^{\alpha_i, \beta_i; \psi} f_i$, $i = 1, ..., m$.

Proof From (9.32) we get that

$$w_i(x) = I_{b-}^{\mu_i; \psi} \, {}^H\mathbb{D}_{b-}^{\alpha_i, \beta_i; \psi} f_i(x), \quad (9.73)$$

$\forall \, x \in [a, b]$, $i = 1, ..., m$.

Then we apply Theorem 9.7, along with [5], pp. 51–53, see the relations there (2.132)–(2.143). ■

We continue with

Theorem 9.16 *Here $i = 1, ..., m$. Let $n_i - 1 < \alpha_i < n_i$, $n_i \in \mathbb{N}$, $I = [a, b] \subset \mathbb{R}$ and $f_i \in C^{n_i}([a, b])$, $\theta := \max\{n_1, ..., n_m\}$, $\psi \in C^{\theta}([a, b])$, ψ is increasing and $\psi'(x) \neq 0$, for all $x \in I$. Here $0 \leq \beta_i \leq 1$ and $\gamma_i = \alpha_i + \beta_i(n_i - \alpha_i)$. Assume that $\Delta_{a+}^{\gamma_i; \psi} f_i, \Delta_{b-}^{\gamma_i; \psi} f_i \in C([a, b])$, $i = 1, ..., m$.*

Let also $\Phi_i : \mathbb{R}_+ \to \mathbb{R}_+$, $i = 1, ..., m$, be convex and increasing functions; $u \geq 0$ is a weight measurable function on $[a, b]$. We assume that

$$\lambda_{m+}(y) = \left(\prod_{i=1}^{m} (\gamma_i - \alpha_i)\right) \left(\psi'(y)\right)^m \int_y^b u(x) \frac{(\psi(x) - \psi(y))^{\sum_{i=1}^{m} (\gamma_i - \alpha_i) - m}}{(\psi(x) - \psi(a))^{\sum_{i=1}^{m} (\gamma_i - \alpha_i)}} dx < \infty, \quad (9.74)$$

for all $a < y < b$ and λ_{m+}^{ψ} is Lebesgue integrable; and

$$\lambda_{m-}(y)=\left(\prod_{i=1}^{m}(\gamma_i-\alpha_i)\right)\left(\psi'(y)\right)^m\int_a^y u(x)\frac{(\psi(y)-\psi(x))^{\sum_{i=1}^{m}(\gamma_i-\alpha_i)-m}}{(\psi(b)-\psi(x))^{\sum_{i=1}^{m}(\gamma_i-\alpha_i)}}dx<\infty, \tag{9.75}$$

for all $a<y<b$ and λ_{m-}^{ψ} is Lebesgue integrable.

Then

i)

$$\int_a^b u(x)\prod_{i=1}^{m}\Phi_i\left(\frac{\left|{}^{H}\mathbb{D}_{a+}^{\alpha_i,\beta_i;\psi}f_i(x)\right|}{(\psi(x)-\psi(a))^{\gamma_i-\alpha_i}}\Gamma(\gamma_i-\alpha_i+1)\right)dx\leq$$

$$\left(\prod_{\substack{i=1\\ i\neq j}}^{m}\int_a^b\Phi_i\left(\left|\left(\Delta_{a+}^{\gamma_i;\psi}f_i\right)(y)\right|\right)dy\right)\left(\int_a^b\Phi_j\left(\left|\left(\Delta_{a+}^{\gamma_j;\psi}f_j\right)(y)\right|\right)\lambda_{m+}^{\psi}(y)\,dy\right), \tag{9.76}$$

with $j\in\{1,...,m\}$,

and

ii)

$$\int_a^b u(x)\prod_{i=1}^{m}\Phi_i\left(\frac{\left|{}^{H}\mathbb{D}_{b-}^{\alpha_i,\beta_i;\psi}f_i(x)\right|}{(\psi(b)-\psi(x))^{\gamma_i-\alpha_i}}\Gamma(\gamma_i-\alpha_i+1)\right)dx\leq$$

$$\left(\prod_{\substack{i=1\\ i\neq j}}^{m}\int_a^b\Phi_i\left(\left|\left(\Delta_{b-}^{\gamma_i;\psi}f_i\right)(y)\right|\right)dy\right)\left(\int_a^b\Phi_j\left(\left|\left(\Delta_{b-}^{\gamma_j;\psi}f_j\right)(y)\right|\right)\lambda_{m-}^{\psi}(y)\,dy\right), \tag{9.77}$$

with $j\in\{1,...,m\}$.

Proof By (9.19) and (9.20), respectively, we have that

$${}^{H}\mathbb{D}_{a+}^{\alpha_i,\beta_i;\psi}f_i(x)=I_{a+}^{\gamma_i-\alpha_i;\psi}\Delta_{a+}^{\gamma_i;\psi}f_i(x), \tag{9.78}$$

and

$${}^{H}\mathbb{D}_{b-}^{\alpha_i,\beta_i;\psi}f_i(x)=I_{b-}^{\gamma_i-\alpha_i;\psi}\Delta_{b-}^{\gamma_i;\psi}f_i(x), \tag{9.79}$$

$\forall\, x\in[a,b]$, $i=1,...,m$.

Then, we apply Theorem 9.7 twice, along with [5], pp. 47–49, see there (2.107)–(2.120), and [5], pp. 51–53, see there (2.132)–(2.143), respectively. ■

We make

Remark 9.17 We pick $\Omega_1=\Omega_2=(a,b)$, $d\mu_1(x)=dx$, $d\mu_2(y)=dy$, the Lebesgue measure. Here $i=1,...,m$. Let $\rho_i,\mu_i,\gamma_i,\omega_i>0$, and $f_i\in C([a,b])$, with $\psi\in C^1([a,b])$ which is increasing.

We have that ($x \in [a, b]$):

$$\left(e^{\gamma_i;\psi}_{\rho_i,\mu_i,\omega_i,a+} f_i\right)(x) =$$

$$\int_a^x \psi'(t)(\psi(x) - \psi(t))^{\mu_i - 1} E^{\gamma_i}_{\rho_i,\mu_i}\left[\omega_i (\psi(x) - \psi(t))^{\rho_i}\right] f_i(t)\,dt \tag{9.80}$$

$$= \int_a^x \chi_{(a,x]}(t)\,\psi'(t)(\psi(x) - \psi(t))^{\mu_i - 1} E^{\gamma_i}_{\rho_i,\mu_i}\left[\omega_i (\psi(x) - \psi(t))^{\rho_i}\right] f_i(t)\,dt,$$

where χ is the characteristic function.

So, we choose here

$$k_i(x,t) := \chi_{(a,x]}(t)\,\psi'(t)(\psi(x) - \psi(t))^{\mu_i - 1} E^{\gamma_i}_{\rho_i,\mu_i}\left[\omega_i (\psi(x) - \psi(t))^{\rho_i}\right], \tag{9.81}$$

$i = 1, ..., m$.

That is

$$k_i(x,y) = \begin{cases} \psi'(y)(\psi(x) - \psi(y))^{\mu_i - 1} E^{\gamma_i}_{\rho_i,\mu_i}\left[\omega_i (\psi(x) - \psi(y))^{\rho_i}\right], & a < y \leq x, \\ 0, \quad x < y < b, \end{cases} \tag{9.82}$$

$i = 1, ..., m$.

Therefore we obtain

$$K_i(x) = \int_a^b \chi_{(a,x]}(y)\,\psi'(y)(\psi(x) - \psi(y))^{\mu_i - 1} E^{\gamma_i}_{\rho_i,\mu_i}\left[\omega_i (\psi(x) - \psi(y))^{\rho_i}\right] dy =$$

$$\int_a^x \psi'(y)(\psi(x) - \psi(y))^{\mu_i - 1} E^{\gamma_i}_{\rho_i,\mu_i}\left[\omega_i (\psi(x) - \psi(y))^{\rho_i}\right] dy =$$

(by [4])

$$\sum_{k_i=0}^{\infty} \frac{(\gamma_i)_{k_i}\,\omega_i^{k_i}}{k_i!\Gamma(\rho_i k_i + \mu_i)} \int_a^x \psi'(y)(\psi(x) - \psi(y))^{(\rho_i k_i + \mu_i) - 1} dy =$$

$$\sum_{k_i=0}^{\infty} \frac{(\gamma_i)_{k_i}}{k_i!\Gamma(\rho_i k_i + \mu_i)} \frac{\omega_i^{k_i}(\psi(x) - \psi(a))^{(\rho_i k_i + \mu_i)}}{(\rho_i k_i + \mu_i)} = \tag{9.83}$$

$$(\psi(x) - \psi(a))^{\mu_i} \sum_{k_i=0}^{\infty} \frac{(\gamma_i)_{k_i}}{k_i!\Gamma(\rho_i k_i + \mu_i + 1)} \left(\omega_i (\psi(x) - \psi(a))^{\rho_i}\right)^{k_i} =$$

$$(\psi(x) - \psi(a))^{\mu_i} E^{\gamma_i}_{\rho_i,\mu_i+1}\left[\omega_i (\psi(x) - \psi(a))^{\rho_i}\right].$$

That is

$$K_i(x) = (\psi(x) - \psi(a))^{\mu_i} E^{\gamma_i}_{\rho_i,\mu_i+1}\left[\omega_i (\psi(x) - \psi(a))^{\rho_i}\right], \tag{9.84}$$

$\forall\, x \in [a, b], i = 1, ..., m.$

Notice that

$$\prod_{i=1}^{m} \frac{k_i(x, y)}{K_i(x)} = \prod_{i=1}^{m} \frac{\chi_{(a,x]}(y)\, \psi'(y)\, (\psi(x) - \psi(y))^{\mu_i - 1} E^{\gamma_i}_{\rho_i,\mu_i}\left[\omega_i (\psi(x) - \psi(y))^{\rho_i}\right]}{(\psi(x) - \psi(a))^{\mu_i} E^{\gamma_i}_{\rho_i,\mu_i+1}\left[\omega_i (\psi(x) - \psi(a))^{\rho_i}\right]} \tag{9.85}$$

$$= \left(\chi_{(a,x]}(y) \left(\psi'(y)\right)^m \frac{(\psi(x) - \psi(y))^{\sum_{i=1}^{m} \mu_i - m}}{(\psi(x) - \psi(a))^{\sum_{i=1}^{m} \mu_i}} \right)$$

$$\prod_{i=1}^{m} \left(\frac{E^{\gamma_i}_{\rho_i,\mu_i}\left[\omega_i (\psi(x) - \psi(y))^{\rho_i}\right]}{E^{\gamma_i}_{\rho_i,\mu_i+1}\left[\omega_i (\psi(x) - \psi(a))^{\rho_i}\right]} \right),$$

$\forall\, x, y \in [a, b]$.

Therefore for (9.55), we get for appropriate weight $u \geq 0$ that (denote this particular λ_m by $\overline{\lambda}^{\psi}_{m+}$):

$$\overline{\lambda}^{\psi}_{m+}(y) = \left(\psi'(y)\right)^m \int_y^b u(x) \left(\frac{(\psi(x) - \psi(y))^{\sum_{i=1}^{m} \mu_i - m}}{(\psi(x) - \psi(a))^{\sum_{i=1}^{m} \mu_i}} \right)$$

$$\prod_{i=1}^{m} \left(\frac{E^{\gamma_i}_{\rho_i,\mu_i}\left[\omega_i (\psi(x) - \psi(y))^{\rho_i}\right]}{E^{\gamma_i}_{\rho_i,\mu_i+1}\left[\omega_i (\psi(x) - \psi(a))^{\rho_i}\right]} \right) dx < \infty, \tag{9.86}$$

for all $a < y < b$.

Based on Theorem 9.7 and the above, we have established the following generalized Prabhakar left fractional Hardy type inequality:

Theorem 9.18 *Here $i = 1, ..., m$. Let $\rho_i, \mu_i, \gamma_i, \omega_i > 0$, and $f_i \in C([a, b])$, with $\psi \in C^1([a, b])$ which is increasing. The function $\overline{\lambda}^{\psi}_{m+}(y) \in \mathbb{R}$ by assumption, $\forall$ $y \in [a, b]$, is given by (9.86). Here $\Phi_i : \mathbb{R}_+ \to \mathbb{R}_+$, $i = 1, ..., m$, are convex and increasing functions. Then*

$$\int_a^b u(x) \prod_{i=1}^{m} \Phi_i \left(\frac{\left| \left(e^{\gamma_i;\psi}_{\rho_i,\mu_i,\omega_i,a+} f_i \right)(x) \right|}{(\psi(x) - \psi(a))^{\mu_i} E^{\gamma_i}_{\rho_i,\mu_i+1}\left[\omega_i (\psi(x) - \psi(a))^{\rho_i}\right]} \right) dx \leq$$

$$\left(\prod_{\substack{i=1\\i\neq j}}^{m}\int_a^b \Phi_i\left(|f_i(y)|\right)dy\right)\left(\int_a^b \Phi_j\left(\left|f_j(y)\right|\right)\overline{\lambda}_{m+}^{\psi}(y)\,dy\right). \tag{9.87}$$

We make

Remark 9.19 We pick $\Omega_1 = \Omega_2 = (a,b)$, $d\mu_1(x) = dx$, $d\mu_2(y) = dy$, the Lebesgue measure. Here $i = 1, ..., m$. Let $\rho_i, \mu_i, \gamma_i, \omega_i > 0$, and $f_i \in C([a,b])$, with $\psi \in C^1([a,b])$ which is increasing.

We have that ($x \in [a,b]$):

$$\left(e_{\rho_i,\mu_i,\omega_i,b-}^{\gamma_i;\psi} f_i\right)(x) =$$

$$\int_x^b \psi'(t)\left(\psi(t)-\psi(x)\right)^{\mu_i-1} E_{\rho_i,\mu_i}^{\gamma_i}\left[\omega_i\left(\psi(t)-\psi(x)\right)^{\rho_i}\right] f_i(t)\,dt \tag{9.88}$$

$$= \int_a^x \chi_{[x,b)}(t)\,\psi'(t)\left(\psi(t)-\psi(x)\right)^{\mu_i-1} E_{\rho_i,\mu_i}^{\gamma_i}\left[\omega_i\left(\psi(t)-\psi(x)\right)^{\rho_i}\right] f_i(t)\,dt,$$

where χ is the characteristic function.

So, we choose here

$$k_i(x,t) := \chi_{[x,b)}(t)\,\psi'(t)\left(\psi(t)-\psi(x)\right)^{\mu_i-1} E_{\rho_i,\mu_i}^{\gamma_i}\left[\omega_i\left(\psi(t)-\psi(x)\right)^{\rho_i}\right], \tag{9.89}$$

$i = 1, ..., m$.

That is

$$k_i(x,y) = \begin{cases} \psi'(y)\left(\psi(y)-\psi(x)\right)^{\mu_i-1} E_{\rho_i,\mu_i}^{\gamma_i}\left[\omega_i\left(\psi(y)-\psi(x)\right)^{\rho_i}\right], & x \leq y < b, \\ 0, & a < y < x, \end{cases} \tag{9.90}$$

$i = 1, ..., m$.

Therefore we obtain

$$K_i(x) = \int_a^b \chi_{[x,b)}(y)\,\psi'(y)\left(\psi(y)-\psi(x)\right)^{\mu_i-1} E_{\rho_i,\mu_i}^{\gamma_i}\left[\omega_i\left(\psi(y)-\psi(x)\right)^{\rho_i}\right]dy =$$

$$\int_x^b \psi'(y)\left(\psi(y)-\psi(x)\right)^{\mu_i-1} E_{\rho_i,\mu_i}^{\gamma_i}\left[\omega_i\left(\psi(y)-\psi(x)\right)^{\rho_i}\right]dy =$$

(by [4])

$$\sum_{k_i=0}^{\infty}\frac{(\gamma_i)_{k_i}\,\omega_i^{k_i}}{k_i!\,\Gamma(\rho_i k_i+\mu_i)}\int_x^b \psi'(y)\left(\psi(y)-\psi(x)\right)^{(\rho_i k_i+\mu_i)-1}dy =$$

$$\sum_{k_i=0}^{\infty} \frac{(\gamma_i)_{k_i}}{k_i!\Gamma(\rho_i k_i + \mu_i)} \frac{\omega_i^{k_i} (\psi(b) - \psi(x))^{(\rho_i k_i + \mu_i)}}{(\rho_i k_i + \mu_i)} = \tag{9.91}$$

$$(\psi(b) - \psi(x))^{\mu_i} \sum_{k_i=0}^{\infty} \frac{(\gamma_i)_{k_i}}{k_i!\Gamma(\rho_i k_i + \mu_i + 1)} (\omega_i (\psi(b) - \psi(x))^{\rho_i})^{k_i} =$$

$$(\psi(b) - \psi(x))^{\mu_i} E_{\rho_i,\mu_i+1}^{\gamma_i} \left[\omega_i (\psi(b) - \psi(x))^{\rho_i}\right].$$

That is

$$K_i(x) = (\psi(b) - \psi(x))^{\mu_i} E_{\rho_i,\mu_i+1}^{\gamma_i} \left[\omega_i (\psi(b) - \psi(x))^{\rho_i}\right], \tag{9.92}$$

$\forall\, x \in [a, b]$, $i = 1, ..., m$.

Notice that

$$\prod_{i=1}^{m} \frac{k_i(x, y)}{K_i(x)} = \prod_{i=1}^{m} \frac{\chi_{[x,b)}(y)\, \psi'(y) (\psi(y) - \psi(x))^{\mu_i - 1} E_{\rho_i,\mu_i}^{\gamma_i} \left[\omega_i (\psi(y) - \psi(x))^{\rho_i}\right]}{(\psi(b) - \psi(x))^{\mu_i} E_{\rho_i,\mu_i+1}^{\gamma_i} \left[\omega_i (\psi(b) - \psi(x))^{\rho_i}\right]}$$

$$= \left(\chi_{[x,b)}(y) \left(\psi'(y)\right)^m \frac{(\psi(y) - \psi(x))^{\sum_{i=1}^{m} \mu_i - m}}{(\psi(b) - \psi(x))^{\sum_{i=1}^{m} \mu_i}} \right)$$

$$\prod_{i=1}^{m} \left(\frac{E_{\rho_i,\mu_i}^{\gamma_i} \left[\omega_i (\psi(y) - \psi(x))^{\rho_i}\right]}{E_{\rho_i,\mu_i+1}^{\gamma_i} \left[\omega_i (\psi(b) - \psi(x))^{\rho_i}\right]} \right), \tag{9.93}$$

$\forall\, x, y \in [a, b]$.

Therefore for (9.55), we get for appropiate weight $u \geq 0$ that (denote this particular λ_m by $\overline{\lambda}_{m-}^{\psi}$):

$$\overline{\lambda}_{m-}^{\psi}(y) = \left(\psi'(y)\right)^m \int_a^y u(x) \left(\frac{(\psi(y) - \psi(x))^{\sum_{i=1}^{m} \mu_i - m}}{(\psi(b) - \psi(x))^{\sum_{i=1}^{m} \mu_i}} \right)$$

$$\prod_{i=1}^{m} \left(\frac{E_{\rho_i,\mu_i}^{\gamma_i} \left[\omega_i (\psi(y) - \psi(x))^{\rho_i}\right]}{E_{\rho_i,\mu_i+1}^{\gamma_i} \left[\omega_i (\psi(b) - \psi(x))^{\rho_i}\right]} \right) dx < \infty, \tag{9.94}$$

for all $a < y < b$.

Based on Theorem 9.7 and the above, we have established the following generalized Prabhakar right fractional Hardy type inequality:

Theorem 9.20 *Here $i = 1, ..., m$. Let $\rho_i, \mu_i, \gamma_i, \omega_i > 0$, and $f_i \in C([a, b])$, with $\psi \in C^1([a, b])$ which is increasing. The function $\overline{\lambda}_{m-}^{\psi}(y) \in \mathbb{R}$ by assumption, $\forall$ $y \in [a, b]$, is given by (9.94). Here $\Phi_i : \mathbb{R}_+ \to \mathbb{R}_+$, $i = 1, ..., m$, are convex and increasing functions. Then*

$$\int_a^b u(x) \prod_{i=1}^m \Phi_i \left(\frac{\left| \left(e_{\rho_i,\mu_i,\omega_i,b-}^{\gamma_i;\psi} f_i \right)(x) \right|}{(\psi(b) - \psi(x))^{\mu_i} E_{\rho_i,\mu_i+1}^{\gamma_i} \left[\omega_i (\psi(b) - \psi(x))^{\rho_i} \right]} \right) dx \leq$$

$$\left(\prod_{\substack{i=1 \\ i \neq j}}^m \int_a^b \Phi_i(|f_i(y)|) \, dy \right) \left(\int_a^b \Phi_j(|f_j(y)|) \, \overline{\lambda}_{m-}^{\psi}(y) \, dy \right). \quad (9.95)$$

We continue with left and right ψ-Prabhakar-Caputo Hardy fractional inequalities:

Theorem 9.21 *Here $i = 1, ..., m$. Let $\rho_i, \mu_i, \omega_i > 0$, $\gamma_i < 0$, and $f_i \in C^{N_i}([a, b])$, $N_i = \lceil \mu_i \rceil$, $\mu_i \notin \mathbb{N}$; $\theta := \max(N_1, ..., N_m)$, $\psi \in C^{\theta}([a, b])$, ψ is increasing with $\psi'(x) \neq 0$ over $[a, b]$. Set $f_{i\psi}^{[N_i]}(x) = \left(\frac{1}{\psi'(x)} \frac{d}{dx} \right)^{N_i} f_i(x)$, $x \in [a, b]$. We assume that the weight function $u \geq 0$ is such that*

$$^{C}\lambda_{m+}^{\psi}(y) := (\psi'(y))^m \int_y^b u(x) \left(\frac{(\psi(x) - \psi(y))^{\sum_{i=1}^m (N_i - \mu_i) - m}}{(\psi(x) - \psi(a))^{\sum_{i=1}^m (N_i - \mu_i)}} \right)$$

$$\prod_{i=1}^m \left(\frac{E_{\rho_i,N_i-\mu_i}^{-\gamma_i} \left[\omega_i (\psi(x) - \psi(y))^{\rho_i} \right]}{E_{\rho_i,N_i-\mu_i+1}^{-\gamma_i} \left[\omega_i (\psi(x) - \psi(a))^{\rho_i} \right]} \right) dx < \infty, \quad (9.96)$$

for all $a < y < b$.

Here $\Phi_i : \mathbb{R}_+ \to \mathbb{R}_+$, $i = 1, ..., m$, are convex and increasing functions. Then

$$\int_a^b u(x) \prod_{i=1}^m \Phi_i \left(\frac{\left| \left({}^C D_{\rho_i,\mu_i,\omega_i,a+}^{\gamma_i;\psi} f_i \right)(x) \right|}{(\psi(x) - \psi(a))^{N_i-\mu_i} E_{\rho_i,N_i-\mu_i+1}^{-\gamma_i} \left[\omega_i (\psi(x) - \psi(a))^{\rho_i} \right]} \right) dx \leq$$

$$\left(\prod_{\substack{i=1 \\ i \neq j}}^m \int_a^b \Phi_i\left(\left| f_{i\psi}^{[N_i]}(y) \right| \right) dy \right) \left(\int_a^b \Phi_j\left(\left| f_{j\psi}^{[N_j]}(y) \right| \right) {}^C\lambda_{m+}^{\psi}(y) \, dy \right). \quad (9.97)$$

Proof By (9.42) we have that

$$\left({}^C D_{\rho_i,\mu_i,\omega_i,a+}^{\gamma_i;\psi} f_i \right)(x) = \left(e_{\rho_i,N_i-\mu_i,\omega_i,a+}^{-\gamma_i;\psi} f_{i\psi}^{[N_i]} \right)(x), \quad (9.98)$$

$\forall\, x \in [a,b], i = 1, ..., m.$

We apply Theorem 9.18. ■

Theorem 9.22 *Here* $i = 1, ..., m$. *Let* $\rho_i, \mu_i, \omega_i > 0$, $\gamma_i < 0$, *and* $f_i \in C^{N_i}([a,b])$, $N_i = \lceil \mu_i \rceil$, $\mu_i \notin \mathbb{N}$; $\theta := \max(N_1, ..., N_m)$, $\psi \in C^{\theta}([a,b])$, ψ *is increasing with* $\psi'(x) \neq 0$ *over* $[a,b]$. *Set* $f_{i\psi}^{[N_i]}(x) = \left(\frac{1}{\psi'(x)}\frac{d}{dx}\right)^{N_i} f_i(x)$, $x \in [a,b]$. *We assume that the weight function* $u \geq 0$ *is such that*

$$^{C}\lambda_{m-}^{\psi}(y) := \left(\psi'(y)\right)^m \int_a^y u(x) \left(\frac{(\psi(y) - \psi(x))^{\sum_{i=1}^{m}(N_i-\mu_i)-m}}{(\psi(b) - \psi(x))^{\sum_{i=1}^{m}(N_i-\mu_i)}} \right)$$

$$\prod_{i=1}^{m} \left(\frac{E_{\rho_i, N_i-\mu_i}^{-\gamma_i}\left[\omega_i (\psi(y) - \psi(x))^{\rho_i}\right]}{E_{\rho_i, N_i-\mu_i+1}^{-\gamma_i}\left[\omega_i (\psi(b) - \psi(x))^{\rho_i}\right]} \right) dx < \infty, \tag{9.99}$$

for all $a < y < b$.

Here $\Phi_i : \mathbb{R}_+ \to \mathbb{R}_+$, $i = 1, ..., m$, *are convex and increasing functions. Then*

$$\int_a^b u(x) \prod_{i=1}^{m} \Phi_i \left(\frac{\left| \left(^{C}D_{\rho_i,\mu_i,\omega_i,b-}^{\gamma_i;\psi} f_i \right)(x) \right|}{(\psi(b) - \psi(x))^{N_i-\mu_i} E_{\rho_i, N_i-\mu_i+1}^{-\gamma_i}\left[\omega_i (\psi(b) - \psi(x))^{\rho_i}\right]} \right) dx \leq$$

$$\left(\prod_{\substack{i=1 \\ i \neq j}}^{m} \int_a^b \Phi_i \left(\left| f_{i\psi}^{[N_i]}(y) \right| \right) dy \right) \left(\int_a^b \Phi_j \left(\left| f_{j\psi}^{[N_j]}(y) \right| \right) {^{C}\lambda_{m-}^{\psi}}(y)\, dy \right). \tag{9.100}$$

Proof By (9.43) we have that

$$\left(^{C}D_{\rho_i,\mu_i,\omega_i,b-}^{\gamma_i;\psi} f_i \right)(x) = (-1)^{N_i} \left(e_{\rho_i, N_i-\mu_i,\omega_i,b-}^{-\gamma_i;\psi} f_{i\psi}^{[N_i]} \right)(x), \tag{9.101}$$

$\forall\, x \in [a,b], i = 1, ..., m.$

We apply Theorem 9.20. ■

Next we present left and right ψ-Hilfer-Prabhakar Hardy fractional inequalities:

Theorem 9.23 *Here* $i = 1, ..., m$. *Let* $\rho_i, \mu_i, \omega_i > 0$, $\gamma_i < 0$, *and* $f_i \in C^{N_i}([a,b])$, $N_i = \lceil \mu_i \rceil$, $\mu_i \notin \mathbb{N}$; $\theta := \max(N_1, ..., N_m)$, $\psi \in C^{\theta}([a,b])$, ψ *is increasing with* $\psi'(x) \neq 0$ *over* $[a,b]$. *Here* $0 \leq \beta_i \leq 1$ *and* $\xi_i = \mu_i + \beta_i(N_i - \mu_i)$. *We assume that* $^{RL}D_{\rho_i,\xi_i,\omega_i,a+}^{\gamma_i(1-\beta_i);\psi} f_i \in C([a,b])$, $i = 1, ..., m$. *We assume further that the weight function* $u \geq 0$ *is such that*

$$
{}^{P}\lambda_{m+}^{\psi}(y) := \left(\psi'(y)\right)^{m} \int_{y}^{b} u(x) \left(\frac{(\psi(x)-\psi(y))^{\sum_{i=1}^{m}(\xi_i-\mu_i)-m}}{(\psi(x)-\psi(a))^{\sum_{i=1}^{m}(\xi_i-\mu_i)}} \right)
$$

$$
\prod_{i=1}^{m} \left(\frac{E_{\rho_i,\xi_i-\mu_i}^{-\gamma_i\beta_i}\left[\omega_i(\psi(x)-\psi(y))^{\rho_i}\right]}{E_{\rho_i,\xi_i-\mu_i+1}^{-\gamma_i\beta_i}\left[\omega_i(\psi(x)-\psi(a))^{\rho_i}\right]} \right) dx < \infty, \tag{9.102}
$$

for all $a < y < b$.

Here $\Phi_i : \mathbb{R}_+ \to \mathbb{R}_+$, $i = 1, ..., m$, are convex and increasing functions. Then

$$
\int_a^b u(x) \prod_{i=1}^{m} \Phi_i \left(\frac{\left| \left({}^{H}\mathbb{D}_{\rho_i,\mu_i,\omega_i,a+}^{\gamma_i,\beta_i;\psi} f_i \right)(x) \right|}{(\psi(x)-\psi(a))^{\xi_i-\mu_i} E_{\rho_i,\xi_i-\mu_i+1}^{-\gamma_i\beta_i}\left[\omega_i(\psi(x)-\psi(a))^{\rho_i}\right]} \right) dx \leq
$$

$$
\left(\prod_{\substack{i=1 \\ i\neq j}}^{m} \int_a^b \Phi_i \left(\left| {}^{RL}D_{\rho_i,\xi_i,\omega_i,a+}^{\gamma_i(1-\beta_i);\psi} f_i(y) \right| \right) dy \right)
$$

$$
\left(\int_a^b \Phi_j \left(\left| {}^{RL}D_{\rho_j,\xi_j,\omega_j,a+}^{\gamma_j(1-\beta_j);\psi} f_j(y) \right| \right) {}^{P}\lambda_{m+}^{\psi}(y)\, dy \right). \tag{9.103}
$$

Proof By (9.51) we have that

$$
\left({}^{H}\mathbb{D}_{\rho_i,\mu_i,\omega_i,a+}^{\gamma_i,\beta_i;\psi} f_i \right)(x) = e_{\rho_i,\xi_i-\mu_i,\omega_i,a+}^{-\gamma_i\beta_i;\psi} \; {}^{RL}D_{\rho_i,\xi_i,\omega_i,a+}^{\gamma_i(1-\beta_i);\psi} f_i(x), \tag{9.104}
$$

$\forall\, x \in [a, b]$, $i = 1, ..., m$.

We apply Theorem 9.18. ■

Theorem 9.24 *Here $i = 1, ..., m$. Let $\rho_i, \mu_i, \omega_i > 0$, $\gamma_i < 0$, and $f_i \in C^{N_i}([a, b])$, $N_i = \lceil \mu_i \rceil$, $\mu_i \notin \mathbb{N}$; $\theta := \max(N_1, ..., N_m)$, $\psi \in C^{\theta}([a, b])$, ψ is increasing with $\psi'(x) \neq 0$ over $[a, b]$. Here $0 \leq \beta_i \leq 1$ and $\xi_i = \mu_i + \beta_i(N_i - \mu_i)$. We assume that ${}^{RL}D_{\rho_i,\xi_i,\omega_i,b-}^{\gamma_i(1-\beta_i);\psi} f_i \in C([a, b])$, $i = 1, ..., m$. We assume further that the weight function $u \geq 0$ is such that*

$$
{}^{P}\lambda_{m-}^{\psi}(y) := \left(\psi'(y)\right)^{m} \int_{a}^{y} u(x) \left(\frac{(\psi(y)-\psi(x))^{\sum_{i=1}^{m}(\xi_i-\mu_i)-m}}{(\psi(b)-\psi(x))^{\sum_{i=1}^{m}(\xi_i-\mu_i)}} \right)
$$

$$
\prod_{i=1}^{m} \left(\frac{E_{\rho_i,\xi_i-\mu_i}^{-\gamma_i\beta_i}\left[\omega_i(\psi(y)-\psi(x))^{\rho_i}\right]}{E_{\rho_i,\xi_i-\mu_i+1}^{-\gamma_i\beta_i}\left[\omega_i(\psi(b)-\psi(x))^{\rho_i}\right]} \right) dx < \infty, \tag{9.105}
$$

for all $a < y < b$.

Here $\Phi_i : \mathbb{R}_+ \to \mathbb{R}_+$, $i = 1, ..., m$, *are convex and increasing functions. Then*

$$\int_a^b u(x) \prod_{i=1}^m \Phi_i \left(\frac{\left| \left({}^H\mathbb{D}^{\gamma_i, \beta_i; \psi}_{\rho_i, \mu_i, \omega_i, b-} f_i \right)(x) \right|}{(\psi(b) - \psi(x))^{\xi_i - \mu_i} E^{-\gamma_i \beta_i}_{\rho_i, \xi_i - \mu_i + 1} \left[\omega_i (\psi(b) - \psi(x))^{\rho_i} \right]} \right) dx \leq$$

$$\left(\prod_{\substack{i=1 \\ i \neq j}}^m \int_a^b \Phi_i \left(\left| {}^{RL}D^{\gamma_i(1-\beta_i); \psi}_{\rho_i, \xi_i, \omega_i, b-} f_i(y) \right| \right) dy \right)$$

$$\left(\int_a^b \Phi_j \left(\left| {}^{RL}D^{\gamma_j(1-\beta_j); \psi}_{\rho_j, \xi_j, \omega_j, b-} f_j(y) \right| \right) {}^P\lambda^{\psi}_{m-}(y) \, dy \right). \quad (9.106)$$

Proof By (9.52) we have that

$$\left({}^H\mathbb{D}^{\gamma_i, \beta_i; \psi}_{\rho_i, \mu_i, \omega_i, b-} f_i \right)(x) = e^{-\gamma_i \beta_i; \psi}_{\rho_i, \xi_i - \mu_i, \omega_i, b-} \; {}^{RL}D^{\gamma_i(1-\beta_i); \psi}_{\rho_i, \xi_i, \omega_i, b-} f_i(x), \quad (9.107)$$

$\forall\, x \in [a, b]$, $i = 1, ..., m$.

We apply Theorem 9.20. ■

We need

Remark 9.25 In the next we consider for $i = 1, ..., m$, the measurable functions $f_{1i}, f_{2i} : \Omega_2 \to \mathbb{R}$, and

$$g_{1i}(x) = \int_{\Omega_2} k_i(x, y) f_{1i}(y) \, d\mu_2(y), \quad (9.108)$$

and

$$g_{2i}(x) = \int_{\Omega_2} k_i(x, y) f_{2i}(y) \, d\mu_2(y), \quad (9.109)$$

these are now instead of (9.54).

Again here $u \geq 0$ is a weight measurable function on Ω_1.

We will use the following rational result:

Theorem 9.26 *([6], p. 405) Here* $0 < f_{2i}(y) < \infty$, *a.e.,* $i = 1, ..., m$. *Let* $\overline{j} \in \{1, ..., m\}$ *be fixed. Assume that the function* $x \to \left(\frac{u(x) \prod_{i=1}^m k_i(x,y) f_{2i}(y)}{\prod_{i=1}^m g_{2i}(x)} \right)$ *is integrable on* Ω_1, *for each* $y \in \Omega_2$. *Define* λ^{**}_m *on* Ω_2 *by*

$$\lambda_m^{**}(y) := \left(\prod_{i=1}^{m} f_{2i}(y)\right) \int_{\Omega_1} \left(\frac{u(x) \prod_{i=1}^{m} k_i(x,y)}{\prod_{i=1}^{m} g_{2i}(x)} \right) d\mu_1(x) < \infty. \tag{9.110}$$

Here $\Phi_i : \mathbb{R}_+ \to \mathbb{R}_+$, $i = 1, ..., m$, are convex and increasing functions.

Then

$$\int_{\Omega_1} u(x) \prod_{i=1}^{m} \Phi_i \left(\left| \frac{g_{1i}(x)}{g_{2i}(x)} \right| \right) d\mu_1(x) \leq$$

$$\left(\prod_{\substack{i=1 \\ i \neq \bar{j}}}^{m} \int_{\Omega_2} \Phi_i \left(\left| \frac{f_{1i}(y)}{f_{2i}(y)} \right| \right) d\mu_2(y) \right) \left(\int_{\Omega_2} \Phi_{\bar{j}} \left(\left| \frac{f_{1\bar{j}}(y)}{f_{2\bar{j}}(y)} \right| \right) \lambda_m^{**}(y) \, d\mu_2(y) \right), \tag{9.111}$$

true for all measurable functions, $i = 1, ..., m$, $f_{1i}, f_{2i} : \Omega_2 \to \mathbb{R}$ such that:

(i) $\frac{f_{1i}(y)}{f_{2i}(y)}$, $\Phi_i \left(\left| \frac{f_{1i}(y)}{f_{2i}(y)} \right| \right)$, are both $k_i(x,y) f_{2i}(y) d\mu_2(y)$ - integrable, μ_1-a.e. in $x \in \Omega_1$,

(ii) $\lambda_m^{**} \Phi_{\bar{j}} \left(\left| \frac{f_{1\bar{j}}(y)}{f_{2\bar{j}}(y)} \right| \right)$, and $\Phi_i \left(\left| \frac{f_{1i}(y)}{f_{2i}(y)} \right| \right)$, for $i = \{1, ..., m\} - \{\bar{j}\}$, are all μ_2-integrable;

and for all corresponding g_{1i} given by (9.108), and g_{2i} given by (9.109).

We present the following left ψ-Hilfer Hardy rational fractional inequality:

Theorem 9.27 *Here $i = 1, ..., m$; $j = 1, 2$. Let $n_i - 1 < \alpha_i < n_i$, $n_i \in \mathbb{N}$, $f_{ji} \in C^{n_i}([a,b])$; $\theta := \max\{n_1, ..., n_m\}$, $\psi \in C^{\theta}([a,b])$, ψ is increasing and $\psi'(x) \neq 0$, for all $x \in [a,b]$; $0 \leq \beta_i \leq 1$ and $\gamma_i = \alpha_i + \beta_i(n_i - \alpha_i)$. Assume that $\Delta_{a+}^{\gamma_i;\psi} f_{ji} \in C([a,b])$ and $0 < \Delta_{a+}^{\gamma_i;\psi} f_{2i}(y) < \infty$, for all $y \in [a,b]$. Define λ_{m+}^{H} so that $(a < y < b)$*

$$\lambda_{m+}^{H}(y) := \frac{\left(\prod_{i=1}^{m} \Delta_{a+}^{\gamma_i;\psi} f_{2i}(y) \right) (\psi'(y))^m}{\left(\prod_{i=1}^{m} \Gamma(\gamma_i - \alpha_i) \right)}$$

$$\int_y^b \frac{u(x) (\psi(x) - \psi(y))^{\sum_{i=1}^{m} (\gamma_i - \alpha_i) - m}}{\prod_{i=1}^{m} \left({}^{H}\mathbb{D}_{a+}^{\alpha_i,\beta_i;\psi} f_{2i} \right)(x)} dx < \infty. \tag{9.112}$$

Here $\Phi_i : \mathbb{R}_+ \to \mathbb{R}_+$, $i = 1, ..., m$, are convex and increasing functions. Then

$$\int_a^b u(x) \prod_{i=1}^{m} \Phi_i \left(\left| \frac{{}^{H}\mathbb{D}_{a+}^{\alpha_i,\beta_i;\psi} f_{1i}(x)}{{}^{H}\mathbb{D}_{a+}^{\alpha_i,\beta_i;\psi} f_{2i}(x)} \right| \right) dx \leq$$

$$\left(\prod_{\substack{i=1 \\ i \neq \overline{j}}}^{m} \int_a^b \Phi_i \left(\left| \frac{\Delta_{a+}^{\gamma_i;\psi} f_{1i}(y)}{\Delta_{a+}^{\gamma_i;\psi} f_{2i}(y)} \right| \right) dy \right) \left(\int_a^b \Phi_{\overline{j}} \left(\left| \frac{\Delta_{a+}^{\gamma_{\overline{j}};\psi} f_{1\overline{j}}(y)}{\Delta_{a+}^{\gamma_{\overline{j}};\psi} f_{2\overline{j}}(y)} \right| \right) \lambda_{m+}^{H}(y)\, dy \right). \tag{9.113}$$

Proof By Theorem 9.26, see also [6], pp. 431–432, and (9.19). ■

The right side analog of (9.113) follows:

Theorem 9.28 *Here* $i = 1, ..., m;\ j = 1, 2$. *Let* $n_i - 1 < \alpha_i < n_i$, $n_i \in \mathbb{N}$, $f_{ji} \in C^{n_i}([a, b])$; $\theta := \max\{n_1, ..., n_m\}$, $\psi \in C^{\theta}([a, b])$, ψ *is increasing and* $\psi'(x) \neq 0$, *for all* $x \in [a, b]$; $0 \leq \beta_i \leq 1$ *and* $\gamma_i = \alpha_i + \beta_i(n_i - \alpha_i)$. *Assume that* $\Delta_{b-}^{\gamma_i;\psi} f_{ji} \in C([a, b])$ *and* $0 < \Delta_{b-}^{\gamma_i;\psi} f_{2i}(y) < \infty$, *for all* $y \in [a, b]$. *Define* λ_{m-}^{H} *so that* $(a < y < b)$

$$\lambda_{m-}^{H}(y) := \frac{\left(\prod_{i=1}^{m} \Delta_{b-}^{\gamma_i;\psi} f_{2i}(y) \right) (\psi'(y))^m}{\left(\prod_{i=1}^{m} \Gamma(\gamma_i - \alpha_i) \right)}$$

$$\int_a^y \frac{u(x)(\psi(y) - \psi(x))^{\sum_{i=1}^{m}(\gamma_i - \alpha_i) - m}}{\prod_{i=1}^{m} \left({}^{H}\mathbb{D}_{b-}^{\alpha_i,\beta_i;\psi} f_{2i} \right)(x)} dx < \infty. \tag{9.114}$$

Here $\Phi_i : \mathbb{R}_+ \to \mathbb{R}_+$, $i = 1, ..., m$, *are convex and increasing functions. Then*

$$\int_a^b u(x) \prod_{i=1}^{m} \Phi_i \left(\left| \frac{{}^{H}\mathbb{D}_{b-}^{\alpha_i,\beta_i;\psi} f_{1i}(x)}{{}^{H}\mathbb{D}_{b-}^{\alpha_i,\beta_i;\psi} f_{2i}(x)} \right| \right) dx \leq$$

$$\left(\prod_{\substack{i=1 \\ i \neq \overline{j}}}^{m} \int_a^b \Phi_i \left(\left| \frac{\Delta_{b-}^{\gamma_i;\psi} f_{1i}(y)}{\Delta_{b-}^{\gamma_i;\psi} f_{2i}(y)} \right| \right) dy \right) \left(\int_a^b \Phi_{\overline{j}} \left(\left| \frac{\Delta_{b-}^{\gamma_{\overline{j}};\psi} f_{1\overline{j}}(y)}{\Delta_{b-}^{\gamma_{\overline{j}};\psi} f_{2\overline{j}}(y)} \right| \right) \lambda_{m-}^{H}(y)\, dy \right). \tag{9.115}$$

Proof By Theorem 9.26, see also [6], pp. 433–434, and (9.20). ■

Next come the ψ-Prabhakar-Riemann-Liouville left side Hardy rational fractional inequality:

Theorem 9.29 *Here* $i = 1, ..., m;\ j = 1, 2$. *Let* $f_{ji} \in C([a, b])$, $\psi \in C^1([a, b])$ *which is increasing, and* $\gamma_i, \rho_i, \mu_i, \omega_i > 0$. *Assume* $0 < f_{2i}(y) < \infty$, $\overline{j} \in \{1, ..., m\}$ *is fixed. Define* λ_{m+}^{e} *so that* $(a < y < b)$

$$\lambda_{m+}^{e}(y) := \left(\prod_{i=1}^{m} f_{2i}(y)\right)\left(\psi'(y)\right)^{m} \tag{9.116}$$

$$\int_{y}^{b}\left(\frac{u(x)\left(\psi(x)-\psi(y)\right)^{\sum_{i=1}^{m}\mu_i-m}\prod_{i=1}^{m}E_{\rho_i,\mu_i}^{\gamma_i}\left[\omega_i\left(\psi(x)-\psi(y)\right)^{\rho_i}\right]}{\prod_{i=1}^{m}e_{\rho_i,\mu_i,\omega_i,a+}^{\gamma_i;\psi}f_{2i}(x)}\right)dx < \infty.$$

Here $\Phi_i : \mathbb{R}_+ \to \mathbb{R}_+$, $i = 1, ..., m$, are convex and increasing functions. Then

$$\int_a^b u(x)\prod_{i=1}^{m}\Phi_i\left(\left|\frac{e_{\rho_i,\mu_i,\omega_i,a+}^{\gamma_i;\psi}f_{1i}(x)}{e_{\rho_i,\mu_i,\omega_i,a+}^{\gamma_i;\psi}f_{2i}(x)}\right|\right)dx \le$$

$$\left(\prod_{\substack{i=1\\ i\neq \overline{j}}}^{m}\int_a^b \Phi_i\left(\left|\frac{f_{1i}(y)}{f_{2i}(y)}\right|\right)dy\right)\left(\int_a^b \Phi_{\overline{j}}\left(\left|\frac{f_{1\overline{j}}(y)}{f_{2\overline{j}}(y)}\right|\right)\lambda_{m+}^{e}(y)\,dy\right). \tag{9.117}$$

Proof See (9.82) and apply Theorem 9.26. ■

The right side analog follows:

Theorem 9.30 *Here $i = 1, ..., m$; $j = 1, 2$. Let $f_{ji} \in C([a, b])$, $\psi \in C^1([a, b])$ which is increasing, and $\gamma_i, \rho_i, \mu_i, \omega_i > 0$. Assume $0 < f_{2i}(y) < \infty$, $\overline{j} \in \{1, ..., m\}$ is fixed. Define λ_{m-}^{e} so that $(a < y < b)$*

$$\lambda_{m-}^{e}(y) := \left(\prod_{i=1}^{m} f_{2i}(y)\right)\left(\psi'(y)\right)^{m} \tag{9.118}$$

$$\int_{a}^{y}\left(\frac{u(x)\left(\psi(y)-\psi(x)\right)^{\sum_{i=1}^{m}\mu_i-m}\prod_{i=1}^{m}E_{\rho_i,\mu_i}^{\gamma_i}\left[\omega_i\left(\psi(y)-\psi(x)\right)^{\rho_i}\right]}{\prod_{i=1}^{m}e_{\rho_i,\mu_i,\omega_i,b-}^{\gamma_i;\psi}f_{2i}(x)}\right)dx < \infty.$$

Here $\Phi_i : \mathbb{R}_+ \to \mathbb{R}_+$, $i = 1, ..., m$, are convex and increasing functions. Then

$$\int_a^b u(x)\prod_{i=1}^{m}\Phi_i\left(\left|\frac{e_{\rho_i,\mu_i,\omega_i,b-}^{\gamma_i;\psi}f_{1i}(x)}{e_{\rho_i,\mu_i,\omega_i,b-}^{\gamma_i;\psi}f_{2i}(x)}\right|\right)dx \le$$

$$\left(\prod_{\substack{i=1\\ i\neq\overline{j}}}^{m}\int_a^b \Phi_i\left(\left|\frac{f_{1i}(y)}{f_{2i}(y)}\right|\right)dy\right)\left(\int_a^b \Phi_{\overline{j}}\left(\left|\frac{f_{1\overline{j}}(y)}{f_{2\overline{j}}(y)}\right|\right)\lambda_{m-}^{e}(y)\,dy\right). \tag{9.119}$$

Proof See (9.90) and apply Theorem 9.26. ■

Rational left and right ψ-Prabhakar-Caputo Hardy fractional inequalities follow:

Theorem 9.31 *Here* $i = 1, ..., m$; $j = 1, 2$; $\rho_i, \mu_i, \omega_i > 0$, $\gamma_i < 0$. *Let* $f_{ji} \in C^{N_i}([a, b])$, $N_i = \lceil \mu_i \rceil$, $\mu_i \notin \mathbb{N}$; $\theta := \max(N_1, ..., N_m)$, $\psi \in C^{\theta}([a, b])$, ψ *is increasing with* $\psi'(x) \neq 0$ *over* $[a, b]$. *Set* $f_{ji\psi}^{[N_i]}(x) = \left(\frac{1}{\psi'(x)}\frac{d}{dx}\right)^{N_i} f_{ji}(x)$, $x \in [a, b]$. *Assume that* $0 < f_{2i\psi}^{[N_i]}(y) < \infty$, $\overline{j} \in \{1, ..., m\}$ *is fixed. Define* ${}^{C}\lambda_{m+}^{R}$ *so that* $(a < y < b)$

$${}^{C}\lambda_{m+}^{R}(y) := \left(\prod_{i=1}^{m} f_{2i\psi}^{[N_i]}(y)\right)\left(\psi'(y)\right)^{m} \tag{9.120}$$

$$\int_y^b\left(\frac{u(x)\left(\psi(x)-\psi(y)\right)^{\sum_{i=1}^{m}(N_i-\mu_i)-m}\prod_{i=1}^{m}E_{\rho_i,N_i-\mu_i}^{-\gamma_i}\left[\omega_i\left(\psi(x)-\psi(y)\right)^{\rho_i}\right]}{\prod_{i=1}^{m}{}^{C}D_{\rho_i,\mu_i,\omega_i,a+}^{\gamma_i;\psi}f_{2i}(x)}\right)dx<\infty.$$

Here $\Phi_i : \mathbb{R}_+ \to \mathbb{R}_+$, $i = 1, ..., m$, *are convex and increasing functions. Then*

$$\int_a^b u(x)\prod_{i=1}^{m}\Phi_i\left(\left|\frac{{}^{C}D_{\rho_i,\mu_i,\omega_i,a+}^{\gamma_i;\psi}f_{1i}(x)}{{}^{C}D_{\rho_i,\mu_i,\omega_i,a+}^{\gamma_i;\psi}f_{2i}(x)}\right|\right)dx\leq$$

$$\left(\prod_{\substack{i=1\\ i\neq\overline{j}}}^{m}\int_a^b \Phi_i\left(\left|\frac{f_{1i\psi}^{[N_i]}(y)}{f_{2i\psi}^{[N_i]}(y)}\right|\right)dy\right)\left(\int_a^b \Phi_{\overline{j}}\left(\left|\frac{f_{1\overline{j}\psi}^{\left[N_{\overline{j}}\right]}(y)}{f_{2\overline{j}\psi}^{\left[N_{\overline{j}}\right]}(y)}\right|\right){}^{C}\lambda_{m+}^{R}(y)\,dy\right). \tag{9.121}$$

Proof By (9.42) and Theorem 9.29. ■

Theorem 9.32 *Here* $i = 1, ..., m$; $j = 1, 2$; $\rho_i, \mu_i, \omega_i > 0$, $\gamma_i < 0$. *Let* $f_{ji} \in C^{N_i}([a, b])$, $N_i = \lceil \mu_i \rceil$, $\mu_i \notin \mathbb{N}$; $\theta := \max(N_1, ..., N_m)$, $\psi \in C^{\theta}([a, b])$, ψ *is increasing with* $\psi'(x) \neq 0$ *over* $[a, b]$. *Set* $f_{ji\psi}^{[N_i]}(x) = \left(\frac{1}{\psi'(x)}\frac{d}{dx}\right)^{N_i} f_{ji}(x)$, $x \in [a, b]$. *Assume that* $0 < f_{2i\psi}^{[N_i]}(y) < \infty$, $\overline{j} \in \{1, ..., m\}$ *is fixed. Define* ${}^{C}\lambda_{m-}^{R}$ *so that* $(a < y < b)$

$${}^{C}\lambda_{m-}^{R}(y) := \left(\prod_{i=1}^{m} f_{2i\psi}^{[N_i]}(y)\right)\left(\psi'(y)\right)^{m} \tag{9.122}$$

$$\int_a^y \left(\frac{u(x)(\psi(y)-\psi(x))^{\sum_{i=1}^{m}(N_i-\mu_i)-m} \prod_{i=1}^{m} E_{\rho_i,N_i-\mu_i}^{-\gamma_i}\left[\omega_i(\psi(y)-\psi(x))^{\rho_i}\right]}{\left(\prod_{i=1}^{m} {}^{C}D_{\rho_i,\mu_i,\omega_i,b-}^{\gamma_i;\psi} f_{2i}(x)\right)(-1)^{\sum_{i=1}^{m} N_i}} \right) dx < \infty.$$

Here $\Phi_i : \mathbb{R}_+ \to \mathbb{R}_+$, $i = 1, ..., m$, *are convex and increasing functions. Then*

$$\int_a^b u(x) \prod_{i=1}^{m} \Phi_i \left(\left| \frac{{}^{C}D_{\rho_i,\mu_i,\omega_i,b-}^{\gamma_i;\psi} f_{1i}(x)}{{}^{C}D_{\rho_i,\mu_i,\omega_i,b-}^{\gamma_i;\psi} f_{2i}(x)} \right| \right) dx \leq$$

$$\left(\prod_{\substack{i=1 \\ i \neq \bar{j}}}^{m} \int_a^b \Phi_i \left(\left| \frac{f_{1i\psi}^{[N_i]}(y)}{f_{2i\psi}^{[N_i]}(y)} \right| \right) dy \right) \left(\int_a^b \Phi_{\bar{j}} \left(\left| \frac{f_{1\bar{j}\psi}^{\left[N_{\bar{j}}\right]}(y)}{f_{2\bar{j}\psi}^{\left[N_{\bar{j}}\right]}(y)} \right| \right) {}^{C}\lambda_{m-}^{R}(y)\, dy \right). \tag{9.123}$$

Proof By (9.43) and Theorem 9.30. ■

We finish with the following rational left and right side ψ-Hilfer-Prabhakar Hardy type fractional inequalities.

Theorem 9.33 *Here* $i = 1, ..., m$; $j = 1, 2$; $\rho_i, \mu_i, \omega_i > 0$, $\gamma_i < 0$, $0 \leq \beta_i \leq 1$, $\xi_i = \mu_i + \beta_i (N_i - \mu_i)$, $N_i = \lceil \mu_i \rceil$, $\mu_i \notin \mathbb{N}$; $\theta := \max(N_1, ..., N_m)$. *Let* $f_{ji} \in C^{N_i}([a, b])$, $\psi \in C^{\theta}([a, b])$, ψ *is increasing with* $\psi'(x) \neq 0$ *over* $[a, b]$. *We assume that* ${}^{RL}D_{\rho_i,\xi_i,\omega_i,a+}^{\gamma_i(1-\beta_i);\psi} f_{ji} \in C([a, b])$ *and that* $0 < {}^{RL}D_{\rho_i,\xi_i,\omega_i,a+}^{\gamma_i(1-\beta_i);\psi} f_{2i}(y) < \infty$, $\bar{j} \in \{1, ..., m\}$ *is fixed. Define* ${}^{H}\lambda_{m+}^{P}$ *so that* $(a < y < b)$

$${}^{H}\lambda_{m+}^{P}(y) = \left(\prod_{i=1}^{m} {}^{RL}D_{\rho_i,\xi_i,\omega_i,a+}^{\gamma_i(1-\beta_i);\psi} f_{2i}(y) \right) \left(\psi'(y)\right)^m$$

$$\int_y^b \left(\frac{u(x)(\psi(x)-\psi(y))^{\sum_{i=1}^{m}(\xi_i-\mu_i)-m} \prod_{i=1}^{m} E_{\rho_i,\xi_i-\mu_i}^{-\gamma_i\beta_i}\left[\omega_i(\psi(x)-\psi(y))^{\rho_i}\right]}{\prod_{i=1}^{m} {}^{H}\mathbb{D}_{\rho_i,\mu_i,\omega_i,a+}^{\gamma_i,\beta_i;\psi} f_{2i}(x)} \right) dx < \infty. \tag{9.124}$$

Here $\Phi_i : \mathbb{R}_+ \to \mathbb{R}_+$, $i = 1, ..., m$, *are convex and increasing functions. Then*

$$\int_a^b u(x) \prod_{i=1}^{m} \Phi_i \left(\left| \frac{{}^{H}\mathbb{D}_{\rho_i,\mu_i,\omega_i,a+}^{\gamma_i,\beta_i;\psi} f_{1i}(x)}{{}^{H}\mathbb{D}_{\rho_i,\mu_i,\omega_i,a+}^{\gamma_i,\beta_i;\psi} f_{2i}(x)} \right| \right) dx \leq$$

$$\left(\prod_{\substack{i=1\\i\neq\overline{j}}}^{m}\int_a^b\Phi_i\left(\left|\frac{{}^{RL}D^{\gamma_i(1-\beta_i);\psi}_{\rho_i,\xi_i,\omega_i,a+}f_{1i}(y)}{{}^{RL}D^{\gamma_i(1-\beta_i);\psi}_{\rho_i,\xi_i,\omega_i,a+}f_{2i}(y)}\right|\right)dy\right)$$

$$\left(\int_a^b\Phi_{\overline{j}}\left(\left|\frac{{}^{RL}D^{\gamma_{\overline{j}}(1-\beta_{\overline{j}});\psi}_{\rho_{\overline{j}},\xi_{\overline{j}},\omega_{\overline{j}},a+}f_{1\overline{j}}(y)}{{}^{RL}D^{\gamma_{\overline{j}}(1-\beta_{\overline{j}});\psi}_{\rho_{\overline{j}},\xi_{\overline{j}},\omega_{\overline{j}},a+}f_{2\overline{j}}(y)}\right|\right){}^{H}\lambda^{P}_{m+}(y)\,dy\right). \tag{9.125}$$

Proof By (9.51) and Theorem 9.29. ∎

Theorem 9.34 *Here* $i=1,...,m;\ j=1,2;\ \rho_i,\mu_i,\omega_i>0,\ \gamma_i<0,\ 0\le\beta_i\le1,$ $\xi_i=\mu_i+\beta_i(N_i-\mu_i),\ N_i=\lceil\mu_i\rceil,\ \mu_i\notin\mathbb{N};\ \theta:=\max(N_1,...,N_m)$. *Let* $f_{ji}\in C^{N_i}([a,b])$, $\psi\in C^{\theta}([a,b])$, ψ *is increasing with* $\psi'(x)\neq0$ *over* $[a,b]$. *We assume that* ${}^{RL}D^{\gamma_i(1-\beta_i);\psi}_{\rho_i,\xi_i,\omega_i,b-}f_{ji}\in C([a,b])$ *and that* $0<{}^{RL}D^{\gamma_i(1-\beta_i);\psi}_{\rho_i,\xi_i,\omega_i,b-}f_{2i}(y)<\infty$, $\overline{j}\in\{1,...,m\}$ *is fixed. Define* ${}^{H}\lambda^{P}_{m-}$ *so that* $(a<y<b)$

$${}^{H}\lambda^{P}_{m-}(y)=\left(\prod_{i=1}^{m}{}^{RL}D^{\gamma_i(1-\beta_i);\psi}_{\rho_i,\xi_i,\omega_i,b-}f_{2i}(y)\right)\left(\psi'(y)\right)^m$$

$$\int_a^y\left(\frac{u(x)\,(\psi(y)-\psi(x))^{\sum_{i=1}^{m}(\xi_i-\mu_i)-m}\prod_{i=1}^{m}E^{-\gamma_i\beta_i}_{\rho_i,\xi_i-\mu_i}\left[\omega_i(\psi(y)-\psi(x))^{\rho_i}\right]}{\prod_{i=1}^{m}{}^{H}\mathbb{D}^{\gamma_i,\beta_i;\psi}_{\rho_i,\mu_i,\omega_i,b-}f_{2i}(x)}\right)dx<\infty. \tag{9.126}$$

Here $\Phi_i:\mathbb{R}_+\to\mathbb{R}_+$, $i=1,...,m$, *are convex and increasing functions. Then*

$$\int_a^b u(x)\prod_{i=1}^{m}\Phi_i\left(\left|\frac{{}^{H}\mathbb{D}^{\gamma_i,\beta_i;\psi}_{\rho_i,\mu_i,\omega_i,b-}f_{1i}(x)}{{}^{H}\mathbb{D}^{\gamma_i,\beta_i;\psi}_{\rho_i,\mu_i,\omega_i,b-}f_{2i}(x)}\right|\right)dx\le$$

$$\left(\prod_{\substack{i=1\\i\neq\overline{j}}}^{m}\int_a^b\Phi_i\left(\left|\frac{{}^{RL}D^{\gamma_i(1-\beta_i);\psi}_{\rho_i,\xi_i,\omega_i,b-}f_{1i}(y)}{{}^{RL}D^{\gamma_i(1-\beta_i);\psi}_{\rho_i,\xi_i,\omega_i,b-}f_{2i}(y)}\right|\right)dy\right)$$

$$\left(\int_a^b\Phi_{\overline{j}}\left(\left|\frac{{}^{RL}D^{\gamma_{\overline{j}}(1-\beta_{\overline{j}});\psi}_{\rho_{\overline{j}},\xi_{\overline{j}},\omega_{\overline{j}},b-}f_{1\overline{j}}(y)}{{}^{RL}D^{\gamma_{\overline{j}}(1-\beta_{\overline{j}});\psi}_{\rho_{\overline{j}},\xi_{\overline{j}},\omega_{\overline{j}},b-}f_{2\overline{j}}(y)}\right|\right){}^{H}\lambda^{P}_{m-}(y)\,dy\right). \tag{9.127}$$

Proof By (9.52) and Theorem 9.30. ∎

References

1. Almeida, R.: A Caputo fractional derivative of a function with respect to another function. Commun. Nonlinear Sci. Numer. Simul. **44**, 460–481 (2017)
2. Anastassiou, G.A.: Advanced Hilfer-Prabhakar-hardy type fractional inequalities. Bull. Allahabad Math. Soc. (2021). [Accepted]
3. Anastassiou, G.: Advancements on ψ-Hilfer fractional calculus and fractional integral inequalities. Nonlinear Complex. (2021). [Accepted, Discontinuity]
4. Anastassiou, G.A.: Foundations of Generalized Prabhakar-Hilfer fractional Calculus with Applications (2021). [Submitted]
5. Anastassiou, G.A.: Univariate Hardy-type fractional inequalities, Chap. 2. In: Anastassiou, G., Duman, O. (eds) Advances in Applied Mathematics and Approximation Theory, Contributions from AMAT 2012, pp. 21–56. Springer, New York (2013)
6. Anastassiou, G.A.: Intelligent Comparisons: Analytic Inequalities. Springer, Heidelberg, New York (2016)
7. da Sousa, J.V.C., de Oliveira, E.C.: On the ψ-Hilfer fractional derivative. Commun. Nonlinear Sci. Numer. Simul. **60**, 72–91 (2018)
8. Giusti, A., et al.: A practical guide to Prabhakar fractional calculus. Fractional Calc. Appl. Anal. **23**(1), 9–54 (2020)
9. Gorenflo, R., Kilbas, A., Mainardi, F., Rogosin, S.: Mittag-Leffler functions. Related Topics and Applications. Springer, Heidelberg, New York (2014)
10. Hardy, H.G.: Notes on some points in the integral calculus. Messenger Math. **47**(10), 145–150 (1918)
11. Kilbas, A.A., Srivastava, H.M., Trujillo, J.J.: Theory and Applications of Fractional Differentiation Equations. North Holland, Amsterdam, New York (2006)
12. Polito, F., Tomovski, Z.: Some properties of Prabhakar-type fractional calculus operators. Fractional Differ. Calc. **6**(1), 73–94 (2016)
13. Prabhakar, T.R.: A singular integral equation with a generalized Mittag Leffler function in the kernel. Yokohama Math. J. **19**, 7–15 (1971)
14. Samko, S.G., Kilbas, A.A., Marichev, O.I.: Fractional Integrals and Derivatives: Theory and Applications. Gordon and Breach Science Publishers, Switzerland (1993)
15. Tomovski, Ž, Hilfer, R., Srivastava, H.M.: Fractional and operational calculus with generalized fractional derivative operators and Mittag-Leffler type functions. Integr. Transforms Spec. Funct. **21**(11), 797–814 (2010)

Chapter 10
Vectorial Advanced Hilfer-Prabhakar-Hardy Fractional Inequalities

We present a variety of univariate and multivariate left and right side Hardy type fractional inequalities, many of them under convexity, and other also of L_p type, $p \geq 1$, in the setting of generalized Hilfer and Prabhakar fractional Calculi. It follows [5].

10.1 Background

Let $-\infty < a < b < \infty$, the left and right Riemann-Liouville fractional integrals of order $\alpha \in \mathbb{C}$ $(\mathcal{R}(\alpha) > 0)$ are defined by

$$\left(I_{a+}^{\alpha} f\right)(x) = \frac{1}{\Gamma(\alpha)} \int_a^x (x-t)^{\alpha-1} f(t)\, dt, \tag{10.1}$$

$x > a$; where Γ stands for the gamma function,

and

$$\left(I_{b-}^{\alpha} f\right)(x) = \frac{1}{\Gamma(\alpha)} \int_x^b (t-x)^{\alpha-1} f(t)\, dt, \tag{10.2}$$

$x < b$.

The Riemann-Liouville left and right fractional derivatives of order $\alpha \in \mathbb{C}$ $(\mathcal{R}(\alpha) \geq 0)$ are defined by

$$\left(\Delta_{a+}^{\alpha} y\right)(x) = \left(\frac{d}{dx}\right)^n \left(I_{a+}^{n-\alpha} y\right)(x) = \frac{1}{\Gamma(n-\alpha)} \left(\frac{d}{dx}\right)^n \int_a^x (x-t)^{n-\alpha-1} y(t)\, dt \tag{10.3}$$

$(n = \lceil \mathcal{R}(\alpha) \rceil$, $\lceil \cdot \rceil$ means ceiling of the number; $x > a)$

G. A. Anastassiou, *Unification of Fractional Calculi with Applications*,
Studies in Systems, Decision and Control 398,
https://doi.org/10.1007/978-3-030-86920-5_10

$$\left(\Delta_{b-}^{\alpha} y\right)(x) = (-1)^n \left(\frac{d}{dx}\right)^n \left(I_{b-}^{n-\alpha} y\right)(x) =$$

$$\frac{(-1)^n}{\Gamma(n-\alpha)} \left(\frac{d}{dx}\right)^n \int_x^b (t-x)^{n-\alpha-1} y(t)\, dt \tag{10.4}$$

($n = \lceil \mathcal{R}(\alpha) \rceil$; $x < b$), respectively, where $\mathcal{R}(\alpha)$ is the real part of α.

In particular, when $\alpha = n \in \mathbb{Z}_+$, then

$$\left(\Delta_{a+}^{0} y\right)(x) = \left(\Delta_{b-}^{0} y\right)(x) = y(x);$$
$$\left(\Delta_{a+}^{n} y\right)(x) = y^{(n)}(x), \text{ and } \left(\Delta_{b-}^{n} y\right)(x) = (-1)^n y^{(n)}(x), \ n \in \mathbb{N}, \tag{10.5}$$

see [14].

Let $\alpha > 0$, $I = [a, b] \subset \mathbb{R}$, f an integrable function defined on I and $\psi \in C^1(I)$ an increasing function such that $\psi'(x) \neq 0$, for all $x \in I$. Left fractional integrals and left Riemann-Liouville fractional derivatives of a function f with respect to another function ψ are defined as ([11, 14])

$$I_{a+}^{\alpha,\psi} f(x) = \frac{1}{\Gamma(\alpha)} \int_a^x \psi'(t)(\psi(x) - \psi(t))^{\alpha-1} f(t)\, dt, \tag{10.6}$$

and

$$\Delta_{a+}^{\alpha,\psi} f(x) = \left(\frac{1}{\psi'(x)} \frac{d}{dx}\right)^n I_{a+}^{n-\alpha,\psi} f(x) = \tag{10.7}$$

$$\frac{1}{\Gamma(n-\alpha)} \left(\frac{1}{\psi'(x)} \frac{d}{dx}\right)^n \int_a^x \psi'(t)(\psi(x) - \psi(t))^{n-\alpha-1} f(t)\, dt,$$

respectively, where $n = \lceil \alpha \rceil$.

Similarly, we define the right ones:

$$I_{b-}^{\alpha,\psi} f(x) = \frac{1}{\Gamma(\alpha)} \int_x^b \psi'(t)(\psi(t) - \psi(x))^{\alpha-1} f(t)\, dt, \tag{10.8}$$

and

$$\Delta_{b-}^{\alpha,\psi} f(x) = \left(-\frac{1}{\psi'(x)} \frac{d}{dx}\right)^n I_{b-}^{n-\alpha,\psi} f(x) =$$

$$\frac{1}{\Gamma(n-\alpha)} \left(-\frac{1}{\psi'(x)} \frac{d}{dx}\right)^n \int_x^b \psi'(t)(\psi(t) - \psi(x))^{n-\alpha-1} f(t)\, dt. \tag{10.9}$$

The following semigroup property holds; if $\alpha, \beta > 0$, $f \in C(I)$, then

$$I_{a+}^{\alpha,\psi} I_{a+}^{\beta,\psi} f = I_{a+}^{\alpha+\beta,\psi} f \text{ and } I_{b-}^{\alpha,\psi} I_{b-}^{\beta,\psi} f = I_{b-}^{\alpha+\beta,\psi} f.$$

Next let again $\alpha > 0$, $n = \lceil \alpha \rceil$, $I = [a, b]$, $f, \psi \in C^n(I) : \psi$ is increasing and $\psi'(x) \neq 0$, for all $x \in I$. The left ψ-Caputo fractional derivative of f of order α is given by ([1])

$$^{C}D_{a+}^{\alpha,\psi} f(x) = I_{a+}^{n-\alpha,\psi} \left(\frac{1}{\psi'(x)} \frac{d}{dx} \right)^n f(x), \qquad (10.10)$$

and the right ψ-Caputo fractional derivative ([1])

$$^{C}D_{b-}^{\alpha,\psi} f(x) = I_{b-}^{n-\alpha,\psi} \left(-\frac{1}{\psi'(x)} \frac{d}{dx} \right)^n f(x). \qquad (10.11)$$

We set

$$f_{\psi}^{[n]}(x) := f_{\psi}^{(n)} f(x) := \left(\frac{1}{\psi'(x)} \frac{d}{dx} \right)^n f(x). \qquad (10.12)$$

Clearly, when $\alpha = m \in \mathbb{N}$ we have

$$^{C}D_{a+}^{\alpha,\psi} f(x) = f_{\psi}^{[m]}(x) \quad \text{and} \quad ^{C}D_{b-}^{\alpha,\psi} f(x) = (-1)^m f_{\psi}^{[m]}(x),$$

and if $\alpha \notin \mathbb{N}$, then

$$^{C}D_{a+}^{\alpha,\psi} f(x) = \frac{1}{\Gamma(n-\alpha)} \int_a^x \psi'(t) (\psi(x) - \psi(t))^{n-\alpha-1} f_{\psi}^{[n]}(t)\, dt, \qquad (10.13)$$

and

$$^{C}D_{b-}^{\alpha,\psi} f(x) = \frac{(-1)^n}{\Gamma(n-\alpha)} \int_x^b \psi'(t) (\psi(t) - \psi(x))^{n-\alpha-1} f_{\psi}^{[n]}(t)\, dt. \qquad (10.14)$$

If $\psi(x) = x$, then we get the usual left and right Caputo fractional derivatives

$$^{C}D_{a+}^{m} f(x) = f^{(m)}(x), \; ^{C}D_{b-}^{m} f(x) = (-1)^m f^{(m)}(x),$$

for $m \in \mathbb{N}$, and ($\alpha \notin \mathbb{N}$)

$$D_{*a}^{\alpha} f(x) = {}^{C}D_{a+}^{\alpha} f(x) = \frac{1}{\Gamma(n-\alpha)} \int_a^x (x-t)^{n-\alpha-1} f^{(n)}(t)\, dt, \qquad (10.15)$$

$$D_{b-}^{\alpha}(x) = {}^{C}D_{b-}^{\alpha} f(x) = \frac{(-1)^n}{\Gamma(n-\alpha)} \int_x^b (t-x)^{n-\alpha-1} f^{(n)}(t)\, dt. \qquad (10.16)$$

Also we set

$$^{C}D_{a+}^{0,\psi} f(x) = {}^{C}D_{b-}^{0,\psi} f(x) = f(x).$$

Next we talk about the ψ-Hilfer fractional derivative.

Definition 10.1 ([7]) Let $n-1<\alpha<n$, $n\in\mathbb{N}$, $I=[a,b]\subset\mathbb{R}$ and $f,\psi\in C^{n}([a,b])$, ψ is increasing and $\psi'(x)\neq 0$, for all $x\in I$. The ψ-Hilfer fractional derivative (left-sided and right-sided) ${}^{H}\mathbb{D}_{a+(b-)}^{\alpha,\beta;\psi}f$ of order α and type $0\leq\beta\leq 1$, respectively, are defined by

$$ {}^{H}\mathbb{D}_{a+}^{\alpha,\beta;\psi}f(x)=I_{a+}^{\beta(n-\alpha);\psi}\left(\frac{1}{\psi'(x)}\frac{d}{dx}\right)^{n}I_{a+}^{(1-\beta)(n-\alpha);\psi}f(x), \tag{10.17} $$

and

$$ {}^{H}\mathbb{D}_{b-}^{\alpha,\beta;\psi}f(x)=I_{b-}^{\beta(n-\alpha);\psi}\left(-\frac{1}{\psi'(x)}\frac{d}{dx}\right)^{n}I_{b-}^{(1-\beta)(n-\alpha);\psi}f(x),\quad x\in[a,b]. \tag{10.18} $$

The original Hilfer fractional derivatives ([15]) come from $\psi(x)=x$, and are denoted by ${}^{H}\mathbb{D}_{a+}^{\alpha,\beta}f(x)$ and ${}^{H}\mathbb{D}_{b-}^{\alpha,\beta}f(x)$.

When $\beta=0$, we get Riemann-Liouville fractional derivatives, while when $\beta=1$ we have Caputo type fractional derivatives.

We define $\gamma=\alpha+\beta(n-\alpha)$. We notice that $n-1<\alpha\leq\alpha+\beta(n-\alpha)\leq\alpha+n-\alpha=n$, hence $\lceil\gamma\rceil=n$. We can easily write that ([7])

$$ {}^{H}\mathbb{D}_{a+}^{\alpha,\beta;\psi}f(x)=I_{a+}^{\gamma-\alpha;\psi}\Delta_{a+}^{\gamma;\psi}f(x), \tag{10.19} $$

and

$$ {}^{H}\mathbb{D}_{b-}^{\alpha,\beta;\psi}f(x)=I_{b-}^{\gamma-\alpha;\psi}\Delta_{b-}^{\gamma;\psi}f(x),\quad x\in[a,b]. \tag{10.20} $$

We have that ([7])

$$ \Delta_{a+}^{\gamma,\psi}f(x)=\left(\frac{1}{\psi'(x)}\frac{d}{dx}\right)^{n}I_{a+}^{(1-\beta)(n-\alpha);\psi}f(x), \tag{10.21} $$

and

$$ \Delta_{b-}^{\gamma,\psi}f(x)=\left(-\frac{1}{\psi'(x)}\frac{d}{dx}\right)^{n}I_{b-}^{(1-\beta)(n-\alpha);\psi}f(x). \tag{10.22} $$

In particular, when $0<\alpha<1$ and $0\leq\beta\leq 1$; $\gamma=\alpha+\beta(1-\alpha)$, we have that

$$ {}^{H}\mathbb{D}_{a+}^{\alpha,\beta;\psi}f(x)=\frac{1}{\Gamma(\gamma-\alpha)}\int_{a}^{x}\psi'(t)(\psi(x)-\psi(t))^{\gamma-\alpha-1}\Delta_{a+}^{\gamma;\psi}f(t)\,dt, \tag{10.23} $$

and

$$ {}^{H}\mathbb{D}_{b-}^{\alpha,\beta;\psi}f(x)=\frac{1}{\Gamma(\gamma-\alpha)}\int_{x}^{b}\psi'(t)(\psi(t)-\psi(x))^{\gamma-\alpha-1}\Delta_{b-}^{\gamma;\psi}f(t)\,dt, \tag{10.24} $$

$x\in[a,b]$.

Remark 10.2 ([7]) Let $\mu = n(1-\beta) + \beta\alpha$, then $\lceil \mu \rceil = n$.

Assume that $g(x) = I_{a+}^{(1-\beta)(n-\alpha);\psi} f(x) \in C^n([a,b])$, we have that

$$ {}^{H}\mathbb{D}_{a+}^{\alpha,\beta;\psi} f(x) = I_{a+}^{n-\mu;\psi} \left(\frac{1}{\psi'(x)} \frac{d}{dx} \right)^n g(x). \tag{10.25} $$

Thus

$$ {}^{H}\mathbb{D}_{a+}^{\alpha,\beta;\psi} f = {}^{C}D_{a+}^{\mu;\psi} g(x) = {}^{C}D_{a+}^{\mu;\psi} \left[I_{a+}^{(1-\beta)(n-\alpha);\psi} f(x) \right]. \tag{10.26} $$

Assume that $w(x) = I_{b-}^{(1-\beta)(n-\alpha);\psi} f(x) \in C^n([a,b])$. Hence

$$ {}^{H}\mathbb{D}_{b-}^{\alpha,\beta;\psi} f(x) = I_{b-}^{\beta(n-\alpha);\psi} \left(-\frac{1}{\psi'(x)} \frac{d}{dx} \right)^n w(x) = I_{b-}^{n-\mu;\psi} \left(-\frac{1}{\psi'(x)} \frac{d}{dx} \right)^n w(x). \tag{10.27} $$

Thus

$$ {}^{H}\mathbb{D}_{b-}^{\alpha,\beta;\psi} f = {}^{C}D_{b-}^{\mu;\psi} w(x) = {}^{C}D_{b-}^{\mu;\psi} \left(I_{b-}^{(1-\beta)(n-\alpha);\psi} f(x) \right). \tag{10.28} $$

We mention the simplified ψ-Hilfer fractional Taylor formulae:

Theorem 10.3 *(see also [7]) Let $\psi, f \in C^n([a,b])$, with ψ being increasing such that $\psi'(x) \neq 0$ over $[a,b]$, where $n-1 < \alpha < n$, $0 \leq \beta \leq 1$, and $\gamma = \alpha + \beta(n-\alpha)$, $x \in [a,b]$. Then*

$$ f(x) - \sum_{k=1}^{n-1} \frac{(\psi(x) - \psi(a))^{\gamma-k}}{\Gamma(\gamma-k+1)} f_{\psi}^{[n-k]} \left(I_{a+}^{(1-\beta)(n-\alpha);\psi} f \right)(a) = $$

$$ \frac{1}{\Gamma(\alpha)} \int_a^x \psi'(t) (\psi(x) - \psi(t))^{\alpha-1} \, {}^{H}\mathbb{D}_{a+}^{\alpha,\beta;\psi} f(t) \, dt, \tag{10.29} $$

and

$$ f(x) - \sum_{k=1}^{n-1} \frac{(-1)^k (\psi(b) - \psi(x))^{\gamma-k}}{\Gamma(\gamma-k+1)} f_{\psi}^{[n-k]} \left(I_{b-}^{(1-\beta)(n-\alpha);\psi} f \right)(b) = $$

$$ \frac{1}{\Gamma(\alpha)} \int_x^b \psi'(t) (\psi(t) - \psi(x))^{\alpha-1} \, {}^{H}\mathbb{D}_{b-}^{\alpha,\beta;\psi} f(t) \, dt. \tag{10.30} $$

Here notice that $\left(I_{a+}^{(1-\beta)(n-\alpha);\psi} f \right)(a) = \left(I_{b-}^{(1-\beta)(n-\alpha);\psi} f \right)(b) = 0$.

We also mention the following alternative ψ-Hilfer fractional Taylor formulae:

Theorem 10.4 *([2]) Let $f, \psi \in C^n([a,b])$, with ψ being increasing, $\psi'(x) \neq 0$ over $[a,b] \subset \mathbb{R}$, $\alpha > 0 : \lceil \alpha \rceil = n$, $0 \leq \beta \leq 1$, $\mu = n(1-\beta) + \beta\alpha$. Assume that $g(x) = I_{a+}^{(1-\beta)(n-\alpha);\psi} f(x)$, $w(x) = I_{b-}^{(1-\beta)(n-\alpha);\psi} f(x) \in C^n([a,b])$.*

Then

1)

$$I_{a+}^{\mu;\psi} \, {}^H\mathbb{D}_{a+}^{\alpha,\beta;\psi} f(x) = g(x) - \sum_{k=0}^{n-1} \frac{g_{\psi}^{[k]}(a)}{k!} (\psi(x) - \psi(a))^k, \tag{10.31}$$

where

$$g_{\psi}^{[k]}(x) = \left(\frac{1}{\psi'(x)} \frac{d}{dx} \right)^k g(x), \quad k = 0, 1, ..., n-1,$$

and

2)

$$I_{b-}^{\mu;\psi} \, {}^H\mathbb{D}_{b-}^{\alpha,\beta;\psi} f(x) = w(x) - \sum_{k=0}^{n-1} \frac{(-1)^k w_{\psi}^{[k]}(b)}{k!} (\psi(b) - \psi(x))^k, \tag{10.32}$$

where

$$w_{\psi}^{[k]}(x) = \left(\frac{1}{\psi'(x)} \frac{d}{dx} \right)^k w(x), \quad k = 0, 1, ..., n-1; \; x \in [a,b].$$

Next we list two Hilfer fractional derivatives representation formulae:

Theorem 10.5 *([2]) Let $\alpha > 0$, $\alpha \notin \mathbb{N}$, $\lceil \alpha \rceil = n$, $0 < \beta < 1$; $f \in C^n([a,b])$, $[a,b] \subset \mathbb{R}$; and set $\gamma = \alpha + \beta(n-\alpha)$. Assume further that $\Delta_{a+}^{\gamma} f \in C([a,b]) : \Delta_{a+}^{\gamma - j} f(a) = 0$, for $j = 1, ..., n$. Let also $\overline{\alpha} > 0 : \lceil \overline{\alpha} \rceil = \overline{n}$, with $\overline{\gamma} = \overline{\alpha} + \beta(\overline{n} - \overline{\alpha})$, and assume that $\alpha > \overline{\alpha}$ and $\gamma > \overline{\gamma}$. Then*

$${}^H\mathbb{D}_{a+}^{\overline{\alpha},\beta} f(x) = \frac{1}{\Gamma(\alpha - \overline{\alpha})} \int_a^x (x-t)^{\alpha - \overline{\alpha} - 1} \, {}^H\mathbb{D}_{a+}^{\alpha,\beta} f(t) \, dt, \tag{10.33}$$

$\forall x \in [a,b]$,

furthermore ${}^H\mathbb{D}_{a+}^{\overline{\alpha},\beta} f \in AC([a,b])$ (absolutely continuous functions) if $\alpha - \overline{\alpha} \geq 1$ and ${}^H\mathbb{D}_{a+}^{\overline{\alpha},\beta} f \in C([a,b])$ if $\alpha - \overline{\alpha} \in (0,1)$.

Theorem 10.6 *([2]) Let $\alpha > 0$, $\alpha \notin \mathbb{N}$, $\lceil \alpha \rceil = n$, $0 < \beta < 1$; $f \in C^n([a,b])$, $[a,b] \subset \mathbb{R}$; and set $\gamma = \alpha + \beta(n-\alpha)$. Assume further that $\Delta_{b-}^{\gamma} f \in C([a,b]) : \Delta_{b-}^{\gamma - j} f(b) = 0$, $j = 1, ..., n$. Let also $\overline{\alpha} > 0 : \lceil \overline{\alpha} \rceil = \overline{n}$, with $\overline{\gamma} = \overline{\alpha} + \beta(\overline{n} - \overline{\alpha})$, and assume that $\alpha > \overline{\alpha}$ and $\gamma > \overline{\gamma}$. Then*

$${}^H\mathbb{D}_{b-}^{\overline{\alpha},\beta} f(x) = \frac{1}{\Gamma(\alpha - \overline{\alpha})} \int_x^b (t-x)^{\alpha - \overline{\alpha} - 1} \, {}^H\mathbb{D}_{b-}^{\alpha,\beta} f(t) \, dt, \tag{10.34}$$

$\forall\, x \in [a, b]$,

furthermore ${}^{H}\mathbb{D}_{b-}^{\overline{\alpha},\beta} f \in AC([a, b])$ *if* $\alpha - \overline{\alpha} \geq 1$ *and* ${}^{H}\mathbb{D}_{b-}^{\overline{\alpha},\beta} f \in C([a, b])$ *if* $\alpha - \overline{\alpha} \in (0, 1)$.

The fractional integral operator $I_{a+}^{\alpha} f$ and $I_{b-}^{\alpha} f$, $\alpha > 0$, are bounded in $L_p(a, b)$, $1 \leq p \leq \infty$, that is

$$\left\| I_{a+}^{\alpha} f \right\|_p \leq K \|f\|_p, \quad \left\| I_{b-}^{\alpha} f \right\|_p \leq K \|f\|_p, \tag{10.35}$$

where

$$K = \frac{(b-a)^{\alpha}}{\Gamma(\alpha+1)}. \tag{10.36}$$

The left inequality (10.35) was proved by H.G. Hardy in one of his first papers, see [10].

We continue this Background section with the following material from [3], where the author introduced the genralized ψ-Prabhakar type of fractional calculus and mixed it with the ψ-Hilfer fractional calculus.

So we consider the Prabhakar function (also known as the three parameter Mittag-Laffler function), (see [9], p. 97; [8])

$$E_{\alpha,\beta}^{\gamma}(z) = \sum_{k=0}^{\infty} \frac{(\gamma)_k}{k!\Gamma(\alpha k + \beta)} z^k, \tag{10.37}$$

where Γ is the gamma function; $\alpha, \beta, \gamma \in \mathbb{R} : \alpha, \beta > 0, z \in \mathbb{R}$, and $(\gamma)_k = \gamma(\gamma+1) \ldots (\gamma+k-1)$. It is $E_{\alpha,\beta}^{0}(z) = \frac{1}{\Gamma(\beta)}$.

Let $a, b \in \mathbb{R}$, $a < b$ and $x \in [a, b]$; $f \in C([a, b])$. Let also $\psi \in C^1([a, b])$ which is increasing. The left and right Prabhakar fractional integrals with respect to ψ are defined as follows:

$$\left(e_{\rho,\mu,\omega,a+}^{\gamma;\psi} f\right)(x) = \int_a^x \psi'(t) (\psi(x) - \psi(t))^{\mu-1} E_{\rho,\mu}^{\gamma} \left[\omega (\psi(x) - \psi(t))^{\rho}\right] f(t)\, dt, \tag{10.38}$$

and

$$\left(e_{\rho,\mu,\omega,b-}^{\gamma;\psi} f\right)(x) = \int_x^b \psi'(t) (\psi(t) - \psi(x))^{\mu-1} E_{\rho,\mu}^{\gamma} \left[\omega (\psi(t) - \psi(x))^{\rho}\right] f(t)\, dt, \tag{10.39}$$

where $\rho, \mu > 0$; $\gamma, \omega \in \mathbb{R}$.

Functions (10.38) and (10.39) are continuous ([3]).

Next, additionally, assume that $\psi'(x) \neq 0$ over $[a, b]$.

Let $\psi, f \in C^N([a, b])$, where $N = \lceil \mu \rceil$, ($\lceil \cdot \rceil$ is the ceiling of the number), $0 < \mu \notin \mathbb{N}$. We define the ψ-Prabhakar-Caputo left and right fractional derivatives of order μ as follows ($x \in [a, b]$):

$$\left({}^{C}D^{\gamma;\psi}_{\rho,\mu,\omega,a+}f\right)(x) = \int_a^x \psi'(t)\left(\psi(x)-\psi(t)\right)^{N-\mu-1}$$

$$E^{-\gamma}_{\rho,N-\mu}\left[\omega\left(\psi(x)-\psi(t)\right)^{\rho}\right]\left(\frac{1}{\psi'(t)}\frac{d}{dt}\right)^{N} f(t)\,dt, \tag{10.40}$$

and

$$\left({}^{C}D^{\gamma;\psi}_{\rho,\mu,\omega,b-}f\right)(x) = (-1)^N \int_x^b \psi'(t)\left(\psi(t)-\psi(x)\right)^{N-\mu-1}$$

$$E^{-\gamma}_{\rho,N-\mu}\left[\omega\left(\psi(t)-\psi(x)\right)^{\rho}\right]\left(\frac{1}{\psi'(t)}\frac{d}{dt}\right)^{N} f(t)\,dt. \tag{10.41}$$

One can write these [(see (10.40), (10.41)] as

$$\left({}^{C}D^{\gamma;\psi}_{\rho,\mu,\omega,a+}f\right)(x) = \left(e^{-\gamma;\psi}_{\rho,N-\mu,\omega,a+}f^{[N]}_{\psi}\right)(x), \tag{10.42}$$

and

$$\left({}^{C}D^{\gamma;\psi}_{\rho,\mu,\omega,b-}f\right)(x) = (-1)^N\left(e^{-\gamma;\psi}_{\rho,N-\mu,\omega,b-}f^{[N]}_{\psi}\right)(x), \tag{10.43}$$

where

$$f^{[N]}_{\psi}(x) = f^{(N)}_{\psi}f(x) := \left(\frac{1}{\psi'(x)}\frac{d}{dx}\right)^{N} f(x), \tag{10.44}$$

$\forall\, x \in [a,b]$.

Functions (10.42) and (10.43) are continuous on $[a,b]$.

Next we define the ψ-Prabhakar-Riemann Liouville left and right fractional derivatives of order μ as follows ($x \in [a,b]$):

$$\left({}^{RL}D^{\gamma;\psi}_{\rho,\mu,\omega,a+}f\right)(x) = \left(\frac{1}{\psi'(x)}\frac{d}{dx}\right)^{N}\int_a^x \psi'(t)\left(\psi(x)-\psi(t)\right)^{N-\mu-1}$$

$$E^{-\gamma}_{\rho,N-\mu}\left[\omega\left(\psi(x)-\psi(t)\right)^{\rho}\right]f(t)\,dt, \tag{10.45}$$

and

$$\left({}^{RL}D^{\gamma;\psi}_{\rho,\mu,\omega,b-}f\right)(x) = \left(-\frac{1}{\psi'(x)}\frac{d}{dx}\right)^{N}\int_x^b \psi'(t)\left(\psi(t)-\psi(x)\right)^{N-\mu-1}$$

$$E^{-\gamma}_{\rho,N-\mu}\left[\omega\left(\psi(t)-\psi(x)\right)^{\rho}\right]f(t)\,dt. \tag{10.46}$$

That is we have

$$\left({}^{RL}D^{\gamma;\psi}_{\rho,\mu,\omega,a+}f\right)(x)=\left(\frac{1}{\psi'(x)}\frac{d}{dx}\right)^{N}\left(e^{-\gamma;\psi}_{\rho,N-\mu,\omega,a+}f\right)(x), \tag{10.47}$$

and

$$\left({}^{RL}D^{\gamma;\psi}_{\rho,\mu,\omega,b-}f\right)(x)=\left(-\frac{1}{\psi'(x)}\frac{d}{dx}\right)^{N}\left(e^{-\gamma;\psi}_{\rho,N-\mu,\omega,b-}f\right)(x), \tag{10.48}$$

$\forall\, x\in[a,b]$.

We define also the ψ-Hilfer-Prabhakar left and right fractional derivatives of order μ and type $0\leq\beta\leq1$, as follows

$$\left({}^{H}\mathbb{D}^{\gamma,\beta;\psi}_{\rho,\mu,\omega,a+}f\right)(x)=e^{-\gamma\beta;\psi}_{\rho,\beta(N-\mu),\omega,a+}\left(\frac{1}{\psi'(x)}\frac{d}{dx}\right)^{N}e^{-\gamma(1-\beta);\psi}_{\rho,(1-\beta)(N-\mu),\omega,a+}f(x), \tag{10.49}$$

and

$$\left({}^{H}\mathbb{D}^{\gamma,\beta;\psi}_{\rho,\mu,\omega,b-}f\right)(x)=e^{-\gamma\beta;\psi}_{\rho,\beta(N-\mu),\omega,b-}\left(-\frac{1}{\psi'(x)}\frac{d}{dx}\right)^{N}e^{-\gamma(1-\beta);\psi}_{\rho,(1-\beta)(N-\mu),\omega,b-}f(x), \tag{10.50}$$

$\forall\, x\in[a,b]$.

When $\beta=0$, we get the Riemann-Liouville version, and when $\beta=1$, we get the Caputo version.

We call $\xi=\mu+\beta(N-\mu)$, we have that $N-1<\mu\leq\mu+\beta(N-\mu)\leq\mu+N-\mu=N$, hence $\lceil\xi\rceil=N$.

We can easily write that

$$\left({}^{H}\mathbb{D}^{\gamma,\beta;\psi}_{\rho,\mu,\omega,a+}f\right)(x)=e^{-\gamma\beta;\psi}_{\rho,\xi-\mu,\omega,a+}\ {}^{RL}D^{\gamma(1-\beta);\psi}_{\rho,\xi,\omega,a+}f(x), \tag{10.51}$$

and

$$\left({}^{H}\mathbb{D}^{\gamma,\beta;\psi}_{\rho,\mu,\omega,b-}f\right)(x)=e^{-\gamma\beta;\psi}_{\rho,\xi-\mu,\omega,b-}\ {}^{RL}D^{\gamma(1-\beta);\psi}_{\rho,\xi,\omega,b-}f(x), \tag{10.52}$$

$\forall\, x\in[a,b]$.

In this chapter we prove univariate and multivariate Hardy type inequalities based on the above mentioned fractional background and convexity of functions. Our chapter is inspired by [4, 6, 10, 12, 13].

10.2 Prerequisites

(I) Here we follow [6], p. 441, see Chap. 22.

Let $(\Omega_1,\Sigma_1,\mu_1)$ and $(\Omega_2,\Sigma_2,\mu_2)$ be measure spaces with positive σ-finite measures, and let $k_i:\Omega_1\times\Omega_2\to\mathbb{R}$ be nonnegative measurable functions, $k_i(x,\cdot)$ measurable on Ω_2, and

$$K_i(x) = \int_{\Omega_2} k_i(x, y)\, d\mu_2(y), \text{ for any } x \in \Omega_1, \tag{10.53}$$

$i = 1, ..., m \in \mathbb{N}$. We assume that $K_i(x) > 0$ a.e. on Ω_1 and the weight functions are nonnegative measurable functions on the related set.

We consider measurable functions $g_i : \Omega_1 \to \mathbb{R}$ with the representation

$$g_i(x) = \int_{\Omega_2} k_i(x, y)\, f_i(y)\, d\mu_2(y), \tag{10.54}$$

where $f_i : \Omega_2 \to \mathbb{R}$ are measurable functions, $i = 1, ..., m$.

Here u stands for a weight function on Ω_1 ($u \geq 0$, which is measurable).

We will use the following general result:

Theorem 10.7 *([6], p. 442) Assume that the functions $(i = 1, 2, ..., m \in \mathbb{N})$ $x \to \left(u(x)\frac{k_i(x,y)}{K_i(x)}\right)$ are integrable on Ω_1, for each fixed $y \in \Omega_2$. Define u_i on Ω_2 by*

$$u_i(y) := \int_{\Omega_1} u(x)\frac{k_i(x, y)}{K_i(x)}\, d\mu_1(x) < \infty. \tag{10.55}$$

Let $p_i > 1 : \sum_{i=1}^{m} \frac{1}{p_i} = 1$. Let the functions $\Phi_i : \mathbb{R}_+ \to \mathbb{R}_+$, $i = 1, ..., m$, be convex and increasing.

Then

$$\int_{\Omega_1} u(x) \prod_{i=1}^{m} \Phi_i\left(\left|\frac{g_i(x)}{K_i(x)}\right|\right) d\mu_1(x) \leq$$

$$\prod_{i=1}^{m} \left(\int_{\Omega_2} u_i(y)\, \Phi_i(|f_i(y)|)^{p_i}\, d\mu_2(y)\right)^{\frac{1}{p_i}}, \tag{10.56}$$

for all measurable functions $f_i : \Omega_2 \to \mathbb{R}$ $(i = 1, ..., m)$ such that

(i) f_i, $\Phi_i(|f_i|)^{p_i}$, are both $k_i(x, y)\, d\mu_2(y)$ - integrable, μ_1-a.e. in $x \in \Omega_1$, $i = 1, ..., m$,

(ii) $u_i \Phi_i(|f_i|)^{p_i}$ is μ_2-integrable, $i = 1, ..., m$,

and for all corresponding functions g_i $(i = 1, ..., m)$ given by (10.54).

(II) Here we follow [6], Chap. 27.

The basic setting follows:

Let $(\Omega_1, \Sigma_1, \mu_1)$ and $(\Omega_2, \Sigma_2, \mu_2)$ be measure spaces with positive σ-finite measures, and let $k : \Omega_1 \times \Omega_2 \to \mathbb{R}$ be nonnegative measurable functions, $k(x, \cdot)$ measurable on Ω_2, and

$$K(x) = \int_{\Omega_2} k(x, y)\, d\mu_2(y), \text{ for any } x \in \Omega_1, \tag{10.57}$$

$i = 1, \ldots, m \in \mathbb{N}$. We assume that $K(x) > 0$ a.e. on Ω_1 and the weight functions are nonnegative measurable functions on the related set.

We consider measurable functions $g_i : \Omega_1 \to \mathbb{R}$ with the representation

$$g_i(x) = \int_{\Omega_2} k(x, y) f_i(y) \, d\mu_2(y), \tag{10.58}$$

where $f_i : \Omega_2 \to \mathbb{R}$ are measurable functions, $i = 1, \ldots, n$.

Denote by $\vec{x} = x := (x_1, \ldots, x_n) \in \mathbb{R}^n$, $\vec{g} := (g_1, \ldots, g_n)$ and $\vec{f} := (f_1, \ldots, f_n)$.

We consider here $\Phi : \mathbb{R}_+^n \to \mathbb{R}$ a convex function, which is increasing per coordinate, i.e. if $x_i \leq y_i$, $i = 1, \ldots, n$, then $\Phi(x_1, \ldots, x_n) \leq \Phi(y_1, \ldots, y_n)$.

Next we may write

$$\vec{g}(x) = \int_{\Omega_2} k(x, y) \vec{f}(y) \, d\mu_2(y), \tag{10.59}$$

which means

$$(g_1(x), \ldots, g_n(x)) = \left(\int_{\Omega_2} k(x, y) f_1(y) \, d\mu_2(y), \ldots, \int_{\Omega_2} k(x, y) f_1(y) \, d\mu_2(y) \right). \tag{10.60}$$

Similarly, we may write

$$|\vec{g}(x)| = \left| \int_{\Omega_2} k(x, y) \vec{f}(y) \, d\mu_2(y) \right|, \tag{10.61}$$

and we mean

$$(|g_1(x)|, \ldots, |g_n(x)|) = \left(\left| \int_{\Omega_2} k(x, y) f_1(y) \, d\mu_2(y) \right|, \ldots, \left| \int_{\Omega_2} k(x, y) f_n(y) \, d\mu_2(y) \right| \right). \tag{10.62}$$

We also can write that

$$|\vec{g}(x)| \leq \int_{\Omega_2} k(x, y) \left| \vec{f}(y) \right| d\mu_2(y), \tag{10.63}$$

and we mean the fact that

$$|g_i(x)| \leq \int_{\Omega_2} k(x, y) |f_i(y)| \, d\mu_2(y), \tag{10.64}$$

for all $i = 1, \ldots, n$, etc.

More precisely here we follow:

Let $(\Omega_1, \Sigma_1, \mu_1)$ and $(\Omega_2, \Sigma_2, \mu_2)$ be measure spaces with positive σ-finite measures, and let $k_j : \Omega_1 \times \Omega_2 \to \mathbb{R}$ be nonnegative measurable functions, $k_j(x, \cdot)$ measurable on Ω_2, and

$$K_j(x) = \int_{\Omega_2} k_j(x, y)\, d\mu_2(y), \quad x \in \Omega_1, \ j = 1, ..., m. \tag{10.65}$$

We suppose that $K_j(x) > 0$ a.e. on Ω_1. Let the measurable functions $g_{ji} : \Omega_1 \to \mathbb{R}$ with the representation

$$g_{ji}(x) = \int_{\Omega_2} k_j(x, y)\, f_{ji}(y)\, d\mu_2(y),$$

written also as

$$\vec{g}_j(x) = \int_{\Omega_2} k_j(x, y)\, \vec{f}_j(y)\, d\mu_2(y), \tag{10.66}$$

where $f_{ji} : \Omega_2 \to \mathbb{R}$ are measurable functions, $i = 1, ..., n$ and $j = 1, ..., m$.

We denote above the function vectors $\vec{g}_j := (g_{j1}, g_{j2}, ..., g_{jn})$ and $\vec{f}_j := (f_{j1}, ..., f_{jn})$, $j = 1, ..., m$.

We say $\vec{f}_j$ is integrable with respect to measure μ, iff all f_{ji} are integrable with respect to μ.

We also consider here $\Phi_j : \mathbb{R}_+^n \to \mathbb{R}$, $j = 1, ..., m$, convex functions that are increasing per coordinate. Again u is a weight function on Ω_1.

We will use the following theorem

Theorem 10.8 *([6], p. 628) Assume that the functions ($j = 1, 2, ..., m \in \mathbb{N}$) $x \to \left(u(x) \frac{k_j(x,y)}{K_j(x)}\right)$ are integrable on Ω_1, for each fixed $y \in \Omega_2$. Define u_j on Ω_2 by*

$$u_j(y) := \int_{\Omega_1} u(x) \frac{k_j(x, y)}{K_j(x)} d\mu_1(x) < \infty. \tag{10.67}$$

Let $p_j > 1 : \sum_{j=1}^{m} \frac{1}{p_j} = 1$. Let the functions $\Phi_j : \mathbb{R}_+^n \to \mathbb{R}_+$, $j = 1, ..., m$, be convex and increasing per coordinate.

Then

$$\int_{\Omega_1} u(x) \prod_{i=1}^{m} \Phi_j\left(\left|\frac{\vec{g}_j(x)}{K_j(x)}\right|\right) d\mu_1(x) \leq$$

$$\prod_{j=1}^{m} \left(\int_{\Omega_2} u_j(y)\, \Phi_j\left(\left|\vec{f}_j(y)\right|\right)^{p_j} d\mu_2(y)\right)^{\frac{1}{p_j}}, \tag{10.68}$$

under the assumptions:

(i) $\vec{f}_j, \Phi_j\left(\left|\vec{f}_j\right|\right)^{p_j}$, *are both* $k_j(x,y)\,d\mu_2(y)$ *- integrable,* μ_1*-a.e. in* $x \in \Omega_1$, $j = 1, ..., m$,

(ii) $u_j \Phi_j\left(\left|\vec{f}_j\right|\right)^{p_j}$ *is* μ_2*-integrable,* $j = 1, ..., m$.

(III) We will also use from [6], Chap. 26, the following theorem:

Theorem 10.9 *([6], p. 598) Let* $\rho \in \{1, ..., m\}$ *be fixed. Assume that the function* $x \to \left(\frac{u(x)\prod_{j=1}^{m} k_j(x,y)}{\prod_{j=1}^{m} K_j(x)}\right)$ *is integrable on* Ω_1*, for each* $y \in \Omega_2$*. Define* λ_m *on* Ω_2 *by*

$$\lambda_m(y) := \int_{\Omega_1} \left(\frac{u(x)\prod_{j=1}^{m} k_j(x,y)}{\prod_{j=1}^{m} K_j(x)}\right) d\mu_1(x) < \infty. \tag{10.69}$$

Let the functions $\Phi_j : \mathbb{R}_+^n \to \mathbb{R}_+$, $j = 1, ..., m$, *be convex and increasing per coordinate. Then*

$$\int_{\Omega_1} u(x) \prod_{j=1}^{m} \Phi_j\left(\left|\frac{\vec{g}_j(x)}{K_j(x)}\right|\right) d\mu_1(x) \leq \tag{10.70}$$

$$\left(\prod_{\substack{j=1\\ j\neq\rho}}^{m} \int_{\Omega_2} \Phi_j\left(\left|\vec{f}_j(y)\right|\right) d\mu_2(y)\right)\left(\int_{\Omega_2} \Phi_\rho\left(\left|\vec{f}_\rho(y)\right|\right) \lambda_m(y)\, d\mu_2(y)\right),$$

under the assumptions:

(i) $\vec{f}_j, \Phi_j\left(\left|\vec{f}_j\right|\right)$, *are* $k_j(x,y)\,d\mu_2(y)$ *- integrable,* μ_1*-a.e. in* $x \in \Omega_1$, $j = 1, ..., m$,

(ii) $\lambda_m \Phi_\rho\left(\left|\vec{f}_\rho\right|\right)$; $\Phi_1\left(\left|\vec{f}_1\right|\right)$, $\Phi_2\left(\left|\vec{f}_2\right|\right)$, $\Phi_3\left(\left|\vec{f}_3\right|\right)$, ..., $\widehat{\Phi_\rho\left(\left|\vec{f}_\rho\right|\right)}$, ..., $\Phi_m\left(\left|\vec{f}_m\right|\right)$, *are all* μ_2*-integrable,*

and for all corresponding functions g_i *given by (10.54). Above* $\widehat{\Phi_\rho\left(\left|\vec{f}_\rho\right|\right)}$ *means a missing item.*

Above all symbols are as in (II).

10.3 Main Results

(I)' Here we apply Theorem 10.7.

Let here $p_i > 1 : \sum_{i=1}^{m} \frac{1}{p_i} = 1$.

We present

Theorem 10.10 *Here $i = 1, ..., m$. Let $\alpha_i > 0$, $\alpha_i \notin \mathbb{N}$, $\lceil \alpha_i \rceil = n_i$, $0 < \beta_i < 1$; $f_i \in C^{n_i}([a,b])$, $[a,b] \subset \mathbb{R}$; and set $\gamma_i = \alpha_i + \beta_i(n_i - \alpha_i)$. Assume further that $\Delta_{a+}^{\gamma_i} f_i \in C([a,b]) : \Delta_{a+}^{\gamma_i - j_i} f_i(a) = 0$, for $j_i = 1, ..., n_i$. Let also $\overline{\alpha}_i > 0 : \lceil \overline{\alpha}_i \rceil = \overline{n}_i$, with $\overline{\gamma}_i = \overline{\alpha}_i + \beta_i(\overline{n}_i - \overline{\alpha}_i)$, and assume that $\alpha_i > \overline{\alpha}_i$ and $\gamma_i > \overline{\gamma}_i$.*

Let also $\Phi_i : \mathbb{R}_+ \to \mathbb{R}_+$, $i = 1, ..., m$, be convex and increasing functions; $u \geq 0$ is a weight measurable function on $[a,b]$. We assume that

$$u_i(y) = (\alpha_i - \overline{\alpha}_i) \int_y^b u(x) \frac{(x-y)^{(\alpha_i - \overline{\alpha}_i)-1}}{(x-a)^{(\alpha_i - \overline{\alpha}_i)}} dx < \infty, \tag{10.71}$$

for all $a < y < b$ and u_i is Lebesgue integrable.

Then

$$\int_a^b u(x) \prod_{i=1}^m \Phi_i \left(\frac{\left| {}^H\mathbb{D}_{a+}^{\overline{\alpha}_i, \beta_i} f_i(x) \right|}{(x-a)^{\alpha_i - \overline{\alpha}_i}} \Gamma(\alpha_i - \overline{\alpha}_i + 1) \right) dx \leq$$

$$\prod_{i=1}^m \left(\int_a^b u_i(y) \left(\Phi_i \left(\left| {}^H\mathbb{D}_{a+}^{\alpha_i, \beta_i} f_i(y) \right| \right) \right)^{p_i} dy \right)^{\frac{1}{p_i}}. \tag{10.72}$$

Proof By Theorems 10.5, 10.7 and from [4], pp. 31–33, see relations there (2.40)–(2.47). ∎

Remark 10.11 (to Theorem 10.10) One can have Φ_i =identity map or e^x, or $\Phi_i(x) = x^{\overline{p}_i}$, $x \in \mathbb{R}_+$, $\overline{p}_i > 1$, etc.

To save space in this chapter we skip these interesting applications here and later.

We continue with

Theorem 10.12 *Here $i = 1, ..., m$. Let $\alpha_i > 0$, $\alpha_i \notin \mathbb{N}$, $\lceil \alpha_i \rceil = n_i$, $0 < \beta_i < 1$; $f_i \in C^{n_i}([a,b])$, $[a,b] \subset \mathbb{R}$; and set $\gamma_i = \alpha_i + \beta_i(n_i - \alpha_i)$. Assume further that $\Delta_{b-}^{\gamma_i} f_i \in C([a,b]) : \Delta_{b-}^{\gamma_i - j_i} f_i(b) = 0$, $j_i = 1, ..., n_i$. Let also $\overline{\alpha}_i > 0 : \lceil \overline{\alpha}_i \rceil = \overline{n}_i$, with $\overline{\gamma}_i = \overline{\alpha}_i + \beta_i(\overline{n}_i - \overline{\alpha}_i)$, and assume that $\alpha_i > \overline{\alpha}_i$ and $\gamma_i > \overline{\gamma}_i$. Let also $\Phi_i : \mathbb{R}_+ \to \mathbb{R}_+$, $i = 1, ..., m$, be convex and increasing functions; $u \geq 0$ is a weight measurable function on $[a,b]$. We assume that*

$$\overline{u}_i(y) := (\alpha_i - \overline{\alpha}_i) \int_a^y u(x) \frac{(y-x)^{(\alpha_i - \overline{\alpha}_i)-1}}{(b-x)^{(\alpha_i - \overline{\alpha}_i)}} dx < \infty, \tag{10.73}$$

for all $a < y < b$ and $\overline{u}_i$ is Lebesgue integrable.

Then

$$\int_a^b u(x) \prod_{i=1}^m \Phi_i \left(\frac{\left| {}^H\mathbb{D}_{b-}^{\overline{\alpha}_i, \beta_i} f_i(x) \right|}{(b-x)^{\alpha_i - \overline{\alpha}_i}} \Gamma(\alpha_i - \overline{\alpha}_i + 1) \right) dx \leq$$

$$\prod_{i=1}^{m}\left(\int_a^b \overline{u}_i(y)\left(\Phi_i\left(\left|{}^H\mathbb{D}_{b-}^{\alpha_i,\beta_i} f_i(y)\right|\right)\right)^{p_i} dy\right)^{\frac{1}{p_i}}. \tag{10.74}$$

Proof By Theorems 10.6, 10.7 and from [4], pp. 35–37, see relations there (2.58)–(2.67). ∎

We continue with

Theorem 10.13 *Here* $i = 1, ..., m$. *Let* $f_i \in C^{n_i}([a,b])$, $\theta := \max\{n_1, ..., n_m\}$, $\psi \in C^{\theta}([a,b])$, *with* ψ *being increasing:* $\psi'(x) \neq 0$ *over* $[a,b]$, *where* $n_i - 1 < \alpha_i < n_i$, $0 \leq \beta_i \leq 1$, *and* $\gamma_i = \alpha_i + \beta_i(n_i - \alpha_i)$.

Assume that $f_{i\psi}^{[n_i-k_i]}\left(I_{a+}^{(1-\beta_i)(n_i-\alpha_i);\psi} f_i\right)(a) = 0$, $k_i = 1, ..., n_i - 1$; $i = 1, ..., m$.

Let also $\Phi_i : \mathbb{R}_+ \to \mathbb{R}_+$, $i = 1, ..., m$, *be convex and increasing functions;* $u \geq 0$ *is a weight measurable function on* $[a,b]$. *We assume that*

$$u_i^{\psi}(y) = \alpha_i \psi'(y) \int_y^b u(x) \frac{(\psi(x) - \psi(y))^{\alpha_i - 1}}{(\psi(x) - \psi(a))^{\alpha_i}} dx < \infty, \tag{10.75}$$

for all $a < y < b$ *and* u_i^{ψ} *is Lebesgue integrable.*

Then

$$\int_a^b u(x) \prod_{i=1}^{m} \Phi_i\left(\frac{|f_i(x)|}{(\psi(x) - \psi(a))^{\alpha_i}} \Gamma(\alpha_i + 1)\right) dx \leq$$

$$\prod_{i=1}^{m}\left(\int_a^b u_i^{\psi}(y)\, \Phi_i\left(\left|\left({}^H\mathbb{D}_{a+}^{\alpha_i,\beta_i;\psi} f_i\right)(y)\right|\right)^{p_i} dy\right)^{\frac{1}{p_i}}, \tag{10.76}$$

true for continuous ${}^H\mathbb{D}_{a+}^{\alpha_i,\beta_i;\psi} f_i$, $i = 1, ..., m$.

Proof From (10.29) we get:

$$f_i(x) = \frac{1}{\Gamma(\alpha_i)} \int_a^x \psi'(t)(\psi(x) - \psi(t))^{\alpha_i - 1}\, {}^H\mathbb{D}_{a+}^{\alpha_i,\beta_i;\psi} f_i(t)\, dt, \tag{10.77}$$

$\forall\, x \in [a,b]$; $i = 1, ..., m$.

Then we apply Theorem 10.7, along with [4], pp. 47–49, see the relations there (2.107)–(2.119). ∎

We also give

Theorem 10.14 *Here* $i = 1, ..., m$. *Let* $f_i \in C^{n_i}([a,b])$, $\theta := \max\{n_1, ..., n_m\}$, $\psi \in C^{\theta}([a,b])$, *with* ψ *being increasing:* $\psi'(x) \neq 0$ *over* $[a,b]$, *where* $n_i - 1 < \alpha_i < n_i$, $0 \leq \beta_i \leq 1$, *and* $\gamma_i = \alpha_i + \beta_i(n_i - \alpha_i)$.

Assume that $f_{i\psi}^{[n_i-k_i]}\left(I_{b-}^{(1-\beta_i)(n_i-\alpha_i);\psi} f_i\right)(b) = 0$, $k_i = 1, ..., n_i - 1$; $i = 1, ..., m$.

Let also $\Phi_i : \mathbb{R}_+ \to \mathbb{R}_+$, $i = 1, ..., m$, *be convex and increasing functions;* $u \geq 0$ *is a weight measurable function on* $[a, b]$. *We assume that*

$$\overline{u}_i^{\psi}(y) = \alpha_i \psi'(y) \int_a^y u(x) \frac{(\psi(y) - \psi(x))^{\alpha_i - 1}}{(\psi(b) - \psi(x))^{\alpha_i}} dx < \infty, \tag{10.78}$$

for all $a < y < b$ *and* $\overline{u}_i^{\psi}$ *is Lebesgue integrable.*

Then

$$\int_a^b u(x) \prod_{i=1}^m \Phi_i \left(\frac{|f_i(x)|}{(\psi(b) - \psi(x))^{\alpha_i}} \Gamma(\alpha_i + 1) \right) dx \leq$$

$$\prod_{i=1}^m \left(\int_a^b \overline{u}_i^{\psi}(y) \Phi_i \left(\left| \left({}^H\mathbb{D}_{b-}^{\alpha_i, \beta_i; \psi} f_i \right)(y) \right| \right)^{p_i} dy \right)^{\frac{1}{p_i}}, \tag{10.79}$$

true for continuous ${}^H\mathbb{D}_{b-}^{\alpha_i, \beta_i; \psi} f_i$, $i = 1, ..., m$.

Proof From (10.30) we get:

$$f_i(x) = \frac{1}{\Gamma(\alpha_i)} \int_x^b \psi'(t) (\psi(t) - \psi(x))^{\alpha_i - 1} \, {}^H\mathbb{D}_{b-}^{\alpha_i, \beta_i; \psi} f_i(t) \, dt, \tag{10.80}$$

$\forall \, x \in [a, b]$; $i = 1, ..., m$.

Then we apply Theorem 10.7, along with [4], pp. 51–53, see the relations there (2.132)–(2.142). ■

We present

Theorem 10.15 *Here* $i = 1, ..., m$. *Let* $f_i \in C^{n_i}([a, b])$, $\theta := \max\{n_1, ..., n_m\}$, $\psi \in C^{\theta}([a, b])$, *with* ψ *being increasing:* $\psi'(x) \neq 0$ *over* $[a, b] \subset \mathbb{R}$, $\alpha_i > 0 : \lceil \alpha_i \rceil = n_i$, $0 \leq \beta_i \leq 1$, $\mu_i = n_i(1 - \beta_i) + \beta_i \alpha_i$. *Assume that* $g_i(x) := I_{a+}^{(1-\beta_i)(n_i - \alpha_i); \psi} f_i(x) \in C^{n_i}([a, b])$, *and* $g_{i\psi}^{[k_i]}(a) = 0$, $k_i = 0, ..., n_i - 1$, *where* $g_{i\psi}^{[k_i]}(x) = \left(\frac{1}{\psi'(x)} \frac{d}{dx} \right)^{k_i} g_i(x)$, $k_i = 0, 1, ..., n_i - 1$.

Let also $\Phi_i : \mathbb{R}_+ \to \mathbb{R}_+$, $i = 1, ..., m$, *be convex and increasing functions;* $u \geq 0$ *is a weight measurable function on* $[a, b]$. *We assume that*

$$\lambda_i^{\psi}(y) = \mu_i \psi'(y) \int_y^b u(x) \frac{(\psi(x) - \psi(y))^{\mu_i - 1}}{(\psi(x) - \psi(a))^{\mu_i}} dx < \infty, \tag{10.81}$$

for all $a < y < b$ *and* λ_i^{ψ} *is Lebesgue integrable.*

Then

$$\int_a^b u(x) \prod_{i=1}^m \Phi_i \left(\frac{|g_i(x)|}{(\psi(x) - \psi(a))^{\mu_i}} \Gamma(\mu_i + 1) \right) dx \leq$$

$$\prod_{i=1}^{m}\left(\int_a^b \lambda_i^{\psi}(y)\,\Phi_i\left(\left|\left({}^{H}\mathbb{D}_{a+}^{\alpha_i,\beta_i;\psi} f_i\right)(y)\right|\right)^{p_i} dy\right)^{\frac{1}{p_i}}, \tag{10.82}$$

true for continuous ${}^{H}\mathbb{D}_{a+}^{\alpha_i,\beta_i;\psi} f_i$, $i = 1, ..., m$.

Proof From (10.31) we get that

$$g_i(x) = I_{a+}^{\mu_i;\psi}\ {}^{H}\mathbb{D}_{a+}^{\alpha_i,\beta_i;\psi} f_i(x), \tag{10.83}$$

$\forall\, x \in [a, b]\,;\ i = 1, ..., m.$

Then we apply Theorem 10.7, along with [4], pp. 47–49, see the relations there (2.107)–(2.119). ■

We also give

Theorem 10.16 *Here* $i = 1, ..., m$. *Let* $f_i \in C^{n_i}([a, b])$, $\theta := \max\{n_1, ..., n_m\}$, $\psi \in C^{\theta}([a, b])$, *with* ψ *being increasing,* $\psi'(x) \neq 0$ *over* $[a, b] \subset \mathbb{R}$, $\alpha_i > 0 : \lceil \alpha_i \rceil = n_i$, $0 \leq \beta_i \leq 1$, $\mu_i = n_i(1 - \beta_i) + \beta_i\alpha_i$. *Assume that* $w_i(x) := I_{b-}^{(1-\beta_i)(n_i-\alpha_i);\psi} f_i(x) \in C^{n_i}([a, b])$, *and* $w_{i\psi}^{[k_i]}(b) = 0$, $k_i = 0, ..., n_i - 1$, *where* $w_{i\psi}^{[k_i]}(x) = \left(\frac{1}{\psi'(x)}\frac{d}{dx}\right)^{k_i} w_i(x)$, $k_i = 0, 1, ..., n_i - 1$.

Let also $\Phi_i : \mathbb{R}_+ \to \mathbb{R}_+$, $i = 1, ..., m$, *be convex and increasing functions;* $u \geq 0$ *is a weight measurable function on* $[a, b]$. *We assume that*

$$\overline{\lambda}_i^{\psi}(y) = \mu_i\psi'(y)\int_a^y u(x)\frac{(\psi(y) - \psi(x))^{\mu_i-1}}{(\psi(b) - \psi(x))^{\mu_i}}dx < \infty, \tag{10.84}$$

for all $a < y < b$ *and* $\overline{\lambda}_i^{\psi}$ *is Lebesgue integrable.*

Then

$$\int_a^b u(x)\prod_{i=1}^{m}\Phi_i\left(\frac{|w_i(x)|}{(\psi(b) - \psi(x))^{\mu_i}}\Gamma(\mu_i + 1)\right)dx \leq$$

$$\prod_{i=1}^{m}\left(\int_a^b \overline{\lambda}_i^{\psi}(y)\,\Phi_i\left(\left|\left({}^{H}\mathbb{D}_{b-}^{\alpha_i,\beta_i;\psi} f_i\right)(y)\right|\right)^{p_i} dy\right)^{\frac{1}{p_i}}, \tag{10.85}$$

true for continuous ${}^{H}\mathbb{D}_{b-}^{\alpha_i,\beta_i;\psi} f_i$, $i = 1, ..., m$.

Proof From (10.32) we get that

$$w_i(x) = I_{b-}^{\mu_i;\psi}\ {}^{H}\mathbb{D}_{b-}^{\alpha_i,\beta_i;\psi} f_i(x), \tag{10.86}$$

$\forall\, x \in [a, b]\,,\ i = 1, ..., m.$

Then we apply Theorem 10.7, along with [4], pp. 51–53, see the relations there (2.132)–(2.142). ■

We continue with

Theorem 10.17 *Here $i = 1, ..., m$. Let $n_i - 1 < \alpha_i < n_i$, $n_i \in \mathbb{N}$, $I = [a, b] \subset \mathbb{R}$ and $f_i \in C^{n_i}([a, b])$, $\theta := \max\{n_1, ..., n_m\}$, $\psi \in C^{\theta}([a, b])$, ψ is increasing and $\psi'(x) \neq 0$, for all $x \in I$. Here $0 \leq \beta_i \leq 1$ and $\gamma_i = \alpha_i + \beta_i (n_i - \alpha_i)$. Assume that $\Delta_{a+}^{\gamma_i;\psi} f_i, \Delta_{b-}^{\gamma_i;\psi} f_i \in C([a, b])$, $i = 1, ..., m$.*

Let also $\Phi_i : \mathbb{R}_+ \to \mathbb{R}_+$, $i = 1, ..., m$, be convex and increasing functions; $u \geq 0$ is a weight measurable function on $[a, b]$. We assume that

$$\lambda_{i+}^{\psi}(y) = (\gamma_i - \alpha_i)\,\psi'(y) \int_y^b u(x) \frac{(\psi(x) - \psi(y))^{(\gamma_i - \alpha_i) - 1}}{(\psi(x) - \psi(a))^{(\gamma_i - \alpha_i)}} dx < \infty, \quad (10.87)$$

for all $a < y < b$ and λ_{i+}^{ψ} is Lebesgue integrable; and

$$\lambda_{i-}^{\psi}(y) = (\gamma_i - \alpha_i)\,\psi'(y) \int_a^y u(x) \frac{(\psi(y) - \psi(x))^{(\gamma_i - \alpha_i) - 1}}{(\psi(b) - \psi(x))^{(\gamma_i - \alpha_i)}} dx < \infty, \quad (10.88)$$

for all $a < y < b$ and λ_{i-}^{ψ} is Lebesgue integrable.

Then

(i)

$$\int_a^b u(x) \prod_{i=1}^m \Phi_i \left(\frac{\left| {}^H\mathbb{D}_{a+}^{\alpha_i,\beta_i;\psi} f_i(x) \right|}{(\psi(x) - \psi(a))^{\gamma_i - \alpha_i}} \Gamma(\gamma_i - \alpha_i + 1) \right) dx \leq$$

$$\prod_{i=1}^m \left(\int_a^b \lambda_{i+}^{\psi}(y)\, \Phi_i \left(\left| \left(\Delta_{a+}^{\gamma_i;\psi} f_i \right)(y) \right| \right)^{p_i} dy \right)^{\frac{1}{p_i}}, \quad (10.89)$$

and

(ii)

$$\int_a^b u(x) \prod_{i=1}^m \Phi_i \left(\frac{\left| {}^H\mathbb{D}_{b-}^{\alpha_i,\beta_i;\psi} f_i(x) \right|}{(\psi(b) - \psi(x))^{\gamma_i - \alpha_i}} \Gamma(\gamma_i - \alpha_i + 1) \right) dx \leq$$

$$\prod_{i=1}^m \left(\int_a^b \lambda_{i-}^{\psi}(y)\, \Phi_i \left(\left| \left(\Delta_{b-}^{\gamma_i;\psi} f_i \right)(y) \right| \right)^{p_i} dy \right)^{\frac{1}{p_i}}. \quad (10.90)$$

Proof By (10.19) and (10.20), respectively, we have that

$${}^H\mathbb{D}_{a+}^{\alpha_i,\beta_i;\psi} f_i(x) = I_{a+}^{\gamma_i - \alpha_i;\psi}\, \Delta_{a+}^{\gamma_i;\psi} f_i(x), \quad (10.91)$$

and

$$ {}^{H}\mathbb{D}_{b-}^{\alpha_i,\beta_i;\psi} f_i(x) = I_{b-}^{\gamma_i-\alpha_i;\psi} \Delta_{b-}^{\gamma_i;\psi} f_i(x), \tag{10.92} $$

$\forall\, x \in [a,b]$, $i = 1, ..., m$.

Then, we apply Theorem 10.7 twice, along with [4], pp. 47–49, see there (2.107)–(2.119), and [4], pp. 51–53, see there (2.132)–(2.142), respectively. ■

We make

Remark 10.18 We pick $\Omega_1 = \Omega_2 = (a,b)$, $d\mu_1(x) = dx$, $d\mu_2(y) = dy$, the Lebesgue measure. Here $i = 1, ..., m$. Let $\rho_i, \mu_i, \gamma_i, \omega_i > 0$, and $f_i \in C([a,b])$, with $\psi \in C^1([a,b])$ which is increasing.

We have that ($x \in [a,b]$):

$$ \left(e_{\rho_i,\mu_i,\omega_i,a+}^{\gamma_i;\psi} f_i\right)(x) = $$

$$ \int_a^x \psi'(t)(\psi(x)-\psi(t))^{\mu_i-1} E_{\rho_i,\mu_i}^{\gamma_i}\left[\omega_i(\psi(x)-\psi(t))^{\rho_i}\right] f_i(t)\,dt \tag{10.93} $$

$$ = \int_a^b \chi_{(a,x]}(t)\,\psi'(t)(\psi(x)-\psi(t))^{\mu_i-1} E_{\rho_i,\mu_i}^{\gamma_i}\left[\omega_i(\psi(x)-\psi(t))^{\rho_i}\right] f_i(t)\,dt, $$

where χ is the characteristic function.

So, we choose here

$$ k_i(x,t) := \chi_{(a,x]}(t)\,\psi'(t)(\psi(x)-\psi(t))^{\mu_i-1} E_{\rho_i,\mu_i}^{\gamma_i}\left[\omega_i(\psi(x)-\psi(t))^{\rho_i}\right], \tag{10.94} $$

$i = 1, ..., m$.

That is

$$ k_i(x,y) = \begin{cases} \psi'(y)(\psi(x)-\psi(y))^{\mu_i-1} E_{\rho_i,\mu_i}^{\gamma_i}\left[\omega_i(\psi(x)-\psi(y))^{\rho_i}\right], & a < y \leq x, \\ 0, \quad x < y < b, \end{cases} \tag{10.95} $$

$i = 1, ..., m$.

Therefore we obtain

$$ K_i(x) = \int_a^b \chi_{(a,x]}(y)\,\psi'(y)(\psi(x)-\psi(y))^{\mu_i-1} E_{\rho_i,\mu_i}^{\gamma_i}\left[\omega_i(\psi(x)-\psi(y))^{\rho_i}\right] dy = $$

$$ \int_a^x \psi'(y)(\psi(x)-\psi(y))^{\mu_i-1} E_{\rho_i,\mu_i}^{\gamma_i}\left[\omega_i(\psi(x)-\psi(y))^{\rho_i}\right] dy = $$

(by [3])

$$\sum_{k_i=0}^{\infty} \frac{(\gamma_i)_{k_i} \omega_i^{k_i}}{k_i!\Gamma(\rho_i k_i + \mu_i)} \int_a^x \psi'(y)(\psi(x) - \psi(y))^{(\rho_i k_i + \mu_i)-1} dy =$$

$$\sum_{k_i=0}^{\infty} \frac{(\gamma_i)_{k_i}}{k_i!\Gamma(\rho_i k_i + \mu_i)} \frac{\omega_i^{k_i} (\psi(x) - \psi(a))^{(\rho_i k_i + \mu_i)}}{(\rho_i k_i + \mu_i)} = \tag{10.96}$$

$$(\psi(x) - \psi(a))^{\mu_i} \sum_{k_i=0}^{\infty} \frac{(\gamma_i)_{k_i}}{k_i!\Gamma(\rho_i k_i + \mu_i + 1)} (\omega_i (\psi(x) - \psi(a))^{\rho_i})^{k_i} =$$

$$(\psi(x) - \psi(a))^{\mu_i} E_{\rho_i,\mu_i+1}^{\gamma_i} \left[\omega_i (\psi(x) - \psi(a))^{\rho_i}\right].$$

That is

$$K_i(x) = (\psi(x) - \psi(a))^{\mu_i} E_{\rho_i,\mu_i+1}^{\gamma_i} \left[\omega_i (\psi(x) - \psi(a))^{\rho_i}\right], \tag{10.97}$$

$\forall\, x \in [a, b]$, $i = 1, ..., m$.

Notice that

$$\frac{k_i(x, y)}{K_i(x)} = \frac{\chi_{(a,x]}(y)\, \psi'(y)(\psi(x) - \psi(y))^{\mu_i - 1} E_{\rho_i,\mu_i}^{\gamma_i} \left[\omega_i (\psi(x) - \psi(y))^{\rho_i}\right]}{(\psi(x) - \psi(a))^{\mu_i} E_{\rho_i,\mu_i+1}^{\gamma_i} \left[\omega_i (\psi(x) - \psi(a))^{\rho_i}\right]} \tag{10.98}$$

$$= \left(\chi_{(a,x]}(y) \left(\psi'(y)\right) \frac{(\psi(x) - \psi(y))^{\mu_i - 1}}{(\psi(x) - \psi(a))^{\mu_i}}\right)$$

$$\left(\frac{E_{\rho_i,\mu_i}^{\gamma_i} \left[\omega_i (\psi(x) - \psi(y))^{\rho_i}\right]}{E_{\rho_i,\mu_i+1}^{\gamma_i} \left[\omega_i (\psi(x) - \psi(a))^{\rho_i}\right]}\right),$$

$\forall\, x, y \in [a, b]$.

Therefore for (10.55), we get for appropriate weight $u \geq 0$ that the Lebesgue integrable

$$u_{i+}^{\psi}(y) = \psi'(y) \int_y^b u(x) \left(\frac{(\psi(x) - \psi(y))^{\mu_i - 1}}{(\psi(x) - \psi(a))^{\mu_i}}\right)$$

$$\left(\frac{E_{\rho_i,\mu_i}^{\gamma_i} \left[\omega_i (\psi(x) - \psi(y))^{\rho_i}\right]}{E_{\rho_i,\mu_i+1}^{\gamma_i} \left[\omega_i (\psi(x) - \psi(a))^{\rho_i}\right]}\right) dx < \infty, \tag{10.99}$$

for all $a < y < b$.

Based on Theorem 10.7 and the above, we have established the following generalized Prabhakar left fractional Hardy type inequality:

Theorem 10.19 *Here $i = 1, ..., m$. Let $\rho_i, \mu_i, \gamma_i, \omega_i > 0$, and $f_i \in C([a, b])$, with $\psi \in C^1([a, b])$ which is increasing. The function $u_{i+}^{\psi}(y) \in \mathbb{R}$ by assumption, $\forall$*

$y \in [a, b]$, *is given by (10.99). Here* $\Phi_i : \mathbb{R}_+ \to \mathbb{R}_+$, $i = 1, ..., m$, *are convex and increasing functions. Then*

$$\int_a^b u(x) \prod_{i=1}^m \Phi_i \left(\frac{\left| \left(e^{\gamma_i;\psi}_{\rho_i,\mu_i,\omega_i,a+} f_i \right)(x) \right|}{(\psi(x) - \psi(a))^{\mu_i} E^{\gamma_i}_{\rho_i,\mu_i+1} \left[\omega_i (\psi(x) - \psi(a))^{\rho_i} \right]} \right) dx \le$$

$$\prod_{i=1}^m \left(\int_a^b u^{\psi}_{i+}(u) \Phi_i (|f_i(y)|)^{p_i} dy \right)^{\frac{1}{p_i}}. \tag{10.100}$$

We make

Remark 10.20 We pick $\Omega_1 = \Omega_2 = (a, b)$, $d\mu_1(x) = dx$, $d\mu_2(y) = dy$, the Lebesgue measure. Here $i = 1, ..., m$. Let $\rho_i, \mu_i, \gamma_i, \omega_i > 0$, and $f_i \in C([a, b])$, with $\psi \in C^1([a, b])$ which is increasing.

We have that ($x \in [a, b]$):

$$\left(e^{\gamma_i;\psi}_{\rho_i,\mu_i,\omega_i,b-} f_i \right)(x) =$$

$$\int_x^b \psi'(t) (\psi(t) - \psi(x))^{\mu_i - 1} E^{\gamma_i}_{\rho_i,\mu_i} \left[\omega_i (\psi(t) - \psi(x))^{\rho_i} \right] f_i(t) dt \tag{10.101}$$

$$= \int_a^b \chi_{[x,b)}(t) \psi'(t) (\psi(t) - \psi(x))^{\mu_i - 1} E^{\gamma_i}_{\rho_i,\mu_i} \left[\omega_i (\psi(t) - \psi(x))^{\rho_i} \right] f_i(t) dt,$$

where χ is the characteristic function.

So, we choose here

$$k_i(x, t) := \chi_{[x,b)}(t) \psi'(t) (\psi(t) - \psi(x))^{\mu_i - 1} E^{\gamma_i}_{\rho_i,\mu_i} \left[\omega_i (\psi(t) - \psi(x))^{\rho_i} \right], \tag{10.102}$$

$i = 1, ..., m$.

That is

$$k_i(x, y) = \begin{cases} \psi'(y) (\psi(y) - \psi(x))^{\mu_i - 1} E^{\gamma_i}_{\rho_i,\mu_i} \left[\omega_i (\psi(y) - \psi(x))^{\rho_i} \right], & x \le y < b, \\ 0, \quad a < y < x. \end{cases} \tag{10.103}$$

$i = 1, ..., m$.

Therefore we obtain

$$K_i(x) = \int_a^b \chi_{[x,b)}(y) \psi'(y) (\psi(y) - \psi(x))^{\mu_i - 1} E^{\gamma_i}_{\rho_i,\mu_i} \left[\omega_i (\psi(y) - \psi(x))^{\rho_i} \right] dy =$$

$$\int_x^b \psi'(y) (\psi(y) - \psi(x))^{\mu_i - 1} E^{\gamma_i}_{\rho_i,\mu_i} \left[\omega_i (\psi(y) - \psi(x))^{\rho_i} \right] dy =$$

(by [3])

$$\sum_{k_i=0}^{\infty} \frac{(\gamma_i)_{k_i} \omega_i^{k_i}}{k_i!\Gamma(\rho_i k_i + \mu_i)} \int_x^b \psi'(y)(\psi(y) - \psi(x))^{(\rho_i k_i + \mu_i)-1} dy =$$

$$\sum_{k_i=0}^{\infty} \frac{(\gamma_i)_{k_i}}{k_i!\Gamma(\rho_i k_i + \mu_i)} \frac{\omega_i^{k_i}(\psi(b) - \psi(x))^{(\rho_i k_i + \mu_i)}}{(\rho_i k_i + \mu_i)} = \tag{10.104}$$

$$(\psi(b) - \psi(x))^{\mu_i} \sum_{k_i=0}^{\infty} \frac{(\gamma_i)_{k_i}}{k_i!\Gamma(\rho_i k_i + \mu_i + 1)} (\omega_i (\psi(b) - \psi(x))^{\rho_i})^{k_i} =$$

$$(\psi(b) - \psi(x))^{\mu_i} E_{\rho_i,\mu_i+1}^{\gamma_i} \left[\omega_i (\psi(b) - \psi(x))^{\rho_i}\right].$$

That is

$$K_i(x) = (\psi(b) - \psi(x))^{\mu_i} E_{\rho_i,\mu_i+1}^{\gamma_i} \left[\omega_i (\psi(b) - \psi(x))^{\rho_i}\right], \tag{10.105}$$

$\forall\, x \in [a, b]$, $i = 1, ..., m$.

Notice that

$$\frac{k_i(x, y)}{K_i(x)} = \frac{\chi_{[x,b)}(y)\psi'(y)(\psi(y) - \psi(x))^{\mu_i-1} E_{\rho_i,\mu_i}^{\gamma_i}\left[\omega_i(\psi(y) - \psi(x))^{\rho_i}\right]}{(\psi(b) - \psi(x))^{\mu_i} E_{\rho_i,\mu_i+1}^{\gamma_i}\left[\omega_i(\psi(b) - \psi(x))^{\rho_i}\right]}$$

$$= \left(\chi_{[x,b)}(y)\left(\psi'(y)\right)\frac{(\psi(y) - \psi(x))^{\mu_i-1}}{(\psi(b) - \psi(x))^{\mu_i}}\right)$$

$$\left(\frac{E_{\rho_i,\mu_i}^{\gamma_i}\left[\omega_i(\psi(y) - \psi(x))^{\rho_i}\right]}{E_{\rho_i,\mu_i+1}^{\gamma_i}\left[\omega_i(\psi(b) - \psi(x))^{\rho_i}\right]}\right), \tag{10.106}$$

$\forall\, x, y \in [a, b]$.

Therefore for (10.55), we get for appropriate weight $u \geq 0$ that the Lebesgue integrable:

$$u_{i-}^{\psi}(y) = \psi'(y) \int_a^y u(x) \left(\frac{(\psi(y) - \psi(x))^{\mu_i-1}}{(\psi(b) - \psi(x))^{\mu_i}}\right)$$

$$\left(\frac{E_{\rho_i,\mu_i}^{\gamma_i}\left[\omega_i(\psi(y) - \psi(x))^{\rho_i}\right]}{E_{\rho_i,\mu_i+1}^{\gamma_i}\left[\omega_i(\psi(b) - \psi(x))^{\rho_i}\right]}\right) dx < \infty, \tag{10.107}$$

for all $a < y < b$.

Based on Theorem 10.7 and the above, we have established the following generalized Prabhakar right fractional Hardy type inequality:

Theorem 10.21 *Here $i = 1, ..., m$. Let $\rho_i, \mu_i, \gamma_i, \omega_i > 0$, and $f_i \in C([a, b])$, with $\psi \in C^1([a, b])$ which is increasing. The function $u_{i-}^{\psi}(y) \in \mathbb{R}$ by assumption, $\forall$ $y \in [a, b]$, is given by (10.107). Here $\Phi_i : \mathbb{R}_+ \to \mathbb{R}_+$, $i = 1, ..., m$, are convex and increasing functions. Then*

$$\int_a^b u(x) \prod_{i=1}^m \Phi_i \left(\frac{\left| \left(e_{\rho_i,\mu_i,\omega_i,b-}^{\gamma_i;\psi} f_i \right)(x) \right|}{(\psi(b) - \psi(x))^{\mu_i} E_{\rho_i,\mu_i+1}^{\gamma_i} \left[\omega_i (\psi(b) - \psi(x))^{\rho_i} \right]} \right) dx \leq$$

$$\prod_{i=1}^m \left(\int_a^b u_{i-}^{\psi}(y) \Phi_i (|f_i(y)|)^{p_i} dy \right)^{\frac{1}{p_i}}. \tag{10.108}$$

We continue with left and right ψ-Prabhakar-Caputo Hardy fractional inequalities:

Theorem 10.22 *Here $i = 1, ..., m$. Let $\rho_i, \mu_i, \omega_i > 0, \gamma_i < 0$, and $f_i \in C^{N_i}([a, b])$, $N_i = \lceil \mu_i \rceil$, $\mu_i \notin \mathbb{N}$; $\theta := \max(N_1, ..., N_m)$, $\psi \in C^{\theta}([a, b])$, ψ is increasing with $\psi'(x) \neq 0$ over $[a, b]$. Set $f_{i\psi}^{[N_i]}(x) = \left(\frac{1}{\psi'(x)} \frac{d}{dx} \right)^{N_i} f_i(x)$, $x \in [a, b]$. We assume that the weight function $u \geq 0$ is such that*

$$^C\lambda_{i+}^{\psi}(y) := \psi'(y) \int_y^b u(x) \left(\frac{(\psi(x) - \psi(y))^{(N_i - \mu_i) - 1}}{(\psi(x) - \psi(a))^{(N_i - \mu_i)}} \right)$$

$$\left(\frac{E_{\rho_i,N_i-\mu_i}^{-\gamma_i} \left[\omega_i (\psi(x) - \psi(y))^{\rho_i} \right]}{E_{\rho_i,N_i-\mu_i+1}^{-\gamma_i} \left[\omega_i (\psi(x) - \psi(a))^{\rho_i} \right]} \right) dx < \infty, \tag{10.109}$$

for all $a < y < b$, which is a Lebesgue integrable function.

Here $\Phi_i : \mathbb{R}_+ \to \mathbb{R}_+$, $i = 1, ..., m$, are convex and increasing functions. Then

$$\int_a^b u(x) \prod_{i=1}^m \Phi_i \left(\frac{\left| \left({}^C D_{\rho_i,\mu_i,\omega_i,a+}^{\gamma_i;\psi} f_i \right)(x) \right|}{(\psi(x) - \psi(a))^{N_i - \mu_i} E_{\rho_i,N_i-\mu_i+1}^{-\gamma_i} \left[\omega_i (\psi(x) - \psi(a))^{\rho_i} \right]} \right) dx \leq$$

$$\prod_{i=1}^m \left(\int_a^b {}^C\lambda_{i+}^{\psi}(y) \Phi_i \left(\left| f_{i\psi}^{[N_i]}(y) \right| \right)^{p_i} dy \right)^{\frac{1}{p_i}}. \tag{10.110}$$

Proof By (10.42) we have that

$$\left({}^{C}D^{\gamma_i;\psi}_{\rho_i,\mu_i,\omega_i,a+}f_i\right)(x) = \left(e^{-\gamma_i;\psi}_{\rho_i,N_i-\mu_i,\omega_i,a+}f^{[N_i]}_{i\psi}\right)(x), \tag{10.111}$$

$\forall\, x \in [a,b]$, $i = 1, ..., m$.

We apply Theorem 10.19. ■

Theorem 10.23 *Here $i = 1, ..., m$. Let $\rho_i, \mu_i, \omega_i > 0$, $\gamma_i < 0$, and $f_i \in C^{N_i}([a,b])$, $N_i = \lceil \mu_i \rceil$, $\mu_i \notin \mathbb{N}$; $\theta := \max(N_1, ..., N_m)$, $\psi \in C^{\theta}([a,b])$, ψ is increasing with $\psi'(x) \neq 0$ over $[a,b]$. Set $f^{[N_i]}_{i\psi}(x) = \left(\frac{1}{\psi'(x)}\frac{d}{dx}\right)^{N_i} f_i(x)$, $x \in [a,b]$. We assume that the weight function $u \geq 0$ is such that*

$$ {}^{C}\lambda^{\psi}_{i-}(y) := \psi'(y) \int_a^y u(x) \left(\frac{(\psi(y) - \psi(x))^{(N_i-\mu_i)-1}}{(\psi(b) - \psi(x))^{(N_i-\mu_i)}} \right)$$

$$\left(\frac{E^{-\gamma_i}_{\rho_i,N_i-\mu_i}\left[\omega_i(\psi(y) - \psi(x))^{\rho_i}\right]}{E^{-\gamma_i}_{\rho_i,N_i-\mu_i+1}\left[\omega_i(\psi(b) - \psi(x))^{\rho_i}\right]} \right) dx < \infty, \tag{10.112}$$

for all $a < y < b$, which is integrable.

Here $\Phi_i : \mathbb{R}_+ \to \mathbb{R}_+$, $i = 1, ..., m$, are convex and increasing functions. Then

$$\int_a^b u(x) \prod_{i=1}^m \Phi_i \left(\frac{\left|\left({}^{C}D^{\gamma_i;\psi}_{\rho_i,\mu_i,\omega_i,b-}f_i\right)(x)\right|}{(\psi(b) - \psi(x))^{N_i-\mu_i} E^{-\gamma_i}_{\rho_i,N_i-\mu_i+1}\left[\omega_i(\psi(b) - \psi(x))^{\rho_i}\right]} \right) dx \leq$$

$$\prod_{i=1}^m \left(\int_a^b {}^{C}\lambda^{\psi}_{i-}(y)\, \Phi_i\left(\left|f^{[N_i]}_{i\psi}(y)\right|\right)^{p_i} dy \right)^{\frac{1}{p_i}}. \tag{10.113}$$

Proof By (10.43) we have that

$$\left({}^{C}D^{\gamma_i;\psi}_{\rho_i,\mu_i,\omega_i,b-}f_i\right)(x) = (-1)^{N_i}\left(e^{-\gamma_i;\psi}_{\rho_i,N_i-\mu_i,\omega_i,b-}f^{[N_i]}_{i\psi}\right)(x), \tag{10.114}$$

$\forall\, x \in [a,b]$, $i = 1, ..., m$.

We apply Theorem 10.21. ■

Next we present left and right ψ-Hilfer-Prabhakar Hardy fractional inequalities:

Theorem 10.24 *Here $i = 1, ..., m$. Let $\rho_i, \mu_i, \omega_i > 0$, $\gamma_i < 0$, and $f_i \in C^{N_i}([a,b])$, $N_i = \lceil \mu_i \rceil$, $\mu_i \notin \mathbb{N}$; $\theta := \max(N_1, ..., N_m)$, $\psi \in C^{\theta}([a,b])$, ψ is increasing with $\psi'(x) \neq 0$ over $[a,b]$. Here $0 \leq \beta_i \leq 1$ and $\xi_i = \mu_i + \beta_i(N_i - \mu_i)$. We assume that ${}^{RL}D^{\gamma_i(1-\beta_i);\psi}_{\rho_i,\xi_i,\omega_i,a+}f_i \in C([a,b])$, $i = 1, ..., m$. We assume further that the weight function $u \geq 0$ is such that*

$$^{P}\lambda_{i+}^{\psi}(y) := \psi'(y)\int_y^b u(x)\left(\frac{(\psi(x)-\psi(y))^{(\xi_i-\mu_i)-1}}{(\psi(x)-\psi(a))^{(\xi_i-\mu_i)}}\right)$$

$$\left(\frac{E_{\rho_i,\xi_i-\mu_i}^{-\gamma_i\beta_i}\left[\omega_i(\psi(x)-\psi(y))^{\rho_i}\right]}{E_{\rho_i,\xi_i-\mu_i+1}^{-\gamma_i\beta_i}\left[\omega_i(\psi(x)-\psi(a))^{\rho_i}\right]}\right)dx<\infty, \tag{10.115}$$

for all $a < y < b$, which is integrable.

Here $\Phi_i : \mathbb{R}_+ \to \mathbb{R}_+$, $i = 1, ..., m$, are convex and increasing functions. Then

$$\int_a^b u(x)\prod_{i=1}^m \Phi_i\left(\frac{\left|\left({}^{H}\mathbb{D}_{\rho_i,\mu_i,\omega_i,a+}^{\gamma_i,\beta_i;\psi}f_i\right)(x)\right|}{(\psi(x)-\psi(a))^{\xi_i-\mu_i}E_{\rho_i,\xi_i-\mu_i+1}^{-\gamma_i\beta_i}\left[\omega_i(\psi(x)-\psi(a))^{\rho_i}\right]}\right)dx \leq$$

$$\prod_{i=1}^m\left(\int_a^b {}^{P}\lambda_{i+}^{\psi}(y)\,\Phi_i\left(\left|{}^{RL}D_{\rho_i,\xi_i,\omega_i,a+}^{\gamma_i(1-\beta_i);\psi}f_i(y)\right|\right)^{p_i}dy\right)^{\frac{1}{p_i}}. \tag{10.116}$$

Proof By (10.51) we have that

$$\left({}^{H}\mathbb{D}_{\rho_i,\mu_i,\omega_i,a+}^{\gamma_i,\beta_i;\psi}f_i\right)(x) = e_{\rho_i,\xi_i-\mu_i,\omega_i,a+}^{-\gamma_i\beta_i;\psi}\ {}^{RL}D_{\rho_i,\xi_i,\omega_i,a+}^{\gamma_i(1-\beta_i);\psi}f_i(x), \tag{10.117}$$

$\forall\, x \in [a, b]$, $i = 1, ..., m$.

We apply Theorem 10.19. ■

Theorem 10.25 *Here $i = 1, ..., m$. Let $\rho_i, \mu_i, \omega_i > 0$, $\gamma_i < 0$, and $f_i \in C^{N_i}([a, b])$, $N_i = \lceil \mu_i \rceil$, $\mu_i \notin \mathbb{N}$; $\theta := \max(N_1, ..., N_m)$, $\psi \in C^{\theta}([a, b])$, ψ is increasing with $\psi'(x) \neq 0$ over $[a, b]$. Here $0 \leq \beta_i \leq 1$ and $\xi_i = \mu_i + \beta_i(N_i - \mu_i)$. We assume that ${}^{RL}D_{\rho_i,\xi_i,\omega_i,b-}^{\gamma_i(1-\beta_i);\psi}f_i \in C([a, b])$, $i = 1, ..., m$. We assume further that the weight function $u \geq 0$ is such that*

$$^{P}\lambda_{i-}^{\psi}(y) := \psi'(y)\int_a^y u(x)\left(\frac{(\psi(y)-\psi(x))^{(\xi_i-\mu_i)-1}}{(\psi(b)-\psi(x))^{(\xi_i-\mu_i)}}\right)$$

$$\left(\frac{E_{\rho_i,\xi_i-\mu_i}^{-\gamma_i\beta_i}\left[\omega_i(\psi(y)-\psi(x))^{\rho_i}\right]}{E_{\rho_i,\xi_i-\mu_i+1}^{-\gamma_i\beta_i}\left[\omega_i(\psi(b)-\psi(x))^{\rho_i}\right]}\right)dx<\infty, \tag{10.118}$$

for all $a < y < b$, which is integrable.

Here $\Phi_i : \mathbb{R}_+ \to \mathbb{R}_+$, $i = 1, ..., m$, are convex and increasing functions. Then

$$\int_a^b u(x)\prod_{i=1}^m \Phi_i\left(\frac{\left|\left({}^{H}\mathbb{D}_{\rho_i,\mu_i,\omega_i,b-}^{\gamma_i,\beta_i;\psi}f_i\right)(x)\right|}{(\psi(b)-\psi(x))^{\xi_i-\mu_i}E_{\rho_i,\xi_i-\mu_i+1}^{-\gamma_i\beta_i}\left[\omega_i(\psi(b)-\psi(x))^{\rho_i}\right]}\right)dx \leq$$

$$\prod_{i=1}^{m}\left(\int_a^b {}^{P}\lambda_{i-}^{\psi}(y)\,\Phi_i\left(\left|{}^{RL}D_{\rho_i,\xi_i,\omega_i,b-}^{\gamma_i(1-\beta_i);\psi}f_i(y)\right|^{p_i}\right)dy\right)^{\frac{1}{p_i}}. \tag{10.119}$$

Proof By (10.52) we have that

$$\left({}^{H}\mathbb{D}_{\rho_i,\mu_i,\omega_i,b-}^{\gamma_i,\beta_i;\psi}f_i\right)(x)=e_{\rho_i,\xi_i-\mu_i,\omega_i,b-}^{-\gamma_i\beta_i;\psi}\ {}^{RL}D_{\rho_i,\xi_i,\omega_i,b-}^{\gamma_i(1-\beta_i);\psi}f_i(x), \tag{10.120}$$

$\forall\, x\in[a,b],\ i=1,...,m.$

We apply Theorem 10.21. ∎

(II)' Next we apply Theorem 10.8.

We present the following result.

Theorem 10.26 *Here $j=1,...,m$. Let $\rho_j,\mu_j,\gamma_j,\omega_j>0$, and $f_{ji}\in C([a,b])$, $i=1,...,n$; with $\psi\in C^1([a,b])$, which is increasing. For appropiate weight $u\geq 0$, we assume that*

$$u_{j+}^{\psi}(y):=\psi'(y)\int_y^b u(x)\left(\frac{(\psi(x)-\psi(y))^{\mu_j-1}}{(\psi(x)-\psi(a))^{\mu_j}}\right)$$

$$\left(\frac{E_{\rho_j,\mu_j}^{\gamma_j}\left[\omega_j(\psi(x)-\psi(y))^{\rho_j}\right]}{E_{\rho_j,\mu_j+1}^{\gamma_j}\left[\omega_j(\psi(x)-\psi(a))^{\rho_j}\right]}\right)dx<\infty, \tag{10.121}$$

for all $a<y<b$, which is integrable.

Let $p_j>1:\sum_{j=1}^{m}\frac{1}{p_j}=1$. Let also the functions $\Phi_j:\mathbb{R}_+^n\to\mathbb{R}_+$, $j=1,...,m$, be convex and increasing per coordinate. Then

$$\int_a^b u(x)\prod_{i=1}^{m}\Phi_j\left(\frac{\left|\left(e_{\rho_j,\mu_j,\omega_j,a+}^{\gamma_j;\psi}\overrightarrow{f_j}\right)(x)\right|}{(\psi(x)-\psi(a))^{\mu_j}E_{\rho_j,\mu_j+1}^{\gamma_j}\left[\omega_j(\psi(x)-\psi(a))^{\rho_j}\right]}\right)dx\leq$$

$$\prod_{j=1}^{m}\left(\int_a^b u_{j+}^{\psi}(y)\,\Phi_j\left(\left|\overrightarrow{f_j}(y)\right|\right)^{p_j}dy\right)^{\frac{1}{p_j}}. \tag{10.122}$$

Proof By Theorem 10.8, see also Remark 10.18. ∎

We continue with

Theorem 10.27 *Here $j=1,...,m$. Let $\rho_j,\mu_j,\gamma_j,\omega_j>0$, and $f_{ji}\in C([a,b])$, $i=1,...,n$; with $\psi\in C^1([a,b])$, which is increasing. For appropiate weight $u\geq 0$, we assume that*

$$u_{j-}^{\psi}(y) := \psi'(y) \int_a^y u(x) \left(\frac{(\psi(y) - \psi(x))^{\mu_j - 1}}{(\psi(b) - \psi(x))^{\mu_j}} \right)$$

$$\left(\frac{E_{\rho_j,\mu_j}^{\gamma_j} \left[\omega_j (\psi(y) - \psi(x))^{\rho_j} \right]}{E_{\rho_j,\mu_j+1}^{\gamma_j} \left[\omega_j (\psi(b) - \psi(x))^{\rho_j} \right]} \right) dx < \infty, \tag{10.123}$$

for all $a < y < b$, which is integrable.

Let $p_j > 1 : \sum_{j=1}^{m} \frac{1}{p_j} = 1$. Let also the functions $\Phi_j : \mathbb{R}_+^n \to \mathbb{R}_+$, $j = 1, ..., m$, be convex and increasing per coordinate. Then

$$\int_a^b u(x) \prod_{i=1}^{m} \Phi_j \left(\frac{\left| \left(\overrightarrow{e_{\rho_j,\mu_j,\omega_j,b-}^{\gamma_j:\psi} f_j} \right)(x) \right|}{(\psi(b) - \psi(x))^{\mu_j} E_{\rho_j,\mu_j+1}^{\gamma_j} \left[\omega_j (\psi(b) - \psi(x))^{\rho_j} \right]} \right) dx \le$$

$$\prod_{j=1}^{m} \left(\int_a^b u_{j-}^{\psi}(y) \Phi_j \left(\left| \overrightarrow{f_j}(y) \right| \right)^{p_j} dy \right)^{\frac{1}{p_j}}. \tag{10.124}$$

Proof By Theorem 10.8, see also Remark 10.20. ■

We also give

Theorem 10.28 *Here $j = 1, ..., m$. Let $\rho_j, \mu_j, \omega_j > 0$, $\gamma_j < 0$, and $f_{ji} \in C^{N_j}([a, b])$, $N_j = \lceil \mu_j \rceil$, $\mu_j \notin \mathbb{N}$; $\theta := \max(N_1, ..., N_m)$, $\psi \in C^{\theta}([a, b])$, ψ is increasing with $\psi'(x) \neq 0$ over $[a, b]$; $i = 1, ..., n$. Set $f_{ji\psi}^{[N_j]}(x) = \left(\frac{1}{\psi'(x)} \frac{d}{dx} \right)^{N_j} f_{ji}(x)$, $x \in [a, b]$. We assume that the weight function $u \ge 0$ is such that*

$${}^{C}\lambda_{j+}^{\psi}(y) := \psi'(y) \int_y^b u(x) \left(\frac{(\psi(x) - \psi(y))^{(N_j - \mu_j) - 1}}{(\psi(x) - \psi(a))^{(N_j - \mu_j)}} \right)$$

$$\left(\frac{E_{\rho_j,N_j-\mu_j}^{-\gamma_j} \left[\omega_j (\psi(x) - \psi(y))^{\rho_j} \right]}{E_{\rho_j,N_j-\mu_j+1}^{-\gamma_j} \left[\omega_j (\psi(x) - \psi(a))^{\rho_j} \right]} \right) dx < \infty, \tag{10.125}$$

for all $a < y < b$, which is integrable.

Let $p_j > 1 : \sum_{j=1}^{m} \frac{1}{p_j} = 1$. Let also the functions $\Phi_j : \mathbb{R}_+^n \to \mathbb{R}_+$, $j = 1, ..., m$, be convex and increasing per coordinate. Then

$$\int_a^b u(x) \prod_{i=1}^m \Phi_i \left(\frac{\left| \left(\overrightarrow{{}^C D^{\gamma_j;\psi}_{\rho_j,\mu_j,\omega_j,a+} f_j} \right)(x) \right|}{(\psi(x) - \psi(a))^{N_j-\mu_j} E^{-\gamma_j}_{\rho_j,N_j-\mu_j+1} \left[\omega_j (\psi(x) - \psi(a))^{\rho_j} \right]} \right) dx \leq$$

$$\prod_{j=1}^m \left(\int_a^b {}^C\lambda^{\psi}_{j+}(y) \Phi_j \left(\left| \overrightarrow{f^{[N_j]}_{j\psi}}(y) \right| \right)^{p_j} dy \right)^{\frac{1}{p_j}}. \tag{10.126}$$

Proof By Theorem 10.26 and (10.42), see also (10.111). ■

We continue with

Theorem 10.29 *Here $j = 1, ..., m$. Let $\rho_j, \mu_j, \omega_j > 0$, $\gamma_j < 0$, and $f_{ji} \in C^{N_j}([a, b])$, $N_j = \lceil \mu_j \rceil$, $\mu_j \notin \mathbb{N}$; $\theta := \max(N_1, ..., N_m)$, $\psi \in C^{\theta}([a, b])$, ψ is increasing with $\psi'(x) \neq 0$ over $[a, b]$; $i = 1, ..., n$. Set $f^{[N_j]}_{ji\psi}(x) = \left(\frac{1}{\psi'(x)} \frac{d}{dx} \right)^{N_j} f_{ji}(x)$, $x \in [a, b]$. We assume that the weight function $u \geq 0$ is such that*

$${}^C\lambda^{\psi}_{j-}(y) := \psi'(y) \int_a^y u(x) \left(\frac{(\psi(y) - \psi(x))^{(N_j-\mu_j)-1}}{(\psi(b) - \psi(x))^{(N_j-\mu_j)}} \right)$$

$$\left(\frac{E^{-\gamma_j}_{\rho_j,N_j-\mu_j} \left[\omega_j (\psi(y) - \psi(x))^{\rho_j} \right]}{E^{-\gamma_j}_{\rho_j,N_j-\mu_j+1} \left[\omega_j (\psi(b) - \psi(x))^{\rho_j} \right]} \right) dx < \infty, \tag{10.127}$$

for all $a < y < b$, which is integrable.

Let $p_j > 1 : \sum_{j=1}^m \frac{1}{p_j} = 1$. Let also the functions $\Phi_j : \mathbb{R}^n_+ \to \mathbb{R}_+$, $j = 1, ..., m$, be convex and increasing per coordinate. Then

$$\int_a^b u(x) \prod_{i=1}^m \Phi_i \left(\frac{\left| \left(\overrightarrow{{}^C D^{\gamma_j;\psi}_{\rho_j,\mu_j,\omega_j,b-} f_j} \right)(x) \right|}{(\psi(b) - \psi(x))^{N_j-\mu_j} E^{-\gamma_j}_{\rho_j,N_j-\mu_j+1} \left[\omega_j (\psi(b) - \psi(x))^{\rho_j} \right]} \right) dx \leq$$

$$\prod_{j=1}^m \left(\int_a^b {}^C\lambda^{\psi}_{j-}(y) \Phi_j \left(\left| \overrightarrow{f^{[N_j]}_{j\psi}}(y) \right| \right)^{p_j} dy \right)^{\frac{1}{p_j}}. \tag{10.128}$$

Proof By Theorem 10.27 and (10.43), see also (10.114). ■

We continue with

Theorem 10.30 *Here $j = 1, ..., m$. Let $\rho_j, \mu_j, \omega_j > 0$, $\gamma_j < 0$, and $f_{ji} \in C^{N_j}([a, b])$, $N_j = \lceil \mu_j \rceil$, $\mu_j \notin \mathbb{N}$; $\theta := \max(N_1, ..., N_m)$, $\psi \in C^{\theta}([a, b])$, ψ is increasing with $\psi'(x) \neq 0$ over $[a, b]$; $i = 1, ..., n$. Here $0 \leq \beta_j \leq 1$ and $\xi_j = \mu_j +$*

$\beta_j\left(N_j-\mu_j\right)$. *We assume that* ${}^{RL}D^{\gamma_j(1-\beta_j);\psi}_{\rho_j,\xi_j,\omega_j,a+} f_{ji} \in C([a,b])$, $j=1,...,m$ *and* $i=1,...,n$. *We assume further that the weight function* $u \geq 0$ *is such that*

$$^{P}\lambda^{\psi}_{j+}(y) := \psi'(y)\int_y^b u(x)\left(\frac{(\psi(x)-\psi(y))^{(\xi_j-\mu_j)-1}}{(\psi(x)-\psi(a))^{(\xi_j-\mu_j)}}\right)$$

$$\left(\frac{E^{-\gamma_j\beta_j}_{\rho_j,\xi_j-\mu_j}\left[\omega_j(\psi(x)-\psi(y))^{\rho_j}\right]}{E^{-\gamma_j\beta_j}_{\rho_j,\xi_j-\mu_j+1}\left[\omega_j(\psi(x)-\psi(a))^{\rho_j}\right]}\right)dx < \infty, \tag{10.129}$$

for all $a<y<b$, *which is integrable.*

Let $p_j>1: \sum\limits_{j=1}^{m}\frac{1}{p_j}=1$. *Let also the functions* $\Phi_j:\mathbb{R}^n_+\to\mathbb{R}_+$, $j=1,...,m$, *be convex and increasing per coordinate. Then*

$$\int_a^b u(x)\prod_{j=1}^m \Phi_j\left(\frac{\left|\left(\overrightarrow{{}^{H}\mathbb{D}^{\gamma_j,\beta_j;\psi}_{\rho_j,\mu_j,\omega_j,a+}f_j}\right)(x)\right|}{(\psi(x)-\psi(a))^{\xi_j-\mu_j}E^{-\gamma_j\beta_j}_{\rho_j,\xi_j-\mu_j+1}\left[\omega_j(\psi(x)-\psi(a))^{\rho_j}\right]}\right)dx \leq$$

$$\prod_{i=1}^m\left(\int_a^b {}^{P}\lambda^{\psi}_{j+}(y)\,\Phi_j\left(\left\|\left(\overrightarrow{{}^{RL}D^{\gamma_j(1-\beta_j);\psi}_{\rho_j,\xi_j,\omega_j,a+}f_j}\right)(y)\right\|\right)^{p_j}dy\right)^{\frac{1}{p_j}}. \tag{10.130}$$

Proof By Theorem 10.26 and (10.51), see also (10.117). ■

The counter part of the last theorem follows:

Theorem 10.31 *Here* $j=1,...,m$. *Let* $\rho_j,\mu_j,\omega_j>0$, $\gamma_j<0$, *and* $f_{ji}\in C^{N_j}$ $([a,b])$, $N_j=\lceil\mu_j\rceil$, $\mu_j\notin\mathbb{N}$; $\theta:=\max(N_1,...,N_m)$, $\psi\in C^{\theta}([a,b])$, ψ *is increasing with* $\psi'(x)\neq 0$ *over* $[a,b]$; $i=1,...,n$. *Here* $0\leq\beta_j\leq 1$ *and* $\xi_j=\mu_j+\beta_j\left(N_j-\mu_j\right)$. *We assume that* ${}^{RL}D^{\gamma_j(1-\beta_j);\psi}_{\rho_j,\xi_j,\omega_j,b-} f_{ji}\in C([a,b])$, $j=1,...,m$ *and* $i=1,...,n$. *We assume further that the weight function* $u\geq 0$ *is such that*

$$^{P}\lambda^{\psi}_{j-}(y) := \psi'(y)\int_a^y u(x)\left(\frac{(\psi(y)-\psi(x))^{(\xi_j-\mu_j)-1}}{(\psi(b)-\psi(x))^{(\xi_j-\mu_j)}}\right)$$

$$\left(\frac{E^{-\gamma_j\beta_j}_{\rho_j,\xi_j-\mu_j}\left[\omega_j(\psi(y)-\psi(x))^{\rho_j}\right]}{E^{-\gamma_j\beta_j}_{\rho_j,\xi_j-\mu_j+1}\left[\omega_j(\psi(b)-\psi(x))^{\rho_j}\right]}\right)dx < \infty, \tag{10.131}$$

for all $a<y<b$, *which is integrable.*

Let $p_j > 1 : \sum_{j=1}^{m} \frac{1}{p_j} = 1$. *Let also the functions* $\Phi_j : \mathbb{R}_+^n \to \mathbb{R}_+$, $j = 1, ..., m$, *be convex and increasing per coordinate. Then*

$$\int_a^b u(x) \prod_{j=1}^m \Phi_j \left(\frac{\left| \left(\overrightarrow{{}^H\mathbb{D}_{\rho_j,\mu_j,\omega_j,b-}^{\gamma_j,\beta_j;\psi} f_j} \right)(x) \right|}{(\psi(b) - \psi(x))^{\xi_j - \mu_j} E_{\rho_j,\xi_j-\mu_j+1}^{-\gamma_j\beta_j} \left[\omega_j (\psi(b) - \psi(x))^{\rho_j} \right]} \right) dx \leq$$

$$\prod_{i=1}^m \left(\int_a^b {}^P\lambda_{j-}^{\psi}(y) \Phi_j \left(\left| \left(\overrightarrow{{}^{RL}D_{\rho_j,\xi_j,\omega_j,b-}^{\gamma_j(1-\beta_j);\psi} f_j} \right)(y) \right|^{p_j} \right) dy \right)^{\frac{1}{p_j}}. \tag{10.132}$$

Proof By Theorem 10.27 and (10.52), see also (10.120). ■

(III)' Here we apply Theorem 10.9.

Based on (10.69) and Remark 10.18, we get for appropiate weight $u \geq 0$ that (denote this particular λ_m by $\overline{\lambda}_{m+}^{\psi}$) the integrable function:

$$\overline{\lambda}_{m+}^{\psi}(y) = \left(\psi'(y) \right)^m \int_y^b u(x) \left(\frac{(\psi(x) - \psi(y))^{\sum_{j=1}^m \mu_j - m}}{(\psi(x) - \psi(a))^{\sum_{j=1}^m \mu_j}} \right)$$

$$\prod_{j=1}^m \left(\frac{E_{\rho_j,\mu_j}^{\gamma_j} \left[\omega_j (\psi(x) - \psi(y))^{\rho_j} \right]}{E_{\rho_j,\mu_j+1}^{\gamma_j} \left[\omega_j (\psi(x) - \psi(a))^{\rho_j} \right]} \right) dx < \infty, \tag{10.133}$$

for all $a < y < b$.

By Theorem 10.9 and the above, we have established the following multivariate generalized Prabhakar left fractional Hardy type inequality:

Theorem 10.32 *Here* $j = 1, ..., m$. *Let* $\rho_j, \mu_j, \gamma_j, \omega_j > 0$, *and* $f_{ji} \in C([a, b])$, $i = 1, ..., n$, *with* $\psi \in C^1([a, b])$ *which is increasing. The function* $\overline{\lambda}_{m+}^{\psi}(y) \in \mathbb{R}$ *by assumption,* $\forall\, y \in [a, b]$, *is given by (10.133). Here* $\Phi_j : \mathbb{R}_+^n \to \mathbb{R}_+$, $j = 1, ..., m$, *are convex and increasing per coordinate functions. Then*

$$\int_a^b u(x) \prod_{j=1}^m \Phi_j \left(\frac{\left| \left(\overrightarrow{e_{\rho_j,\mu_j,\omega_j,a+}^{\gamma_j;\psi} f_j} \right)(x) \right|}{(\psi(x) - \psi(a))^{\mu_j} E_{\rho_j,\mu_j+1}^{\gamma_j} \left[\omega_j (\psi(x) - \psi(a))^{\rho_j} \right]} \right) dx \leq$$

$$\left(\prod_{\substack{j=1\\j\neq\rho}}^{m}\int_a^b \Phi_j\left(\left|\overrightarrow{f_j}(y)\right|\right)dy\right)\left(\int_a^b \Phi_\rho\left(\left|\overrightarrow{f_\rho}(y)\right|\right)\overline{\lambda}_{m+}^{\psi}(y)\,dy\right). \tag{10.134}$$

Based on (10.69) and Remark 10.20, we get for appropiate weight $u \geq 0$ that (denote this particular λ_m by $\overline{\lambda}_{m-}^{\psi}$) the integrable function:

$$\overline{\lambda}_{m-}^{\psi}(y) = \left(\psi'(y)\right)^m \int_a^y u(x)\left(\frac{(\psi(y)-\psi(x))^{\sum_{j=1}^{m}\mu_j - m}}{(\psi(b)-\psi(x))^{\sum_{j=1}^{m}\mu_j}}\right)$$

$$\prod_{j=1}^{m}\left(\frac{E_{\rho_j,\mu_j}^{\gamma_j}\left[\omega_j(\psi(y)-\psi(x))^{\rho_j}\right]}{E_{\rho_j,\mu_j+1}^{\gamma_j}\left[\omega_j(\psi(b)-\psi(x))^{\rho_j}\right]}\right)dx < \infty, \tag{10.135}$$

for all $a < y < b$.

By Theorem 10.9 and the above, we have established the following multivariate generalized Prabhakar right fractional Hardy type inequality:

Theorem 10.33 *Here $j = 1, ..., m$; i=1,...,n. Let $\rho_j, \mu_j, \gamma_j, \omega_j > 0$, and $f_{ji} \in C([a,b])$, $i = 1, ..., n$, with $\psi \in C^1([a,b])$ which is increasing. The function $\overline{\lambda}_{m-}^{\psi}(y) \in \mathbb{R}$ by assumption, $\forall\, y \in [a,b]$, is given by (10.135). Here $\Phi_j : \mathbb{R}_+^n \to \mathbb{R}_+$, $j = 1, ..., m$, are convex and increasing per coordinate functions. Then*

$$\int_a^b u(x)\prod_{j=1}^{m}\Phi_j\left(\frac{\left|\left(\overrightarrow{e_{\rho_j,\mu_j,\omega_j,b-}^{\gamma_j;\psi} f_j}\right)(x)\right|}{(\psi(b)-\psi(x))^{\mu_j}E_{\rho_j,\mu_j+1}^{\gamma_j}\left[\omega_j(\psi(b)-\psi(x))^{\rho_j}\right]}\right)dx \leq$$

$$\left(\prod_{\substack{j=1\\j\neq\rho}}^{m}\int_a^b \Phi_j\left(\left|\overrightarrow{f_j}(y)\right|\right)dy\right)\left(\int_a^b \Phi_\rho\left(\left|\overrightarrow{f_\rho}(y)\right|\right)\overline{\lambda}_{m-}^{\psi}(y)\,dy\right). \tag{10.136}$$

We continue with multivariate left and right ψ-Prabhakar-Caputo Hardy fractional inequalities:

Theorem 10.34 *Here $j = 1, ..., m$; $i = 1, ..., n$. Let $\rho_j, \mu_j, \omega_j > 0$, $\gamma_j < 0$, and $f_{ji} \in C^{N_j}([a,b])$, $N_j = \lceil \mu_j \rceil$, $\mu_j \notin \mathbb{N}$; $\theta := \max(N_1, ..., N_m)$, $\psi \in C^{\theta}([a,b])$, ψ is increasing with $\psi'(x) \neq 0$ over $[a,b]$. Set $f_{ji\psi}^{[N_j]}(x) = \left(\frac{1}{\psi'(x)}\frac{d}{dx}\right)^{N_j} f_{ji}(x)$, $x \in [a,b]$. We assume that the weight function $u \geq 0$ is such that*

$$
{}_{C}\lambda_{m+}^{\psi}(y) := \left(\psi'(y)\right)^{m} \int_{y}^{b} u(x) \left(\frac{(\psi(x) - \psi(y))^{\sum_{j=1}^{m}(N_j - \mu_j) - m}}{(\psi(x) - \psi(a))^{\sum_{j=1}^{m}(N_j - \mu_j)}} \right)
$$

$$
\prod_{j=1}^{m} \left(\frac{E_{\rho_j, N_j - \mu_j}^{-\gamma_j}\left[\omega_j (\psi(x) - \psi(y))^{\rho_j}\right]}{E_{\rho_j, N_j - \mu_j + 1}^{-\gamma_j}\left[\omega_j (\psi(x) - \psi(a))^{\rho_j}\right]} \right) dx < \infty, \tag{10.137}
$$

for all $a < y < b$, which is integrable.

Here $\Phi_j : \mathbb{R}_+^n \to \mathbb{R}_+$, $j = 1, ..., m$, are convex and increasing per coordinate functions. Then

$$
\int_a^b u(x) \prod_{j=1}^{m} \Phi_j \left(\frac{\left| \left(\overrightarrow{{}^{C}D_{\rho_j, \mu_j, \omega_j, a+}^{\gamma_j; \psi} f_j} \right)(x) \right|}{(\psi(x) - \psi(a))^{N_j - \mu_j} E_{\rho_j, N_j - \mu_j + 1}^{-\gamma_j}\left[\omega_j (\psi(x) - \psi(a))^{\rho_j}\right]} \right) dx \leq
$$

$$
\left(\prod_{\substack{j=1 \\ j \neq \rho}}^{m} \int_a^b \Phi_j \left(\left\| \overrightarrow{f_{j\psi}^{[N_j]}}(y) \right\| \right) dy \right) \left(\int_a^b \Phi_\rho \left(\left\| \overrightarrow{f_{\rho\psi}^{[N_\rho]}}(y) \right\| \right) {}_{C}\lambda_{m+}^{\psi}(y) \, dy \right). \tag{10.138}
$$

Proof By (10.42) we have that

$$
\left({}^{C}D_{\rho_j, \mu_j, \omega_j, a+}^{\gamma_j; \psi} f_{ji} \right)(x) = \left(e_{\rho_j, N_j - \mu_j, \omega_j, a+}^{-\gamma_j; \psi} f_{ji\psi}^{[N_j]} \right)(x), \tag{10.139}
$$

$\forall\, x \in [a, b]$, $j = 1, ..., m$, $i = 1, ..., n$.

We apply Theorem 10.32. ∎

Theorem 10.35 *Here $j = 1, ..., m$; $i = 1, ..., n$. Let $\rho_j, \mu_j, \omega_j > 0$, $\gamma_j < 0$, and $f_{ji} \in C^{N_j}([a, b])$, $N_j = \lceil \mu_j \rceil$, $\mu_j \notin \mathbb{N}$; $\theta := \max(N_1, ..., N_m)$, $\psi \in C^{\theta}([a, b])$, ψ is increasing with $\psi'(x) \neq 0$ over $[a, b]$. Set $f_{ji\psi}^{[N_j]}(x) = \left(\frac{1}{\psi'(x)} \frac{d}{dx} \right)^{N_i} f_{ji}(x)$, $x \in [a, b]$. We assume that the weight function $u \geq 0$ is such that*

$$
{}_{C}\lambda_{m-}^{\psi}(y) := \left(\psi'(y)\right)^{m} \int_{a}^{y} u(x) \left(\frac{(\psi(y) - \psi(x))^{\sum_{j=1}^{m}(N_j - \mu_j) - m}}{(\psi(b) - \psi(x))^{\sum_{j=1}^{m}(N_j - \mu_j)}} \right)
$$

$$
\prod_{j=1}^{m} \left(\frac{E_{\rho_j, N_j - \mu_j}^{-\gamma_j}\left[\omega_j (\psi(y) - \psi(x))^{\rho_j}\right]}{E_{\rho_j, N_j - \mu_j + 1}^{-\gamma_j}\left[\omega_j (\psi(b) - \psi(x))^{\rho_j}\right]} \right) dx < \infty, \tag{10.140}
$$

for all $a < y < b$, which is integrable.

Here $\Phi_j : \mathbb{R}_+^n \to \mathbb{R}_+$, $j = 1, ..., m$, are convex and increasing per coordinate functions. Then

$$\int_a^b u(x) \prod_{j=1}^m \Phi_j \left(\frac{\left| \left(\overrightarrow{{}^C D^{\gamma_j;\psi}_{\rho_j,\mu_j,\omega_j,b-} f_j} \right)(x) \right|}{(\psi(b) - \psi(x))^{N_j-\mu_j} E^{-\gamma_j}_{\rho_j,N_j-\mu_j+1} \left[\omega_j (\psi(b) - \psi(x))^{\rho_j} \right]} \right) dx \leq$$

$$\left(\prod_{\substack{j=1 \\ j\neq\rho}}^m \int_a^b \Phi_j \left(\left| \overrightarrow{f^{[N_j]}_{j\psi}}(y) \right| \right) dy \right) \left(\int_a^b \Phi_\rho \left(\left| \overrightarrow{f^{[N_\rho]}_{\rho\psi}}(y) \right| \right) {}^C\lambda^{\psi}_{m-}(y)\, dy \right). \tag{10.141}$$

Proof By (10.43) we have that

$$\left({}^C D^{\gamma_j;\psi}_{\rho_j,\mu_j,\omega_j,b-} f_{ji} \right)(x) = (-1)^{N_j} \left(e^{-\gamma_j;\psi}_{\rho_j,N_j-\mu_j,\omega_j,b-} f^{[N_j]}_{ji\psi} \right)(x), \tag{10.142}$$

$\forall\, x \in [a, b]$, $j = 1, ..., m$, $i = 1, ..., n$..

We apply Theorem 10.33. ■

Next we present multivariate left and right ψ-Hilfer-Prabhakar Hardy fractional inequalities:

Theorem 10.36 *Here $j = 1, ..., m$, $i = 1, ..., n$. Let $\rho_j, \mu_j, \omega_j > 0$, $\gamma_j < 0$, and $f_{ji} \in C^{N_j}([a, b])$, $N_j = \lceil \mu_j \rceil$, $\mu_j \notin \mathbb{N}$; $\theta := \max(N_1, ..., N_m)$, $\psi \in C^\theta([a, b])$, ψ is increasing with $\psi'(x) \neq 0$ over $[a, b]$. Here $0 \leq \beta_j \leq 1$ and $\xi_j = \mu_j + \beta_j (N_j - \mu_j)$. We assume that ${}^{RL}D^{\gamma_j(1-\beta_j);\psi}_{\rho_j,\xi_j,\omega_j,a+} f_{ji} \in C([a, b])$, $j = 1, ..., m$, $i = 1, ..., n$. We assume further that the weight function $u \geq 0$ is such that*

$${}^P\lambda^{\psi}_{m+}(y) := (\psi'(y))^m \int_y^b u(x) \left(\frac{(\psi(x) - \psi(y))^{\sum_{j=1}^m (\xi_j-\mu_j)-m}}{(\psi(x) - \psi(a))^{\sum_{j=1}^m (\xi_j-\mu_j)}} \right)$$

$$\prod_{j=1}^m \left(\frac{E^{-\gamma_j\beta_j}_{\rho_j,\xi_j-\mu_j} \left[\omega_j (\psi(x) - \psi(y))^{\rho_j} \right]}{E^{-\gamma_j\beta_j}_{\rho_j,\xi_j-\mu_j+1} \left[\omega_j (\psi(x) - \psi(a))^{\rho_j} \right]} \right) dx < \infty, \tag{10.143}$$

for all $a < y < b$, which is integrable.

Here $\Phi_j : \mathbb{R}_+^n \to \mathbb{R}_+$, $j = 1, ..., m$, are convex and increasing per coordinate functions. Then

$$\int_a^b u(x) \prod_{j=1}^m \Phi_j \left(\frac{\left| \overrightarrow{\left({}^H\mathbb{D}^{\gamma_j,\beta_j;\psi}_{\rho_j,\mu_j,\omega_j,a+} f_j \right)(x)} \right|}{(\psi(x)-\psi(a))^{\xi_j-\mu_j} E^{-\gamma_j\beta_j}_{\rho_j,\xi_j-\mu_j+1}\left[\omega_j(\psi(x)-\psi(a))^{\rho_j}\right]} \right) dx \leq$$

$$\left(\prod_{\substack{j=1 \\ j\neq\overline{\rho}}}^m \int_a^b \Phi_j \left(\overrightarrow{\left| {}^{RL}D^{\gamma_j(1-\beta_j);\psi}_{\rho_j,\xi_j,\omega_j,a+} f_j(y) \right|} \right) dy \right)$$

$$\left(\int_a^b \Phi_{\overline{\rho}} \left(\left\| \overrightarrow{{}^{RL}D^{\gamma_{\overline{\rho}}(1-\beta_{\overline{\rho}});\psi}_{\rho_{\overline{\rho}},\xi_{\overline{\rho}},\omega_{\overline{\rho}},a+} f_{\overline{\rho}}(y)} \right\| \right) {}^P\lambda^{\psi}_{m+}(y)\, dy \right). \tag{10.144}$$

Proof By (10.51) we have that

$$\left({}^H\mathbb{D}^{\gamma_j,\beta_j;\psi}_{\rho_j,\mu_j,\omega_j,a+} f_{ji} \right)(x) = e^{-\gamma_j\beta_j;\psi}_{\rho_j,\xi_j-\mu_j,\omega_j,a+} \; {}^{RL}D^{\gamma_j(1-\beta_j);\psi}_{\rho_j,\xi_j,\omega_j,a+} f_{ji}(x), \tag{10.145}$$

$\forall\, x \in [a,b]$, $j = 1,...,m$, $i = 1,...,n$.

We apply Theorem 10.32. ■

Theorem 10.37 *Here $j = 1,...,m$, $i = 1,...,n$. Let $\rho_j, \mu_j, \omega_j > 0$, $\gamma_j < 0$, and $f_{ji} \in C^{N_j}([a,b])$, $N_j = \lceil \mu_j \rceil$, $\mu_j \notin \mathbb{N}$; $\theta := \max(N_1,...,N_m)$, $\psi \in C^{\theta}([a,b])$, ψ is increasing with $\psi'(x) \neq 0$ over $[a,b]$. Here $0 \leq \beta_j \leq 1$ and $\xi_j = \mu_j + \beta_j(N_j - \mu_j)$. We assume that ${}^{RL}D^{\gamma_j(1-\beta_j);\psi}_{\rho_j,\xi_j,\omega_j,b-} f_{ji} \in C([a,b])$, $j = 1,...,m$, $i = 1,...,n$. We assume further that the weight function $u \geq 0$ is such that*

$${}^P\lambda^{\psi}_{m-}(y) := (\psi'(y))^m \int_a^y u(x) \left(\frac{(\psi(y)-\psi(x))^{\sum_{j=1}^m(\xi_j-\mu_j)-m}}{(\psi(b)-\psi(x))^{\sum_{j=1}^m(\xi_j-\mu_j)}} \right)$$

$$\prod_{j=1}^m \left(\frac{E^{-\gamma_j\beta_j}_{\rho_j,\xi_j-\mu_j}\left[\omega_j(\psi(y)-\psi(x))^{\rho_j}\right]}{E^{-\gamma_j\beta_j}_{\rho_j,\xi_j-\mu_j+1}\left[\omega_j(\psi(b)-\psi(x))^{\rho_j}\right]} \right) dx < \infty, \tag{10.146}$$

for all $a < y < b$, which is integrable.

Here $\Phi_j : \mathbb{R}^n_+ \to \mathbb{R}_+$, $j = 1,...,m$, are convex and increasing per coordinate functions. Then

$$\int_a^b u(x) \prod_{j=1}^m \Phi_j \left(\frac{\left| \left(\overrightarrow{{}^H\mathbb{D}^{\gamma_j,\beta_j;\psi}_{\rho_j,\mu_j,\omega_j,b-} f_j} \right)(x) \right|}{(\psi(b)-\psi(x))^{\xi_j-\mu_j} E^{-\gamma_j\beta_j}_{\rho_j,\xi_j-\mu_j+1}\left[\omega_j(\psi(b)-\psi(x))^{\rho_j}\right]} \right) dx \leq$$

$$\left(\prod_{\substack{j=1 \\ j \neq \overline{\rho}}}^{m} \int_a^b \Phi_j \left(\overrightarrow{\left| {}^{RL} D_{\rho_j,\xi_j,\omega_j,b-}^{\gamma_j(1-\beta_j);\psi} f_j(y) \right|} \right) dy \right)$$

$$\left(\int_a^b \Phi_{\overline{\rho}} \left(\overrightarrow{\left| {}^{RL} D_{\rho_{\overline{\rho}},\xi_{\overline{\rho}},\omega_{\overline{\rho}},b-}^{\gamma_{\overline{\rho}}(1-\beta_{\overline{\rho}});\psi} f_{\overline{\rho}}(y) \right|} \right) {}^{P}\lambda_{m-}^{\psi}(y)\, dy \right). \tag{10.147}$$

Proof By (10.52) we have that

$$\left({}^{H}\mathbb{D}_{\rho_j,\mu_j,\omega_j,b-}^{\gamma_j,\beta_j;\psi} f_{ji} \right)(x) = e_{\rho_j,\xi_j-\mu_j,\omega_j,b-}^{-\gamma_j\beta_j;\psi} \; {}^{RL} D_{\rho_j,\xi_j,\omega_j,b-}^{\gamma_j(1-\beta_j);\psi} f_{ji}(x), \tag{10.148}$$

$\forall\, x \in [a,b]$, $j = 1, ..., m$, $i = 1, ..., n$.

We apply Theorem 10.33. ■

References

1. Almeida, R.: A Caputo fractional derivative of a function with respect to another function. Commun. Nonlinear Sci. Numer. Simul. **44**, 460–481 (2017)
2. Anastassiou, G.: Advancements on ψ-Hilfer fractional calculus and fractional integral inequalities. Nonlinear and Complexity (2021). [Accepted, Discontinuity]
3. Anastassiou, G.A.: Foundations of Generalized Prabhakar-Hilfer Fractional Calculus with Applications (2021). [Submitted]
4. Anastassiou, G.A.: Univariate Hardy-type fractional inequalities, Chap. 2. In: Anastassiou, G., Duman, O. (eds.) Advances in Applied Mathematics and Approximation Theory, Contributions from AMAT 2012, pp. 21–56. Springer, New York (2013)
5. Anastassiou, G.A.: Vectorial Hilfer-Prabhakar-Hardy type fractional inequalities (2021). [Submitted]
6. Anastassiou, G.A.: Intelligent Comparisons: Analytic Inequalities. Springer, Heidelberg, New York (2016)
7. da Sousa, J.C.V, Oliveira, D.C.: On the ψ-Hilfer fractional derivative. Commun. Nonlinear Sci. Numer. Simul. **60**, 72–91 (2018)
8. Giusti, A., et al.: A practical guide to Prabhakar factional calculus. Fractional Calc. Appl. Anal. **23**(1), 9–54 (2020)
9. Gorenflo, R., Kilbas, A., Mainardi, F., Rogosin, S.: Mittag-Leffler functions. Related Topics and Applications, Springer, Heidelberg, New York (2014)
10. Hardy, H.G.: Notes on some points in the integral calculus. Messenger Math. **47**(10), 145–150 (1918)
11. Kilbas, A.A., Srivastava, H.M., Trujillo, J.J.: Theory and Applications of Fractional Differentiation Equations. North Holland, Amsterdam, New York (2006)
12. Polito, F., Tomovski, Z.: Some properties of Prabhakar-type fractional calculus operators. Fractional Differ. Calc. **6**(1), 73–94 (2016)
13. Prabhakar, T.R.: A singular integral equation with a generalized Mittag Leffler function in the kernel. Yokohama Math. J. **19**, 7–15 (1971)

14. Samko, S.G., Kilbas, A.A., Marichev, O.I.: Fractional Integrals and Derivatives: Theory and Applications. Gordon and Breach Science Publishers, Switzerland (1993)
15. Tomovski, Ž, Hilfer, R., Srivastava, H.M.: Fractional and operational calculus with generalized fractional derivative operators and Mittag-Leffler type functions. Integral Trans. Spec. Funct. **21**(11), 797–814 (2010)

Chapter 11
Vectorial Prabhakar Hardy Advanced Fractional Inequalities Under Convexity

We present a detailed great variety of Hardy type fractional inequalities under convexity and L_p norm in the setting of generalized Prabhakar and Hilfer fractional calculi of left and right integrals and derivatives. The radial multivariate case of the above over a spherical shell is developed in detail to all directions. Many inequalities are of vectorial splitting rational L_p type or of separating rational L_p type, others involve ratios of functions and of fractional integral operators. It follows [5].

11.1 Background

This chapter is inspired by Anastassiou [3, 4], Giusti et al. [6], Goreno et al. [7], Hardy [8], Iqbal et al. [9, 10], Polito and Tomovski [11], Prabhakar [12].

Here we consider the Prabhakar function (also known as the three parameter Mittag-Leffler function), (see [6], [7, p. 97])

$$E_{\alpha,\beta}^{\gamma}(z) = \sum_{k=0}^{\infty} \frac{(\gamma)_k}{k!\Gamma(\alpha k + \beta)} z^k, \tag{11.1}$$

where Γ is the gamma function; $\alpha, \beta, \gamma \in \mathbb{R} : \alpha, \beta > 0, z \in \mathbb{R}$, and $(\gamma)_k = \gamma(\gamma + 1) \ldots (\gamma + k - 1)$. It is $E_{\alpha,\beta}^{0}(z) = \frac{1}{\Gamma(\beta)}$.

Here we follow [4].

Let $a, b \in \mathbb{R}, a < b$ and $x \in [a, b]; f \in C([a, b])$. Let also $\psi \in C^1([a, b])$ which is increasing. The left and right Prabhakar fractional integrals with respect to ψ are defined as follows:

$$\left(e_{\rho,\mu,\omega,a+}^{\gamma;\psi} f\right)(x) = \int_a^x \psi'(t)(\psi(x) - \psi(t))^{\mu-1} E_{\rho,\mu}^{\gamma}\left[\omega(\psi(x) - \psi(t))^{\rho}\right] f(t)\,dt, \tag{11.2}$$

G. A. Anastassiou, *Unification of Fractional Calculi with Applications*,
Studies in Systems, Decision and Control 398,
https://doi.org/10.1007/978-3-030-86920-5_11

and

$$\left(e^{\gamma;\psi}_{\rho,\mu,\omega,b-}f\right)(x)=\int_x^b \psi'(t)\left(\psi(t)-\psi(x)\right)^{\mu-1}E^{\gamma}_{\rho,\mu}\left[\omega\left(\psi(t)-\psi(x)\right)^{\rho}\right]f(t)\,dt, \tag{11.3}$$

where $\rho,\mu>0$; $\gamma,\omega\in\mathbb{R}$.

Functions (11.2) and (11.3) are continuous ([4]).

Next, additionally, assume that $\psi'(x)\neq 0$ over $[a,b]$ and let $\psi,f\in C^N([a,b])$, where $N=\lceil\mu\rceil$, ($\lceil\cdot\rceil$ is the ceiling of the number), $0<\mu\notin\mathbb{N}$. We define the ψ-Prabhakar-Caputo left and right fractional derivatives of order μ ([4]) as follows ($x\in[a,b]$):

$$\left({}^{C}D^{\gamma;\psi}_{\rho,\mu,\omega,a+}f\right)(x)=\int_a^x \psi'(t)\left(\psi(x)-\psi(t)\right)^{N-\mu-1}$$

$$E^{-\gamma}_{\rho,N-\mu}\left[\omega\left(\psi(x)-\psi(t)\right)^{\rho}\right]\left(\frac{1}{\psi'(t)}\frac{d}{dt}\right)^{N}f(t)\,dt, \tag{11.4}$$

and

$$\left({}^{C}D^{\gamma;\psi}_{\rho,\mu,\omega,b-}f\right)(x)=(-1)^N\int_x^b \psi'(t)\left(\psi(t)-\psi(x)\right)^{N-\mu-1}$$

$$E^{-\gamma}_{\rho,N-\mu}\left[\omega\left(\psi(t)-\psi(x)\right)^{\rho}\right]\left(\frac{1}{\psi'(t)}\frac{d}{dt}\right)^{N}f(t)\,dt. \tag{11.5}$$

One can write these (see (11.4), (11.5)) as

$$\left({}^{C}D^{\gamma;\psi}_{\rho,\mu,\omega,a+}f\right)(x)=\left(e^{-\gamma;\psi}_{\rho,N-\mu,\omega,a+}f^{[N]}_{\psi}\right)(x), \tag{11.6}$$

and

$$\left({}^{C}D^{\gamma;\psi}_{\rho,\mu,\omega,b-}f\right)(x)=(-1)^N\left(e^{-\gamma;\psi}_{\rho,N-\mu,\omega,b-}f^{[N]}_{\psi}\right)(x), \tag{11.7}$$

where

$$f^{[N]}_{\psi}(x)=f^{(N)}_{\psi}f(x):=\left(\frac{1}{\psi'(x)}\frac{d}{dx}\right)^{N}f(x), \tag{11.8}$$

$\forall\, x\in[a,b]$.

Functions (11.6) and (11.7) are continuous on $[a,b]$.

Next we define the ψ-Prabhakar-Riemann Liouville left and right fractional derivatives of order μ ([4]) as follows ($x\in[a,b], f\in C([a,b])$):

$$\left({}^{RL}D^{\gamma;\psi}_{\rho,\mu,\omega,a+}f\right)(x)=\left(\frac{1}{\psi'(x)}\frac{d}{dx}\right)^{N}\int_a^x \psi'(t)\left(\psi(x)-\psi(t)\right)^{N-\mu-1}$$

$$E^{-\gamma}_{\rho,N-\mu}\left[\omega\left(\psi(x)-\psi(t)\right)^{\rho}\right]f(t)\,dt, \tag{11.9}$$

and

$$\left({}^{RL}D^{\gamma;\psi}_{\rho,\mu,\omega,b-}f\right)(x) = \left(-\frac{1}{\psi'(x)}\frac{d}{dx}\right)^N \int_x^b \psi'(t)\left(\psi(t)-\psi(x)\right)^{N-\mu-1}$$

$$E^{-\gamma}_{\rho,N-\mu}\left[\omega\left(\psi(t)-\psi(x)\right)^{\rho}\right] f(t)\,dt. \tag{11.10}$$

That is we have

$$\left({}^{RL}D^{\gamma;\psi}_{\rho,\mu,\omega,a+}f\right)(x) = \left(\frac{1}{\psi'(x)}\frac{d}{dx}\right)^N \left(e^{-\gamma;\psi}_{\rho,N-\mu,\omega,a+}f\right)(x), \tag{11.11}$$

and

$$\left({}^{RL}D^{\gamma;\psi}_{\rho,\mu,\omega,b-}f\right)(x) = \left(-\frac{1}{\psi'(x)}\frac{d}{dx}\right)^N \left(e^{-\gamma;\psi}_{\rho,N-\mu,\omega,b-}f\right)(x), \tag{11.12}$$

$\forall\, x \in [a,b]$.

We define also the ψ-Hilfer-Prabhakar left and right fractional derivatives of order μ and type $0 \leq \beta \leq 1$ ([4]), as follows

$$\left({}^{H}\mathbb{D}^{\gamma,\beta;\psi}_{\rho,\mu,\omega,a+}f\right)(x) = e^{-\gamma\beta;\psi}_{\rho,\beta(N-\mu),\omega,a+}\left(\frac{1}{\psi'(x)}\frac{d}{dx}\right)^N e^{-\gamma(1-\beta);\psi}_{\rho,(1-\beta)(N-\mu),\omega,a+}f(x), \tag{11.13}$$

and

$$\left({}^{H}\mathbb{D}^{\gamma,\beta;\psi}_{\rho,\mu,\omega,b-}f\right)(x) = e^{-\gamma\beta;\psi}_{\rho,\beta(N-\mu),\omega,b-}\left(-\frac{1}{\psi'(x)}\frac{d}{dx}\right)^N e^{-\gamma(1-\beta);\psi}_{\rho,(1-\beta)(N-\mu),\omega,b-}f(x), \tag{11.14}$$

$\forall\, x \in [a,b]$.

When $\beta = 0$, we get the Riemann-Liouville version, and when $\beta = 1$, we get the Caputo version.

We call $\xi = \mu + \beta(N-\mu)$, we have that $N-1 < \mu \leq \mu + \beta(N-\mu) \leq \mu + N - \mu = N$, hence $\lceil \xi \rceil = N$.

We can easily write that

$$\left({}^{H}\mathbb{D}^{\gamma,\beta;\psi}_{\rho,\mu,\omega,a+}f\right)(x) = e^{-\gamma\beta;\psi}_{\rho,\xi-\mu,\omega,a+}\ {}^{RL}D^{\gamma(1-\beta);\psi}_{\rho,\xi,\omega,a+}f(x), \tag{11.15}$$

and

$$\left({}^{H}\mathbb{D}^{\gamma,\beta;\psi}_{\rho,\mu,\omega,b-}f\right)(x) = e^{-\gamma\beta;\psi}_{\rho,\xi-\mu,\omega,b-}\ {}^{RL}D^{\gamma(1-\beta);\psi}_{\rho,\xi,\omega,b-}f(x), \tag{11.16}$$

$\forall\, x \in [a,b]$.

In this chapter we develop a great variety of fractional inequalities of Hardy type invovling convexity and engaging the above exposed: ψ-Prabhakar fractional left and right fractional integrals, the ψ-Prabhakar-Caputo left and right fractional derivatives, the ψ-Riemann-Liouville left and right fractional derivatives, and the ψ-Hilfer-Prabhakar left and right fractional derivatives. The radial multivariate case of all of the above over a spherical shell is studied in full detail. We involve ratios of functions and of integral operators and we produce among others vectorial splitting rational L_p inequalities, as well as separating rational L_p inequalities.

11.2 Prerequisites

Let $(\Omega_1, \Sigma_1, \mu_1)$ and $(\Omega_2, \Sigma_2, \mu_2)$ be measure spaces with positive σ-finite measures, and let $k : \Omega_1 \times \Omega_2 \to \mathbb{R}$ be nonnegative measurable functions, $k(x, \cdot)$ measurable on Ω_2, and

$$K(x) = \int_{\Omega_2} k(x, y)\, d\mu_2(y), \text{ for any } x \in \Omega_1. \tag{11.17}$$

7We suppose that $K(x) > 0$ a.e. on Ω_1 and by a weight function (shortly: a weight), we mean a nonnegative measurable function on the actual set. Let the measurable functions $g_i : \Omega_1 \to \mathbb{R}$, $i = 1, ..., n$, with the representation

$$g_i(x) = \int_{\Omega_2} k(x, y) f_i(y)\, d\mu_2(y), \tag{11.18}$$

where $f_i : \Omega_2 \to \mathbb{R}$ are measurable functions, $i = 1, ..., n$.

Denote by $\mathbf{x} = x := (x_1, ..., x_n) \in \mathbb{R}^n$, $\mathbf{g} := (g_1, ..., g_n)$ and $\mathbf{f} := (f_1, ..., f_n)$.

We consider here $\Phi : \mathbb{R}^n_+ \to \mathbb{R}$ a convex function, which is increasing per coordinate, i.e. if $x_i \leq y_i$, $i = 1, ..., n$, then

$$\Phi(x_1, ..., x_n) \leq \Phi(y_1, ..., y_n).$$

In [3], p. 588, we proved that

Theorem 11.1 *Let u be a weight function on Ω_1, and k, K, g_i, f_i, $i = 1, ..., n \in \mathbb{N}$, and Φ defined as above. Assume that the function $x \to u(x) \frac{k(x,y)}{K(x)}$ is integrable on Ω_1 for each fixed $y \in \Omega_2$. Define v on Ω_2 by*

$$v(y) := \int_{\Omega_1} u(x) \frac{k(x, y)}{K(x)} d\mu_1(x) < \infty. \tag{11.19}$$

Then

$$\int_{\Omega_1} u(x)\, \Phi\left(\frac{|g_1(x)|}{K(x)}, ..., \frac{|g_n(x)|}{K(x)}\right) d\mu_1(x) \leq$$

$$\int_{\Omega_2} v(y)\, \Phi(|f_1(y)|, ..., |f_n(y)|)\, d\mu_2(y), \tag{11.20}$$

under the assumptions:

(i) f_i, $\Phi(|f_1|, ..., |f_n|)$, are $k(x, y)\, d\mu_2(y)$ -integrable, μ_1 -a.e. in $x \in \Omega_1$, for all $i = 1, ..., n$,
(ii) $v(y)\, \Phi(|f_1(y)|, ..., |f_n(y)|)$ is μ_2 -integrable.

Notation 11.2 From now on we may write

$$\overrightarrow{g}(x) = \int_{\Omega_2} k(x, y) \overrightarrow{f}(y)\, d\mu_2(y), \tag{11.21}$$

which means

$$(g_1(x), \ldots, g_n(x)) = \left(\int_{\Omega_2} k(x,y) f_1(y)\, d\mu_2(y), \ldots, \int_{\Omega_2} k(x,y) f_n(y)\, d\mu_2(y)\right). \tag{11.22}$$

Similarly, we may write

$$\left|\overrightarrow{g}(x)\right| = \left|\int_{\Omega_2} k(x,y) \overrightarrow{f}(y)\, d\mu_2(y)\right|, \tag{11.23}$$

and we mean

$$(|g_1(x)|, \ldots, |g_n(x)|) =$$

$$\left(\left|\int_{\Omega_2} k(x,y) f_1(y)\, d\mu_2(y)\right|, \ldots, \left|\int_{\Omega_2} k(x,y) f_n(y)\, d\mu_2(y)\right|\right). \tag{11.24}$$

We also can write that

$$\left|\overrightarrow{g}(x)\right| \leq \int_{\Omega_2} k(x,y) \left|\overrightarrow{f}(y)\right| d\mu_2(y), \tag{11.25}$$

and we mean the fact that

$$|g_i(x)| \leq \int_{\Omega_2} k(x,y) |f_i(y)|\, d\mu_2(y), \tag{11.26}$$

for all $i = 1, \ldots, n$, etc.

Notation 11.3 Next let $(\Omega_1, \Sigma_1, \mu_1)$ and $(\Omega_2, \Sigma_2, \mu_2)$ be measure spaces with positive σ-finite measures, and let $k_j : \Omega_1 \times \Omega_2 \to \mathbb{R}$ be a nonnegative measurable function, $k_j(x, \cdot)$ measurable on Ω_2 and

$$K_j(x) = \int_{\Omega_2} k_j(x,y)\, d\mu_2(y), \quad x \in \Omega_1, j = 1, \ldots, m. \tag{11.27}$$

We suppose that $K_j(x) > 0$ a.e. on Ω_1. Let the measurable functions $g_{ji} : \Omega_1 \to \mathbb{R}$ with the representation

$$g_{ji}(x) = \int_{\Omega_2} k_j(x,y) f_{ji}(y)\, d\mu_2(y), \tag{11.28}$$

where $f_{ji} : \Omega_2 \to \mathbb{R}$ are measurable functions, $i = 1, \ldots, n$ and $j = 1, \ldots, m$.

Denote the function vectors $\overrightarrow{g_j} := (g_{j1}, g_{j2}, \ldots, g_{jn})$ and $\overrightarrow{f_j} := (f_{j1}, \ldots, f_{jn})$, $j = 1, \ldots, m$.

We say $\overrightarrow{f_j}$ is integrable with respect to measure μ, iff all f_{ji} are integrable with respect to μ.

We also consider here $\Phi_j : \mathbb{R}_+^n \to \mathbb{R}_+, j = 1, \ldots, m$, convex functions that are increasing per coordinate. Again u is a weight function on Ω_1.

We make

Remark 11.4 Following Notation 11.3, let $F_j : \Omega_2 \rightarrow \mathbb{R} \cup \{\pm\infty\}$ be measurable functions, $j = 1, ..., m$, with $0 < F_j(y) < \infty$ on Ω_2. In (11.27) we replace $k_j(x, y)$ by $k_j(x, y) F_j(y)$, $j = 1, ..., m$, and we have the modified $K_j(x)$ as

$$L_j(x) := \int_{\Omega_2} k_j(x, y) F_j(y) \, d\mu_2(y), \quad x \in \Omega_1. \tag{11.29}$$

We assume $L_j(x) > 0$ a.e. on Ω_1.

As new $\overrightarrow{f_j}$ we consider now $\overrightarrow{\gamma_j} := \frac{\overrightarrow{f_j}}{F_j}$, $j = 1, ..., m$, where $\overrightarrow{f_j} = (f_{j1}, ..., f_{jn})$; $\overrightarrow{\gamma_j} = \left(\frac{f_{j1}}{F_j}, ..., \frac{f_{jn}}{F_j}\right)$.

Notice that

$$g_{ji}(x) = \int_{\Omega_2} k_j(x, y) f_{ji}(y) \, d\mu_2(y) \tag{11.30}$$

$$= \int_{\Omega_2} \left(k_j(x, y) F_j(y)\right) \left(\frac{f_{ji}(y)}{F_j(y)}\right) d\mu_2(y),$$

$x \in \Omega_1$, all $j = 1, ..., m$; $i = 1, ..., n$.

So we can write

$$\overrightarrow{g_j}(x) = \int_{\Omega_2} \left(k_j(x, y) F_j(y)\right) \overrightarrow{\gamma_j}(y) \, d\mu_2(y), \quad j = 1, ..., m. \tag{11.31}$$

We mention

Theorem 11.5 ([3, p. 481]) *Here we follow Remark 11.4. Let $\rho \in \{1, ..., m\}$ be fixed. Assume that the function*

$$x \mapsto \left(\frac{u(x) \left(\prod_{j=1}^{m} F_j(y)\right) \left(\prod_{j=1}^{m} k_j(x, y)\right)}{\prod_{j=1}^{m} L_j(x)}\right)$$

is integrable on Ω_1, for each $y \in \Omega_2$. Define U_m on Ω_2 by

$$U_m(y) := \left(\prod_{j=1}^{m} F_j(y)\right) \int_{\Omega_1} \frac{u(x) \prod_{j=1}^{m} k_j(x, y)}{\prod_{j=1}^{m} L_j(x)} d\mu_1(x) < \infty. \tag{11.32}$$

Then

$$\int_{\Omega_1} u(x) \prod_{j=1}^{m} \Phi_j \left(\left\| \frac{\overrightarrow{g_j}(x)}{L_j(x)} \right\|\right) d\mu_1(x) \leq \tag{11.33}$$

$$\left(\prod_{\substack{j=1\\ j\neq\rho}}^{m}\int_{\Omega_2}\Phi_j\left(\left|\frac{\overrightarrow{f_j}(y)}{F_j(y)}\right|\right)d\mu_2(y)\right)\cdot\left(\int_{\Omega_2}\Phi_\rho\left(\left|\frac{\overrightarrow{f_\rho}(y)}{F_\rho(y)}\right|\right)U_m(y)\,d\mu_2(y)\right),$$

under the assumptions:

(i) $\frac{\overrightarrow{f_j}}{F_j}$, $\Phi_j\left(\frac{\left|\overrightarrow{f_j}\right|}{F_j}\right)$ *are both* $k_j(x,y)F_j(y)\,d\mu_2(y)$ *-integrable,* μ_1 *-a.e. in* $x\in\Omega_1$, $j=1,\ldots,m$,

(ii) $U_m\Phi_\rho\left(\frac{\left|\overrightarrow{f_\rho}\right|}{F_\rho}\right)$; $\Phi_1\left(\frac{\left|\overrightarrow{f_1}\right|}{F_1}\right), \Phi_2\left(\frac{\left|\overrightarrow{f_2}\right|}{F_2}\right), \ldots, \widehat{\Phi_\rho\left(\frac{\left|\overrightarrow{f_\rho}\right|}{F_\rho}\right)}, \ldots, \Phi_m\left(\frac{\left|\overrightarrow{f_m}\right|}{F_m}\right)$, *are* μ_2 *-integrable, where* $\widehat{\Phi_\rho\left(\frac{\left|\overrightarrow{f_\rho}\right|}{F_\rho}\right)}$ *is absent.*

We also mention

Theorem 11.6 ([3, p. 519]) *Here all as in Notation 11.3 and Remark 11.4. Assume that the functions* ($j=1,2,\ldots,m\in\mathbb{N}$)

$$x\mapsto\left(\frac{u(x)\,k_j(x,y)\,F_j(y)}{K_j(x)}\right)$$

are integrable on Ω_1*, for each fixed* $y\in\Omega_2$*. Define* W_j *on* Ω_2 *by*

$$W_j(y):=\left(\int_{\Omega_1}\frac{u(x)\,k_j(x,y)}{K_j(x)}d\mu_1(x)\right)F_j(y)<\infty, \tag{11.34}$$

on Ω_2.

Let $p_j>1:\sum_{j=1}^{m}\frac{1}{p_j}=1$*. Let the functions* $\Phi_j:\mathbb{R}_+^n\to\mathbb{R}_+$, $j=1,\ldots,m$, *be convex and increasing per coordinate.*

Then

$$\int_{\Omega_1}u(x)\prod_{j=1}^{m}\Phi_j\left(\left|\frac{\overrightarrow{g_j}(x)}{L_j(x)}\right|\right)d\mu_1(x)\leq \tag{11.35}$$

$$\prod_{j=1}^{m}\left(\int_{\Omega_2}W_j(y)\,\Phi_j\left(\left|\frac{\overrightarrow{f_j}(y)}{F_j(y)}\right|\right)^{p_j}d\mu_2(y)\right)^{\frac{1}{p_j}},$$

under the assumptions:

(i) $\frac{\overrightarrow{f_j}}{F_j}$, $\Phi_j\left(\frac{\left|\overrightarrow{f_j}\right|}{F_j}\right)^{p_j}$ *are both* $k_j(x,y)F_j(y)\,d\mu_2(y)$ *-integrable,* μ_1 *-a.e. in* $x\in\Omega_1$, $j=1,\ldots,m$, *(ii)* $W_j \Phi_j \left(\frac{|\overrightarrow{f_j}|}{F_j} \right)^{p_j}$ *is* μ_2 *-integrable,* $j = 1, ..., m$.

We make

Remark 11.7 Let $(\Omega_1, \Sigma_1, \mu_1)$ and $(\Omega_2, \Sigma_2, \mu_2)$ be measure spaces with positive σ-finite measures, and let $k : \Omega_1 \times \Omega_2 \to \mathbb{R}$ be nonnegative measurable functions, $k(x, \cdot)$ measurable on Ω_2, and

$$K(x) = \int_{\Omega_2} k(x, y)\, d\mu_2(y), \text{ for any } x \in \Omega_1.$$

We assume $K(x) > 0$ a.e. on Ω_1 and the weight functions are nonnegative functions on the related set. We consider measurable functions $g_i : \Omega_1 \to \mathbb{R}$, with the representation

$$g_i(x) = \int_{\Omega_2} k(x, y) f_i(y)\, d\mu_2(y),$$

where $f_i : \Omega_2 \to \mathbb{R}$ are measurable functions, $i = 1, ..., n$. Here u stands for a weight function on Ω_1. So we follow Notation 11.3 for $j = m = 1$. We write here $\mathbf{g} := (g_1, ..., g_n)$, $\mathbf{f} := (f_1, ..., f_n)$.

We set

$$\left\| \overrightarrow{f}(y) \right\|_\infty := \max \{|f_1(y)|, ..., |f_n(y)|\},$$

and

$$\left\| \overrightarrow{f}(y) \right\|_q := \left(\sum_{i=1}^n |f_i(y)|^q \right)^{\frac{1}{q}}, \quad q \geq 1. \tag{11.36}$$

We assume that

$$0 < \left\| \overrightarrow{f}(y) \right\|_q < \infty, \text{ a.e. on } (a, b), \tag{11.37}$$

$1 \leq q \leq \infty$ fixed.

Let

$$L_q(x) := \int_{\Omega_2} k(x, y) \left\| \overrightarrow{f}(y) \right\|_q d\mu_2(y), \ x \in \Omega_1, \tag{11.38}$$

$1 \leq q \leq \infty$ fixed.

We assume $L_q(x) > 0$ a.e. on Ω_1.

We furher assume that the function

$$x \mapsto \left(\frac{u(x) k(x, y) \left\| \overrightarrow{f}(y) \right\|_q}{L_q(x)} \right) \tag{11.39}$$

is integrable on Ω_1, for almost each fixed $y \in \Omega_2$.

Define W_q on Ω_2 by

$$W_q(y) := \left(\int_{\Omega_1} \frac{u(x) k(x, y)}{L_q(x)} d\mu_1(x) \right) \left\| \overrightarrow{f}(y) \right\|_q < \infty, \tag{11.40}$$

a.e. on Ω_2.

Let

$$\overrightarrow{\gamma} := \left(\frac{f_1}{\left\| \overrightarrow{f}(y) \right\|_q}, \frac{f_2}{\left\| \overrightarrow{f}(y) \right\|_q}, \ldots, \frac{f_n}{\left\| \overrightarrow{f}(y) \right\|_q} \right), \tag{11.41}$$

i.e. $\overrightarrow{\gamma} = \frac{\overrightarrow{f}}{\left\| \overrightarrow{f}(y) \right\|_q}$.

Here $\Phi : \mathbb{R}_+^n \to \mathbb{R}$ is a convex and increasing per coordinate function.

We mention

Theorem 11.8 ([3, p. 536]) *Let all here as in Remark 11.7. Then*

$$\int_{\Omega_1} u(x)\, \Phi\left(\left| \frac{\overrightarrow{g}(x)}{L_q(x)} \right| \right) d\mu_1(x) \le \tag{11.42}$$

$$\int_{\Omega_2} W_q(y)\, \Phi\left(\frac{\left| \overrightarrow{f}(y) \right|}{\left\| \overrightarrow{f}(y) \right\|_q} \right) d\mu_2(y),$$

under the assumptions:

(i) $\frac{\overrightarrow{f}(y)}{\left\| \overrightarrow{f}(y) \right\|_q}, \Phi\left(\frac{\left| \overrightarrow{f}(y) \right|}{\left\| \overrightarrow{f}(y) \right\|_q} \right)$ *are both* $k(x,y) \left\| \overrightarrow{f}(y) \right\|_q d\mu_2(y)$ *-integrable,* μ_1 *-a.e. in* $x \in \Omega_1$,

(ii) $W_q(y)\, \Phi\left(\frac{\left| \overrightarrow{f}(y) \right|}{\left\| \overrightarrow{f}(y) \right\|_q} \right)$ *is* μ_2 *-integrable.*

Theorem 11.8 comes directly from Theorem 11.1.

We will also use:

Let $(\Omega_1, \Sigma_1, \mu_1)$, $(\Omega_2, \Sigma_2, \mu_2)$ measure spaces with positive σ-finite measures, and $k_i : \Omega_1 \times \Omega_2 \to \mathbb{R}$ are nonnegative measurable functions, with $k_i(x, \cdot)$ measurable on Ω_2, and measurable functions $g_{ji} : \Omega_1 \to \mathbb{R}$:

$$g_{ji}(x) = \int_{\Omega_2} k_i(x,y) f_{ji}(y)\, d\mu_2(y),$$

where $f_{ji} : \Omega_2 \to \mathbb{R}$ are measurable functions, for all $j = 1, 2$; $i = 1, \ldots, m$.

Theorem 11.9 ([3, p. 552]) *Here* $0 < f_{2i}(y) < \infty$, *a.e.,* $i = 1, \ldots, m$. *Assume that the functions* ($i = 1, \ldots, m \in \mathbb{N}$)

$$x \mapsto \left(\frac{u(x) k_i(x,y) f_{2i}(y)}{g_{2i}(x)} \right)$$

are integrable on Ω_1, *for each fixed* $y \in \Omega_2$; *with* $g_{2i}(x) > 0$, *a.e. on* Ω_1.

Define ψ_i *on* Ω_2 *by*

$$\psi_i(y) := f_{2i}(y) \int_{\Omega_1} u(x) \frac{k_i(x,y)}{g_{2i}(x)} d\mu_1(x) < \infty, \tag{11.43}$$

a.e. on Ω_2.

Let $p_i > 1 : \sum_{i=1}^{m} \frac{1}{p_i} = 1$. *Let the functions* $\Phi_i : \mathbb{R}_+ \to \mathbb{R}_+$, $i = 1, ..., m$, *be convex and increasing. Then*

$$\int_{\Omega_1} u(x) \prod_{i=1}^{m} \Phi_i \left(\left| \frac{g_{1i}(x)}{g_{2i}(x)} \right| \right) d\mu_1(x) \leq \tag{11.44}$$

$$\prod_{i=1}^{m} \left(\int_{\Omega_2} \psi_i(y) \Phi_i \left(\left| \frac{f_{1i}(y)}{f_{2i}(y)} \right| \right)^{p_i} d\mu_2(y) \right)^{\frac{1}{p_i}},$$

under the assumptions:

(i) $\frac{f_{1i}(y)}{f_{2i}(y)}$, $\Phi_i \left(\left| \frac{f_{1i}(y)}{f_{2i}(y)} \right| \right)^{p_i}$ *are both* $k_i(x,y) f_{2i}(y) d\mu_2(y)$ *-integrable,* μ_1 *-a.e. in* $x \in \Omega_1$,
(ii) $\psi_i(y) \Phi_i \left(\left| \frac{f_{1i}(y)}{f_{2i}(y)} \right| \right)^{p_i}$ *is* μ_2*-integrable,* $i = 1, ..., m$.

11.3 Main Results

We make

Remark 11.10 Here $\rho_j, \mu_j, \gamma_j, \omega_j > 0$; $f_{ji} \in C([a,b])$ and $\psi \in C^1([a,b])$ which is increasing; $j = 1, ..., m$ and $i = 1, ..., n$. Set

$$_{\infty}\varphi_{j+}(y) := \left\| \overrightarrow{e^{\gamma_j;\psi}_{\rho_j,\mu_j,\omega_j,a+} f_j(y)} \right\|_{\infty} := \max_{\substack{j=1,...,m \\ i=1,...,n}} \left\{ \left| e^{\gamma_j;\psi}_{\rho_j,\mu_j,\omega_j,a+} f_{ji}(y) \right| \right\}, \tag{11.45}$$

and

$$_{q}\varphi_{j+}(y) := \left\| \overrightarrow{e^{\gamma_j;\psi}_{\rho_j,\mu_j,\omega_j,a+} f_j(y)} \right\|_{q} := \left(\sum_{i=1}^{n} \left| e^{\gamma_j;\psi}_{\rho_j,\mu_j,\omega_j,a+} f_{ji}(y) \right|^q \right)^{\frac{1}{q}}, \quad q \geq 1; \tag{11.46}$$

$y \in [a,b]$, which $_{q}\varphi_{j+}$ are continuous functions, $j = 1, ..., m$. We have that

$$0 <_q \varphi_{j+}(y) < \infty \text{ in } [a,b], \tag{11.47}$$

$j = 1, ..., m$; where $1 \leq q \leq \infty$ is fixed.

Here it is

$$k_j^+(x,y) := k_j(x,y) =$$

$$\begin{cases} \psi'(y)(\psi(x)-\psi(y))^{\mu_j-1} E^{\gamma_j}_{\rho_j,\mu_j}\left[\omega_j(\psi(x)-\psi(y))^{\rho_j}\right], & a<y\leq x, \\ 0, \ x<y<b, \end{cases} \tag{11.48}$$

$j=1,...,m$, and

$$L^{+}_{jq}(x) := \int_a^x \psi'(y)(\psi(x)-\psi(y))^{\mu_j-1} E^{\gamma_j}_{\rho_j,\mu_j}\left[\omega_j(\psi(x)-\psi(y))^{\rho_j}\right] {}_q\varphi_{j+}(y)\,dy, \tag{11.49}$$

$\forall\, x\in[a,b]$, $1\leq q\leq\infty$.

We have that $L^{+}_{jq}(x)>0$ on $[a,b]$.

Let $\rho\in\{1,...,m\}$ be fixed. The weight function u is chosen so that

$$U^{+}_{m}(y) := \left(\prod_{j=1}^{m} {}_q\varphi_{j+}(y)\right)\int_y^b \frac{u(x)\prod_{j=1}^{m}k^{+}_{j}(x,y)}{\prod_{j=1}^{m}L^{+}_{jq}(x)}dx<\infty, \tag{11.50}$$

$\forall\, y\in[a,b]$, and that U^{+}_{m} is integrable on $[a,b]$.

A direct application of Theorem 11.5 gives:

Theorem 11.11 *It is all as in Remark 11.10. Here $\Phi_j:\mathbb{R}^n_+\to\mathbb{R}_+$, $j=1,...,m$, are convex functions increasing per coordinate. Then*

$$\int_a^b u(x)\prod_{j=1}^{m}\Phi_j\left(\frac{\left|\overrightarrow{e^{\gamma_j;\psi}_{\rho_j,\mu_j,\omega_j,a+}f_j(x)}\right|}{L^{+}_{jq}(x)}\right)dx\leq$$

$$\left(\prod_{\substack{j=1\\ j\neq\rho}}^{m}\int_a^b\Phi_j\left(\frac{\left|\overrightarrow{f_j(y)}\right|}{{}_q\varphi_{j+}(y)}\right)dy\right)\left(\int_a^b\Phi_\rho\left(\frac{\left|\overrightarrow{f_\rho(y)}\right|}{{}_q\varphi_{\rho+}(y)}\right)U^{+}_{m}(y)\,dy\right). \tag{11.51}$$

We make

Remark 11.12 Here $\rho_j,\mu_j,\gamma_j,\omega_j>0$; $f_{ji}\in C([a,b])$ and $\psi\in C^1([a,b])$ which is increasing; $j=1,...,m$ and $i=1,...,n$. Set

$${}_\infty\varphi_{j-}(y) := \left\|\overrightarrow{e^{\gamma_j;\psi}_{\rho_j,\mu_j,\omega_j,b-}f_j(y)}\right\|_\infty := \max_{\substack{j=1,...,m\\ i=1,...,n}}\left\{\left|e^{\gamma_j;\psi}_{\rho_j,\mu_j,\omega_j,b-}f_{ji}(y)\right|\right\}, \tag{11.52}$$

and

$${}_q\varphi_{j-}(y) := \left\|\overrightarrow{e^{\gamma_j;\psi}_{\rho_j,\mu_j,\omega_j,b-}f_j(y)}\right\|_q := \left(\sum_{i=1}^{n}\left|e^{\gamma_j;\psi}_{\rho_j,\mu_j,\omega_j,b-}f_{ji}(y)\right|^q\right)^{\frac{1}{q}},\quad q\geq1; \tag{11.53}$$

$y \in [a, b]$, which ${}_q\varphi_{j-}$ are continuous functions, $j = 1, ..., m$. We have also that

$$0 <_q \varphi_{j-}(y) < \infty \text{ in } [a, b], \tag{11.54}$$

$j = 1, ..., m$; where $1 \leq q \leq \infty$ is fixed.

Here it is

$$k_j^-(x, y) := k_j(x, y) =$$

$$\begin{cases} \psi'(y)(\psi(y) - \psi(x))^{\mu_j - 1} E_{\rho_j, \mu_j}^{\gamma_j}\left[\omega_j(\psi(y) - \psi(x))^{\rho_j}\right], & x \leq y < b, \\ 0, \quad a < y < x, \end{cases} \tag{11.55}$$

$j = 1, ..., m$, and

$$L_{jq}^-(x) := \int_x^b \psi'(y)(\psi(y) - \psi(x))^{\mu_j - 1} E_{\rho_j, \mu_j}^{\gamma_j}\left[\omega_j(\psi(y) - \psi(x))^{\rho_j}\right] {}_q\varphi_{j-}(y)\, dy, \tag{11.56}$$

$\forall\, x \in [a, b]$, $1 \leq q \leq \infty$.

We have that $L_{jq}^-(x) > 0$ on $[a, b]$.

Let $\rho \in \{1, ..., m\}$ be fixed. The weight function u is chosen so that

$$U_m^-(y) := \left(\prod_{j=1}^m {}_q\varphi_{j-}(y)\right) \int_a^y \frac{u(x) \prod_{j=1}^m k_j^-(x, y)}{\prod_{j=1}^m L_{jq}^-(x)} dx < \infty, \tag{11.57}$$

$\forall\, y \in [a, b]$, and that U_m^- is integrable on $[a, b]$.

A direct application of Theorem 11.5 gives:

Theorem 11.13 *It is all as in Remark 11.12. Here $\Phi_j : \mathbb{R}_+^n \to \mathbb{R}_+$, $j = 1, ..., m$, are convex functions increasing per coordinate. Then*

$$\int_a^b u(x) \prod_{j=1}^m \Phi_j\left(\frac{\left|\overrightarrow{e_{\rho_j, \mu_j, \omega_j, b-}^{\gamma_j; \psi} f_j(x)}\right|}{L_{jq}^-(x)}\right) dx \leq$$

$$\left(\prod_{\substack{j=1 \\ j \neq \rho}}^m \int_a^b \Phi_j\left(\frac{\left|\overrightarrow{f_j(y)}\right|}{{}_q\varphi_{j-}(y)}\right) dy\right)\left(\int_a^b \Phi_\rho\left(\frac{\left|\overrightarrow{f_\rho(y)}\right|}{{}_q\varphi_{\rho-}(y)}\right) U_m^-(y)\, dy\right). \tag{11.58}$$

We make

Remark 11.14 Here $j = 1, ..., m;$ $i = 1, ..., n.$ Let $\rho_j, \mu_j, \omega_j > 0,$ $\gamma_j < 0,$ and $f_{ji} \in C^{N_j}([a,b]), N_j = \lceil \mu_j \rceil, \mu_j \notin \mathbb{N}; \theta := \max(N_1, ..., N_m), \psi \in C^{\theta}([a,b]), \psi$ is increasing with $\psi'(x) \neq 0$ over $[a,b]$. Set $f_{ji\psi}^{[N_j]}(x) = \left(\frac{1}{\psi'(x)}\frac{d}{dx}\right)^{N_j} f_{ji}(x), x \in [a,b]$. Set

$$ {}_{\infty}\lambda_{j+}(y) := \left\| \overrightarrow{{}^{C}D^{\gamma_j;\psi}_{\rho_j,\mu_j,\omega_j,a+}f_j(y)} \right\|_{\infty} := \max_{\substack{j=1,...,m \\ i=1,...,n}} \left\{ \left| {}^{C}D^{\gamma_j;\psi}_{\rho_j,\mu_j,\omega_j,a+}f_{ji}(y) \right| \right\}, \tag{11.59} $$

and

$$ {}_{q}\lambda_{j+}(y) := \left\| \overrightarrow{{}^{C}D^{\gamma_j;\psi}_{\rho_j,\mu_j,\omega_j,a+}f_j(y)} \right\|_{q} := \left(\sum_{i=1}^{n} \left| {}^{C}D^{\gamma_j;\psi}_{\rho_j,\mu_j,\omega_j,a+}f_{ji}(y) \right|^{q} \right)^{\frac{1}{q}}, \quad q \geq 1; \tag{11.60} $$

$y \in [a,b]$, which all ${}_q\lambda_{j+}$ are continuous functions, $j = 1, ..., m$. We also have that

$$ 0 <_q \lambda_{j+}(y) < \infty \text{ in } [a,b], \tag{11.61} $$

$j = 1, ..., m$; where $1 \leq q \leq \infty$ is fixed.

Here it is

$$ {}^{C}k_j^{+}(x,y) := k_j(x,y) = $$

$$ \begin{cases} \psi'(y)(\psi(x) - \psi(y))^{N_j-\mu_j-1} E^{-\gamma_j}_{\rho_j,N_j-\mu_j}\left[\omega_j(\psi(x) - \psi(y))^{\rho_j}\right], & a < y \leq x, \\ 0, \quad x < y < b, \end{cases} \tag{11.62} $$

$j = 1, ..., m$, and

$$ {}^{C}L^{+}_{jq}(x) := \int_a^x \psi'(y)(\psi(x) - \psi(y))^{N_j-\mu_j-1} $$

$$ E^{-\gamma_i}_{\rho_j,N_j-\mu_j}\left[\omega_j(\psi(x) - \psi(y))^{\rho_j}\right] {}_q\lambda_{j+}(y)\,dy, \tag{11.63} $$

$\forall\, x \in [a,b]$, $1 \leq q \leq \infty$, $j = 1, ..., m$.

We have that ${}^{C}L^{+}_{jq}(x) > 0$ on $[a,b]$.

Let $\rho \in \{1, ..., m\}$ be fixed. The weight function u is chosen so that

$$ {}^{C}U^{+}_{m}(y) := \left(\prod_{j=1}^{m} {}_q\lambda_{j+}(y) \right) \int_y^b \frac{u(x) \prod_{j=1}^{m} {}^{C}k_j^{+}(x,y)}{\prod_{j=1}^{m} {}^{C}L^{+}_{jq}(x)} dx < \infty, \tag{11.64} $$

$\forall\, y \in [a,b]$, and that ${}^{C}U^{+}_{m}$ is integrable on $[a,b]$.

A direct application of Theorem 11.11, see also (11.6), gives:

Theorem 11.15 *It is all as in Remark 11.14. Here $\Phi_j : \mathbb{R}_+^n \to \mathbb{R}_+$, $j = 1, ..., m$, are convex functions increasing per coordinate. Then*

$$\int_a^b u(x) \prod_{j=1}^m \Phi_j \left(\frac{\left| \overrightarrow{{}^C D_{\rho_j,\mu_j,\omega_j,a+}^{\gamma_j;\psi} f_j(x)} \right|}{{}^C L_{jq}^+(x)} \right) dx \leq$$

$$\left(\prod_{\substack{j=1 \\ j \neq \rho}}^m \int_a^b \Phi_j \left(\frac{\left| \overrightarrow{f_{j\psi}^{[N_j]}(y)} \right|}{{}_q\lambda_{j+}(y)} \right) dy \right) \left(\int_a^b \Phi_\rho \left(\frac{\left| \overrightarrow{f_{\rho\psi}^{[N_\rho]}(y)} \right|}{{}_q\lambda_{\rho+}(y)} \right) {}^C U_m^+(y) \, dy \right). \quad (11.65)$$

We make

Remark 11.16 Here $j = 1, ..., m$; $i = 1, ..., n$. Let $\rho_j, \mu_j, \omega_j > 0$, $\gamma_j < 0$, and $f_{ji} \in C^{N_j}([a, b])$, $N_j = \lceil \mu_j \rceil$, $\mu_j \notin \mathbb{N}$; $\theta := \max(N_1, ..., N_m)$, $\psi \in C^\theta([a, b])$, ψ is increasing with $\psi'(x) \neq 0$ over $[a, b]$. Set $f_{ji\psi}^{[N_j]}(x) = \left(\frac{1}{\psi'(x)} \frac{d}{dx} \right)^{N_j} f_{ji}(x)$, $x \in [a, b]$. Set

$${}_\infty\lambda_{j-}(y) := \left\| \overrightarrow{{}^C D_{\rho_j,\mu_j,\omega_j,b-}^{\gamma_j;\psi} f_j(y)} \right\|_\infty := \max_{\substack{j=1,...,m \\ i=1,...,n}} \left\{ \left| {}^C D_{\rho_j,\mu_j,\omega_j,b-}^{\gamma_j;\psi} f_{ji}(y) \right| \right\}, \quad (11.66)$$

and

$${}_q\lambda_{j-}(y) := \left\| \overrightarrow{{}^C D_{\rho_j,\mu_j,\omega_j,b-}^{\gamma_j;\psi} f_j(y)} \right\|_q := \left(\sum_{i=1}^n \left| {}^C D_{\rho_j,\mu_j,\omega_j,b-}^{\gamma_j;\psi} f_{ji}(y) \right|^q \right)^{\frac{1}{q}}, \quad q \geq 1; \quad (11.67)$$

$y \in [a, b]$, which all ${}_q\lambda_{j-}$ are continuous functions, $j = 1, ..., m$. We also have that

$$0 <_q \lambda_{j-}(y) < \infty \text{ in } [a, b], \quad (11.68)$$

$j = 1, ..., m$; where $1 \leq q \leq \infty$ is fixed.

Here it is

$${}^C k_j^-(x, y) := k_j(x, y) =$$

$$\begin{cases} \psi'(y)(\psi(y) - \psi(x))^{N_j-\mu_j-1} E_{\rho_j,N_j-\mu_j}^{-\gamma_j} \left[\omega_j (\psi(y) - \psi(x))^{\rho_j} \right], & x \leq y < b, \\ 0, \quad a < y < x, \end{cases} \quad (11.69)$$

$j = 1, ..., m$, and

$${}^C L_{jq}^-(x) := \int_x^b \psi'(y)(\psi(y) - \psi(x))^{N_j-\mu_j-1}$$

$$E^{-\gamma_j}_{\rho_j,N_j-\mu_j}\left[\omega_j\left(\psi(y)-\psi(x)\right)^{\rho_j}\right]{}_q\lambda_{j-}(y)\,dy, \tag{11.70}$$

$\forall\, x \in [a,b]$, $1 \le q \le \infty$, $j = 1, ..., m$.

We have that ${}^C L^-_{jq}(x) > 0$ on $[a,b]$.

Let $\rho \in \{1, ..., m\}$ be fixed. The weight function u is chosen so that

$${}^C U^-_m(y) := \left(\prod_{j=1}^m {}_q\lambda_{j-}(y)\right)\int_a^y \frac{u(x)\prod_{j=1}^m {}^C k^-_j(x,y)}{\prod_{j=1}^m {}^C L^-_{jq}(x)}\,dx < \infty, \tag{11.71}$$

$\forall\, y \in [a,b]$, and that ${}^C U^-_m$ is integrable on $[a,b]$.

A direct application of Theorem 11.13, see also (11.7), gives:

Theorem 11.17 *It is all as in Remark 11.16. Here $\Phi_j : \mathbb{R}^n_+ \to \mathbb{R}_+$, $j = 1, ..., m$, are convex functions increasing per coordinate. Then*

$$\int_a^b u(x)\prod_{j=1}^m \Phi_j\left(\frac{\left|\overrightarrow{{}^C D^{\gamma_j;\psi}_{\rho_j,\mu_j,\omega_j,b-} f_j(x)}\right|}{{}^C L^-_{jq}(x)}\right)dx \le$$

$$\left(\prod_{\substack{j=1\\ j\ne\rho}}^m \int_a^b \Phi_j\left(\frac{\left|\overrightarrow{f^{[N_j]}_{j\psi}(y)}\right|}{{}_q\lambda_{j-}(y)}\right)dy\right)\left(\int_a^b \Phi_\rho\left(\frac{\left|\overrightarrow{f^{[N_\rho]}_{\rho\psi}(y)}\right|}{{}_q\lambda_{\rho-}(y)}\right){}^C U^-_m(y)\,dy\right). \tag{11.72}$$

We make

Remark 11.18 Here $j = 1, ..., m$; $i = 1, ..., n$. Let $\rho_j, \mu_j, \omega_j > 0$, $\gamma_j < 0$, and $f_{ji} \in C([a,b])$, $N_j = \lceil \mu_j \rceil$, $\mu_j \notin \mathbb{N}$; $\theta := \max(N_1, ..., N_m)$, $\psi \in C^\theta([a,b])$, ψ is increasing with $\psi'(x) \ne 0$ over $[a,b]$. Here $0 \le \beta_j \le 1$ and $\xi_j = \mu_j + \beta_j\left(N_j - \mu_j\right)$. We assume that ${}^{RL}D^{\gamma_j(1-\beta_j);\psi}_{\rho_j,\xi_j,\omega_j,a+} f_{ji} \in C([a,b])$, $j = 1, ..., m$, $i = 1, ..., n$. Set

$${}_\infty M_{j+}(y) := \left\|\overrightarrow{{}^H\mathbb{D}^{\gamma_j,\beta_j;\psi}_{\rho_j,\mu_j,\omega_j,a+} f_j(y)}\right\|_\infty := \max_{\substack{j=1,...,m\\ i=1,...,n}}\left\{\left|{}^H\mathbb{D}^{\gamma_j,\beta_j;\psi}_{\rho_j,\mu_j,\omega_j,a+} f_{ji}(y)\right|\right\}, \tag{11.73}$$

and

$${}_q M_{j+}(y) := \left\|\overrightarrow{{}^H\mathbb{D}^{\gamma_j,\beta_j;\psi}_{\rho_j,\mu_j,\omega_j,a+} f_j(y)}\right\|_q := \left(\sum_{i=1}^n \left|{}^H\mathbb{D}^{\gamma_j,\beta_j;\psi}_{\rho_j,\mu_j,\omega_j,a+} f_{ji}(y)\right|^q\right)^{\frac{1}{q}},\quad q \ge 1; \tag{11.74}$$

$y \in [a, b]$, which all ${}_qM_{j+}$ are continuous functions, $j = 1, ..., m$. We also have that

$$0 <_q M_{j+}(y) < \infty \text{ in } [a, b], \tag{11.75}$$

$j = 1, ..., m$; where $1 \leq q \leq \infty$ is fixed.

Here it is

$$^{P}k_j^{+}(x, y) := k_j(x, y) =$$

$$\begin{cases} \psi'(y)(\psi(x) - \psi(y))^{\xi_j - \mu_j - 1} E_{\rho_j, \xi_j - \mu_j}^{-\gamma_j \beta_j}\left[\omega_j (\psi(x) - \psi(y))^{\rho_j}\right], & a < y \leq x, \\ 0, \quad x < y < b, \end{cases} \tag{11.76}$$

$j = 1, ..., m$, and

$$^{P}L_{jq}^{+}(x) := \int_a^x \psi'(y)(\psi(x) - \psi(y))^{\xi_j - \mu_j - 1}$$

$$E_{\rho_j, \xi_j - \mu_j}^{-\gamma_j \beta_j}\left[\omega_j (\psi(x) - \psi(y))^{\rho_j}\right] {}_qM_{j+}(y)\, dy, \tag{11.77}$$

$\forall\, x \in [a, b]$, $1 \leq q \leq \infty$.

We have that ${}^{P}L_{jq}^{+}(x) > 0$ on $[a, b]$.

Let $\overline{\rho} \in \{1, ..., m\}$ be fixed. The weight function u is chosen so that

$$^{P}U_m^{+}(y) := \left(\prod_{j=1}^{m} {}_qM_{j+}(y)\right) \int_y^b \frac{u(x) \prod_{j=1}^{m} {}^{P}k_j^{+}(x, y)}{\prod_{j=1}^{m} {}^{P}L_{jq}^{+}(x)} dx < \infty, \tag{11.78}$$

$\forall\, y \in [a, b]$, and that ${}^{P}U_m^{+}$ is integrable on $[a, b]$.

A direct application of Theorem 11.11, see also (11.15), gives:

Theorem 11.19 *It is all as in Remark 11.18. Here $\Phi_j : \mathbb{R}_+^n \to \mathbb{R}_+$, $j = 1, ..., m$, are convex functions increasing per coordinate. Then*

$$\int_a^b u(x) \prod_{j=1}^{m} \Phi_j \left(\frac{\left| \overrightarrow{{}^{H}\mathbb{D}_{\rho_j, \mu_j, \omega_j, a+}^{\gamma_j, \beta_j; \psi} f_j(x)} \right|}{{}^{P}L_{jq}^{+}(x)} \right) dx \leq$$

$$\left(\prod_{\substack{j=1 \\ j \neq \overline{\rho}}}^{m} \int_a^b \Phi_j \left(\frac{\left| \overrightarrow{{}^{RL}D_{\rho_j, \xi_j, \omega_j, a+}^{\gamma_j (1 - \beta_j); \psi} f_j(y)} \right|}{{}_qM_{j+}(y)} \right) dy \right) \tag{11.79}$$

$$\left(\int_a^b \Phi_{\overline{\rho}}\left(\frac{\left|\overrightarrow{{}^{RL}D^{\gamma_j(1-\beta_j);\psi}_{\rho_j,\xi_j,\omega_j,a+}f_{\overline{\rho}}(y)}\right|}{{}_qM_{\overline{\rho}+}(y)}\right){}^PU_m^+(y)\,dy\right).$$

We make

Remark 11.20 Here $j=1,...,m$; $i=1,...,n$. Let $\rho_j,\mu_j,\omega_j>0$, $\gamma_j<0$, and $f_{ji}\in C([a,b])$, $N_j=\lceil\mu_j\rceil$, $\mu_j\notin\mathbb{N}$; $\theta:=\max(N_1,...,N_m)$, $\psi\in C^{\theta}([a,b])$, ψ is increasing with $\psi'(x)\neq 0$ over $[a,b]$. Here $0\leq\beta_j\leq 1$ and $\xi_j=\mu_j+\beta_j(N_j-\mu_j)$. We assume that ${}^{RL}D^{\gamma_j(1-\beta_j);\psi}_{\rho_j,\xi_j,\omega_j,b-}f_{ji}\in C([a,b]), j=1,...,m, i=1,...,n$. Set

$$_\infty M_{j-}(y):=\left\|\overrightarrow{{}^H\mathbb{D}^{\gamma_j,\beta_j;\psi}_{\rho_j,\mu_j,\omega_j,b-}f_j(y)}\right\|_\infty:=\max_{\substack{j=1,...,m\\ i=1,...,n}}\left\{\left|{}^H\mathbb{D}^{\gamma_j,\beta_j;\psi}_{\rho_j,\mu_j,\omega_j,b-}f_{ji}(y)\right|\right\},\quad (11.80)$$

and

$$_qM_{j-}(y):=\left\|\overrightarrow{{}^H\mathbb{D}^{\gamma_j,\beta_j;\psi}_{\rho_j,\mu_j,\omega_j,b-}f_j(y)}\right\|_q:=\left(\sum_{i=1}^n\left|{}^H\mathbb{D}^{\gamma_j,\beta_j;\psi}_{\rho_j,\mu_j,\omega_j,b-}f_{ji}(y)\right|^q\right)^{\frac{1}{q}},\quad q\geq 1;\quad (11.81)$$

$y\in[a,b]$, which all ${}_qM_{j-}$ are continuous functions, $j=1,...,m$. We also have that

$$0<_q M_{j-}(y)<\infty \text{ in } [a,b],\quad (11.82)$$

$j=1,...,m$; where $1\leq q\leq\infty$ is fixed.

Here it is

$$^Pk_j^-(x,y):=k_j(x,y)=$$

$$\begin{cases}\psi'(y)(\psi(y)-\psi(x))^{\xi_j-\mu_j-1}E^{-\gamma_j\beta_j}_{\rho_j,\xi_j-\mu_j}\left[\omega_j(\psi(y)-\psi(x))^{\rho_j}\right], & x\leq y<b,\\ 0,\ a<y<x,\end{cases}\quad (11.83)$$

$j=1,...,m$, and

$$^PL_{jq}^-(x):=\int_x^b\psi'(y)(\psi(y)-\psi(x))^{\xi_j-\mu_j-1}$$

$$E^{-\gamma_j\beta_j}_{\rho_j,\xi_j-\mu_j}\left[\omega_j(\psi(y)-\psi(x))^{\rho_j}\right]{}_qM_{j-}(y)\,dy,\quad (11.84)$$

$\forall\, x\in[a,b]$, $1\leq q\leq\infty$.

We have that ${}^PL_{jq}^-(x)>0$ on $[a,b]$.

Let $\overline{\rho}\in\{1,...,m\}$ be fixed. The weight function u is chosen so that

$$ {}^{P}U_{m}^{-}(y) := \left(\prod_{j=1}^{m} {}_{q}M_{j-}(y)\right) \int_{a}^{y} \frac{u(x) \prod_{j=1}^{m} {}^{P}k_{j}^{-}(x,y)}{\prod_{j=1}^{m} {}^{P}L_{jq}^{-}(x)} dx < \infty, \tag{11.85} $$

$\forall\, y \in [a,b]$, and that ${}^{P}U_{m}^{-}$ is integrable on $[a,b]$.

A direct application of Theorem 11.13, see also (11.16), gives:

Theorem 11.21 *It is all as in Remark 11.20. Here $\Phi_j : \mathbb{R}_{+}^{n} \to \mathbb{R}_{+}$, $j = 1, ..., m$, are convex functions increasing per coordinate. Then*

$$ \int_{a}^{b} u(x) \prod_{j=1}^{m} \Phi_j \left(\frac{\left| \overrightarrow{{}^{H}\mathbb{D}_{\rho_j,\mu_j,\omega_j,b-}^{\gamma_j,\beta_j;\psi} f_j(x)} \right|}{{}^{P}L_{jq}^{-}(x)} \right) dx \leq $$

$$ \left(\prod_{\substack{j=1 \\ j \neq \overline{\rho}}}^{m} \int_{a}^{b} \Phi_j \left(\frac{\left| \overrightarrow{{}^{RL}D_{\rho_j,\xi_j,\omega_j,b-}^{\gamma_j(1-\beta_j);\psi} f_j(y)} \right|}{{}_{q}M_{j-}(y)} \right) dy \right) \tag{11.86} $$

$$ \left(\int_{a}^{b} \Phi_{\overline{\rho}} \left(\frac{\left| \overrightarrow{{}^{RL}D_{\rho_j,\xi_j,\omega_j,b-}^{\gamma_j(1-\beta_j);\psi} f_{\overline{\rho}}(y)} \right|}{{}_{q}M_{\overline{\rho}-}(y)} \right) {}^{P}U_{m}^{-}(y)\, dy \right). $$

We make

Remark 11.22 The basic background here is as in Remark 11.10. Also ${}_{q}\varphi_{j+}(y)$, $1 \leq q \leq \infty$, $y \in [a,b]$ is as in (11.45)–(11.47); $k_j^{+}(x,y)$ is as (11.48) and $L_{jq}^{+}(x)$ as in (11.49), where $x, y \in [a,b]$. Here it is

$$ K_j^{+}(x) := K_j(x) = (\psi(x) - \psi(a))^{\mu_j} E_{\rho_j,\mu_j+1}^{\gamma_j} \left[\omega_j (\psi(x) - \psi(a))^{\rho_j} \right], \tag{11.87} $$

$\forall\, x \in [a,b]$, $j = 1, ..., m$. Indeed it is

$$ \frac{k_j^{+}(x,y)}{K_j^{+}(x)} = \left(\chi_{(a,x]}(y)\, \psi'(y) \frac{(\psi(x) - \psi(y))^{\mu_j - 1}}{(\psi(x) - \psi(a))^{\mu_j}} \right) \tag{11.88} $$

$$ \left(\frac{E_{\rho_j,\mu_j}^{\gamma_j} \left[\omega_j (\psi(x) - \psi(y))^{\rho_j} \right]}{E_{\rho_j,\mu_j+1}^{\gamma_j} \left[\omega_j (\psi(x) - \psi(a))^{\rho_j} \right]} \right), $$

$\forall\, x, y \in [a, b], j = 1, ..., m$; χ is the characteristic function.

We define ${}_qW_{j+}$ on $[a, b]$, with appropiate choice of weight function u, by

$$ {}_qW_{j+}(y) :=_q \varphi_{j+}(y) \left(\int_y^b \frac{u(x)\, k_j^+(x, y)}{K_j^+(x)} dx \right) < \infty, \tag{11.89} $$

$\forall\, y \in [a, b]$, and that ${}_qW_{j+}$ is integrable on $[a, b]$; $j = 1, ..., m$.

A direct application of Theorem 11.6, see also (11.2), follows:

Theorem 11.23 *It is all as in Remark 11.22. Let* $p_j > 1 : \sum_{j=1}^{m} \frac{1}{p_j} = 1$. *Let the functions* $\Phi_j : \mathbb{R}_+^n \to \mathbb{R}_+$, $j = 1, ..., m$, *be convex and increasing per coordinate. Then*

$$ \int_a^b u(x) \prod_{j=1}^{m} \Phi_j \left(\frac{\left| \overrightarrow{e_{\rho_j,\mu_j,\omega_j,a+}^{\gamma_j;\psi} f_j(x)} \right|}{L_{jq}^+(x)} \right) dx \le $$

$$ \prod_{j=1}^{m} \left(\int_a^b {}_qW_{j+}(y)\, \Phi_j \left(\frac{\left| \overrightarrow{f_j(y)} \right|}{{}_q\varphi_{j+}(y)} \right)^{p_j} dy \right)^{\frac{1}{p_j}}. \tag{11.90} $$

We make

Remark 11.24 The basic background here is as in Remark 11.12. Also ${}_q\varphi_{j-}(y)$, $1 \le q \le \infty$, $y \in [a, b]$ is as in (11.52)–(11.54); $k_j^-(x, y)$ is as (11.55) and $L_{jq}^-(x)$ as in (11.56), where $x, y \in [a, b]$. Here it is

$$ K_j^-(x) := K_j(x) = (\psi(b) - \psi(x))^{\mu_j} E_{\rho_j,\mu_j+1}^{\gamma_j} \left[\omega_j (\psi(b) - \psi(x))^{\rho_j} \right], \tag{11.91} $$

$\forall\, x \in [a, b], j = 1, ..., m$. Indeed it is

$$ \frac{k_j^-(x, y)}{K_j^-(x)} = \left(\chi_{[x,b)}(y)\, \psi'(y) \frac{(\psi(y) - \psi(x))^{\mu_j - 1}}{(\psi(b) - \psi(x))^{\mu_j}} \right) \tag{11.92} $$

$$ \left(\frac{E_{\rho_j,\mu_j}^{\gamma_j} \left[\omega_j (\psi(y) - \psi(x))^{\rho_j} \right]}{E_{\rho_j,\mu_j+1}^{\gamma_j} \left[\omega_j (\psi(b) - \psi(x))^{\rho_j} \right]} \right), $$

$\forall\, x, y \in [a, b], j = 1, ..., m$.

We define ${}_qW_{j-}$ on $[a, b]$, with appropiate choice of weight function u, by

$$ {}_qW_{j-}(y) :=_q \varphi_{j-}(y) \left(\int_a^y \frac{u(x)\, k_j^-(x, y)}{K_j^-(x)} dx \right) < \infty, \tag{11.93} $$

$\forall\, y \in [a, b]$, and that ${}_qW_{j-}$ is integrable on $[a, b]$; $j = 1, ..., m$.

A direct application of Theorem 11.6, see also (11.3), follows:

Theorem 11.25 *It is all as in Remark 11.24. Let $p_j > 1 : \sum_{j=1}^{m} \frac{1}{p_j} = 1$. Let the functions $\Phi_j : \mathbb{R}_+^n \to \mathbb{R}_+$, $j = 1, ..., m$, be convex and increasing per coordinate. Then*

$$\int_a^b u(x) \prod_{j=1}^{m} \Phi_j \left(\frac{\left| \overrightarrow{e^{\gamma_j;\psi}_{\rho_j,\mu_j,\omega_j,b-} f_j(x)} \right|}{L_{jq}^{-}(x)} \right) dx \leq$$

$$\prod_{j=1}^{m} \left(\int_a^b {}_qW_{j-}(y)\, \Phi_j \left(\frac{\left| \overrightarrow{f_j(y)} \right|}{{}_q\varphi_{j-}(y)} \right)^{p_j} dy \right)^{\frac{1}{p_j}}. \tag{11.94}$$

We need

Remark 11.26 The basic background here is as in Remark 11.14. Also ${}_q\lambda_{j+}(y)$, $1 \leq q \leq \infty$, $y \in [a, b]$ is as in (11.59)–(11.61); ${}^Ck_j^+(x, y)$ is as (11.62) and ${}^CL_{jq}^+(x)$ as in (11.63), where $x, y \in [a, b]$. Here it is

$${}^CK_j^+(x) := K_j(x) = (\psi(x) - \psi(a))^{N_j-\mu_j} E^{-\gamma_j}_{\rho_j,N_j-\mu_j+1}\left[\omega_j(\psi(x) - \psi(a))^{\rho_j}\right], \tag{11.95}$$

$\forall\, x \in [a, b]$, $j = 1, ..., m$. Indeed it is

$$\frac{{}^Ck_j^+(x, y)}{{}^CK_j^+(x)} = \left(\chi_{(a,x]}(y)\, \psi'(y)\, \frac{(\psi(x) - \psi(y))^{N_j-\mu_j-1}}{(\psi(x) - \psi(a))^{N_j-\mu_j}} \right)$$

$$\left(\frac{E^{-\gamma_j}_{\rho_j,N_j-\mu_j}\left[\omega_j(\psi(x) - \psi(y))^{\rho_j}\right]}{E^{-\gamma_j}_{\rho_j,N_j-\mu_j+1}\left[\omega_j(\psi(x) - \psi(a))^{\rho_j}\right]} \right), \tag{11.96}$$

$\forall\, x, y \in [a, b]$, $j = 1, ..., m$.

We define ${}_q^CW_{j+}$ on $[a, b]$, with appropiate choice of weight function u, by

$${}_q^CW_{j+}(y) := {}_q\lambda_{j+}(y) \left(\int_y^b \frac{u(x)\, {}^Ck_j^+(x, y)}{{}^CK_j^+(x)} dx \right) < \infty, \tag{11.97}$$

$\forall\, y \in [a, b]$, and that ${}_q^CW_{j+}$ is integrable on $[a, b]$; $j = 1, ..., m$.

A direct application of Theorem 11.23, see also (11.6), follows:

Theorem 11.27 *It is all as in Remark 11.26. Let* $p_j > 1 : \sum_{j=1}^{m} \frac{1}{p_j} = 1$. *Let the functions* $\Phi_j : \mathbb{R}_+^n \to \mathbb{R}_+$, $j = 1, ..., m$, *be convex and increasing per coordinate. Then*

$$\int_a^b u(x) \prod_{j=1}^{m} \Phi_j \left(\frac{\left| \overrightarrow{{}^C D^{\gamma_j;\psi}_{\rho_j,\mu_j,\omega_j,a+} f_j(x)} \right|}{{}^C L^+_{jq}(x)} \right) dx \leq$$

$$\prod_{j=1}^{m} \left(\int_a^b {}^C_q W_{j+}(y) \, \Phi_j \left(\frac{\left| \overrightarrow{f^{[N_j]}_{j\psi}(y)} \right|}{{}_q\lambda_{j+}(y)} \right)^{p_j} dy \right)^{\frac{1}{p_j}}. \quad (11.98)$$

We need

Remark 11.28 The basic background here is as in Remark 11.16. Also ${}_q\lambda_{j-}(y)$, $1 \leq q \leq \infty$, $y \in [a, b]$ is as in (11.66)–(11.68); ${}^C k_j^-(x, y)$ is as (11.69) and ${}^C L^-_{jq}(x)$ as in (11.70), where $x, y \in [a, b]$. Here it is

$${}^C K_j^-(x) := K_j(x) = (\psi(b) - \psi(x))^{N_j - \mu_j} E^{-\gamma_j}_{\rho_j, N_j - \mu_j + 1} \left[\omega_j (\psi(b) - \psi(x))^{\rho_j} \right], \quad (11.99)$$

$\forall\, x \in [a, b]$, $j = 1, ..., m$. Indeed it is

$$\frac{{}^C k_j^-(x, y)}{{}^C K_j^-(x)} = \left(\chi_{[x,b)}(y) \, \psi'(y) \frac{(\psi(y) - \psi(x))^{N_j - \mu_j - 1}}{(\psi(b) - \psi(x))^{N_j - \mu_j}} \right)$$

$$\left(\frac{E^{-\gamma_j}_{\rho_j, N_j - \mu_j} \left[\omega_j (\psi(y) - \psi(x))^{\rho_j} \right]}{E^{-\gamma_j}_{\rho_j, N_j - \mu_j + 1} \left[\omega_j (\psi(b) - \psi(x))^{\rho_j} \right]} \right), \quad (11.100)$$

$\forall\, x, y \in [a, b]$, $j = 1, ..., m$.

We define ${}^C_q W_{j-}$ on $[a, b]$, with appropiate choice of weight function u, by

$${}^C_q W_{j-}(y) :=_q \lambda_{j-}(y) \left(\int_a^y \frac{u(x) {}^C k_j^-(x, y)}{{}^C K_j^-(x)} dx \right) < \infty, \quad (11.101)$$

$\forall\, y \in [a, b]$, and that ${}^C_q W_{j-}$ is integrable on $[a, b]$; $j = 1, ..., m$.

A direct application of Theorem 11.25, see also (11.7), follows:

Theorem 11.29 *It is all as in Remark 11.28. Let* $p_j > 1 : \sum_{j=1}^{m} \frac{1}{p_j} = 1$. *Let the functions* $\Phi_j : \mathbb{R}_+^n \to \mathbb{R}_+$, $j = 1, ..., m$, *be convex and increasing per coordinate. Then*

$$\int_a^b u(x) \prod_{j=1}^m \Phi_j \left(\frac{\left| \overrightarrow{{}^C D^{\gamma_j;\psi}_{\rho_j,\mu_j,\omega_j,b-} f_j(x)} \right|}{{}^C L^-_{jq}(x)} \right) dx \leq$$

$$\prod_{j=1}^m \left(\int_a^b {}^C_q W_{j-}(y) \, \Phi_j \left(\frac{\left| \overrightarrow{f^{[N_j]}_{j\psi}(y)} \right|}{{}_q\lambda_{j-}(y)} \right)^{p_j} dy \right)^{\frac{1}{p_j}}. \tag{11.102}$$

We need

Remark 11.30 The basic background here is as in Remark 11.18. Also ${}_qM_{j+}(y)$, $1 \leq q \leq \infty$, $y \in [a,b]$ is as in (11.73)–(11.75); ${}^P k_j^+(x,y)$ is as (11.76) and ${}^P L^+_{jq}(x)$ as in (11.77), where $x, y \in [a,b]$. Here it is

$${}^P K_j^+(x) := K_j(x) = (\psi(x) - \psi(a))^{\xi_j - \mu_j} E^{-\gamma_j \beta_j}_{\rho_j, \xi_j - \mu_j + 1} \left[\omega_j (\psi(x) - \psi(a))^{\rho_j} \right], \tag{11.103}$$

$\forall\, x \in [a,b]$, $j = 1, ..., m$. Indeed it is

$$\frac{{}^P k_j^+(x,y)}{{}^P K_j^+(x)} = \left(\chi_{(a,x]}(y)\, \psi'(y) \frac{(\psi(x) - \psi(y))^{\xi_j - \mu_j - 1}}{(\psi(x) - \psi(a))^{\xi_j - \mu_j}} \right.$$

$$\left. \left(\frac{E^{-\gamma_j \beta_j}_{\rho_j, \xi_j - \mu_j} \left[\omega_j (\psi(x) - \psi(y))^{\rho_j} \right]}{E^{-\gamma_j \beta_j}_{\rho_j, \xi_j - \mu_j + 1} \left[\omega_j (\psi(x) - \psi(a))^{\rho_j} \right]} \right) \right), \tag{11.104}$$

$\forall\, x, y \in [a,b]$, $j = 1, ..., m$.

We define ${}^P_q W_{j+}$ on $[a,b]$, with appropiate choice of weight function u,

$${}^P_q W_{j+}(y) :=_q M_{j+}(y) \left(\int_y^b \frac{u(x)\, {}^P k_j^+(x,y)}{{}^P K_j^+(x)} dx \right) < \infty, \tag{11.105}$$

$\forall\, y \in [a,b]$, and that ${}^P_q W_{j+}$ is integrable on $[a,b]$; $j = 1, ..., m$.

A direct application of Theorem 11.23, see also (11.15), follows:

Theorem 11.31 *It is all as in Remark 11.30. Here $\Phi_j : \mathbb{R}^n_+ \to \mathbb{R}_+$, $j = 1, ..., m$, are convex functions increasing per coordinate. Then*

$$\int_a^b u(x) \prod_{j=1}^m \Phi_j \left(\frac{\left| \overrightarrow{{}^H\mathbb{D}^{\gamma_j,\beta_j;\psi}_{\rho_j,\mu_j,\omega_j,a+} f_j(x)} \right|}{{}^P L^+_{jq}(x)} \right) dx \leq$$

$$\prod_{j=1}^m \left(\int_a^b {}^P_q W_{j+}(y)\, \Phi_j \left(\frac{\left| \overrightarrow{{}^{RL}D^{\gamma_j(1-\beta_j);\psi}_{\rho_j,\xi_j,\omega_j,a+} f_j(y)} \right|}{{}_qM_{j+}(y)} \right)^{p_j} dy \right)^{\frac{1}{p_j}}. \tag{11.106}$$

We need

Remark 11.32 The basic background here is as in Remark 11.20. Also ${}_qM_{j-}(y)$, $1 \leq q \leq \infty$, $y \in [a, b]$ is as in (11.80)–(11.82); ${}^Pk_j^-(x, y)$ is as in (11.83) and ${}^PL^-_{jq}(x)$ as in (11.84), where $x, y \in [a, b]$. Here it is

$${}^PK_j^-(x) := K_j(x) = (\psi(b) - \psi(x))^{\xi_j-\mu_j} E^{-\gamma_j\beta_j}_{\rho_j,\xi_j-\mu_j+1}\left[\omega_j(\psi(b) - \psi(x))^{\rho_j}\right], \tag{11.107}$$

$\forall\, x \in [a, b], j = 1, ..., m$. Indeed it is

$$\frac{{}^Pk_j^-(x, y)}{{}^PK_j^-(x)} = \left(\chi_{[x,b)}(y)\, \psi'(y) \frac{(\psi(y) - \psi(x))^{\xi_j-\mu_j-1}}{(\psi(b) - \psi(x))^{\xi_j-\mu_j}} \right.$$

$$\left. \left(\frac{E^{-\gamma_j\beta_j}_{\rho_j,\xi_j-\mu_j}\left[\omega_j(\psi(y) - \psi(x))^{\rho_j}\right]}{E^{-\gamma_j\beta_j}_{\rho_j,\xi_j-\mu_j+1}\left[\omega_j(\psi(b) - \psi(x))^{\rho_j}\right]} \right) \right), \tag{11.108}$$

$\forall\, x, y \in [a, b], j = 1, ..., m$.

We define ${}^P_qW_{j-}$ on $[a, b]$, with appropiate choice of weight function u,

$${}^P_qW_{j-}(y) :=_q M_{j-}(y) \left(\int_a^y \frac{u(x)\, {}^Pk_j^-(x, y)}{{}^PK_j^-(x)} dx \right) < \infty, \tag{11.109}$$

$\forall\, y \in [a, b]$, and that ${}^P_qW_{j-}$ is integrable on $[a, b]$; $j = 1, ..., m$.

A direct application of Theorem 11.25, see also (11.16), follows:

Theorem 11.33 *It is all as in Remark 11.32. Here $\Phi_j : \mathbb{R}^n_+ \to \mathbb{R}_+$, $j = 1, ..., m$, are convex functions increasing per coordinate. Then*

$$\int_a^b u(x) \prod_{j=1}^{m} \Phi_j \left(\frac{\left| \overrightarrow{{}^{H}\mathbb{D}^{\gamma_j, \beta_j; \psi}_{\rho_j, \mu_j, \omega_j, b-} f_j (x)} \right|}{{}^{P}L^{-}_{jq} (x)} \right) dx \leq$$

$$\prod_{j=1}^{m} \left(\int_a^b {}^{P}_{q} W_{j-} (y) \, \Phi_j \left(\frac{\left| \overrightarrow{{}^{RL}D^{\gamma_j (1-\beta_j); \psi}_{\rho_j, \xi_j, \omega_j, b-} f_j (y)} \right|}{{}_{q}M_{j-} (y)} \right)^{p_j} dy \right)^{\frac{1}{p_j}}. \quad (11.110)$$

We make

Remark 11.34 Let $f_i \in C([a, b])$, $i = 1, ..., n$, and $\overrightarrow{f} = (f_1, ..., f_n)$. We set

$$\left\| \overrightarrow{f} (y) \right\|_{\infty} := \max \{ |f_1 (y)|, ..., |f_n (y)| \},$$

and

$$\left\| \overrightarrow{f} (y) \right\|_{q} := \left(\sum_{i=1}^{n} |f_i (y)|^q \right)^{\frac{1}{q}}, \quad q \geq 1; \; y \in [a, b]. \quad (11.111)$$

Clearly it is $\left\| \overrightarrow{f} (y) \right\|_{q} \in C([a, b])$, for all $1 \leq q \leq \infty$. We assume that $\left\| \overrightarrow{f} (y) \right\|_{q} > 0$, a.e. on (a, b), for $q \in [1, \infty]$ being fixed.

Let

$$L_q^{+} (x) := \int_a^x k^{+} (x, y) \left\| \overrightarrow{f} (y) \right\|_{q} dy, \; x \in [a, b], \quad (11.112)$$

$1 \leq q \leq \infty$ fixed.

We assume $L_q^{+} (x) > 0$ a.e. on (a, b).

Here we considered

$$k^{+} (x, y) := k (x, y) :=$$

$$\begin{cases} \psi' (y) (\psi (x) - \psi (y))^{\mu - 1} E^{\gamma}_{\rho, \mu} \left[\omega (\psi (x) - \psi (y))^{\rho} \right], & a < y \leq x, \\ 0, \; x < y < b, \end{cases} \quad (11.113)$$

where $\rho, \mu, \gamma, \omega > 0$; $\psi \in C^1([a, b])$ which is increasing.

The weight function u is chosen so that

$$W_q^{+} (y) := \left\| \overrightarrow{f} (y) \right\|_{q} \left(\int_y^b \frac{u (x) k^{+} (x, y)}{L_q^{+} (x)} dx \right) < \infty, \quad (11.114)$$

a.e. on (a, b) and that W_q^{+} is integrable on $[a, b]$.

A direct application of Theorem 11.8 produces:

Theorem 11.35 *Let all as in Remark 11.34. Here $\Phi : \mathbb{R}_{+}^{n} \to \mathbb{R}$ is a convex and increasing per coordinate function. Then*

$$\int_a^b u(x)\,\Phi\left(\frac{\left|\overrightarrow{e_{\rho,\mu,\omega,a+}^{\gamma;\psi} f}(x)\right|}{L_q^{+}(x)}\right)dx \leq \int_a^b W_q^{+}(y)\,\Phi\left(\frac{\left|\overrightarrow{f}(y)\right|}{\left\|\overrightarrow{f}(y)\right\|_q}\right)dy. \tag{11.115}$$

We make

Remark 11.36 Let $f_i \in C([a,b])$, $i = 1, \ldots, n$, and $\overrightarrow{f} = (f_1, \ldots, f_n)$. We set

$$\left\|\overrightarrow{f}(y)\right\|_\infty := \max\{|f_1(y)|, \ldots, |f_n(y)|\},$$
and
$$\left\|\overrightarrow{f}(y)\right\|_q := \left(\sum_{i=1}^n |f_i(y)|^q\right)^{\frac{1}{q}}, \quad q \geq 1;\ y \in [a,b]. \tag{11.116}$$

Clearly it is $\left\|\overrightarrow{f}(y)\right\|_q \in C([a,b])$, for all $1 \leq q \leq \infty$. We assume that $\left\|\overrightarrow{f}(y)\right\|_q > 0$, a.e. on (a,b), for $q \in [1,\infty]$ being fixed.

Let

$$L_q^{-}(x) := \int_x^b k^{-}(x,y)\left\|\overrightarrow{f}(y)\right\|_q dy,\ x \in [a,b], \tag{11.117}$$

$1 \leq q \leq \infty$ fixed.

We assume $L_q^{-}(x) > 0$ a.e. on (a,b).

Here we considered

$$k^{-}(x,y) := k(x,y) :=$$

$$\begin{cases} \psi'(y)(\psi(y)-\psi(x))^{\mu-1} E_{\rho,\mu}^{\gamma}\left[\omega(\psi(y)-\psi(x))^{\rho}\right], & x \leq y < b, \\ 0, \quad a \prec y \prec x, \end{cases} \tag{11.118}$$

where $\rho, \mu, \gamma, \omega > 0$; $\psi \in C^1([a,b])$ which is increasing.

The weight function u is chosen so that

$$W_q^{-}(y) := \left\|\overrightarrow{f}(y)\right\|_q \left(\int_a^y \frac{u(x)\,k^{-}(x,y)}{L_q^{-}(x)}dx\right) < \infty, \tag{11.119}$$

a.e. on (a,b) and that W_q^{-} is integrable on $[a,b]$.

A direct application of Theorem 11.8 produces:

Theorem 11.37 *Let all as in Remark 11.36. Here $\Phi : \mathbb{R}_{+}^{n} \to \mathbb{R}$ is a convex and increasing per coordinate function. Then*

$$\int_a^b u(x)\,\Phi\left(\frac{\left|\overrightarrow{e^{\gamma;\psi}_{\rho,\mu,\omega,b-}f}(x)\right|}{L_q^-(x)}\right)dx \le \int_a^b W_q^-(y)\,\Phi\left(\frac{\left|\overrightarrow{f}(y)\right|}{\left\|\overrightarrow{f}(y)\right\|_q}\right)dy. \tag{11.120}$$

Next we deal with the spherical shell:

Background 11.38 We need:

Let $N \ge 2$, $S^{N-1} := \{x \in \mathbb{R}^N : |x| = 1\}$ the unit sphere on $\mathbb{R}^N$, where $|\cdot|$ stands for the Euclidean norm in $\mathbb{R}^N$. Also denote the ball $B(0,R) := \{x \in \mathbb{R}^N : |x| < R\} \subseteq \mathbb{R}^N$, $R > 0$, and the spherical shell

$$A := B(0,R_2) - \overline{B(0,R_1)},\ \ 0 < R_1 < R_2. \tag{11.121}$$

For the following see [13, pp. 149–150], and [14, pp. 87–88].

For $x \in \mathbb{R}^N - \{0\}$ we can write uniquely $x = r\omega$, where $r = |x| > 0$, and $\omega = \frac{x}{r} \in S^{N-1}$, $|\omega| = 1$.

Clearly here

$$\mathbb{R}^N - \{0\} = (0,\infty) \times S^{N-1}, \tag{11.122}$$

and

$$\overline{A} = [R_1, R_2] \times S^{N-1}. \tag{11.123}$$

We will be using

Theorem 11.39 ([1, p. 322]) *Let $f : A \to \mathbb{R}$ be a Lebesgue integrable function. Then*

$$\int_A f(x)\,dx = \int_{S^{N-1}} \left(\int_{R_1}^{R_2} f(r\omega)\, r^{N-1} dr\right) d\omega. \tag{11.124}$$

So we are able to write an integral on the shell in polar form using the polar coordinates (r,ω).

We need

Definition 11.40 Let $\rho, \mu, \gamma, w > 0$; $f \in C\left(\overline{A}\right)$ and $\psi \in C^1([R_1, R_2])$ which is increasing. The left and right radial Prabhakar fractional integrals with respect to ψ are defined as follows:

$$\left(e^{\gamma;\psi}_{\rho,\mu,w,R_1+}f\right)(x) = \int_{R_1}^{r} \psi'(t)\,(\psi(r) - \psi(t))^{\mu-1}\, E^{\gamma}_{\rho,\mu}\left[w\,(\psi(r) - \psi(t))^{\rho}\right] f(t\omega)\,dt, \tag{11.125}$$

and

$$\left(e^{\gamma;\psi}_{\rho,\mu,w,R_2-}f\right)(x) = \int_{r}^{R_2} \psi'(t)\,(\psi(t) - \psi(r))^{\mu-1}\, E^{\gamma}_{\rho,\mu}\left[w\,(\psi(t) - \psi(r))^{\rho}\right] f(t\omega)\,dt, \tag{11.126}$$

where $x \in \overline{A}$, that is $x = r\omega$, $r \in [R_1, R_2]$, $\omega \in S^{N-1}$.

Based on [1, p. 288], and [2, 4], we have that (11.125), (11.126) are continuous functions over $\overline{A}$ when $\mu \ge 1$.

We make

Remark 11.41 Let $f_i \in C(\overline{A})$, where the shell A is as in (11.121), $i = 1, ..., n$, and $\overrightarrow{f} = (f_1, ..., f_n)$. We set

$$\left\| \overrightarrow{f}(y) \right\|_\infty := \max\{|f_1(y)|, ..., |f_n(y)|\},$$

and

$$\left\| \overrightarrow{f}(y) \right\|_q := \left(\sum_{i=1}^n |f_i(y)|^q \right)^{\frac{1}{q}}, \quad q \ge 1; \ y \in \overline{A}. \tag{11.127}$$

Clearly it is $\left\| \overrightarrow{f}(y) \right\|_q \in C(\overline{A})$, $1 \le q \le \infty$. One can write that

$$\left\| \overrightarrow{f}(y) \right\|_q = \left\| \overrightarrow{f}(t\omega) \right\|_q, \quad 1 \le q \le \infty, \tag{11.128}$$

where $t \in [R_1, R_2]$, $\omega \in S^{N-1}$; $y = t\omega$, by Background 11.38.

We assume that $\left\| \overrightarrow{f}(y) \right\|_q > 0$ on $\overline{A}$, $1 \le q \le \infty$ fixed.

Consider the kernel

$$k_*^+(r, t) := k(r, t) := \chi_{(R_1, r]}(t)\psi'(t)(\psi(r) - \psi(t))^{\mu-1} E_{\rho,\mu}^\gamma\left[w(\psi(r) - \psi(t))^\rho\right], \tag{11.129}$$

where $\rho, \mu, \gamma, w > 0$; $\psi \in C^1([R_1, R_2])$ which is increasing.

Let

$$L_{q*}^+(x) = L_{q*}^+(r\omega) = \int_{R_1}^{R_2} k_*^+(r, t) \left\| \overrightarrow{f(t\omega)} \right\|_q dt, \tag{11.130}$$

$x = r\omega \in \overline{A}$, $1 \le q \le \infty$ fixed; $r \in [R_1, R_2]$, $\omega \in S^{N-1}$.

We have that $L_{q*}^+(r\omega) > 0$ for $r \in (R_1, R_2]$, for every $\omega \in S^{N-1}$.

Here we choose the weight $u(x) = u(r\omega) = L_{q*}^+(r\omega)$.

Consider the function

$$W_{q*}^+(y) = W_{q*}^+(t\omega) = \left\| \overrightarrow{f(t\omega)} \right\|_q \left(\int_{R_1}^{R_2} k_*^+(r, t)\, dr \right) < \infty, \tag{11.131}$$

$\forall\, t \in [R_1, R_2]$, $\omega \in S^{N-1}$; and $W_{q*}^+(t\omega)$ is integrable over $[R_1, R_2]$, $\forall\, \omega \in S^{N-1}$.

Here $\Phi : \mathbb{R}_+^n \to \mathbb{R}$ is a convex and increasing per coordinate function. By (11.115) we obtain

$$\int_{R_1}^{R_2} L_{q*}^+(r\omega)\, \Phi\left(\frac{\left| \overrightarrow{e_{\rho,\mu,w,R_1+}^{\gamma;\psi} f(r\omega)} \right|}{L_{q*}^+(r\omega)} \right) dr \le \int_{R_1}^{R_2} W_{q*}^+(t\omega)\, \Phi\left(\frac{\left| \overrightarrow{f(t\omega)} \right|}{\left\| \overrightarrow{f(t\omega)} \right\|_q} \right) dt, \tag{11.132}$$

$\forall\, \omega \in S^{N-1}$.

Here we have $R_1 \le r \le R_2$, and $R_1^{N-1} \le r^{N-1} \le R_2^{N-1}$, and $R_2^{1-N} \le r^{1-N} \le R_1^{1-N}$, also $r^{N-1} r^{1-N} = 1$. Thus by (11.132), we have

$$\int_{R_1}^{R_2} L_{q*}^{+}(r\omega)\,\Phi\left(\frac{\left|\overrightarrow{e_{\rho,\mu,w,R_1+}^{\gamma;\psi}f\,(r\omega)}\right|}{L_{q*}^{+}(r\omega)}\right) r^{N-1}dr \leq$$

$$\left(\frac{R_2}{R_1}\right)^{N-1} \int_{R_1}^{R_2} W_{q*}^{+}(r\omega)\,\Phi\left(\frac{\left|\overrightarrow{f\,(r\omega)}\right|}{\left\|\overrightarrow{f\,(r\omega)}\right\|_q}\right) r^{N-1}dr, \tag{11.133}$$

$\forall\, \omega \in S^{N-1}$.

Therefore it holds

$$\int_{S^{N-1}} \left(\int_{R_1}^{R_2} L_{q*}^{+}(r\omega)\,\Phi\left(\frac{\left|\overrightarrow{e_{\rho,\mu,w,R_1+}^{\gamma;\psi}f\,(r\omega)}\right|}{L_{q*}^{+}(r\omega)}\right) r^{N-1}dr\right) d\omega \leq$$

$$\left(\frac{R_2}{R_1}\right)^{N-1} \int_{S^{N-1}} \left(\int_{R_1}^{R_2} W_{q*}^{+}(r\omega)\,\Phi\left(\frac{\left|\overrightarrow{f\,(r\omega)}\right|}{\left\|\overrightarrow{f\,(r\omega)}\right\|_q}\right) r^{N-1}dr\right) d\omega. \tag{11.134}$$

Using Theorem 11.39 we derive:

Theorem 11.42 *All as in Remark 11.41. Then*

$$\int_{A} L_{q*}^{+}(x)\,\Phi\left(\frac{\left|\overrightarrow{e_{\rho,\mu,w,R_1+}^{\gamma;\psi}f\,(x)}\right|}{L_{q*}^{+}(x)}\right) dx \leq$$

$$\left(\frac{R_2}{R_1}\right)^{N-1} \int_{A} W_{q*}^{+}(x)\,\Phi\left(\frac{\left|\overrightarrow{f\,(x)}\right|}{\left\|\overrightarrow{f\,(x)}\right\|_q}\right) dx, \tag{11.135}$$

where $\overrightarrow{e_{\rho,\mu,w,R_1+}^{\gamma;\psi}f\,(x)} = \left(\left(e_{\rho,\mu,w,R_1+}^{\gamma;\psi}f_1\right)(x), \ldots, \left(e_{\rho,\mu,w,R_1+}^{\gamma;\psi}f_n\right)(x)\right)$ *and coordinates are assumed to be continuous functions on* $\overline{A}$.

We make

Remark 11.43 Let $f_i \in C\left(\overline{A}\right)$, where the shell A is as in (11.121), $i = 1, \ldots, n$, and $\overrightarrow{f} = (f_1, \ldots, f_n)$. We set

$$\left\|\overrightarrow{f}\,(y)\right\|_{\infty} := \max\left\{|f_1(y)|, \ldots, |f_n(y)|\right\},$$

and

$$\left\|\overrightarrow{f}\,(y)\right\|_{q} := \left(\sum_{i=1}^{n} |f_i(y)|^q\right)^{\frac{1}{q}}, \quad q \geq 1;\ y \in \overline{A}. \tag{11.136}$$

Clearly it is $\left\| \overrightarrow{f}(y) \right\|_q \in C\left(\overline{A}\right)$, $1 \leq q \leq \infty$. One can write that

$$\left\| \overrightarrow{f}(y) \right\|_q = \left\| \overrightarrow{f}(t\omega) \right\|_q, \quad 1 \leq q \leq \infty, \tag{11.137}$$

where $t \in [R_1, R_2]$, $\omega \in S^{N-1}$; $y = t\omega$, by Background 11.38.

We assume that $\left\| \overrightarrow{f}(y) \right\|_q > 0$ on $\overline{A}$, $1 \leq q \leq \infty$ fixed.

Consider the kernel

$$k_*^-(r,t) := k(r,t) := \chi_{[r,R_2)}(t)\psi'(t)(\psi(t) - \psi(r))^{\mu-1} E_{\rho,\mu}^{\gamma}\left[w(\psi(t) - \psi(r))^{\rho}\right], \tag{11.138}$$

where $\rho, \mu, \gamma, w > 0$; $\psi \in C^1([R_1, R_2])$ which is increasing.

Let

$$L_{q*}^-(x) = L_{q*}^-(r\omega) = \int_{R_1}^{R_2} k_*^-(r,t) \left\| \overrightarrow{f(t\omega)} \right\|_q dt, \tag{11.139}$$

$x = r\omega \in \overline{A}$, $1 \leq q \leq \infty$ fixed; $r \in [R_1, R_2]$, $\omega \in S^{N-1}$.

We have that $L_{q*}^-(r\omega) > 0$ for $r \in (R_1, R_2]$, for every $\omega \in S^{N-1}$.

Here we choose the weight $u(x) = u(r\omega) = L_{q*}^-(r\omega)$.

Consider the function

$$W_{q*}^-(y) = W_{q*}^-(t\omega) = \left\| \overrightarrow{f(t\omega)} \right\|_q \left(\int_{R_1}^{R_2} k_*^-(r,t)\, dr \right) < \infty, \tag{11.140}$$

$\forall\, t \in [R_1, R_2]$, $\omega \in S^{N-1}$; and $W_{q*}^-(t\omega)$ is integrable over $[R_1, R_2]$, $\forall\, \omega \in S^{N-1}$.

Here $\Phi : \mathbb{R}_+^n \to \mathbb{R}$ is a convex and increasing per coordinate function. By (11.120) we obtain

$$\int_{R_1}^{R_2} L_{q*}^-(r\omega)\, \Phi\left(\frac{\left| \overrightarrow{e_{\rho,\mu,w,R_2-}^{\gamma;\psi} f(r\omega)} \right|}{L_{q*}^-(r\omega)} \right) dr \leq \int_{R_1}^{R_2} W_{q*}^-(t\omega)\, \Phi\left(\frac{\left| \overrightarrow{f(t\omega)} \right|}{\left\| \overrightarrow{f(t\omega)} \right\|_q} \right) dt, \tag{11.141}$$

$\forall\, \omega \in S^{N-1}$.

Here we have $R_1 \leq r \leq R_2$, and $R_1^{N-1} \leq r^{N-1} \leq R_2^{N-1}$, and $R_2^{1-N} \leq r^{1-N} \leq R_1^{1-N}$, also $r^{N-1}r^{1-N} = 1$. Thus by (11.141), we have

$$\int_{R_1}^{R_2} L_{q*}^-(r\omega)\, \Phi\left(\frac{\left| \overrightarrow{e_{\rho,\mu,w,R_2-}^{\gamma;\psi} f(r\omega)} \right|}{L_{q*}^-(r\omega)} \right) r^{N-1} dr \leq$$

$$\left(\frac{R_2}{R_1}\right)^{N-1} \int_{R_1}^{R_2} W_{q*}^-(r\omega)\, \Phi\left(\frac{\left| \overrightarrow{f(r\omega)} \right|}{\left\| \overrightarrow{f(r\omega)} \right\|_q} \right) r^{N-1} dr, \tag{11.142}$$

$\forall\, \omega \in S^{N-1}$.

Therefore it holds

$$\int_{S^{N-1}}\left(\int_{R_1}^{R_2} L_{q*}^{-}(r\omega)\,\Phi\left(\frac{\left|\overrightarrow{e_{\rho,\mu,w,R_2-}^{\gamma;\psi} f}(r\omega)\right|}{L_{q*}^{-}(r\omega)}\right) r^{N-1}dr\right) d\omega \leq$$

$$\left(\frac{R_2}{R_1}\right)^{N-1}\int_{S^{N-1}}\left(\int_{R_1}^{R_2} W_{q*}^{-}(r\omega)\,\Phi\left(\frac{\left|\overrightarrow{f}(r\omega)\right|}{\left\|\overrightarrow{f}(r\omega)\right\|_q}\right) r^{N-1}dr\right) d\omega. \tag{11.143}$$

Using Theorem 11.39 we derive:

Theorem 11.44 *All as in Remark 11.43. Then*

$$\int_A L_{q*}^{-}(x)\,\Phi\left(\frac{\left|\overrightarrow{e_{\rho,\mu,w,R_2-}^{\gamma;\psi} f}(x)\right|}{L_{q*}^{-}(x)}\right) dx \leq$$

$$\left(\frac{R_2}{R_1}\right)^{N-1}\int_A W_{q*}^{-}(x)\,\Phi\left(\frac{\left|\overrightarrow{f}(x)\right|}{\left\|\overrightarrow{f}(x)\right\|_q}\right) dx, \tag{11.144}$$

where $\overrightarrow{e_{\rho,\mu,w,R_2-}^{\gamma;\psi} f}(x) = \left(\left(e_{\rho,\mu,w,R_2-}^{\gamma;\psi} f_1\right)(x), ..., \left(e_{\rho,\mu,w,R_2-}^{\gamma;\psi} f_n\right)(x)\right)$ *and coordinates are assumed to be continuous functions on* $\overline{A}$.

We need

Definition 11.45 Let $\rho, \mu, w > 0$, $\gamma < 0$, $N = \lceil \mu \rceil$, $\mu \notin \mathbb{N}$; $f \in C^N\left(\overline{A}\right)$ and $\psi \in C^N([R_1, R_2])$, $\psi'(r) \neq 0$, $\forall\ r \in [R_1, R_2]$, and ψ is increasing. We define the ψ-Prabhakar-Caputo radial left and right fractional derivatives of order μ as follows ($x \in \overline{A}$; $x = r\omega$, $r \in [R_1, R_2]$, $\omega \in S^{N-1}$)

$$\left({}_R^C D_{\rho,\mu,w,R_1+}^{\gamma;\psi} f\right)(x) = \left({}_R^C D_{\rho,\mu,w,R_1+}^{\gamma;\psi} f\right)(r\omega) :=$$

$$\int_{R_1}^{r} \psi'(t)(\psi(r)-\psi(t))^{N-\mu-1} E_{\rho,N-\mu}^{-\gamma}\left[w(\psi(r)-\psi(t))^{\rho}\right]\left(\frac{1}{\psi'(r)}\frac{d}{dr}\right)^N f(t\omega)\,dt$$

$$\overset{(11.125)}{=} \left(e_{\rho,N-\mu,w,R_1+}^{-\gamma;\psi} f_{\psi}^{[N]}\right)(x), \tag{11.145}$$

where

$$f_{\psi}^{[N]}(x) = f_{\psi}^{[N]}(r\omega) := \left(\frac{1}{\psi'(r)}\frac{d}{dr}\right)^N f(r\omega), \tag{11.146}$$

is the Nth order ψ-radial derivative of f,

and

$$\left({}_{R}^{C}D^{\gamma;\psi}_{\rho,\mu,w,R_2-}f\right)(x) = \left({}_{R}^{C}D^{\gamma;\psi}_{\rho,\mu,w,R_2-}f\right)(r\omega) :=$$

$$(-1)^N \int_r^{R_2} \psi'(t)\left(\psi(t)-\psi(r)\right)^{N-\mu-1} E^{-\gamma}_{\rho,N-\mu}\left[w\left(\psi(t)-\psi(r)\right)^{\rho}\right] \tag{11.147}$$

$$\left(\frac{1}{\psi'(r)}\frac{d}{dr}\right)^N f(t\omega)\,dt \overset{(11.126)}{=} (-1)^N \left(e^{-\gamma;\psi}_{\rho,N-\mu,w,R_2-}f^{[N]}_{\psi}\right)(x),$$

$\forall\, x \in \overline{A}$.

In this chapter we assume that $\left({}_{R}^{C}D^{\gamma;\psi}_{\rho,\mu,w,R_1+}f\right)$ and $\left({}_{R}^{C}D^{\gamma;\psi}_{\rho,\mu,w,R_2-}f\right)$ are continuous functions over $\overline{A}$.

We make

Remark 11.46 Let $\rho, \mu, w > 0$, $\gamma < 0$, $N = \lceil \mu \rceil$, $\mu \notin \mathbb{N}$; $f_i \in C^N\left(\overline{A}\right)$, $i = 1, ..., n$, and $\overrightarrow{f} = (f_1, ..., f_n)$, and $\psi \in C^N([R_1, R_2])$, $\psi'(r) \neq 0$, $\forall\, r \in [R_1, R_2]$, and ψ is increasing. We follow Definition 11.45 and we set:

$$\left\|\overrightarrow{f^{[N]}_{\psi}}(y)\right\|_{\infty} := \max\left\{\left|f^{[N]}_{1\psi}(y)\right|, ..., \left|f^{[N]}_{n\psi}(y)\right|\right\},$$

and

$$\left\|\overrightarrow{f^{[N]}_{\psi}}(y)\right\|_{q} := \left(\sum_{i=1}^{n}\left|f^{[N]}_{i\psi}(y)\right|^q\right)^{\frac{1}{q}}, \quad q \geq 1;\ y \in \overline{A}. \tag{11.148}$$

One can write that

$$\left\|\overrightarrow{f^{[N]}_{\psi}}(y)\right\|_{q} = \left\|\overrightarrow{f^{[N]}_{\psi}}(t\omega)\right\|_{q}, \quad 1 \leq q \leq \infty, \tag{11.149}$$

where $t \in [R_1, R_2]$, $\omega \in S^{N-1}$; $y = t\omega$.

Notice that $\left\|\overrightarrow{f^{[N]}_{\psi}}(y)\right\|_{q} \in C\left(\overline{A}\right)$, $1 \leq q \leq \infty$.

We assume that $\left\|\overrightarrow{f^{[N]}_{\psi}}(y)\right\|_{q} > 0$ on $\overline{A}$, $1 \leq q \leq \infty$ fixed.

Consider the kernel

$${}^{C}k^{+}(r,t) := k(r,t) :=$$

$$\chi_{(R_1,r]}(t)\psi'(t)\left(\psi(r)-\psi(t)\right)^{N-\mu-1} E^{-\gamma}_{\rho,N-\mu}\left[w\left(\psi(r)-\psi(t)\right)^{\rho}\right]. \tag{11.150}$$

Let

$${}^{C}L^{+}_{q}(x) = {}^{C}L^{+}_{q}(r\omega) = \int_{R_1}^{R_2} {}^{C}k^{+}(r,t)\left\|\overrightarrow{f^{[N]}_{\psi}}(t\omega)\right\|_{q} dt, \tag{11.151}$$

$x = r\omega \in \overline{A}$, $1 \leq q \leq \infty$ fixed; $r \in [R_1, R_2]$, $\omega \in S^{N-1}$.

We have that ${}^{C}L^{+}_{q}(r\omega) > 0$ for $r \in (R_1, R_2]$, $\forall\, \omega \in S^{N-1}$.

Here we choose the weight $u(x) = u(r\omega) = {}^{C}L^{+}_{q}(r\omega)$.

Consider the function

$$^{C}W_{q}^{+}(y)={}^{C}W_{q}^{+}(t\omega)=\left\Vert \overrightarrow{f_{\psi}^{[N]}(t\omega)}\right\Vert_{q}\left(\int_{R_{1}}^{R_{2}}{}^{C}k^{+}(r,t)\,dr\right)<\infty, \tag{11.152}$$

$\forall\, t\in[R_1,R_2]$, $\omega\in S^{N-1}$; and $^{C}W_{q}^{+}(t\omega)$ is integrable over $[R_1,R_2]$, $\forall\,\omega\in S^{N-1}$.
Here $\Phi:\mathbb{R}_{+}^{n}\to\mathbb{R}$ is a convex and increasing per coordinate function.

A direct application of Theorem 11.42, along with (11.145) follows:

Theorem 11.47 *All as in Remark 11.46. Then*

$$\int_{A}{}^{C}L_{q}^{+}(x)\,\Phi\left(\frac{\left|\overrightarrow{\left({}_{R}^{C}D_{\rho,\mu,w,R_{1}+}^{\gamma;\psi}f\right)(x)}\right|}{^{C}L_{q}^{+}(x)}\right)dx\leq$$

$$\left(\frac{R_2}{R_1}\right)^{N-1}\int_{A}{}^{C}W_{q}^{+}(x)\,\Phi\left(\frac{\left|\overrightarrow{f_{\psi}^{[N]}(x)}\right|}{\left\Vert \overrightarrow{f_{\psi}^{[N]}(x)}\right\Vert_{q}}\right)dx, \tag{11.153}$$

where $\overrightarrow{\left({}_{R}^{C}D_{\rho,\mu,w,R_{1}+}^{\gamma;\psi}f\right)(x)}=\left(\left({}_{R}^{C}D_{\rho,\mu,w,R_{1}+}^{\gamma;\psi}f_{1}\right)(x),\ldots,\left({}_{R}^{C}D_{\rho,\mu,w,R_{1}+}^{\gamma;\psi}f_{n}\right)(x)\right)$ *and the coordinates are assumed to be continuous on* $\overline{A}$.

We make

Remark 11.48 Let $\rho,\mu,w>0$, $\gamma<0$, $N=\lceil\mu\rceil$, $\mu\notin\mathbb{N}$; $f_i\in C^N\left(\overline{A}\right)$, $i=1,\ldots,n$, and $\overrightarrow{f}=(f_1,\ldots,f_n)$, and $\psi\in C^N([R_1,R_2])$, $\psi'(r)\neq0$, $\forall\, r\in[R_1,R_2]$, and ψ is increasing. We follow Definition 11.45 and we set:

$$\left\Vert \overrightarrow{f_{\psi}^{[N]}(y)}\right\Vert_{\infty}:=\max\left\{\left|f_{1\psi}^{[N]}(y)\right|,\ldots,\left|f_{n\psi}^{[N]}(y)\right|\right\},$$

and

$$\left\Vert \overrightarrow{f_{\psi}^{[N]}(y)}\right\Vert_{q}:=\left(\sum_{i=1}^{n}\left|f_{i\psi}^{[N]}(y)\right|^{q}\right)^{\frac{1}{q}},\quad q\geq1;\ y\in\overline{A}. \tag{11.154}$$

One can write that

$$\left\Vert \overrightarrow{f_{\psi}^{[N]}(y)}\right\Vert_{q}=\left\Vert \overrightarrow{f_{\psi}^{[N]}}(t\omega)\right\Vert_{q},\quad 1\leq q\leq\infty, \tag{11.155}$$

where $t\in[R_1,R_2]$, $\omega\in S^{N-1}$; $y=t\omega$.

Notice that $\left\Vert \overrightarrow{f_{\psi}^{[N]}(y)}\right\Vert_{q}\in C\left(\overline{A}\right)$, $1\leq q\leq\infty$.

We assume that $\left\Vert \overrightarrow{f_{\psi}^{[N]}(y)}\right\Vert_{q}>0$ on $\overline{A}$, $1\leq q\leq\infty$ fixed.

Consider the kernel

$$^{C}k^{-}(r,t) := k(r,t) :=$$

$$\chi_{[r,R_2)}(t)\psi'(t)(\psi(t)-\psi(r))^{N-\mu-1}E^{-\gamma}_{\rho,N-\mu}\left[w(\psi(t)-\psi(r))^{\rho}\right]. \quad (11.156)$$

Let

$$^{C}L_q^{-}(x) = {}^{C}L_q^{-}(r\omega) = \int_{R_1}^{R_2} {}^{C}k^{-}(r,t)\left\|\overrightarrow{f_{\psi}^{[N]}(t\omega)}\right\|_q dt, \quad (11.157)$$

$x = r\omega \in \overline{A}$, $1 \leq q \leq \infty$ fixed; $r \in [R_1, R_2]$, $\omega \in S^{N-1}$.

We have that $^{C}L_q^{-}(r\omega) > 0$ for $r \in (R_1, R_2]$, $\forall\, \omega \in S^{N-1}$.

Here we choose the weight $u(x) = u(r\omega) = {}^{C}L_q^{-}(r\omega)$.

Consider the function

$$^{C}W_q^{-}(y) = {}^{C}W_q^{-}(t\omega) = \left\|\overrightarrow{f_{\psi}^{[N]}(t\omega)}\right\|_q \left(\int_{R_1}^{R_2} {}^{C}k^{-}(r,t)\,dr\right) < \infty, \quad (11.158)$$

$\forall\, t \in [R_1, R_2]$, $\omega \in S^{N-1}$; and $^{C}W_q^{-}(t\omega)$ is integrable over $[R_1, R_2]$, $\forall\, \omega \in S^{N-1}$.

Here $\Phi : \mathbb{R}_{+}^{n} \to \mathbb{R}$ is a convex and increasing per coordinate function.

A direct application of Theorem 11.44, along with (11.147) follows:

Theorem 11.49 *All as in Remark 11.48. Then*

$$\int_A {}^{C}L_q^{-}(x)\,\Phi\left(\frac{\left|\overrightarrow{\left({}_{R}^{C}D^{\gamma;\psi}_{\rho,\mu,w,R_2-}f\right)(x)}\right|}{^{C}L_q^{-}(x)}\right)dx \leq$$

$$\left(\frac{R_2}{R_1}\right)^{N-1}\int_A {}^{C}W_q^{-}(x)\,\Phi\left(\frac{\left|\overrightarrow{f_{\psi}^{[N]}(x)}\right|}{\left\|\overrightarrow{f_{\psi}^{[N]}(x)}\right\|_q}\right)dx, \quad (11.159)$$

where $\overrightarrow{\left({}_{R}^{C}D^{\gamma;\psi}_{\rho,\mu,w,R_2-}f\right)(x)} = \left(\left({}_{R}^{C}D^{\gamma;\psi}_{\rho,\mu,w,R_2-}f_1\right)(x), \ldots, \left({}_{R}^{C}D^{\gamma;\psi}_{\rho,\mu,w,R_2-}f_n\right)(x)\right)$ *and the coordinates are assumed to be continuous on* $\overline{A}$.

We need

Definition 11.50 Let $\rho, \mu, w > 0$, $\gamma < 0$, $N = \lceil \mu \rceil$, $\mu \notin \mathbb{N}$; $f \in C\left(\overline{A}\right)$ and $\psi \in C^N([R_1, R_2])$, $\psi'(r) \neq 0, \forall\, r \in [R_1, R_2]$, and ψ is increasing. The ψ-Prabhakar-Riemann Liouville left and right radial fractional derivatives of order μ are defined as follows (see also Definition 11.40)

$$\left({}_{R}^{RL}D^{\gamma;\psi}_{\rho,\mu,w,R_1+}f\right)(x) = \left({}_{R}^{RL}D^{\gamma;\psi}_{\rho,\mu,w,R_1+}f\right)(r\omega) :=$$

$$\left(\frac{1}{\psi'(r)}\frac{d}{dr}\right)^N \left(e^{-\gamma;\psi}_{\rho,N-\mu,w,R_1+}f\right)(x), \tag{11.160}$$

and

$$\left({}^{RL}_{R}D^{\gamma;\psi}_{\rho,\mu,w,R_2-}f\right)(x) = \left({}^{RL}_{R}D^{\gamma;\psi}_{\rho,\mu,w,R_2-}f\right)(r\omega) :=$$

$$\left(-\frac{1}{\psi'(r)}\frac{d}{dr}\right)^N \left(e^{-\gamma;\psi}_{\rho,N-\mu,w,R_2-}f\right)(x), \tag{11.161}$$

$\forall\, x \in \overline{A}$; where $x = r\omega$, $r \in [R_1, R_2]$, $\omega \in S^{N-1}$.

In this chapter we assume that $\left({}^{RL}_{R}D^{\gamma;\psi}_{\rho,\mu,w,R_1+}f\right), \left({}^{RL}_{R}D^{\gamma;\psi}_{\rho,\mu,w,R_2-}f\right) \in C\left(\overline{A}\right)$.

Next we define the ψ-Hilfer-Prabhakar left and right radial fractional derivatives of order μ and type $\beta \in [0, 1]$, as follows ($\xi := \mu + \beta(N-\mu)$, see also Definition 11.40):

$$\left({}^{H}_{R}\mathbb{D}^{\gamma,\beta;\psi}_{\rho,\mu,w,R_1+}f\right)(x) = \left({}^{H}_{R}\mathbb{D}^{\gamma,\beta;\psi}_{\rho,\mu,w,R_1+}f\right)(r\omega) :=$$

$$e^{-\gamma\beta;\psi}_{\rho,\xi-\mu,w,R_1+}\left({}^{RL}_{R}D^{\gamma(1-\beta);\psi}_{\rho,\xi,w,R_1+}f\right)(x), \tag{11.162}$$

and

$$\left({}^{H}_{R}\mathbb{D}^{\gamma,\beta;\psi}_{\rho,\mu,w,R_2-}f\right)(x) = \left({}^{H}_{R}\mathbb{D}^{\gamma,\beta;\psi}_{\rho,\mu,w,R_2-}f\right)(r\omega) :=$$

$$e^{-\gamma\beta;\psi}_{\rho,\xi-\mu,w,R_2-}\left({}^{RL}_{R}D^{\gamma(1-\beta);\psi}_{\rho,\xi,w,R_2-}f\right)(x), \tag{11.163}$$

$\forall\, x \in \overline{A}$; where $x = r\omega$, $r \in [R_1, R_2]$, $\omega \in S^{N-1}$.

In this chapter we assume that $\left({}^{H}_{R}\mathbb{D}^{\gamma,\beta;\psi}_{\rho,\mu,w,R_1+}f\right), \left({}^{H}_{R}\mathbb{D}^{\gamma,\beta;\psi}_{\rho,\mu,w,R_2-}f\right) \in C\left(\overline{A}\right)$.

We make

Remark 11.51 Let $\rho, \mu, w > 0$, $\gamma < 0$, $N = \lceil \mu \rceil$, $\mu \notin \mathbb{N}$; $0 \le \beta \le 1$, $\xi = \mu + \beta(N-\mu)$, $f_i \in C\left(\overline{A}\right)$, $i = 1, ..., n$, and $\psi \in C^N([R_1, R_2])$, $\psi'(r) \neq 0$, $\forall\ r \in [R_1, R_2]$, and ψ is increasing. We follow Definition 11.50, especially (11.162) and we set:

$$\left\|\overrightarrow{\left({}^{RL}_{R}D^{\gamma(1-\beta);\psi}_{\rho,\xi,w,R_1+}f\right)(y)}\right\|_\infty :=$$
$$\max\left\{\left|\left({}^{RL}_{R}D^{\gamma(1-\beta);\psi}_{\rho,\xi,w,R_1+}f_1\right)(y)\right|, ..., \left|\left({}^{RL}_{R}D^{\gamma(1-\beta);\psi}_{\rho,\xi,w,R_1+}f_n\right)(y)\right|\right\},$$

and

$$\left\|\overrightarrow{\left({}^{RL}_{R}D^{\gamma(1-\beta);\psi}_{\rho,\xi,w,R_1+}f\right)(y)}\right\|_q := \left(\sum_{i=1}^n \left|\left({}^{RL}_{R}D^{\gamma(1-\beta);\psi}_{\rho,\xi,w,R_1+}f_i\right)(y)\right|^q\right)^{\frac{1}{q}}, \quad q \ge 1;\ y \in \overline{A}. \tag{11.164}$$

One can write that

$$\left\|\overrightarrow{\left({}^{RL}_{R}D^{\gamma(1-\beta);\psi}_{\rho,\xi,w,R_1+}f\right)(y)}\right\|_q = \left\|\overrightarrow{\left({}^{RL}_{R}D^{\gamma(1-\beta);\psi}_{\rho,\xi,w,R_1+}f\right)(t\omega)}\right\|_q, \quad 1 \le q \le \infty, \tag{11.165}$$

where $t \in [R_1, R_2]$, $\omega \in S^{N-1}$; $y = t\omega$.

Notice that $\left\| \overrightarrow{\left({}_{R}^{RL}D_{\rho,\xi,w,R_1+}^{\gamma(1-\beta);\psi} f \right)(y)} \right\|_q \in C\left(\overline{A}\right)$, $1 \leq q \leq \infty$.

We assume that $\left\| \overrightarrow{\left({}_{R}^{RL}D_{\rho,\xi,w,R_1+}^{\gamma(1-\beta);\psi} f \right)(y)} \right\|_q > 0$ on $\overline{A}$, $1 \leq q \leq \infty$ fixed.

Consider the kernel

$$
{}^{P}k^{+}(r,t) := k(r,t) :=
$$

$$
\chi_{(R_1,r]}(t)\psi'(t)(\psi(r)-\psi(t))^{\xi-\mu-1} E_{\rho,\xi-\mu}^{-\gamma\beta}\left[w(\psi(r)-\psi(t))^{\rho}\right]. \tag{11.166}
$$

Let

$$
{}^{P}L_q^{+}(x) = {}^{P}L_q^{+}(r\omega) = \int_{R_1}^{R_2} {}^{P}k^{+}(r,t) \left\| \overrightarrow{\left({}_{R}^{RL}D_{\rho,\xi,w,R_1+}^{\gamma(1-\beta);\psi} f \right)(t\omega)} \right\|_q dt, \tag{11.167}
$$

$x = r\omega \in \overline{A}$, $1 \leq q \leq \infty$ fixed; $r \in [R_1, R_2]$, $\omega \in S^{N-1}$.

We have that ${}^{P}L_q^{+}(r\omega) > 0$ for $r \in (R_1, R_2]$, $\forall\, \omega \in S^{N-1}$.

Here we choose the weight $u(x) = u(r\omega) = {}^{P}L_q^{+}(r\omega)$.

Consider the function

$$
{}^{P}W_q^{+}(y) = {}^{P}W_q^{+}(t\omega) = \left\| \overrightarrow{\left({}_{R}^{RL}D_{\rho,\xi,w,R_1+}^{\gamma(1-\beta);\psi} f \right)(t\omega)} \right\|_q \left(\int_{R_1}^{R_2} {}^{P}k^{+}(r,t)\,dr \right) < \infty, \tag{11.168}
$$

$\forall\, t \in [R_1, R_2]$, $\omega \in S^{N-1}$; and ${}^{P}W_q^{+}(t\omega)$ is integrable over $[R_1, R_2]$, $\forall\, \omega \in S^{N-1}$.

Here $\Phi : \mathbb{R}_{+}^{n} \to \mathbb{R}$ is a convex and increasing per coordinate function.

A direct application of Theorem 11.42, along with (11.162) follows:

Theorem 11.52 *All as in Remark 11.51. Then*

$$
\int_A {}^{P}L_q^{+}(x)\,\Phi\left(\frac{\left| \overrightarrow{\left({}_{R}^{H}\mathbb{D}_{\rho,\mu,w,R_1+}^{\gamma,\beta;\psi} f \right)(x)} \right|}{{}^{P}L_q^{+}(x)} \right) dx \leq
$$

$$
\left(\frac{R_2}{R_1}\right)^{N-1} \int_A {}^{P}W_q^{+}(x)\,\Phi\left(\frac{\left| \overrightarrow{\left({}_{R}^{RL}D_{\rho,\xi,w,R_1+}^{\gamma(1-\beta);\psi} f \right)(x)} \right|}{\left\| \overrightarrow{\left({}_{R}^{RL}D_{\rho,\xi,w,R_1+}^{\gamma(1-\beta);\psi} f \right)(x)} \right\|_q} \right) dx, \tag{11.169}
$$

where $\overrightarrow{\left({}_{R}^{H}\mathbb{D}_{\rho,\mu,w,R_1+}^{\gamma,\beta;\psi} f \right)(x)} = \left(\left({}_{R}^{H}\mathbb{D}_{\rho,\mu,w,R_1+}^{\gamma,\beta;\psi} f_1 \right)(x), \ldots, \left({}_{R}^{H}\mathbb{D}_{\rho,\mu,w,R_1+}^{\gamma,\beta;\psi} f_n \right)(x) \right)$ *and the coordinates are assumed to be continuous on* $\overline{A}$.

We make

Remark 11.53 Let $\rho, \mu, w > 0$, $\gamma < 0$, $N = \lceil \mu \rceil$, $\mu \notin \mathbb{N}$; $0 \leq \beta \leq 1$, $\xi = \mu + \beta (N - \mu)$, $f_i \in C(\overline{A})$, $i = 1, ..., n$, and $\psi \in C^N([R_1, R_2])$, $\psi'(r) \neq 0$, $\forall\ r \in [R_1, R_2]$, and ψ is increasing. We follow Definition 11.50, especially (11.163) and we set:

$$\left\| \overrightarrow{\left({}^{RL}_{R} D^{\gamma(1-\beta);\psi}_{\rho,\xi,w,R_2-} f \right)(y)} \right\|_{\infty} := \max\left\{ \left| \left({}^{RL}_{R} D^{\gamma(1-\beta);\psi}_{\rho,\xi,w,R_2-} f_1 \right)(y) \right|, ..., \left| \left({}^{RL}_{R} D^{\gamma(1-\beta);\psi}_{\rho,\xi,w,R_2-} f_n \right)(y) \right| \right\},$$

and

$$\left\| \overrightarrow{\left({}^{RL}_{R} D^{\gamma(1-\beta);\psi}_{\rho,\xi,w,R_2-} f \right)(y)} \right\|_{q} := \left(\sum_{i=1}^{n} \left| \left({}^{RL}_{R} D^{\gamma(1-\beta);\psi}_{\rho,\xi,w,R_2-} f_i \right)(y) \right|^q \right)^{\frac{1}{q}}, \quad q \geq 1;\ y \in \overline{A}. \tag{11.170}$$

One can write that

$$\left\| \overrightarrow{\left({}^{RL}_{R} D^{\gamma(1-\beta);\psi}_{\rho,\xi,w,R_2-} f \right)(y)} \right\|_{q} = \left\| \overrightarrow{\left({}^{RL}_{R} D^{\gamma(1-\beta);\psi}_{\rho,\xi,w,R_2-} f \right)(t\omega)} \right\|_{q}, \quad 1 \leq q \leq \infty, \tag{11.171}$$

where $t \in [R_1, R_2]$, $\omega \in S^{N-1}$; $y = t\omega$.

Notice that $\left\| \overrightarrow{\left({}^{RL}_{R} D^{\gamma(1-\beta);\psi}_{\rho,\xi,w,R_2-} f \right)(y)} \right\|_{q} \in C(\overline{A})$, $1 \leq q \leq \infty$.

We assume that $\left\| \overrightarrow{\left({}^{RL}_{R} D^{\gamma(1-\beta);\psi}_{\rho,\xi,w,R_2-} f \right)(y)} \right\|_{q} > 0$ on $\overline{A}$, $1 \leq q \leq \infty$ fixed.

Consider the kernel

$$^{P}k^{-}(r, t) := k(r, t) :=$$

$$\chi_{[r,R_2)}(t)\psi'(t)(\psi(t) - \psi(r))^{\xi-\mu-1} E^{-\gamma\beta}_{\rho,\xi-\mu}\left[w(\psi(t) - \psi(r))^{\rho} \right]. \tag{11.172}$$

Let

$$^{P}L^{-}_{q}(x) = {}^{P}L^{-}_{q}(r\omega) = \int_{R_1}^{R_2} {}^{P}k^{-}(r, t) \left\| \overrightarrow{\left({}^{RL}_{R} D^{\gamma(1-\beta);\psi}_{\rho,\xi,w,R_2-} f \right)(t\omega)} \right\|_{q} dt, \tag{11.173}$$

$x = r\omega \in \overline{A}$, $1 \leq q \leq \infty$ fixed; $r \in [R_1, R_2]$, $\omega \in S^{N-1}$.

We have that ${}^{P}L^{-}_{q}(r\omega) > 0$ for $r \in (R_1, R_2]$, $\forall\ \omega \in S^{N-1}$.

Here we choose the weight $u(x) = u(r\omega) = {}^{P}L^{-}_{q}(r\omega)$.

Consider the function

$$^{P}W^{-}_{q}(y) = {}^{P}W^{-}_{q}(t\omega) = \left\| \overrightarrow{\left({}^{RL}_{R} D^{\gamma(1-\beta);\psi}_{\rho,\xi,w,R_2-} f \right)(t\omega)} \right\|_{q} \left(\int_{R_1}^{R_2} {}^{P}k^{-}(r, t)\, dr \right) < \infty, \tag{11.174}$$

$\forall\ t \in [R_1, R_2]$, $\omega \in S^{N-1}$; and ${}^{P}W^{-}_{q}(t\omega)$ is integrable over $[R_1, R_2]$, $\forall\ \omega \in S^{N-1}$.

Here $\Phi : \mathbb{R}^n_+ \to \mathbb{R}$ is a convex and increasing per coordinate function.

A direct application of Theorem 11.44, along with (11.163) follows:

Theorem 11.54 *All as in Remark 11.53. Then*

$$\int_A {}^P L_q^- (x)\, \Phi \left(\frac{\left| \overrightarrow{\left({}^H_R \mathbb{D}^{\gamma,\beta;\psi}_{\rho,\mu,w,R_2 -} f \right) (x)} \right|}{{}^P L_q^- (x)} \right) dx \leq$$

$$\left(\frac{R_2}{R_1} \right)^{N-1} \int_A {}^P W_q^- (x)\, \Phi \left(\frac{\left| \overrightarrow{\left({}^{RL}_R D^{\gamma(1-\beta);\psi}_{\rho,\xi,w,R_2 -} f \right) (x)} \right|}{\left\| \overrightarrow{\left({}^{RL}_R D^{\gamma(1-\beta);\psi}_{\rho,\xi,w,R_2 -} f \right) (x)} \right\|_q} \right) dx, \tag{11.175}$$

where $\overrightarrow{\left({}^H_R \mathbb{D}^{\gamma,\beta;\psi}_{\rho,\mu,w,R_2 -} f \right) (x)} = \left(\left({}^H_R \mathbb{D}^{\gamma,\beta;\psi}_{\rho,\mu,w,R_2 -} f_1 \right) (x), ..., \left({}^H_R \mathbb{D}^{\gamma,\beta;\psi}_{\rho,\mu,w,R_2 -} f_n \right) (x) \right)$ *and the coordinates are assumed to be continuous on* $\overline{A}$.

We make

Remark 11.55 Let $f_{ji} \in C([a,b])$, $j = 1, 2$; $i = 1, ..., m$; $\psi \in C^1([a,b])$ which is increasing. Let also $\rho_i, \mu_i, \gamma_i, \omega_i > 0$ and $\left(e^{\gamma_i;\psi}_{\rho_i,\mu_i,\omega_i,a+} f_{ji} \right) (x)$, $x \in [a,b]$ as in (11.2). We assume here that $0 < f_{2i}(y) < \infty$ on $[a,b]$, $i = 1, ..., m$.

Here we consider the kernel

$$k_i^+ (x,y) := k_i (x,y) =$$

$$\begin{cases} \psi'(y) (\psi(x) - \psi(y))^{\mu_i - 1} E^{\gamma_i}_{\rho_i,\mu_i} \left[\omega_i (\psi(x) - \psi(y))^{\rho_i} \right], & a < y \leq x, \\ 0, \quad x < y < b, \end{cases} \tag{11.176}$$

$i = 1, ..., m$.

Choose weight $u(x) \geq 0$, so that

$$\psi_i (y) := f_{2i} (y) \int_y^b u(x) \frac{k_i^+ (x,y)}{\left(e^{\gamma_i;\psi}_{\rho_i,\mu_i,\omega_i,a+} f_{2i} \right) (x)} dx < \infty, \tag{11.177}$$

a.e. on $[a,b]$, and that ψ_i is integrable on $[a,b]$, $i = 1, ..., m$.

Theorem 11.9 immediately implies:

Theorem 11.56 *All as in Remark 11.55. Let* $p_i > 1 : \sum_{i=1}^m \frac{1}{p_i} = 1$. *Let the functions* $\Phi_i : \mathbb{R}_+ \to \mathbb{R}_+$, $i = 1, ..., m$, *be convex and increasing. Then*

$$\int_a^b u(x) \prod_{i=1}^m \Phi_i \left(\frac{\left| \left(e^{\gamma_i;\psi}_{\rho_i,\mu_i,\omega_i,a+} f_{1i} \right) (x) \right|}{\left(e^{\gamma_i;\psi}_{\rho_i,\mu_i,\omega_i,a+} f_{2i} \right) (x)} \right) dx \leq$$

$$\prod_{i=1}^{m}\left(\int_{a}^{b}\psi_i(y)\,\Phi_i\left(\frac{|f_{1i}(y)|}{f_{2i}(y)}\right)^{p_i}dy\right)^{\frac{1}{p_i}}. \tag{11.178}$$

We make

Remark 11.57 Let $f_{ji}\in C([a,b])$, $j=1,2$; $i=1,\ldots,m$; $\psi\in C^1([a,b])$ which is increasing. Let also $\rho_i,\mu_i,\gamma_i,\omega_i>0$ and $\left(e^{\gamma_i;\psi}_{\rho_i,\mu_i,\omega_i,b-}f_{ji}\right)(x)$, $x\in[a,b]$ as in (11.3). We assume here that $0<f_{2i}(y)<\infty$ on $[a,b]$, $i=1,\ldots,m$.

Here we consider the kernel

$$k_i^-(x,y):=k_i(x,y)=$$

$$\begin{cases}\psi'(y)(\psi(y)-\psi(x))^{\mu_i-1}E^{\gamma_i}_{\rho_i,\mu_i}\left[\omega_i(\psi(y)-\psi(x))^{\rho_i}\right], & x\le y<b,\\ 0, & a<y<x,\end{cases} \tag{11.179}$$

$i=1,\ldots,m$.

Choose weight $u(x)\ge 0$, so that

$$\overline{\psi_i}(y):=f_{2i}(y)\int_a^y u(x)\frac{k_i^-(x,y)}{\left(e^{\gamma_i;\psi}_{\rho_i,\mu_i,\omega_i,b-}f_{2i}\right)(x)}dx<\infty, \tag{11.180}$$

a.e. on $[a,b]$, and that $\overline{\psi_i}$ is integrable on $[a,b]$, $i=1,\ldots,m$.

Theorem 11.9 immediately implies:

Theorem 11.58 *All as in Remark 11.57. Let $p_i>1:\sum_{i=1}^{m}\frac{1}{p_i}=1$. Let the functions $\Phi_i:\mathbb{R}_+\to\mathbb{R}_+$, $i=1,\ldots,m$, be convex and increasing. Then*

$$\int_a^b u(x)\prod_{i=1}^{m}\Phi_i\left(\frac{\left|\left(e^{\gamma_i;\psi}_{\rho_i,\mu_i,\omega_i,b-}f_{1i}\right)(x)\right|}{\left(e^{\gamma_i;\psi}_{\rho_i,\mu_i,\omega_i,b-}f_{2i}\right)(x)}\right)dx\le$$

$$\prod_{i=1}^{m}\left(\int_{a}^{b}\overline{\psi_i}(y)\,\Phi_i\left(\frac{|f_{1i}(y)|}{f_{2i}(y)}\right)^{p_i}dy\right)^{\frac{1}{p_i}}. \tag{11.181}$$

We make

Remark 11.59 Let $j=1,2$; $i=1,\ldots,n$; $\rho_i,\mu_i,\omega_i>0$, $\gamma_i<0$, $N_i=\lceil\mu_i\rceil$, $\mu_i\notin\mathbb{N}$; $\theta:=\max(N_1,\ldots,N_m)$, $\psi\in C^\theta([a,b])$, $\psi'(x)\neq 0$ over $[a,b]$, ψ is increasing; $f_{ji}\in C^{N_i}([a,b])$ and $f_{ji\psi}^{[N_i]}(x)=\left(\frac{1}{\psi'(x)}\frac{d}{dx}\right)^{N_i}f_{ji}(x)$, $\forall\,x\in[a,b]$. Here

$$\left({}^{C}D^{\gamma_i;\psi}_{\rho_i,\mu_i,\omega_i,a+}f_{ji}\right)(x)\overset{(11.6)}{=}\left(e^{-\gamma_i;\psi}_{\rho_i,N_i-\mu_i,\omega_i,a+}f_{ji\psi}^{[N_i]}\right)(x), \tag{11.182}$$

$\forall\, x \in [a,b], j = 1,2;\ i = 1,...,m.$

We assume that $0 < f_{2i\psi}^{[N_i]}(y) < \infty$ on $[a,b]$, $i = 1,...,m$.

Here we consider the kernel

$$^{C}k_i^{+}(x,y) := k_i(x,y) =$$

$$\begin{cases} \psi'(y)(\psi(x)-\psi(y))^{N_i-\mu_i-1} E_{\rho_i,N_i-\mu_i}^{-\gamma_i}\left[\omega_i(\psi(x)-\psi(y))^{\rho_i}\right], & a < y \le x, \\ 0, \quad x < y < b, \end{cases} \tag{11.183}$$

$i = 1,...,m.$

Choose weight $u \ge 0$, so that

$$^{C}\psi_i(y) := f_{2i\psi}^{[N_i]}(y) \int_y^b u(x) \frac{^{C}k_i^{+}(x,y)}{\left({}^{C}D_{\rho_i,\mu_i,\omega_i,a+}^{\gamma_i;\psi} f_{2i}\right)(x)} dx < \infty, \tag{11.184}$$

a.e. on $[a,b]$, and that $^{C}\psi_i$ is integrable on $[a,b]$, $i = 1,...,m$.

Theorem 11.56 immediately produces:

Theorem 11.60 *All as in Remark 11.59. Let* $p_i > 1 : \sum_{i=1}^{m} \frac{1}{p_i} = 1$. *Let the functions* $\Phi_i : \mathbb{R}_+ \to \mathbb{R}_+$, $i = 1,...,m$, *be convex and increasing. Then*

$$\int_a^b u(x) \prod_{i=1}^{m} \Phi_i\left(\frac{\left|\left({}^{C}D_{\rho_i,\mu_i,\omega_i,a+}^{\gamma_i;\psi} f_{1i}\right)(x)\right|}{\left({}^{C}D_{\rho_i,\mu_i,\omega_i,a+}^{\gamma_i;\psi} f_{2i}\right)(x)}\right) dx \le$$

$$\prod_{i=1}^{m}\left(\int_a^b {}^{C}\psi_i(y)\, \Phi_i\left(\frac{\left|f_{1i\psi}^{[N_i]}(y)\right|}{f_{2i\psi}^{[N_i]}(y)}\right)^{p_i} dy\right)^{\frac{1}{p_i}}. \tag{11.185}$$

We make

Remark 11.61 Let $j = 1,2;\ i = 1,...,n;\ \rho_i, \mu_i, \omega_i > 0,\ \gamma_i < 0,\ N_i = \lceil \mu_i \rceil,\ \mu_i \notin \mathbb{N};\ \theta := \max(N_1,...,N_m)$, $\psi \in C^{\theta}([a,b])$, $\psi'(x) \ne 0$ over $[a,b]$, ψ is increasing; $f_{ji} \in C^{N_i}([a,b])$ and $f_{ji\psi}^{[N_i]}(x) = \left(\frac{1}{\psi'(x)}\frac{d}{dx}\right)^{N_i} f_{ji}(x)$, $\forall\, x \in [a,b]$. Here

$$\left({}^{C}D_{\rho_i,\mu_i,\omega_i,b-}^{\gamma_i;\psi} f_{ji}\right)(x) \overset{(11.7)}{=} (-1)^{N_i}\left(e_{\rho_i,N_i-\mu_i,\omega_i,b-}^{-\gamma_i;\psi} f_{ji\psi}^{[N_i]}\right)(x), \tag{11.186}$$

$\forall\, x \in [a,b], j = 1,2;\ i = 1,...,m.$

We assume that $0 < f_{2i\psi}^{[N_i]}(y) < \infty$ on $[a,b]$, $i = 1,...,m$.

Here we consider the kernel

$$^{C}k_i^{-}(x,y) := k_i(x,y) =$$

$$\begin{cases} \psi'(y)(\psi(y)-\psi(x))^{N_i-\mu_i-1} E^{-\gamma_i}_{\rho_i,N_i-\mu_i}\left[\omega_i(\psi(y)-\psi(x))^{\rho_i}\right], & x \le y < b, \\ 0, \quad a < y < x, \end{cases} \tag{11.187}$$

$i = 1, ..., m$.

Choose weight $u \ge 0$, so that

$$^{C}\overline{\psi_i}(y) := f^{[N_i]}_{2i\psi}(y) \int_a^y u(x) \frac{^{C}k_i^{-}(x,y)}{\left(^{C}D^{\gamma_i;\psi}_{\rho_i,\mu_i,\omega_i,b-} f_{2i}\right)(x)} dx < \infty, \tag{11.188}$$

a.e. on $[a, b]$, and that $^{C}\overline{\psi_i}$ is integrable on $[a, b]$, $i = 1, ..., m$.

Theorem 11.58 immediately produces:

Theorem 11.62 *All as in Remark 11.61. Let $p_i > 1 : \sum_{i=1}^{m} \frac{1}{p_i} = 1$. Let the functions $\Phi_i : \mathbb{R}_+ \to \mathbb{R}_+$, $i = 1, ..., m$, be convex and increasing. Then*

$$\int_a^b u(x) \prod_{i=1}^{m} \Phi_i \left(\frac{\left| \left(^{C}D^{\gamma_i;\psi}_{\rho_i,\mu_i,\omega_i,b-} f_{1i}\right)(x) \right|}{\left(^{C}D^{\gamma_i;\psi}_{\rho_i,\mu_i,\omega_i,b-} f_{2i}\right)(x)} \right) dx \le$$

$$\prod_{i=1}^{m} \left(\int_a^b {^{C}\overline{\psi_i}}(y)\, \Phi_i \left(\frac{\left| f^{[N_i]}_{1i\psi}(y) \right|}{f^{[N_i]}_{2i\psi}(y)} \right)^{p_i} dy \right)^{\frac{1}{p_i}}. \tag{11.189}$$

We make

Remark 11.63 Let $j = 1, 2$; $i = 1, ..., m$; $\rho_i, \mu_i, \omega_i > 0$, $\gamma_i < 0$, $N_i = \lceil \mu_i \rceil$, $\mu_i \notin \mathbb{N}$; $\theta := \max(N_1, ..., N_m)$, $\psi \in C^{\theta}([a, b])$, $\psi'(x) \neq 0$ over $[a, b]$, ψ is increasing; $f_{ji} \in C([a, b])$. Let $0 \le \beta_i \le 1$ and $\xi_i = \mu_i + \beta_i (N_i - \mu_i)$, $i = 1, ..., m$. We assume that $^{RL}D^{\gamma_i(1-\beta_i);\psi}_{\rho_i,\xi_i,\omega_i,a+} f_{ji} \in C([a, b])$ and $0 <^{RL} D^{\gamma_i(1-\beta_i);\psi}_{\rho_i,\xi_i,\omega_i,a+} f_{2i}(y) < \infty$ on $[a, b]$, $i = 1, ..., m$. Here we have

$$\left(^{H}\mathbb{D}^{\gamma_i,\beta_i;\psi}_{\rho_i,\mu_i,\omega_i,a+} f_{ji}\right)(x) \overset{(11.15)}{=} e^{-\gamma_i\beta_i;\psi}_{\rho_i,\xi_i-\mu_i,\omega_i,a+} \; {^{RL}D^{\gamma_i(1-\beta_i);\psi}_{\rho_i,\xi_i,\omega_i,a+}} f_{ji}(x), \tag{11.190}$$

$\forall\, x \in [a, b]$, $j = 1, 2$; $i = 1, ..., m$.

Here we consider the kernel

$$^{P}k_i^{+}(x, y) := k_i(x, y) =$$

$$\begin{cases} \psi'(y)(\psi(x)-\psi(y))^{\xi_i-\mu_i-1} E^{-\gamma_i\beta_i}_{\rho_i,\xi_i-\mu_i}\left[\omega_i(\psi(x)-\psi(y))^{\rho_i}\right], & a < y \le x, \\ 0, \quad x < y < b, \end{cases} \tag{11.191}$$

$i = 1, ..., m.$

Choose weight $u \geq 0$, so that

$$ {}^{P}\psi_i(y) := \left({}^{RL}D^{\gamma_i(1-\beta_i);\psi}_{\rho_i,\xi_i,\omega_i,a+} f_{2i}(y) \right) \int_y^b \frac{u(x)\, {}^{P}k_i^{+}(x,y)}{\left({}^{H}\mathbb{D}^{\gamma_i,\beta_i;\psi}_{\rho_i,\mu_i,\omega_i,a+} f_{2i} \right)(x)} dx < \infty, \tag{11.192} $$

a.e. on $[a, b]$, and that ${}^{P}\psi_i$ is integrable on $[a, b]$, $i = 1, ..., m$.

Theorem 11.56 immediately produces:

Theorem 11.64 *All as in Remark 11.63. Let $p_i > 1 : \sum_{i=1}^{m} \frac{1}{p_i} = 1$. Let the functions $\Phi_i : \mathbb{R}_+ \to \mathbb{R}_+$, $i = 1, ..., m$, be convex and increasing. Then*

$$ \int_a^b u(x) \prod_{i=1}^{m} \Phi_i \left(\frac{\left| \left({}^{H}\mathbb{D}^{\gamma_i,\beta_i;\psi}_{\rho_i,\mu_i,\omega_i,a+} f_{1i} \right)(x) \right|}{\left({}^{H}\mathbb{D}^{\gamma_i,\beta_i;\psi}_{\rho_i,\mu_i,\omega_i,a+} f_{2i} \right)(x)} \right) dx \leq $$

$$ \prod_{i=1}^{m} \left(\int_a^b {}^{P}\psi_i(y)\, \Phi_i \left(\frac{\left| \left({}^{RL}D^{\gamma_i(1-\beta_i);\psi}_{\rho_i,\xi_i,\omega_i,a+} f_{1i} \right)(y) \right|}{\left({}^{RL}D^{\gamma_i(1-\beta_i);\psi}_{\rho_i,\xi_i,\omega_i,a+} f_{2i} \right)(y)} \right)^{p_i} dy \right)^{\frac{1}{p_i}}. \tag{11.193} $$

We make

Remark 11.65 Let $j = 1, 2$; $i = 1, ..., m$; $\rho_i, \mu_i, \omega_i > 0$, $\gamma_i < 0$, $N_i = \lceil \mu_i \rceil$, $\mu_i \notin \mathbb{N}$; $\theta := \max(N_1, ..., N_m)$, $\psi \in C^{\theta}([a, b])$, $\psi'(x) \neq 0$ over $[a, b]$, ψ is increasing; $f_{ji} \in C([a, b])$. Let $0 \leq \beta_i \leq 1$ and $\xi_i = \mu_i + \beta_i(N_i - \mu_i)$, $i = 1, ..., m$. We assume that ${}^{RL}D^{\gamma_i(1-\beta_i);\psi}_{\rho_i,\xi_i,\omega_i,b-} f_{ji} \in C([a, b])$ and $0 < {}^{RL}D^{\gamma_i(1-\beta_i);\psi}_{\rho_i,\xi_i,\omega_i,b-} f_{2i}(y) < \infty$ on $[a, b]$, $i = 1, ..., m$. Here we have

$$ \left({}^{H}\mathbb{D}^{\gamma_i,\beta_i;\psi}_{\rho_i,\mu_i,\omega_i,b-} f_{ji} \right)(x) \overset{(11.16)}{=} e^{-\gamma_i\beta_i;\psi}_{\rho_i,\xi_i-\mu_i,\omega_i,b-}\; {}^{RL}D^{\gamma_i(1-\beta_i);\psi}_{\rho_i,\xi_i,\omega_i,b-} f_{ji}(x), \tag{11.194} $$

$\forall\, x \in [a, b]$, $j = 1, 2$; $i = 1, ..., m$.

Here we consider the kernel

$$ {}^{P}k_i^{-}(x,y) := k_i(x,y) = $$

$$ \begin{cases} \psi'(y)(\psi(y) - \psi(x))^{\xi_i-\mu_i-1} E^{-\gamma_i\beta_i}_{\rho_i,\xi_i-\mu_i}\left[\omega_i(\psi(y) - \psi(x))^{\rho_i}\right], & x \leq y < b, \\ 0, \quad a < y < x, \end{cases} \tag{11.195} $$

$i = 1, ..., m.$

Choose weight $u \geq 0$, so that

$$ {}^{P}\overline{\psi_i}(y) := \left({}^{RL}D^{\gamma_i(1-\beta_i);\psi}_{\rho_i,\xi_i,\omega_i,b-} f_{2i}(y) \right) \int_a^y \frac{u(x)\, {}^{P}k_i^{-}(x,y)}{\left({}^{H}\mathbb{D}^{\gamma_i,\beta_i;\psi}_{\rho_i,\mu_i,\omega_i,b-} f_{2i} \right)(x)} dx < \infty, \tag{11.196} $$

a.e. on $[a, b]$, and that ${}^{P}\overline{\psi_i}$ is integrable on $[a, b]$, $i = 1, ..., m$.

Theorem 11.58 immediately produces:

Theorem 11.66 *All as in Remark 11.65. Let* $p_i > 1 : \sum_{i=1}^{m} \frac{1}{p_i} = 1$. *Let the functions* $\Phi_i : \mathbb{R}_+ \to \mathbb{R}_+$, $i = 1, ..., m$, *be convex and increasing. Then*

$$\int_a^b u(x) \prod_{i=1}^{m} \Phi_i \left(\frac{\left| \left({}^{H}\mathbb{D}^{\gamma_i, \beta_i; \psi}_{\rho_i, \mu_i, \omega_i, b-} f_{1i} \right)(x) \right|}{\left({}^{H}\mathbb{D}^{\gamma_i, \beta_i; \psi}_{\rho_i, \mu_i, \omega_i, b-} f_{2i} \right)(x)} \right) dx \leq$$

$$\prod_{i=1}^{m} \left(\int_a^b {}^{P}\overline{\psi_i}(y)\, \Phi_i \left(\frac{\left| \left({}^{RL}D^{\gamma_i(1-\beta_i);\psi}_{\rho_i, \xi_i, \omega_i, b-} f_{1i} \right)(y) \right|}{\left({}^{RL}D^{\gamma_i(1-\beta_i);\psi}_{\rho_i, \xi_i, \omega_i, b-} f_{2i} \right)(y)} \right)^{p_i} dy \right)^{\frac{1}{p_i}}. \quad (11.197)$$

References

1. Anastassiou, G.A.: Fractional Differentiation Inequalities. Research Monograph, Springer, New York (2009)
2. Anastassiou, G.A.: Fractional representation formulae and right fractional inequalities. Math. Comput. Model. **54**(11–12), 3098–3115 (2011)
3. Anastassiou, G.A.: Intell. Comparisons: Anal. Inequal. Springer, Heidelberg, New York (2016)
4. Anastassiou, G.A.: Foundations of Generalized Prabhakar-Hilfer Fractional Calculus with Applications (2021) (submitted)
5. Anastassiou, G.A.: Vectorial Prabhakar Hardy type generalized fractional inequalities under convexity (2021) (submitted)
6. Giusti, A., et al.: A practical guide to Prabhakar fractional calculus. Fract. Calculus App. Anal. **23**(1), 9–54 (2020)
7. Gorenflo, R., Kilbas, A., Mainardi, F., Rogosin, S.: Mittag-Leffler Functions. Related Topics and Applications. Springer, Heidelberg, New York (2014)
8. Hardy, H.G.: Notes on some points in the integral calculus. Messenger Math. **47**(10), 145–150 (1918)
9. Iqbal, S., Krulic, K., Pecaric, J.: On an inequality of H.G. Hardy. J. Inequal. Appl. **2010**(Article ID 264347), 23 p
10. Iqbal, S., Krulic, K., Pecaric, J.: On an inequality for convex functions with some applications on fractional derivatives and fractional integrals. J. Math. Inequal. **5**(2), 219–230 (2011)
11. Polito, F., Tomovski, Z.: Some properties of Prabhakar-type fractional calculus operators. Fract. Differ. Calculus **6**(1), 73–94 (2016)
12. Prabhakar, T.R.: A singular integral equation with a generalized Mittag Leffler function in the kernel. Yokohama Math. J. **19**, 7–15 (1971)
13. Rudin, W.: Real and Complex Analysis. International Student Edition, Mc Graw Hill, London, New York (1970)
14. Stroock, D.: A Concise Introduction to the Theory of Integration, 3rd edn. Birkhäuser, Boston, Basel, Berlin (1999)

Chapter 12
Advanced Multivariate Prabhakar fractional integrals and inequalities

We introduce here the mixed generalized multivariate Prabhakar type left and right fractional integrals and study their basic properties, such as preservation of continuity and their boundedness as positive linear operators. Then we produce an interesting variety of related multivariate left and right fractional Hardy type inequalities under convexity. We introduce also other related multivariate fractional integrals. It follows [4].

12.1 Background

This chapter is motivated by Hardy [7], Iqbal et al. [9, 10], Polito and Tomovski [12], Prabhakar [13], Saxena et al. [14], Srivastava and Tomovski [15], Tomovski et al. [16].

Here we consider the Prabhakar function (also known as the three parameters Mittag-Leffler function, an entire function if $z \in \mathbb{C}$), (see [5], [6, p. 97])

$$E_{\alpha,\beta}^{\gamma}(z) = \sum_{k=0}^{\infty} \frac{(\gamma)_k}{k!\Gamma(\alpha k + \beta)} z^k, \tag{12.1}$$

where Γ is the gamma function; $\alpha, \beta, \gamma \in \mathbb{R} : \alpha, \beta > 0, z \in \mathbb{R}$, and $(\gamma)_k = \gamma(\gamma+1) \ldots (\gamma+k-1)$. It is $E_{\alpha,\beta}^{0}(z) = \frac{1}{\Gamma(\beta)}$.

Let $a, b \in \mathbb{R}, a < b$ and $x \in [a,b]$; $f \in C([a,b])$. Let also $\psi \in C^1([a,b])$ which is increasing. In [3] we defined and studied the left and right Prabhakar fractional integrals with respect to ψ as follows:

G. A. Anastassiou, *Unification of Fractional Calculi with Applications*,
Studies in Systems, Decision and Control 398,
https://doi.org/10.1007/978-3-030-86920-5_12

$$\left(e^{\gamma;\psi}_{\rho,\mu,\omega,a+}f\right)(x) = \int_a^x \psi'(t)\left(\psi(x)-\psi(t)\right)^{\mu-1} E^{\gamma}_{\rho,\mu}\left[\omega\left(\psi(x)-\psi(t)\right)^{\rho}\right] f(t)\,dt, \tag{12.2}$$

and

$$\left(e^{\gamma;\psi}_{\rho,\mu,\omega,b-}f\right)(x) = \int_x^b \psi'(t)\left(\psi(t)-\psi(x)\right)^{\mu-1} E^{\gamma}_{\rho,\mu}\left[\omega\left(\psi(t)-\psi(x)\right)^{\rho}\right] f(t)\,dt, \tag{12.3}$$

where $\rho, \mu > 0$; $\gamma, \omega \in \mathbb{R}$, which are continuous functions [3].

In this chapter we define and study the multivariate analogs of (12.2) and (12.3).

Let $\prod_{i=1}^{N}[a_i, b_i] \subset \mathbb{R}^N$, $N > 1$, $a_i < b_i$, $a_i, b_i \in \mathbb{R}$; $f \in C\left(\prod_{i=1}^{N}[a_i, b_i]\right)$. Let also $x_i, t_i \in [a_i, b_i]$; $\rho_i, \mu_i > 0$; $\gamma_i, \omega_i \in \mathbb{R}$; $i = 1, ..., N$. Here $\psi_i \in C^1([a_i, b_i])$ which is increasing, $i = 1, ..., N$. We set $a = (a_1, ..., a_N)$, $b = (b_1, ..., b_N)$, $x = (x_1, ..., x_N)$, $t = (t_1, ..., t_N)$; $\rho = (\rho_1, ..., \rho_N)$, $\mu = (\mu_1, ..., \mu_N)$, $\gamma = (\gamma_1, ..., \gamma_N)$, $\omega = (\omega_1, ..., \omega_N)$, $\psi := (\psi_1, ..., \psi_N)$.

We define the left and right mixed Prabhakar multiple fractional integrals with respect to ψ, respectively, as follows:

$$\left({}^{M}e^{\gamma;\psi}_{\rho,\mu,\omega,a+}f\right)(x) := \int_{a_1}^{x_1} ... \int_{a_N}^{x_N} \prod_{i=1}^{N}\left[\psi_i'(t_i)\left(\psi_i(x_i)-\psi_i(t_i)\right)^{\mu_i-1}\right. \tag{12.4}$$

$$\left. E^{\gamma_i}_{\rho_i,\mu_i}\left[\omega_i\left(\psi_i(x_i)-\psi_i(t_i)\right)^{\rho_i}\right]\right] f(t_1, ..., t_N)\,dt_1...dt_N,$$

with $x_i \geq a_i$, $i = 1, ..., N$;

$$\left({}^{M}e^{\gamma;\psi}_{\rho,\mu,\omega,b-}f\right)(x) := \int_{x_1}^{b_1} ... \int_{x_N}^{b_N} \prod_{i=1}^{N}\left[\psi_i'(t_i)\left(\psi_i(t_i)-\psi_i(x_i)\right)^{\mu_i-1}\right. \tag{12.5}$$

$$\left. E^{\gamma_i}_{\rho_i,\mu_i}\left[\omega_i\left(\psi_i(t_i)-\psi_i(x_i)\right)^{\rho_i}\right]\right] f(t_1, ..., t_N)\,dt_1...dt_N,$$

with $x_i \leq b_i$, $i = 1, ..., N$.

We give

Theorem 12.1 *Let $\rho_i, \gamma_i > 0$, $\mu_i \geq 1$, $\omega_i \in \mathbb{R}$; $i = 1, ..., N$. Then $\left({}^{M}e^{\gamma;\psi}_{\rho,\mu,\omega,a+}f\right)$, $\left({}^{M}e^{\gamma;\psi}_{\rho,\mu,\omega,b-}f\right) \in C\left(\prod_{i=1}^{N}[a_i, b_i]\right)$.*

Proof It is enough to prove that $\left({}^{M}e^{\gamma;\psi}_{\rho,\mu,\omega,a+}f\right) \in C\left(\prod_{i=1}^{N}[a_i, b_i]\right)$. The proof for the second is similar and omitted.

One can write

$$\left({}^{M}e^{\gamma;\psi}_{\rho,\mu,\omega,a+}f\right)(x_0) = \int_{\prod_{i=1}^{N}[a_i,b_i]} \chi_{\prod_{i=1}^{N}[a_i,x_{0i}]}(t) \prod_{i=1}^{N}\left[\psi_i'(t_i)(\psi_i(x_{0i}) - \psi_i(t_i))^{\mu_i-1}\right.$$
$$\left. E^{\gamma_i}_{\rho_i,\mu_i}\left[\omega_i(\psi_i(x_{0i}) - \psi_i(t_i))^{\rho_i}\right]\right] f(t)\,dt, \tag{12.6}$$

where χ is the characteristic function and $x_0 = (x_{01}, ..., x_{0N}) \in \prod_{i=1}^{N}[a_i, b_i]$.

Let $x_m \to x_0$ as $m \to +\infty$, where $x_m \in \prod_{i=1}^{N}[a_i, b_i]$; with $x_m = (x_{m1}, ..., x_{mN})$.

We will prove that $\left({}^{M}e^{\gamma;\psi}_{\rho,\mu,\omega,a+}f\right)(x_m) \to \left({}^{M}e^{\gamma;\psi}_{\rho,\mu,\omega,a+}f\right)(x_0)$, as $m \to +\infty$.

We have that

$$\left({}^{M}e^{\gamma;\psi}_{\rho,\mu,\omega,a+}f\right)(x_m) = \int_{\prod_{i=1}^{N}[a_i,b_i]} \chi_{\prod_{i=1}^{N}[a_i,x_{mi}]}(t) \prod_{i=1}^{N}\left[\psi_i'(t_i)(\psi_i(x_{mi}) - \psi_i(t_i))^{\mu_i-1}\right.$$
$$\left. E^{\gamma_i}_{\rho_i,\mu_i}\left[\omega_i(\psi_i(x_{mi}) - \psi_i(t_i))^{\rho_i}\right]\right] f(t)\,dt. \tag{12.7}$$

We notice that $\chi_{\prod_{i=1}^{N}[a_i,x_{mi}]}(t) \to \chi_{\prod_{i=1}^{N}[a_i,x_{0i}]}(t)$, a.e.; also it holds

$$T_t(x_m) := \prod_{i=1}^{N}\left[\psi_i'(t_i)(\psi_i(x_{mi}) - \psi_i(t_i))^{\mu_i-1} E^{\gamma_i}_{\rho_i,\mu_i}\left[\omega_i(\psi_i(x_{mi}) - \psi_i(t_i))^{\rho_i}\right]\right]$$

$$\to \prod_{i=1}^{N}\left[\psi_i'(t_i)(\psi_i(x_{0i}) - \psi_i(t_i))^{\mu_i-1} E^{\gamma_i}_{\rho_i,\mu_i}\left[\omega_i(\psi_i(x_{0i}) - \psi_i(t_i))^{\rho_i}\right]\right] =: T_t(x_0),$$

as $m \to +\infty$.

Furthermore we obtain

$$\chi_{\prod_{i=1}^{N}[a_i,x_{mi}]}(t)\, T_t(x_m)\, f(t) \to \chi_{\prod_{i=1}^{N}[a_i,x_{0i}]}(t)\, T_t(x_0)\, f(t),$$

a.e. on $\prod_{i=1}^{N}[a_i, b_i]$, as $m \to +\infty$.

However, we have that

$$\chi_{\prod_{i=1}^{N}[a_i,x_{mi}]}(t)\,|T_t(x_m)|\,|f(t)| \le |T_t(x_m)|\,|f(t)| \le \tag{12.8}$$

$$\prod_{i=1}^{N}\left[\left\|\psi_i'\right\|_{\infty,[a_i,b_i]}(\psi_i(b_i)-\psi_i(a_i))^{\mu_i-1}E_{\rho_i,\mu_i}^{\gamma_i}\left[|\omega_i|(\psi_i(b_i)-\psi_i(a_i))^{\rho_i}\right]\right]$$

$$\|f\|_{\infty,\prod_{i=1}^{N}[a_i,b_i]}<+\infty.$$

Thus, by the dominated convergence theorem, we derive

$$\int_{\prod_{i=1}^{N}[a_i,b_i]}\chi_{\prod_{i=1}^{N}[a_i,x_{mi}]}(t)\,T_t(x_m)\,f(t)\,dt\to\int_{\prod_{i=1}^{N}[a_i,b_i]}\chi_{\prod_{i=1}^{N}[a_i,x_{0i}]}(t)\,T_t(x_0)\,f(t)\,dt, \tag{12.9}$$

as $m\to+\infty$, proving the claim. ∎

Conjecture 12.2 Functions (12.4) and (12.5) must be continuous when $\rho_i,\mu_i>0$; $\gamma_i,\omega_i\in\mathbb{R}$; $i=1,...,N$.

We also present the following basic Hardy type inequalities.

Theorem 12.3 *Let $\rho_i,\mu_i,\gamma_i,\omega_i>0$, $i=1,...,N$. Then*

$$\left\{\left\|\left({}^{M}e_{\rho,\mu,\omega,a+}^{\gamma;\psi}f\right)\right\|_{\infty,\prod_{i=1}^{N}[a_i,b_i]},\left\|\left({}^{M}e_{\rho,\mu,\omega,b-}^{\gamma;\psi}f\right)\right\|_{\infty,\prod_{i=1}^{N}[a_i,b_i]}\right\}\le$$

$$\left\{\left\|\left({}^{M}e_{\rho,\mu,\omega,a+}^{\gamma;\psi}|f|\right)\right\|_{\infty,\prod_{i=1}^{N}[a_i,b_i]},\left\|\left({}^{M}e_{\rho,\mu,\omega,b-}^{\gamma;\psi}|f|\right)\right\|_{\infty,\prod_{i=1}^{N}[a_i,b_i]}\right\}\le$$

$$\left[\prod_{i=1}^{N}\left[(\psi_i(b_i)-\psi_i(a_i))^{\mu_i}E_{\rho_i,\mu_i+1}^{\gamma_i}\left[\omega_i(\psi_i(b_i)-\psi_i(a_i))^{\rho_i}\right]\right]\right]\|f\|_{\infty,\prod_{i=1}^{N}[a_i,b_i]}<\infty, \tag{12.10}$$

respectively.

Here (12.4) and (12.5) are bounded and positive linear operators.

Proof It is enough to estimate only $\left\|\left({}^{M}e_{\rho,\mu,\omega,a+}^{\gamma;\psi}f\right)\right\|_{\infty,\prod_{i=1}^{N}[a_i,b_i]}$. The proof of the second as similar is omitted. We have that

$$\left|\left({}^{M}e_{\rho,\mu,\omega,a+}^{\gamma;\psi}f\right)(x)\right|\overset{(12.4)}{\le}\left({}^{M}e_{\rho,\mu,\omega,a+}^{\gamma;\psi}|f|\right)(x)=$$

$$\int_{a_1}^{x_1}...\int_{a_N}^{x_N}\prod_{i=1}^{N}\left[\psi_i'(t_i)(\psi_i(x_i)-\psi_i(t_i))^{\mu_i-1}\right.$$

$$\left.E_{\rho_i,\mu_i}^{\gamma_i}\left[\omega_i(\psi_i(x_i)-\psi_i(t_i))^{\rho_i}\right]\right]|f(t_1,...,t_N)|\,dt_1...dt_N\le$$

$$\|f\|_{\infty,\prod_{i=1}^{N}[a_i,b_i]}\prod_{i=1}^{N}\left(\int_{a_i}^{x_i}\psi_i'(t_i)(\psi_i(x_i)-\psi_i(t_i))^{\mu_i-1}\right.$$

$$\left.E_{\rho_i,\mu_i}^{\gamma_i}\left[\omega_i(\psi_i(x_i)-\psi_i(t_i))^{\rho_i}\right]dt_i\right)= \tag{12.11}$$

$$\|f\|_{\infty,\prod_{i=1}^{N}[a_i,b_i]}\prod_{i=1}^{N}\left(\int_{a_i}^{x_i}\psi_i'(t_i)(\psi_i(x_i)-\psi_i(t_i))^{\mu_i-1}\right.$$

$$\left.\left(\sum_{k_i=0}^{\infty}\frac{(\gamma_i)_{k_i}}{k_i!\Gamma(\rho_ik_i+\mu_i)}(\omega_i(\psi_i(x_i)-\psi_i(t_i))^{\rho_i})^{k_i}\right)dt_i\right)\overset{(7,\text{ p. }175)}{=}$$

$$\|f\|_{\infty,\prod_{i=1}^{N}[a_i,b_i]}\prod_{i=1}^{N}\left[\sum_{k_i=0}^{\infty}\frac{(\gamma_i)_{k_i}\omega_i^{k_i}}{k_i!\Gamma(\rho_ik_i+\mu_i)}\int_{a_i}^{x_i}\psi_i'(t_i)(\psi_i(x_i)-\psi_i(t_i))^{(\rho_ik_i+\mu_i)-1}dt_i\right]=$$

$$\|f\|_{\infty,\prod_{i=1}^{N}[a_i,b_i]}\prod_{i=1}^{N}\left[\sum_{k_i=0}^{\infty}\frac{(\gamma_i)_{k_i}\omega_i^{k_i}}{k_i!\Gamma(\rho_ik_i+\mu_i+1)}(\psi_i(x_i)-\psi_i(a_i))^{\rho_ik_i+\mu_i}\right]=$$

$$\|f\|_{\infty,\prod_{i=1}^{N}[a_i,b_i]}\prod_{i=1}^{N}\left[(\psi_i(x_i)-\psi_i(a_i))^{\mu_i}\sum_{k_i=0}^{\infty}\frac{(\gamma_i)_{k_i}(\omega_i(\psi_i(x_i)-\psi_i(a_i))^{\rho_i})^{k_i}}{k_i!\Gamma(\rho_ik_i+\mu_i+1)}\right]= \tag{12.12}$$

$$\|f\|_{\infty,\prod_{i=1}^{N}[a_i,b_i]}\prod_{i=1}^{N}\left[(\psi_i(x_i)-\psi_i(a_i))^{\mu_i}E_{\rho_i,\mu_i+1}^{\gamma_i}\left[\omega_i(\psi_i(x_i)-\psi_i(a_i))^{\rho_i}\right]\right]\leq$$

$$\|f\|_{\infty,\prod_{i=1}^{N}[a_i,b_i]}\prod_{i=1}^{N}\left[(\psi_i(b_i)-\psi_i(a_i))^{\mu_i}E_{\rho_i,\mu_i+1}^{\gamma_i}\left[\omega_i(\psi_i(b_i)-\psi_i(a_i))^{\rho_i}\right]\right],$$

$\forall\, x\in\prod_{i=1}^{N}[a_i,b_i]$.

The claim is proved. ■

We make

Remark 12.4 Let $\rho_i,\mu_i,\gamma_i,\omega_i>0$, $i=1,...,N$. We define the kernels $\left(y=(y_1,...,y_N)\in\prod_{i=1}^{N}(a_i,b_i)\right)$

$$k_{a+}(x,y)=\chi_{\prod_{i=1}^{N}(a_i,x_i]}(y)\prod_{i=1}^{N}\left[\psi_i'(y_i)(\psi_i(x_i)-\psi_i(y_i))^{\mu_i-1}\right.$$

$$E^{\gamma_i}_{\rho_i,\mu_i}\left[\omega_i\left(\psi_i\left(x_i\right)-\psi_i\left(y_i\right)\right)^{\rho_i}\right]\Big], \tag{12.13}$$

and

$$k_{b-}(x,y)=\chi_{\prod\limits_{i=1}^{N}[x_i,b_i)}(y)\prod_{i=1}^{N}\Big[\psi_i'(y_i)\left(\psi_i(y_i)-\psi_i(x_i)\right)^{\mu_i-1}$$

$$E^{\gamma_i}_{\rho_i,\mu_i}\left[\omega_i\left(\psi_i\left(y_i\right)-\psi_i\left(x_i\right)\right)^{\rho_i}\right]\Big], \tag{12.14}$$

$\forall\, x,y\in\prod\limits_{i=1}^{N}(a_i,b_i)$.

We compute the integral

$$K_{a+}(x):=\int_{\prod\limits_{i=1}^{N}(a_i,b_i)}k_{a+}(x,y)\,dy\overset{(12.4)}{=}\left({}^{M}e^{\gamma;\psi}_{\rho,\mu,\omega,a+}1\right)(x)=$$

$$\int_{a_1}^{x_1}\ldots\int_{a_N}^{x_N}\prod_{i=1}^{N}\Big[\psi_i'(t_i)\left(\psi_i(x_i)-\psi_i(t_i)\right)^{\mu_i-1}$$

$$E^{\gamma_i}_{\rho_i,\mu_i}\left[\omega_i\left(\psi_i\left(x_i\right)-\psi_i\left(t_i\right)\right)^{\rho_i}\right]\Big]dt_1\ldots dt_N= \tag{12.15}$$

$$\prod_{i=1}^{N}\left(\int_{a_i}^{x_i}\psi_i'(t_i)\left(\psi_i(x_i)-\psi_i(t_i)\right)^{\mu_i-1}E^{\gamma_i}_{\rho_i,\mu_i}\left[\omega_i\left(\psi_i(x_i)-\psi_i(t_i)\right)^{\rho_i}\right]dt_i\right)=$$

(by (12.11), (12.12))

$$\prod_{i=1}^{N}\Big[\left(\psi_i(x_i)-\psi_i(a_i)\right)^{\mu_i}E^{\gamma_i}_{\rho_i,\mu_i+1}\left[\omega_i\left(\psi_i(x_i)-\psi_i(a_i)\right)^{\rho_i}\right]\Big].$$

That is

$$K_{a+}(x)=\prod_{i=1}^{N}\Big[\left(\psi_i(x_i)-\psi_i(a_i)\right)^{\mu_i}E^{\gamma_i}_{\rho_i,\mu_i+1}\left[\omega_i\left(\psi_i(x_i)-\psi_i(a_i)\right)^{\rho_i}\right]\Big], \tag{12.16}$$

$\forall\, x\in\prod\limits_{i=1}^{N}(a_i,b_i)$.

Similarly, we compute the integral

$$K_{b-}(x):=\int_{\prod\limits_{i=1}^{N}(a_i,b_i)}k_{b-}(x,y)\,dy\overset{(12.5)}{=}\left({}^{M}e^{\gamma;\psi}_{\rho,\mu,\omega,b-}1\right)(x)=$$

$$\prod_{i=1}^{N}\left[\left(\psi_i(b_i)-\psi_i(x_i)\right)^{\mu_i}E_{\rho_i,\mu_i+1}^{\gamma_i}\left[\omega_i\left(\psi_i(b_i)-\psi_i(x_i)\right)^{\rho_i}\right]\right], \tag{12.17}$$

$\forall\, x\in\prod_{i=1}^{N}(a_i,b_i)$.

Next, we form the ratios

$$\frac{k_{a+}(x,y)}{K_{a+}(x)}=$$

$$\frac{\chi_{\prod_{i=1}^{N}(a_i,x_i]}(y)\prod_{i=1}^{N}\left[\psi_i'(y_i)\left(\psi_i(x_i)-\psi_i(y_i)\right)^{\mu_i-1}E_{\rho_i,\mu_i}^{\gamma_i}\left[\omega_i\left(\psi_i(x_i)-\psi_i(y_i)\right)^{\rho_i}\right]\right]}{\prod_{i=1}^{N}\left[\left(\psi_i(x_i)-\psi_i(a_i)\right)^{\mu_i}E_{\rho_i,\mu_i+1}^{\gamma_i}\left[\omega_i\left(\psi_i(x_i)-\psi_i(a_i)\right)^{\rho_i}\right]\right]}$$

$$=\chi_{\prod_{i=1}^{N}(a_i,x_i]}(y)\prod_{i=1}^{N}\left(\frac{\psi_i'(y_i)\left(\psi_i(x_i)-\psi_i(y_i)\right)^{\mu_i-1}}{\left(\psi_i(x_i)-\psi_i(a_i)\right)^{\mu_i}}\right)$$

$$\prod_{i=1}^{N}\left(\frac{E_{\rho_i,\mu_i}^{\gamma_i}\left[\omega_i\left(\psi_i(x_i)-\psi_i(y_i)\right)^{\rho_i}\right]}{E_{\rho_i,\mu_i+1}^{\gamma_i}\left[\omega_i\left(\psi_i(x_i)-\psi_i(a_i)\right)^{\rho_i}\right]}\right), \tag{12.18}$$

and

$$\frac{k_{b-}(x,y)}{K_{b-}(x)}=\chi_{\prod_{i=1}^{N}[x_i,b_i)}(y)\prod_{i=1}^{N}\left(\frac{\psi_i'(y_i)\left(\psi_i(y_i)-\psi_i(x_i)\right)^{\mu_i-1}}{\left(\psi_i(b_i)-\psi_i(x_i)\right)^{\mu_i}}\right)$$

$$\prod_{i=1}^{N}\left(\frac{E_{\rho_i,\mu_i}^{\gamma_i}\left[\omega_i\left(\psi_i(y_i)-\psi_i(x_i)\right)^{\rho_i}\right]}{E_{\rho_i,\mu_i+1}^{\gamma_i}\left[\omega_i\left(\psi_i(b_i)-\psi_i(x_i)\right)^{\rho_i}\right]}\right), \tag{12.19}$$

$\forall\, x,y\in\prod_{i=1}^{N}(a_i,b_i)$.

In this chapter we prove a variety of interesting generalized multivariate Hardy type fractional inequalities under convexity, related to (12.4) and (12.5), mixed generalized multivariate left and right Prabhakar type fractional integrals.

12.2 Prerequisites

(I) Let $(\Omega_1,\Sigma_1,\mu_1)$ and $(\Omega_2,\Sigma_2,\mu_2)$ be measure spaces with positive σ-finite measures, and let $k_j:\Omega_1\times\Omega_2\to\mathbb{R}$ be nonnegative measurable functions, $k_j(x,\cdot)$ measurable on Ω_2, and

$$K_j(x) = \int_{\Omega_2} k_j(x,y)\, d\mu_2(y), \text{ for any } x \in \Omega_1, \tag{12.20}$$

$j = 1, ..., r \in \mathbb{N}$. We assume that $K_j(x) > 0$ a.e. on Ω_1 and the weight function are nonnegative measurable functions on the related set.

We consider measurable functions $g_j : \Omega_1 \to \mathbb{R}$ with the representation

$$g_j(x) = \int_{\Omega_2} k_j(x,y)\, f_j(y)\, d\mu_2(y), \tag{12.21}$$

where $f_j : \Omega_2 \to \mathbb{R}$ are measurable functions, $j = 1, ..., r$. Here u stands for a weight function on Ω_1 (nonnegative measurable function).

We mention

Theorem 12.5 ([1, p. 29]) *Let $\overline{j} \in \{1, ..., r\}$ be fixed. Assume that the function $x \to \left(\frac{u(x) \prod_{j=1}^{r} k_j(x,y)}{\prod_{j=1}^{r} K_j(x)} \right)$ is integrable on Ω_1, for each $y \in \Omega_2$. Define λ_r on Ω_2 by*

$$\lambda_r(y) := \int_{\Omega_1} \left(\frac{u(x) \prod_{j=1}^{r} k_j(x,y)}{\prod_{j=1}^{r} K_j(x)} \right) d\mu_1(x) < \infty. \tag{12.22}$$

Here $\Phi_j : \mathbb{R}_+ \to \mathbb{R}_+$, $j = 1, ..., r$, are convex and increasing functions.

Then

$$\int_{\Omega_1} u(x) \prod_{j=1}^{r} \Phi_j \left(\left| \frac{g_j(x)}{K_j(x)} \right| \right) d\mu_1(x) \leq$$

$$\left(\prod_{\substack{j=1 \\ j \neq \overline{j}}}^{r} \int_{\Omega_2} \Phi_j \left(\left| f_j(y) \right| \right) d\mu_2(y) \right) \left(\int_{\Omega_2} \Phi_{\overline{j}} \left(\left| f_{\overline{j}}(y) \right| \right) \lambda_r(y)\, d\mu_2(y) \right), \tag{12.23}$$

true for all measurable functions, $j = 1, ..., r$, $f_j : \Omega_2 \to \mathbb{R}$ such that:

(i) f_j, $\Omega_j \left(\left| f_j \right| \right)$, are both $k_j(x,y)\, d\mu_2(y)$ - integrable, μ_1-a.e. in $x \in \Omega_1$,

(ii) $\lambda_r \Phi_{\overline{j}} \left(\left| f_{\overline{j}} \right| \right)$; $\Phi_1(|f_1|)$, $\Phi_2(|f_2|)$, $\Phi_3(|f_3|)$, ..., $\widehat{\Phi_{\overline{j}} \left(\left| f_{\overline{j}} \right| \right)}$, ..., $\Phi_r(|f_r|)$, are all μ_2-integrable,

and for all corresponding functions g_j given by (12.21). Above $\widehat{\Phi_{\overline{j}} \left(\left| f_{\overline{j}} \right| \right)}$ means missing item.

(II) We need

Remark 12.6 In the next we consider for $j = 1, ..., r$, the measurable functions $f_{1j}, f_{2j} : \Omega_2 \to \mathbb{R}$, and

$$g_{1j}(x) = \int_{\Omega_2} k_j(x, y) f_{1j}(y) d\mu_2(y), \tag{12.24}$$

and

$$g_{2j}(x) = \int_{\Omega_2} k_j(x, y) f_{2j}(y) d\mu_2(y), \tag{12.25}$$

these are now instead of (12.21).

Again here $u \geq 0$ is a weight measurable function on Ω_1.

We will use the following rational result:

Theorem 12.7 ([2, p. 405]) *Here $0 < f_{2j}(y) < \infty$, a.e., $j = 1, ..., r$. Let $\overline{j} \in \{1, ..., r\}$ be fixed. Assume that the function $x \to \left(\frac{u(x) \prod_{j=1}^{r} k_j(x,y) f_{2j}(y)}{\prod_{j=1}^{r} g_{2j}(x)} \right)$ is integrable on Ω_1, for each $y \in \Omega_2$. Define λ_r^{**} on Ω_2 by*

$$\lambda_r^{**}(y) := \left(\prod_{j=1}^{r} f_{2j}(y) \right) \int_{\Omega_1} \left(\frac{u(x) \prod_{j=1}^{r} k_j(x, y)}{\prod_{j=1}^{r} g_{2j}(x)} \right) d\mu_1(x) < \infty. \tag{12.26}$$

Here $\Phi_j : \mathbb{R}_+ \to \mathbb{R}_+$, $j = 1, ..., r$, are convex and increasing functions.

Then

$$\int_{\Omega_1} u(x) \prod_{j=1}^{r} \Phi_j \left(\left| \frac{g_{1j}(x)}{g_{2j}(x)} \right| \right) d\mu_1(x) \leq$$

$$\left(\prod_{\substack{j=1 \\ j \neq \overline{j}}}^{r} \int_{\Omega_2} \Phi_j \left(\left| \frac{f_{1j}(y)}{f_{2j}(y)} \right| \right) d\mu_2(y) \right) \left(\int_{\Omega_2} \Phi_{\overline{j}} \left(\left| \frac{f_{1\overline{j}}(y)}{f_{2\overline{j}}(y)} \right| \right) \lambda_r^{**}(y) d\mu_2(y) \right), \tag{12.27}$$

true for all measurable functions, $j = 1, ..., r$, $f_{1j}, f_{2j} : \Omega_2 \to \mathbb{R}$ such that:

(i) $\frac{f_{1j}(y)}{f_{2j}(y)}$, $\Phi_j \left(\left| \frac{f_{1j}(y)}{f_{2j}(y)} \right| \right)$, are both $k_j(x, y) f_{2j}(y) d\mu_2(y)$ - integrable, μ_1-a.e. in $x \in \Omega_1$,

(ii) $\lambda_r^{**} \Phi_{\overline{j}} \left(\left| \frac{f_{1\overline{j}}(y)}{f_{2\overline{j}}(y)} \right| \right)$, and $\Phi_j \left(\left| \frac{f_{1j}(y)}{f_{2j}(y)} \right| \right)$, for $j = \{1, ..., r\} - \{\overline{j}\}$, are all μ_2-integrable;

and for all corresponding g_{1j} given by (12.24), and g_{2j} given by (12.25).

(III) Here we follow [2, p. 441], see Chapter 22.

Let $(\Omega_1, \Sigma_1, \mu_1)$ and $(\Omega_2, \Sigma_2, \mu_2)$ be measure spaces with positive σ-finite measures, and let $k_j : \Omega_1 \times \Omega_2 \to \mathbb{R}$ be nonnegative measurable functions, $k_j(x, \cdot)$ measurable on Ω_2, and

$$K_j(x) = \int_{\Omega_2} k_j(x, y)\, d\mu_2(y), \text{ for any } x \in \Omega_1, \tag{12.28}$$

$j = 1, ..., r \in \mathbb{N}$. We assume that $K_j(x) > 0$ a.e. on Ω_1 and the weight functions are nonnegative measurable functions on the related set.

We consider measurable functions $g_j : \Omega_1 \to \mathbb{R}$ with the representation

$$g_j(x) = \int_{\Omega_2} k_j(x, y)\, f_j(y)\, d\mu_2(y), \tag{12.29}$$

where $f_j : \Omega_2 \to \mathbb{R}$ are measurable functions, $j = 1, ..., r$.

Here u stands for a weight function on Ω_1 ($u \geq 0$, which is measurable).

We will use the following general result:

Theorem 12.8 ([2, p. 442]) *Assume that the functions* $(j = 1, 2, ..., r \in \mathbb{N})$ $x \to \left(u(x) \frac{k_j(x,y)}{K_j(x)}\right)$ *are integrable on* Ω_1, *for each fixed* $y \in \Omega_2$. *Define* u_j *on* Ω_2 *by*

$$u_j(y) := \int_{\Omega_1} u(x) \frac{k_j(x, y)}{K_j(x)} d\mu_1(x) < \infty. \tag{12.30}$$

Let $p_j > 1 : \sum_{j=1}^{r} \frac{1}{p_j} = 1$. *Let the functions* $\Phi_j : \mathbb{R}_+ \to \mathbb{R}_+$, $j = 1, ..., r$, *be convex and increasing.*

Then

$$\int_{\Omega_1} u(x) \prod_{j=1}^{r} \Phi j \left(\left| \frac{g_j(x)}{K_j(x)} \right| \right) d\mu_1(x) \leq$$

$$\prod_{j=1}^{r} \left(\int_{\Omega_2} u_j(y)\, \Phi_j \left(\left| f_j(y) \right| \right)^{p_j} d\mu_2(y) \right)^{\frac{1}{p_j}}, \tag{12.31}$$

for all measurable functions $f_j : \Omega_2 \to \mathbb{R}$ $(j = 1, ..., r)$ *such that*

(i) $f_j, \Phi_j\left(\left|f_j\right|\right)^{p_j}$, *are both* $k_j(x, y)\, d\mu_2(y)$ *- integrable,* μ_1*-a.e. in* $x \in \Omega_1$, $j = 1, ..., r$,

(ii) $u_j \Phi_j\left(\left|f_j\right|\right)^{p_j}$ *is* μ_2*-integrable,* $j = 1, ..., r$,

and for all corresponding functions g_j $(j = 1, ..., r)$ *given by (12.29).*

(IV) Let $(\Omega_1, \Sigma_1, \mu_1)$ and $(\Omega_2, \Sigma_2, \mu_2)$ be measure spaces with positive σ-finite measures, and let $k : \Omega_1 \times \Omega_2 \to \mathbb{R}$ be nonnegative measurable functions, $k(x, \cdot)$

measurable on Ω_2, and

$$K(x) = \int_{\Omega_2} k(x, y)\, d\mu_2(y), \text{ for any } x \in \Omega_1. \tag{12.32}$$

We suppose that $K(x) > 0$ a.e. on Ω_1 and by a weight function u (shortly: a weight), we mean a nonnegative measurable function on the actual set. Let the measurable functions $g_j : \Omega_1 \to \mathbb{R}$, $j = 1, ..., r$, with the representation

$$g_j(x) = \int_{\Omega_2} k(x, y)\, f_j(y)\, d\mu_2(y), \tag{12.33}$$

where $f_j : \Omega_2 \to \mathbb{R}$ are measurable functions, $j = 1, ..., r$.

Denote by $\mathbf{x} = x := (x_1, ..., x_r) \in \mathbb{R}^r$, $\mathbf{g} := (g_1, ..., g_r)$ and $\mathbf{f} := (f_1, ..., f_r)$.

We consider here $\Phi : \mathbb{R}_+^r \to \mathbb{R}$ a convex function, which is increasing per coordinate, i.e. if $x_j \leq y_j$, $j = 1, ..., r$, then

$$\Phi(x_1, ..., x_r) \leq \Phi(y_1, ..., y_r).$$

In [2, p. 588], we proved that

Theorem 12.9 *Let u be a weight function on Ω_1, and k, K, g_j, f_j, $j = 1, ..., r \in \mathbb{N}$, and Φ defined as above. Assume that the function $x \to u(x) \frac{k(x,y)}{K(x)}$ is integrable on Ω_1 for each fixed $y \in \Omega_2$. Define v on Ω_2 by*

$$v(y) := \int_{\Omega_1} u(x) \frac{k(x, y)}{K(x)} d\mu_1(x) < \infty. \tag{12.34}$$

Then

$$\int_{\Omega_1} u(x)\, \Phi\left(\frac{|g_1(x)|}{K(x)}, ..., \frac{|g_r(x)|}{K(x)}\right) d\mu_1(x) \leq$$

$$\int_{\Omega_2} v(y)\, \Phi(|f_1(y)|, ..., |f_r(y)|)\, d\mu_2(y), \tag{12.35}$$

under the assumptions:

(i) f_j, $\Phi(|f_1|, ..., |f_r|)$, *are* $k(x, y)\, d\mu_2(y)$ *-integrable,* μ_1 *-a.e. in* $x \in \Omega_1$, *for all* $j = 1, ..., r$,

(ii) $v(y)\, \Phi(|f_1(y)|, ..., |f_r(y)|)$ *is* μ_2 *-integrable.*

12.3 Main Results

From now on in this chapter we assume $\rho_i, \mu_i, \gamma_i, \omega_i > 0$, for $i = 1, ..., N$. Clearly here $\left({}^{M}e^{\gamma;\psi}_{\rho,\mu,\omega,a+}f\right), \left({}^{M}e^{\gamma;\psi}_{\rho,\mu,\omega,b-}f\right)$ are measurable functions over $\prod_{i=1}^{N}[a_i, b_i]$, where $f \in C\left(\prod_{i=1}^{N}[a_i, b_i]\right)$. Infact by Theorem 12.3, these are integrable functions on $\prod_{i=1}^{N}[a_i, b_i]$.

(I') Here we apply Theorems 12.5 to (12.4), (12.5) for $f_j \in C\left(\prod_{i=1}^{N}[a_i, b_i]\right)$, $j = 1, ..., r$.

Theorem 12.10 *Let $\overline{j} \in \{1, ..., r\}$ be fixed. Assume that the function $x \rightarrow \left(u(x)\left(\frac{k_{a+}(x,y)}{K_{a+}(x)}\right)^r\right)$ is integrable on $\prod_{i=1}^{N}(a_i, b_i)$, for each $y \in \prod_{i=1}^{N}(a_i, b_i)$, see (12.16), (12.18). Here $u \geq 0$ stands for a weight function on $\prod_{i=1}^{N}(a_i, b_i)$. Define λ_r^+ on $\prod_{i=1}^{N}(a_i, b_i)$ by*

$$\lambda_r^+(y) := \int_{\prod_{i=1}^{N}(a_i,b_i)} \left(u(x)\left(\frac{k_{a+}(x,y)}{K_{a+}(x)}\right)^r\right) dx < \infty, \tag{12.36}$$

which is assumed to be integrable on $\prod_{i=1}^{N}(a_i, b_i)$.

Here $\Phi_j : \mathbb{R}_+ \rightarrow \mathbb{R}_+$, $j = 1, ..., r$, are convex and increasing functions. Then

$$\int_{\prod_{i=1}^{N}(a_i,b_i)} u(x) \prod_{j=1}^{r} \Phi_j \left(\frac{\left|\left({}^{M}e^{\gamma;\psi}_{\rho,\mu,\omega,a+}f_j\right)(x)\right|}{\prod_{i=1}^{N}\left[(\psi_i(x_i) - \psi_i(a_i))^{\mu_i} E^{\gamma_i}_{\rho_i,\mu_i+1}\left[\omega_i(\psi_i(x_i) - \psi_i(a_i))^{\rho_i}\right]\right]}\right) dx \tag{12.37}$$

$$\leq \left(\prod_{\substack{j=1 \\ j \neq \overline{j}}}^{r} \int_{\prod_{i=1}^{N}(a_i,b_i)} \Phi_j\left(\left|f_j(y)\right|\right) dy\right)\left(\int_{\prod_{i=1}^{N}(a_i,b_i)} \Phi_{\overline{j}}\left(\left|f_{\overline{j}}(y)\right|\right)\lambda_r^+(y)\, dy\right).$$

We continue with

Theorem 12.11 *Let $\overline{j} \in \{1, ..., r\}$ be fixed. Assume that the function $x \to \left(u(x)\left(\frac{k_{b-}(x,y)}{K_{b-}(x)}\right)^r\right)$ is integrable on $\prod_{i=1}^{N}(a_i, b_i)$, for each $y \in \prod_{i=1}^{N}(a_i, b_i)$, see (12.17), (12.19). Here $u \geq 0$ stands for a weight function on $\prod_{i=1}^{N}(a_i, b_i)$. Define λ_r^- on $\prod_{i=1}^{N}(a_i, b_i)$ by*

$$\lambda_r^-(y) := \int_{\prod_{i=1}^{N}(a_i,b_i)} \left(u(x)\left(\frac{k_{b-}(x,y)}{K_{b-}(x)}\right)^r\right) dx < \infty, \tag{12.38}$$

which is assumed to be integrable on $\prod_{i=1}^{N}(a_i, b_i)$.

Here $\Phi_j : \mathbb{R}_+ \to \mathbb{R}_+$, $j = 1, ..., r$, are convex and increasing functions. Then

$$\int_{\prod_{i=1}^{N}(a_i,b_i)} u(x) \prod_{j=1}^{r} \Phi_j \left(\frac{\left|\left({}^{M}e_{\rho,\mu,\omega,b-}^{\gamma;\psi} f_j\right)(x)\right|}{\prod_{i=1}^{N}\left[(\psi_i(b_i) - \psi_i(x_i))^{\mu_i} E_{\rho_i,\mu_i+1}^{\gamma_i}\left[\omega_i(\psi_i(b_i) - \psi_i(x_i))^{\rho_i}\right]\right]} \right) dx \tag{12.39}$$

$$\leq \left(\prod_{\substack{j=1 \\ j \neq \overline{j}}}^{r} \int_{\prod_{i=1}^{N}(a_i,b_i)} \Phi_j(|f_j(y)|)\, dy \right) \left(\int_{\prod_{i=1}^{N}(a_i,b_i)} \Phi_{\overline{j}}\left(\left|f_{\overline{j}}(y)\right|\right) \lambda_r^-(y)\, dy \right).$$

(II') Here we apply Theorem 12.7 to (12.4), (12.5) for $f_{1j}, f_{2j} \in C\left(\prod_{i=1}^{N}[a_i, b_i]\right)$, $j = 1, ..., r$.

Theorem 12.12 *Consider $f_{2j} > 0$, $j = 1, ..., r$. Let $\overline{j} \in \{1, ..., r\}$ be fixed. Assume that the function $x \to \left(u(x)(k_{a+}(x,y))^r \prod_{j=1}^{r}\left(\frac{f_{2j}(y)}{\left({}^{M}e_{\rho,\mu,\omega,a+}^{\gamma;\psi} f_{2j}\right)(x)}\right)\right)$ (see (12.16)) is integrable on $\prod_{i=1}^{N}(a_i, b_i)$, for each $y \in \prod_{i=1}^{N}(a_i, b_i)$. Define λ_r^{*+} on $\prod_{i=1}^{N}(a_i, b_i)$ by*

$$\lambda_r^{*+}(y) := \left(\prod_{j=1}^{r} f_{2j}(y)\right) \int_{\prod_{i=1}^{N}(a_i,b_i)} \left(\frac{u(x)(k_{a+}(x,y))^r}{\prod_{j=1}^{r}\left({}^{M}e_{\rho,\mu,\omega,a+}^{\gamma;\psi} f_{2j}(x)\right)} \right) dx < \infty, \tag{12.40}$$

for appropiate weight $u \geq 0$, so that λ_r^{*+} is integrable on $\prod_{i=1}^{N}(a_i, b_i)$.

Here $\Phi_j : \mathbb{R}_+ \to \mathbb{R}_+$, $j = 1, ..., r$, *are convex and increasing functions. Then*

$$\int_{\prod_{i=1}^{N}(a_i,b_i)} u(x) \prod_{j=1}^{r} \Phi_j \left(\frac{\left| {}_M e^{\gamma;\psi}_{\rho,\mu,\omega,a+} f_{1j}(x) \right|}{{}_M e^{\gamma;\psi}_{\rho,\mu,\omega,a+} f_{2j}(x)} \right) dx \leq$$

$$\left(\prod_{\substack{j=1 \\ j \neq \overline{j}}}^{r} \int_{\prod_{i=1}^{N}(a_i,b_i)} \Phi_j \left(\frac{|f_{1j}(y)|}{f_{2j}(y)} \right) dy \right) \left(\int_{\prod_{i=1}^{N}(a_i,b_i)} \Phi_{\overline{j}} \left(\frac{|f_{1\overline{j}}(y)|}{f_{2\overline{j}}(y)} \right) \lambda_r^{*+}(y)\, dy \right). \tag{12.41}$$

We continue with

Theorem 12.13 *Consider* $f_{2j} > 0$, $j = 1, ..., r$. *Let* $\overline{j} \in \{1, ..., r\}$ *be fixed. Assume that the function* $x \to \left(u(x)(k_{b-}(x,y))^r \prod_{j=1}^{r} \left(\frac{f_{2j}(y)}{\left({}_M e^{\gamma;\psi}_{\rho,\mu,\omega,b-} f_{2j}\right)(x)} \right) \right)$ *(see (12.17)) is integrable on* $\prod_{i=1}^{N}(a_i,b_i)$, *for each* $y \in \prod_{i=1}^{N}(a_i,b_i)$. *Define* λ_r^{*-} *on* $\prod_{i=1}^{N}(a_i,b_i)$ *by*

$$\lambda_r^{*-}(y) := \left(\prod_{j=1}^{r} f_{2j}(y) \right) \int_{\prod_{i=1}^{N}(a_i,b_i)} \left(\frac{u(x)(k_{b-}(x,y))^r}{\prod_{j=1}^{r} \left({}_M e^{\gamma;\psi}_{\rho,\mu,\omega,b-} f_{2j}(x) \right)} \right) dx < \infty, \tag{12.42}$$

for appropiate weight $u \geq 0$, *so that* λ_r^{*-} *is integrable on* $\prod_{i=1}^{N}(a_i,b_i)$.

Here $\Phi_j : \mathbb{R}_+ \to \mathbb{R}_+$, $j = 1, ..., r$, *are convex and increasing functions. Then*

$$\int_{\prod_{i=1}^{N}(a_i,b_i)} u(x) \prod_{j=1}^{r} \Phi_j \left(\frac{\left| {}_M e^{\gamma;\psi}_{\rho,\mu,\omega,b-} f_{1j}(x) \right|}{{}_M e^{\gamma;\psi}_{\rho,\mu,\omega,b-} f_{2j}(x)} \right) dx \leq$$

$$\left(\prod_{\substack{j=1 \\ j \neq \overline{j}}}^{r} \int_{\prod_{i=1}^{N}(a_i,b_i)} \Phi_j \left(\frac{|f_{1j}(y)|}{f_{2j}(y)} \right) dy \right) \left(\int_{\prod_{i=1}^{N}(a_i,b_i)} \Phi_{\overline{j}} \left(\frac{|f_{1\overline{j}}(y)|}{f_{2\overline{j}}(y)} \right) \lambda_r^{*-}(y)\, dy \right). \tag{12.43}$$

(III') Here we apply Theorem 12.8 to (12.4), (12.5) for $f_j \in C\left(\prod_{i=1}^{N}[a_i,b_i]\right)$, $j = 1, ..., r$.

Theorem 12.14 *Assume that the function* $x \to \left(u(x)\frac{k_{a+}(x,y)}{K_{a+}(x)}\right)$ *is integrable on* $\prod_{i=1}^{N}(a_i,b_i)$, *for each fixed* $y \in \prod_{i=1}^{N}(a_i,b_i)$. *Define* ρ^+ *on* $\prod_{i=1}^{N}(a_i,b_i)$ *by*

$$\rho^+(y) := \int_{\prod_{i=1}^{N}(a_i,b_i)} u(x)\frac{k_{a+}(x,y)}{K_{a+}(x)}dx < \infty, \tag{12.44}$$

(see (12.13), (12.16), (12.18)) for appropiate weight $u \geq 0$, *so that* ρ^+ *is integrable on* $\prod_{i=1}^{N}(a_i,b_i)$.

Let $p_j > 1 : \sum_{j=1}^{r}\frac{1}{p_j} = 1$. *Let the functions* $\Phi_j : \mathbb{R}_+ \to \mathbb{R}_+$, $j = 1, ..., r$, *be convex and increasing.*

Then

$$\int_{\prod_{i=1}^{N}(a_i,b_i)} u(x)\prod_{j=1}^{r}\Phi_j\left(\frac{\left|{}^{M}e^{\gamma;\psi}_{\rho,\mu,\omega,a+}f_j(x)\right|}{K_{a+}(x)}\right)dx \leq$$

$$\prod_{j=1}^{r}\left(\int_{\prod_{i=1}^{N}(a_i,b_i)}\rho^+(y)\,\Phi_j\left(\left|f_j(y)\right|\right)^{p_j}dy\right)^{\frac{1}{p_j}}. \tag{12.45}$$

We continue with

Theorem 12.15 *Assume that the function* $x \to \left(u(x)\frac{k_{b-}(x,y)}{K_{b-}(x)}\right)$ *is integrable on* $\prod_{i=1}^{N}(a_i,b_i)$, *for each fixed* $y \in \prod_{i=1}^{N}(a_i,b_i)$. *Define* ρ^- *on* $\prod_{i=1}^{N}(a_i,b_i)$ *by*

$$\rho^-(y) := \int_{\prod_{i=1}^{N}(a_i,b_i)} u(x)\frac{k_{b-}(x,y)}{K_{b-}(x)}dx < \infty, \tag{12.46}$$

(see (12.14), (12.17), (12.19)) for appropiate weight $u \geq 0$, *so that* ρ^- *is integrable on* $\prod_{i=1}^{N}(a_i,b_i)$.

Let $p_j > 1 : \sum_{j=1}^{r}\frac{1}{p_j} = 1$. *Let the functions* $\Phi_j : \mathbb{R}_+ \to \mathbb{R}_+$, $j = 1, ..., r$, *be convex and increasing.*

Then

$$\int_{\prod_{i=1}^{N}(a_i,b_i)} u(x)\prod_{j=1}^{r}\Phi_j\left(\frac{\left|{}^{M}e^{\gamma;\psi}_{\rho,\mu,\omega,b-}f_j(x)\right|}{K_{b-}(x)}\right)dx \leq$$

$$\prod_{j=1}^{r}\left(\int_{\prod_{i=1}^{N}(a_i,b_i)}\rho^{-}(y)\,\Phi_j\left(\left|f_j(y)\right|\right)^{p_j}dy\right)^{\frac{1}{p_j}}. \tag{12.47}$$

(IV') Here we apply Theorem 12.9 to (12.4), (12.5) for $f_j \in C\left(\prod_{i=1}^{N}[a_i,b_i]\right)$, $j=1,...,r$.

Theorem 12.16 *Consider here $\Phi:\mathbb{R}_{+}^{r}\to\mathbb{R}$ a convex function, which is increasing per coordinate, and $u\geq 0$ the weight. Assume that the function $x\to u(x)\frac{k_{a+}(x,y)}{K_{a+}(x)}$ is integrable on $\prod_{i=1}^{N}(a_i,b_i)$ for each $y\in\prod_{i=1}^{N}(a_i,b_i)$. Define ν^{+} on $\prod_{i=1}^{N}(a_i,b_i)$ (see (12.16), (12.18)) by*

$$\nu^{+}(y):=\int_{\prod_{i=1}^{N}(a_i,b_i)}u(x)\frac{k_{a+}(x,y)}{K_{a+}(x)}dx<\infty, \tag{12.48}$$

for appropiate u, so that ν^{+} is integrable on $\prod_{i=1}^{N}(a_i,b_i)$. Then

$$\int_{\prod_{i=1}^{N}(a_i,b_i)}u(x)\,\Phi\left(\frac{\left|{}^{M}e_{\rho,\mu,\omega,a+}^{\gamma;\psi}f_1(x)\right|}{K_{a+}(x)},...,\frac{\left|{}^{M}e_{\rho,\mu,\omega,a+}^{\gamma;\psi}f_r(x)\right|}{K_{a+}(x)}\right)dx\leq$$

$$\int_{\prod_{i=1}^{N}(a_i,b_i)}\nu^{+}(y)\,\Phi\left(\left|f_1(y)\right|,...,\left|f_r(y)\right|\right)dy. \tag{12.49}$$

We also give

Theorem 12.17 *Consider here $\Phi:\mathbb{R}_{+}^{r}\to\mathbb{R}$ a convex function, which is increasing per coordinate, and $u\geq 0$ the weight. Assume that the function $x\to u(x)\frac{k_{b-}(x,y)}{K_{b-}(x)}$ is integrable on $\prod_{i=1}^{N}(a_i,b_i)$ for each $y\in\prod_{i=1}^{N}(a_i,b_i)$. Define ν^{-} on $\prod_{i=1}^{N}(a_i,b_i)$ (see (12.17), (12.19)) by*

$$\nu^{-}(y):=\int_{\prod_{i=1}^{N}(a_i,b_i)}u(x)\frac{k_{b-}(x,y)}{K_{b-}(x)}dx<\infty, \tag{12.50}$$

for appropiate u, so that ν^{-} *is integrable on* $\prod_{i=1}^{N}(a_i,b_i)$. *Then*

$$\int_{\prod_{i=1}^{N}(a_i,b_i)} u(x)\,\Phi\left(\frac{\left|{}^{M}e^{\gamma;\psi}_{\rho,\mu,\omega,b-}f_1(x)\right|}{K_{b-}(x)},\ldots,\frac{\left|{}^{M}e^{\gamma;\psi}_{\rho,\mu,\omega,b-}f_r(x)\right|}{K_{b-}(x)}\right)dx \leq$$

$$\int_{\prod_{i=1}^{N}(a_i,b_i)} \nu^{-}(y)\,\Phi\left(|f_1(y)|,\ldots,|f_r(y)|\right)dy. \tag{12.51}$$

One can produce a vast wealth of similar results, related to (12.4) and (12.5), by applying the general results of [2], Chaps. 21-27, but we choose here to skip this task.

We mention

Definition 12.18 All as in (12.4) and (12.5), $x \in \prod_{i=1}^{N}[a_i,b_i]$. We define the left and right partial Prabhakar fractional integrals with respect to ψ_k, respectively, for $k = 1,\ldots,N$, as follows:

$$\left({}^{P}e^{\gamma_k;\psi_k}_{\rho_k,\mu_k,\omega_k,a_k+}f\right)(x) := \int_{a_k}^{x_k} \psi_k'(t_k)\left(\psi_k(x_k)-\psi_k(t_k)\right)^{\mu_k-1} \tag{12.52}$$

$$E^{\gamma_k}_{\rho_k,\mu_k}\left[\omega_k\left(\psi_k(x_k)-\psi_k(t_k)\right)^{\rho_k}\right] f(x_1,\ldots,x_{k-1},t_k,x_{k+1},\ldots,x_N)\,dt_k,$$

with $a_k \leq x_k \leq b_k$;

and

$$\left({}^{P}e^{\gamma_k;\psi_k}_{\rho_k,\mu_k,\omega_k,b_k-}f\right)(x) := \int_{x_k}^{b_k} \psi_k'(t_k)\left(\psi_k(t_k)-\psi_k(x_k)\right)^{\mu_k-1} \tag{12.53}$$

$$E^{\gamma_k}_{\rho_k,\mu_k}\left[\omega_k\left(\psi_k(t_k)-\psi_k(x_k)\right)^{\rho_k}\right] f(x_1,\ldots,x_{k-1},t_k,x_{k+1},\ldots,x_N)\,dt_k,$$

with $a_k \leq x_k \leq b_k$.

By [3], (12.52), (12.53) are continuous functions in x_k, $k = 1,\ldots,N$.

We need

Definition 12.19 ([17, p. 35]) Here we consider the Prabhakar function with respect to another function $\alpha(x) > 0$, $x \in [a,b]$, $\alpha \in C([a,b])$, as follows:

$$E^{\gamma}_{\alpha(x),\beta}(z) := \sum_{k=0}^{\infty}\frac{(\gamma)_k}{k!\Gamma(k\alpha(x)+\beta)}z^k, \tag{12.54}$$

where $\beta > 0$, $\gamma \in \mathbb{R}$, $z \in \mathbb{R}$.

We mention

Definition 12.20 All as in (12.4) and (12.5), however now instead of ρ we have $\rho(x) = (\rho_1(x_1), ..., \rho_N(x_N))$, where $0 < \rho_i \in C([a_i, b_i])$, for $i = 1, ..., N$. We define the left and right mixed Prabhakar multiple fractional integrals of variable degree with respect to ψ, respectively, as follows:

$$\left({}_{v}^{M}e_{\rho,\mu,\omega,a+}^{\gamma;\psi} f\right)(x) := \int_{a_1}^{x_1} ... \int_{a_N}^{x_N} \prod_{i=1}^{N} \left[\psi_i'(t_i)(\psi_i(x_i) - \psi_i(t_i))^{\mu_i - 1}\right.$$

$$\left. E_{\rho_i(t_i),\mu_i}^{\gamma_i}\left[\omega_i(\psi_i(x_i) - \psi_i(t_i))^{\rho_i(t_i)}\right]\right] f(t_1, ..., t_N)\, dt_1...dt_N, \tag{12.55}$$

with $x_i \geq a_i$, $i = 1, ..., N$;

$$\left({}_{v}^{M}e_{\rho,\mu,\omega,b-}^{\gamma;\psi} f\right)(x) := \int_{x_1}^{b_1} ... \int_{x_N}^{b_N} \prod_{i=1}^{N} \left[\psi_i'(t_i)(\psi_i(t_i) - \psi_i(x_i))^{\mu_i - 1}\right.$$

$$\left. E_{\rho_i(t_i),\mu_i}^{\gamma_i}\left[\omega_i(\psi_i(t_i) - \psi_i(x_i))^{\rho_i(t_i)}\right]\right] f(t_1, ..., t_N)\, dt_1...dt_N, \tag{12.56}$$

with $x_i \leq b_i$, $i = 1, ..., N$.

One can prove similar results as above and in [2, Chapters 21–27], for the operators (12.52), (12.53), (12.55), (12.56), but we omit here this task.

References

1. Anastassiou, G.A.: Univariate Hardy-type fractional inequalities, Chapter 2. In: Anastassiou, G., Duman, O. (eds.) "Advances in Applied Mathematics and Approximation Theory", Contributions from AMAT 2012, pp. 21–56. Springer, New York (2013)
2. Anastassiou, G.A.: Intelligent Comparisons: Analytic Inequalities. Springer, Heidelberg, New York (2016)
3. Anastassiou, G.A.: Foundations of Generalized Prabhakar-Hilfer fractional Calculus with Applications (2021) (submitted)
4. Anastassiou, G.A.: Generalized Multivariate Prabhakar type fractional integrals and inequalities (2021) (submitted)
5. Giusti, A., et al.: A practical guide to Prabhakar fractional calculus. Fract. Calculus Appl. Anal. **23**(1), 9–54 (2020)
6. Gorenflo, R., Kilbas, A., Mainardi, F., Rogosin, S.: Mittag-Leffler Functions. Related Topics and Applications. Springer, Heidelberg, New York (2014)
7. Hardy, H.G.: Notes on some points in the integral calculus. Messenger Math. **47**(10), 145–150 (1918)
8. Hewith, E., Stromberg, K.: Real and Abstract Analysis. Springer, New York, Heidelberg, Berlin (1965)
9. Iqbal, S., Krulic, K., Pecaric, J.: On an Inequality of H.G. Hardy. J. Inequal. Appl. **2010**(Article ID 264347), 23 p

10. Iqbal, S., Krulic, K., Pecaric, J.: On an inequality for convex functions with some applications on fractional derivatives and fractional integrals. J. Math. Inequal. **5**(2), 219–230 (2011)
11. Kilbas, A.A., Saigo, M., Saxena, R.K.: Generalized Mittag-Leffler function and generalized fractional calculus operators. Integr. Transforms Spec. Functions **15**(1), 31–49 (2004)
12. Polito, F., Tomovski, Z.: Some properties of Prabhakar-type fractional calculus operators. Fract. Differ. Calculus **6**(1), 73–94 (2016)
13. Prabhakar, T.R.: A singular integral equation with a generalized Mittag Leffler function in the kernel. Yokohama Math. J. **19**, 7–15 (1971)
14. Saxena, R.K., Kalla, S.L., Saxena, R.: Multivariate analogue of generalized Mittag-Leffler function. Integr. Transforms Spec. Functions **22**(7), 533–548 (2011)
15. Srivastava, H.M., Tomovski, Z.: Fractional calculus with an integral operator containing a generalized Mittag-Leffler function in the kernel. Appl. Math. Comput. **211**, 198–210 (2009)
16. Tomovski, Ž, Hilfer, R., Srivastava, H.M.: Fractional and Operational Calculus with Generalized Fractional Derivative Operators and Mittag-Leffler Type Functions. Integr. Transforms Spec Functions **21**(11), 797–814 (2010)
17. Yang, X.-J.: General Fractional Derivatives. Theory, Methods and Applications. CRC Press, Taylor & Francis Group, Boca Raton, FL (2019)

Chapter 13
Non Singular Kernel Multiparameter Fractional Differentiation

We introduce here Caputo and Riemann-Liouville type non singular kernel very general multi parameter left and right side fractional derivatives and we prove their continuity. These have the advantage to describe accurately complex situations and phenomena and we can measure their fractional smoothness with memory and non-locality. Then, we derive related left and right fractional integral inequalities of Hardy, Opial and Hilbert-Pachpatte types, also of Hardy type involving convexity. It follows [3].

13.1 Background

Of great inspiration here are the articles [4, 5, 7]. But the most important for the author to write this article has been the work of Saxena et al. [8], about the multivariate analogue of generalized Mittag-Leffler function. Their extension allows at the same time to be treated both the univariate and multivariate level because of the many parameters involved. That gives us the advantage to record and treat lots of various data resulting from complicated natural phenomena and from diverse socioeconomical interactions, as well as mechanics, chemistry, etc. Also, their function serves well as a non singular kernel.

Here we use the multivariate analogue of generalized Mittag-Leffler function, see [8], defined for $\lambda, \gamma_j, \rho_j, z_j \in \mathbb{C}$, $Re\left(\rho_j\right) > 0\ (j = 1, ..., m)$ in terms of a multiple series of the form:

G. A. Anastassiou, *Unification of Fractional Calculi with Applications*,
Studies in Systems, Decision and Control 398,
https://doi.org/10.1007/978-3-030-86920-5_13

$$E_{(\rho_j),\lambda}^{(\gamma_j)}(z_1, ..., z_m) = E_{(\rho_1,...,\rho_m),\lambda}^{(\gamma_1,...,\gamma_m)}(z_1, ..., z_m) =$$

$$\sum_{k_1,...,k_m=0}^{\infty} \frac{(\gamma_1)_{k_1} \cdots (\gamma_m)_{k_m}}{\Gamma\left(\lambda + \sum_{j=1}^{m} k_j \rho_j\right)} \frac{z_1^{k_1} ... z_m^{k_m}}{k_1! ... k_m!}, \tag{13.1}$$

where $(\gamma_j)_{k_j}$ is the Pochhammer symbol, Γ is the gamma function. This is a special case of the generalized Lauricella series in several variables, see [9, p. 454], and [11].

By [10, p. 157], (13.1) converges for $Re(\rho_j) > 0$, $j = 1, ..., m$.

In what follows we will use the particular case of $E_{(\rho,...,\rho),\lambda}^{(\gamma_1,...,\gamma_m)}[\omega_1 t^\rho, ..., \omega_m t^\rho]$, denoted by $E_{(\rho),\lambda}^{(\gamma_j)}[\omega_1 t^\rho, ..., \omega_m t^\rho]$, where $0 < \rho < 1$, $t \geq 0$, $\lambda > 0$, $\gamma_j \in \mathbb{R}$ with $(\gamma_j)_{k_j} := \gamma_j(\gamma_j + 1) ... (\gamma_j + k_j - 1)$, $\omega_j \in \mathbb{R} - \{0\}$, for $j = 1, ..., m$. Let $f \in C^1([a, b])$, we define the following Caputo type generalized left fractional derivative with non singular kernel of order ρ, as

$$D_{a*}^{\rho} f(x) := {}_{(\gamma_j)(\omega_j)}^{CA} D_{a*}^{\rho,\lambda} f(x) :=$$

$$\frac{A(\rho)}{1-\rho} \int_a^x E_{(\rho),\lambda}^{(\gamma_j)} \left[\frac{-\omega_1 \rho}{1-\rho}(x-t)^\rho, ..., \frac{-\omega_m \rho}{1-\rho}(x-t)^\rho\right] f'(t)\, dt, \quad x \in [a, b], \tag{13.2}$$

where $A(\rho)$ is a normalizing constant.

Let now $f \in C^{n+1}([a, b])$, $n \in \mathbb{Z}_+$.

We define the Caputo type generalized left fractional derivative with non singular kernel of order $n + \rho$, as

$$D_{a*}^{n+\rho} f(x) := {}_{(\gamma_j)(\omega_j)}^{CA} D_{a*}^{n+\rho,\lambda} f(x) :=$$

$$\frac{A(\rho)}{1-\rho} \int_a^x E_{(\rho),\lambda}^{(\gamma_j)} \left[\frac{-\omega_1 \rho}{1-\rho}(x-t)^\rho, ..., \frac{-\omega_m \rho}{1-\rho}(x-t)^\rho\right] f^{(n+1)}(t)\, dt, \quad x \in [a, b]. \tag{13.3}$$

Similarly, we define the Caputo type generalized right fractional derivative with non singular kernel of order ρ, as

$$D_{b-}^{\rho} f(x) := {}_{(\gamma_j)(\omega_j)}^{CA} D_{b-}^{\rho,\lambda} f(x) :=$$

$$\frac{-A(\rho)}{1-\rho} \int_x^b E_{(\rho),\lambda}^{(\gamma_j)} \left[\frac{-\omega_1 \rho}{1-\rho}(t-x)^\rho, ..., \frac{-\omega_m \rho}{1-\rho}(t-x)^\rho\right] f'(t)\, dt, \quad x \in [a, b]. \tag{13.4}$$

And, for $f \in C^{n+1}([a,b])$, $n \in \mathbb{Z}_+$, we define

$$D_{b-}^{n+\rho} f(x) := {}_{(\gamma_j)(\omega_j)}^{CA} D_{b-}^{n+\rho,\lambda} f(x) :=$$

$$(-1)^{n+1} \frac{A(\rho)}{1-\rho} \int_x^b E_{(\rho),\lambda}^{(\gamma_j)} \left[\frac{-\omega_1 \rho}{1-\rho} (t-x)^\rho, \ldots, \frac{-\omega_m \rho}{1-\rho} (t-x)^\rho \right] f^{(n+1)}(t)\, dt, \tag{13.5}$$

$x \in [a,b]$.

The above derivatives generalize the Atangana-Baleanu fractional derivative, see [4].

Similarly, we defie the Riemann-Liouville corresponding versions of above fractional derivatives for $f \in C([a,b])$, $n \in \mathbb{Z}_+$, $0 < \rho < 1$, the left one:

$$D_a^{n+\rho} f(x) := {}_{(\gamma_j)(\omega_j)}^{RLA} D_a^{n+\rho,\lambda} f(x) :=$$

$$\frac{A(\rho)}{1-\rho} \frac{d^{n+1}}{dx^{n+1}} \int_a^x E_{(\rho),\lambda}^{(\gamma_j)} \left[\frac{-\omega_1 \rho}{1-\rho} (x-t)^\rho, \ldots, \frac{-\omega_m \rho}{1-\rho} (x-t)^\rho \right] f(t)\, dt, \tag{13.6}$$

$x \in [a,b]$, and right one:

$$D_b^{n+\rho} f(x) := {}_{(\gamma_j)(\omega_j)}^{RLA} D_b^{n+\rho,\lambda} f(x) :=$$

$$\frac{(-1)^{n+1} A(\rho)}{1-\rho} \frac{d^{n+1}}{dx^{n+1}} \int_x^b E_{(\rho),\lambda}^{(\gamma_j)} \left[\frac{-\omega_1 \rho}{1-\rho} (t-x)^\rho, \ldots, \frac{-\omega_m \rho}{1-\rho} (t-x)^\rho \right] f(t)\, dt, \tag{13.7}$$

$x \in [a,b]$.

In this chapter we emphasize on the Caputo version. The advantage of our fractional derivatives here is, that they have non singular and very general kernels able to incorporate lots of various fixed data sets from complicated physical phenomena, carried by the different sets of their parameters.

We present the following basic Hardy type inequalities:

Theorem 13.1 *All as above with $\gamma_j > 0$, $j = 1, \ldots, m$; $\lambda = 1$. Then*

$$\left\{ \left\| D_{a*}^{n+\rho} f \right\|_\infty, \left\| D_{b-}^{n+\rho} f \right\|_\infty \right\} \leq \frac{(b-a)|A(\rho)|}{1-\rho}$$

$$E_{(\rho),2}^{(\gamma_j)} \left[\frac{|\omega_1| \rho}{1-\rho} (b-a)^\rho, \ldots, \frac{|\omega_m| \rho}{1-\rho} (b-a)^\rho \right] \left\| f^{(n+1)} \right\|_\infty < \infty, \tag{13.8}$$

where $n \in \mathbb{Z}_+$.

Proof We prove the first inequality, for the second one as similar the proof is omitted. We have by (13.3) that

$$\left|D_{a*}^{n+\rho} f(x)\right| \leq$$

$$\frac{|A(\rho)|}{1-\rho} \int_a^x \left(E_{(\rho),1}^{(\gamma_j)} \left[\frac{|\omega_1| \rho}{1-\rho} (x-t)^{\rho}, ..., \frac{|\omega_m| \rho}{1-\rho} (x-t)^{\rho} \right] dt \right) \left\| f^{(n+1)} \right\|_{\infty}$$

$$\overset{(5,\text{ p. }175)}{=} \frac{|A(\rho)|}{1-\rho} \left[\sum_{k_1,...,k_m=0}^{\infty} \frac{(\gamma_1)_{k_1} \cdots (\gamma_m)_{k_m}}{\Gamma\left(1+\rho\left(\sum_{i=1}^m k_i\right)\right) \prod_{i=1}^m k_i!} \left(\frac{\rho}{1-\rho}\right)^{\sum_{i=1}^m k_i} \right.$$

$$\left. \left(\prod_{i=1}^m |\omega_i|^{k_i}\right) \left(\int_a^x (x-t)^{\rho\left(\sum_{i=1}^m k_i\right)} dt\right) \right] \left\| f^{(n+1)} \right\|_{\infty} =$$

$$\frac{|A(\rho)|}{1-\rho} \left[\sum_{k_1,...,k_m=0}^{\infty} \frac{(\gamma_1)_{k_1} \cdots (\gamma_m)_{k_m}}{\Gamma\left(1+\rho\left(\sum_{i=1}^m k_i\right)\right) \prod_{i=1}^m k_i!} \left(\frac{\rho}{1-\rho}\right)^{\sum_{i=1}^m k_i} \right.$$

$$\left. \left(\prod_{i=1}^m |\omega_i|^{k_i}\right) \frac{(x-a)^{\rho\left(\sum_{i=1}^m k_i\right)+1}}{\left(\rho\left(\sum_{i=1}^m k_i\right)+1\right)} \right] \left\| f^{(n+1)} \right\|_{\infty} \leq \tag{13.9}$$

$$(b-a) \frac{|A(\rho)|}{1-\rho} \left[\sum_{k_1,...,k_m=0}^{\infty} \frac{(\gamma_1)_{k_1} \cdots (\gamma_m)_{k_m}}{\Gamma\left(2+\rho\left(\sum_{i=1}^m k_i\right)\right) \prod_{i=1}^m k_i!} \left(\frac{\rho}{1-\rho}\right)^{\sum_{i=1}^m k_i} \right.$$

$$\left. \left(\prod_{i=1}^m |\omega_i|^{k_i}\right) (b-a)^{\rho\left(\sum_{i=1}^m k_i\right)} \right] \left\| f^{(n+1)} \right\|_{\infty} =$$

$$\frac{(b-a)|A(\rho)|}{1-\rho} \left[E_{(\rho),2}^{(\gamma_j)} \left[\frac{|\omega_1| \rho}{1-\rho} (b-a)^{\rho}, ..., \frac{|\omega_m| \rho}{1-\rho} (b-a)^{\rho} \right] \right] \left\| f^{(n+1)} \right\|_{\infty} < \infty,$$

proving the claim. ■

We also give

Theorem 13.2 *All as above with* $\gamma_j > 0$, $j = 1, ..., m$, *and* $\lambda > 0$, $0 < \rho < 1$, *etc. Then*

$$D_{a*}^{n+\rho} f,\ D_{b-}^{n+\rho} f \in C([a,b]),\ n \in \mathbb{Z}_+.$$

Proof We prove it only for the first one, the other one as similar is omitted. Let $x_N, x_0 \in [a, b]$ such that $x_N \to x_0$, as $N \to \infty$. Then

$$\chi_{[a,x_N]}(t) \to \chi_{[a,x_0]}(t), \text{ a.e. } (t \in [a, b]),$$

where χ is the characteristic function.

Also it holds

$$E_{(\rho),\lambda}^{(\gamma_j)}\left[\frac{-\omega_1\rho}{1-\rho}(x_N-t)^{\rho}, ..., \frac{-\omega_m\rho}{1-\rho}(x_N-t)^{\rho}\right] \to$$

$$E_{(\rho),\lambda}^{(\gamma_j)}\left[\frac{-\omega_1\rho}{1-\rho}(x_0-t)^{\rho}, ..., \frac{-\omega_m\rho}{1-\rho}(x_0-t)^{\rho}\right]. \tag{13.10}$$

Therefore we get that

$$\chi_{[a,x_N]}(t)\, E_{(\rho),\lambda}^{(\gamma_j)}\left[\frac{-\omega_1\rho}{1-\rho}(x_N-t)^{\rho}, ..., \frac{-\omega_m\rho}{1-\rho}(x_N-t)^{\rho}\right] f^{(n+1)}(t) \to$$

$$\chi_{[a,x_0]}(t)\, E_{(\rho),\lambda}^{(\gamma_j)}\left[\frac{-\omega_1\rho}{1-\rho}(x_0-t)^{\rho}, ..., \frac{-\omega_m\rho}{1-\rho}(x_0-t)^{\rho}\right] f^{(n+1)}(t),$$

a.e. over $t \in [a, b]$, as $N \to \infty$.

Furthermore it holds

$$\chi_{[a,x_N]}(t)\left|E_{(\rho),\lambda}^{(\gamma_j)}\left[\frac{-\omega_1\rho}{1-\rho}(x_N-t)^{\rho}, ..., \frac{-\omega_m\rho}{1-\rho}(x_N-t)^{\rho}\right]\right|\left|f^{(n+1)}(t)\right| \leq$$

$$E_{(\rho),\lambda}^{(\gamma_j)}\left[\frac{|\omega_1|\rho}{1-\rho}(b-a)^{\rho}, ..., \frac{|\omega_m|\rho}{1-\rho}(b-a)^{\rho}\right]\left\|f^{(n+1)}\right\|_{\infty} < \infty. \tag{13.11}$$

Thus, by the denominated convergence theorem, we derive

$$\frac{A(\rho)}{1-\rho}\int_a^b \chi_{[a,x_N]}(t)\, E_{(\rho),\lambda}^{(\gamma_j)}\left[\frac{-\omega_1\rho}{1-\rho}(x_N-t)^{\rho}, ..., \frac{-\omega_m\rho}{1-\rho}(x_N-t)^{\rho}\right] f^{(n+1)}(t)\,dt \to$$

$$\frac{A(\rho)}{1-\rho}\int_a^b \chi_{[a,x_0]}(t)\, E_{(\rho),\lambda}^{(\gamma_j)}\left[\frac{-\omega_1\rho}{1-\rho}(x_0-t)^{\rho}, ..., \frac{-\omega_m\rho}{1-\rho}(x_0-t)^{\rho}\right] f^{(n+1)}(t)\,dt, \tag{13.12}$$

as $N \to \infty$, proving the continuity of $D_{a*}^{n+\rho} f$ over $[a, b]$, for $n \in \mathbb{Z}_+$. ■

In this chapter we derive left and right side fractional inequalities related to the introduced fractional derivatives (13.2)–(13.7) of non singular kernel and multiple parameters. These are of Hardy, Opial and Hilbert-Pachpatte types, also of Hardy type involving convexity.

13.2 Main Results

From now on we denote by

$$\frac{-\boldsymbol{\omega}\rho}{1-\rho}(x-t)^{\rho} := \left(\frac{-\omega_1\rho}{1-\rho}(x-t)^{\rho}, ..., \frac{-\omega_m\rho}{1-\rho}(x-t)^{\rho}\right), \tag{13.13}$$

and

$$\frac{-\boldsymbol{\omega}\rho}{1-\rho}(t-x)^{\rho} := \left(\frac{-\omega_1\rho}{1-\rho}(t-x)^{\rho}, ..., \frac{-\omega_m\rho}{1-\rho}(t-x)^{\rho}\right). \tag{13.14}$$

So, we rewrite

$$D_{a*}^{n+\rho} f(x) = \frac{A(\rho)}{1-\rho}\int_a^x E_{(\rho),\lambda}^{(\gamma_j)}\left[\frac{-\boldsymbol{\omega}\rho}{1-\rho}(x-t)^{\rho}\right] f^{(n+1)}(t)\,dt, \tag{13.15}$$

and

$$D_{b-}^{n+\rho} f(x) = \frac{(-1)^{n+1} A(\rho)}{1-\rho}\int_x^b E_{(\rho),\lambda}^{(\gamma_j)}\left[\frac{-\boldsymbol{\omega}\rho}{1-\rho}(t-x)^{\rho}\right] f^{(n+1)}(t)\,dt, \tag{13.16}$$

$\forall\, x \in [a,b]$, $f \in C^{n+1}([a,b])$, $n \in \mathbb{Z}_+$.

From now on we take $0<\rho<1$, $\lambda>0$, $\gamma_j>0$, $\omega_j \in \mathbb{R}-\{0\}$, for $j=1,...,m$.

We present the following L_p Hardy type inequalities.

Theorem 13.3 *Let* $f_\nu \in C^{n+1}([a,b])$, $n \in \mathbb{Z}_+$, $\nu = 1,...,N$; $p,q>1 : \frac{1}{p}+\frac{1}{q}=1$. *Then*

$$\left\|\prod_{\nu=1}^{N} D_{a*}^{n+\rho} f_\nu\right\|_p \le \left(\frac{A(\rho)}{1-\rho}\right)^N$$

$$\left(\int_a^b\left(\int_a^x \left|E_{(\rho),\lambda}^{(\gamma_j)}\left[\frac{-\boldsymbol{\omega}\rho}{1-\rho}(x-t)^{\rho}\right]\right|^p dt\right)^N dx\right)^{\frac{1}{p}}\left(\prod_{\nu=1}^{N}\left\|f_\nu^{(n+1)}\right\|_q\right), \tag{13.17}$$

and

$$\left\|\prod_{\nu=1}^{N} D_{b-}^{n+\rho} f_\nu\right\|_p \le \left(\frac{A(\rho)}{1-\rho}\right)^N$$

$$\left(\int_a^b\left(\int_x^b \left|E_{(\rho),\lambda}^{(\gamma_j)}\left[\frac{-\boldsymbol{\omega}\rho}{1-\rho}(t-x)^{\rho}\right]\right|^p dt\right)^N dx\right)^{\frac{1}{p}}\left(\prod_{\nu=1}^{N}\left\|f_\nu^{(n+1)}\right\|_q\right). \tag{13.18}$$

Proof We prove only (13.17).

We have that

$$D_{a*}^{n+\rho} f_\nu (x) = \frac{A(\rho)}{1-\rho} \int_a^x E_{(\rho),\lambda}^{(\gamma_j)} \left[\frac{-\omega\rho}{1-\rho} (x-t)^\rho \right] f_\nu^{(n+1)} (t) \, dt, \tag{13.19}$$

$\nu = 1, ..., N$, $\forall \, x \in [a, b]$.

By Hölder's inequality and (13.19) we obtain

$$\left| D_{a*}^{n+\rho} f_\nu (x) \right| \leq \frac{|A(\rho)|}{1-\rho} \int_a^x \left| E_{(\rho),\lambda}^{(\gamma_j)} \left[\frac{-\omega\rho}{1-\rho} (x-t)^\rho \right] \right| \left| f_\nu^{(n+1)} (t) \right| dt \leq$$

$$\frac{|A(\rho)|}{1-\rho} \left(\int_a^x \left| E_{(\rho),\lambda}^{(\gamma_j)} \left[\frac{-\omega\rho}{1-\rho} (x-t)^\rho \right] \right|^p dt \right)^{\frac{1}{p}} \left\| f_\nu^{(n+1)} \right\|_q, \tag{13.20}$$

and

$$\prod_{\nu=1}^N \left| D_{a*}^{n+\rho} f_\nu (x) \right| \leq \left(\frac{|A(\rho)|}{1-\rho} \right)^N \left(\int_a^x \left| E_{(\rho),\lambda}^{(\gamma_j)} \left[\frac{-\omega\rho}{1-\rho} (x-t)^\rho \right] \right|^p dt \right)^{\frac{N}{p}} \prod_{\nu=1}^N \left\| f_\nu^{(n+1)} \right\|_q,$$

$\forall \, x \in [a, b]$.

Hence it holds

$$\left(\prod_{\nu=1}^N \left| D_{a*}^{n+\rho} f_\nu (x) \right| \right)^p \leq \left(\frac{|A(\rho)|}{1-\rho} \right)^{Np}$$

$$\left(\int_a^x \left| E_{(\rho),\lambda}^{(\gamma_j)} \left[\frac{-\omega\rho}{1-\rho} (x-t)^\rho \right] \right|^p dt \right)^N \left(\prod_{\nu=1}^N \left\| f_\nu^{(n+1)} \right\|_q \right)^p, \tag{13.21}$$

and

$$\int_a^b \left(\prod_{\nu=1}^N \left| D_{a*}^{n+\rho} f_\nu (x) \right| \right)^p dx \leq \left(\frac{|A(\rho)|}{1-\rho} \right)^{Np}$$

$$\left(\int_a^b \left(\int_a^x \left| E_{(\rho),\lambda}^{(\gamma_j)} \left[\frac{-\omega\rho}{1-\rho} (x-t)^\rho \right] \right|^p dt \right)^N dx \right) \left(\prod_{\nu=1}^N \left\| f_\nu^{(n+1)} \right\|_q \right)^p, \tag{13.22}$$

proving the claim. ■

We continue with fractional Opial type inequalities:

Theorem 13.4 *Let $f \in C^{n+1}([a, b])$, $n \in \mathbb{Z}_+$; $p, q > 1 : \frac{1}{p} + \frac{1}{q} = 1$. Then*

(i) the left side one:

$$\int_a^x \left| D_{a*}^{n+\rho} f(w) \right| \left| f^{(n+1)} (w) \right| dw \leq 2^{-\frac{1}{q}} \frac{|A(\rho)|}{1-\rho}$$

$$\left(\int_a^x \left(\int_a^w \left| E_{(\rho),\lambda}^{(\gamma_j)} \left[\frac{-\omega\rho}{1-\rho} (w-t)^\rho \right] \right|^p dt \right) dw \right)^{\frac{1}{p}} \left(\int_a^x \left| f^{(n+1)} (w) \right|^q dw \right)^{\frac{2}{q}}, \tag{13.23}$$

(ii) the right side one:

$$\int_x^b \left|D_{b-}^{n+\rho} f(w)\right| \left|f^{(n+1)}(w)\right| dw \leq 2^{-\frac{1}{q}} \frac{|A(\rho)|}{1-\rho}$$

$$\left(\int_x^b \left(\int_w^b \left|E_{(\rho),\lambda}^{(\gamma_j)}\left[\frac{-\omega\rho}{1-\rho}(t-w)^\rho\right]\right|^p dt\right) dw\right)^{\frac{1}{p}} \left(\int_x^b \left|f^{(n+1)}(w)\right|^q dw\right)^{\frac{2}{q}}, \tag{13.24}$$

$\forall\, x \in [a,b]$.

Proof We prove only (13.23). We have that (by Hölder's inequality)

$$\left|D_{a*}^{n+\rho} f(x)\right| \leq \frac{|A(\rho)|}{1-\rho} \int_a^x \left|E_{(\rho),\lambda}^{(\gamma_j)}\left[\frac{-\omega\rho}{1-\rho}(x-t)^\rho\right]\right| \left|f^{(n+1)}(t)\right| dt \leq$$

$$\frac{|A(\rho)|}{1-\rho} \left(\int_a^x \left|E_{(\rho),\lambda}^{(\gamma_j)}\left[\frac{-\omega\rho}{1-\rho}(x-t)^\rho\right]\right|^p dt\right)^{\frac{1}{p}} \left(\int_a^x \left|f^{(n+1)}(t)\right|^q dt\right)^{\frac{1}{q}}. \tag{13.25}$$

Call

$$\Phi(x) := \int_a^x \left|f^{(n+1)}(t)\right|^q dt,\ \Phi(a) = 0. \tag{13.26}$$

Then

$$\Phi'(x) = \left|f^{(n+1)}(x)\right|^q \geq 0, \tag{13.27}$$

and

$$\left(\Phi'(x)\right)^{\frac{1}{q}} = \left|f^{(n+1)}(x)\right| \geq 0,\ \ \forall\, x \in [a,b].$$

Consequently, we get

$$\left|D_{a*}^{n+\rho} f(w)\right| \left|f^{(n+1)}(w)\right| \leq$$

$$\frac{|A(\rho)|}{1-\rho} \left(\int_a^w \left|E_{(\rho),\lambda}^{(\gamma_j)}\left[\frac{-\omega\rho}{1-\rho}(w-t)^\rho\right]\right|^p dt\right)^{\frac{1}{p}} \left(\Phi(w)\,\Phi'(w)\right)^{\frac{1}{q}}, \tag{13.28}$$

$\forall\, w \in [a,b]$.

Thus, by applying again Hölder's inequality:

$$\int_a^x \left|D_{a*}^{n+\rho} f(w)\right| \left|f^{(n+1)}(w)\right| dw \leq$$

$$\frac{|A(\rho)|}{1-\rho} \int_a^x \left(\int_a^w \left|E_{(\rho),\lambda}^{(\gamma_j)}\left[\frac{-\omega\rho}{1-\rho}(w-t)^\rho\right]\right|^p dt\right)^{\frac{1}{p}} \left(\Phi(w)\,\Phi'(w)\right)^{\frac{1}{q}} dw \leq$$

$$\frac{|A(\rho)|}{1-\rho}\left(\int_a^x\left(\int_a^w\left|E_{(\rho),\lambda}^{(\gamma_j)}\left[\frac{-\omega\rho}{1-\rho}(w-t)^\rho\right]\right|^p dt\right)dw\right)^{\frac{1}{p}}\left(\int_a^x\left(\Phi(w)\Phi'(w)\right)dw\right)^{\frac{1}{q}} \tag{13.29}$$

$$=\frac{|A(\rho)|}{1-\rho}\left(\int_a^x\left(\int_a^w\left|E_{(\rho),\lambda}^{(\gamma_j)}\left[\frac{-\omega\rho}{1-\rho}(w-t)^\rho\right]\right|^p dt\right)dw\right)^{\frac{1}{p}}\left(\frac{\Phi^2(x)}{2}\right)^{\frac{1}{q}}=$$

$$2^{-\frac{1}{q}}\frac{|A(\rho)|}{1-\rho}\left(\int_a^x\left(\int_a^w\left|E_{(\rho),\lambda}^{(\gamma_j)}\left[\frac{-\omega\rho}{1-\rho}(w-t)^\rho\right]\right|^p dt\right)dw\right)^{\frac{1}{p}}\left(\int_a^x\left|f^{(n+1)}(w)\right|^q dw\right)^{\frac{2}{q}},$$

proving the claim. ■

We continue with fractional Hilbert-Pachpatte type inequalities. First comes the left side one:

Theorem 13.5 *Let* $p,q>1:\frac{1}{p}+\frac{1}{q}=1$, $\mu=1,2$. *Let* $[a_\mu,b_\mu]\subset\mathbb{R}$, $f_\mu\in C^{n_\mu+1}([a_\mu,b_\mu])$, $n_\mu\in\mathbb{Z}_+$; $0<\rho_\mu<1$, $\lambda_\mu>0$, $\gamma_{j\mu}>0$, $\omega_{j\mu}\in\mathbb{R}-\{0\}$, $j=1,\dots,m$. *Also we denote*

$$\frac{-\boldsymbol{\omega}_\mu\rho_\mu}{1-\rho_\mu}(x_\mu-t_\mu)^{\rho_\mu}:=\left(\frac{-\omega_{1\mu}\rho_\mu}{1-\rho_\mu}(x_\mu-t_\mu)^{\rho_\mu},\dots,\frac{-\omega_{m\mu}\rho_\mu}{1-\rho_\mu}(x_\mu-t_\mu)^{\rho_\mu}\right), \tag{13.30}$$

$\mu=1,2$.

Then

$$\int_{a_1}^{b_1}\int_{a_2}^{b_2}\frac{\left|D_{a_1*}^{n_1+\rho_1}f_1(x_1)\right|\left|D_{a_2*}^{n_2+\rho_2}f_2(x_2)\right|dx_1dx_2}{\left\{\frac{\left(\int_{a_1}^{x_1}\left|E_{(\rho_1),\lambda_1}^{(\gamma_{j1})}\left[\frac{-\boldsymbol{\omega}_1\rho_1}{1-\rho_1}(x_1-t_1)^{\rho_1}\right]\right|^p dt_1\right)}{p}+\frac{\int_{a_2}^{x_2}\left|E_{(\rho_2),\lambda_2}^{(\gamma_{j2})}\left[\frac{-\boldsymbol{\omega}_2\rho_2}{1-\rho_2}(x_2-t_2)^{\rho_2}\right]\right|^q dt_2}{q}\right\}}$$

$$\le(b_1-a_1)(b_2-a_2)\frac{|A_1(\rho_1)||A_2(\rho_2)|}{(1-\rho_1)(1-\rho_2)}\left\|f_1^{(n_1+1)}\right\|_q\left\|f_2^{(n_2+1)}\right\|_p. \tag{13.31}$$

Proof Here we have ($\mu=1,2$)

$$D_{a_\mu*}^{n_\mu+\rho_\mu}f_\mu(x_\mu)\overset{(13.15)}{=}\frac{A_\mu(\rho_\mu)}{1-\rho_\mu}\int_{a_\mu}^{x_\mu}E_{(\rho_\mu),\lambda_\mu}^{(\gamma_{j\mu})}\left[\frac{-\boldsymbol{\omega}_\mu\rho_\mu}{1-\rho_\mu}(x_\mu-t_\mu)^{\rho_\mu}\right]f_\mu^{(n_\mu+1)}(t_\mu)dt_\mu, \tag{13.32}$$

$\forall x_\mu\in[a_\mu,b_\mu]$, $n_\mu\in\mathbb{Z}_+$.

Then

$$\left|D_{a_\mu*}^{n_\mu+\rho_\mu}f_\mu(x_\mu)\right|\le$$

$$\frac{|A_\mu(\rho_\mu)|}{1-\rho_\mu}\int_{a_\mu}^{x_\mu}\left|E_{(\rho_\mu),\lambda_\mu}^{(\gamma_{j\mu})}\left[\frac{-\boldsymbol{\omega}_\mu\rho_\mu}{1-\rho_\mu}(x_\mu-t_\mu)^{\rho_\mu}\right]\right|\left|f_\mu^{(n_\mu+1)}(t_\mu)\right|dt_\mu, \tag{13.33}$$

$\mu=1,2$.

By applying Hölder's inequality twice we get:

$$\left|D_{a_1*}^{n_1+\rho_1} f_1(x_1)\right| \leq \frac{|A_1(\rho_1)|}{1-\rho_1}$$

$$\left(\int_{a_1}^{x_1} \left|E_{(\rho_1),\lambda_1}^{(\gamma_{j1})}\left[\frac{-\omega_1\rho_1}{1-\rho_1}(x_1-t_1)^{\rho_1}\right]\right|^p dt_1\right)^{\frac{1}{p}} \left(\int_{a_1}^{x_1} \left|f_1^{(n_1+1)}(t_1)\right|^q dt_1\right)^{\frac{1}{q}}, \tag{13.34}$$

$\forall\, x_1 \in [a_1, b_1]$, and

$$\left|D_{a_2*}^{n_2+\rho_2} f_2(x_2)\right| \leq \frac{|A_2(\rho_2)|}{1-\rho_2}$$

$$\left(\int_{a_2}^{x_2} \left|E_{(\rho_2),\lambda_2}^{(\gamma_{j2})}\left[\frac{-\omega_2\rho_2}{1-\rho_2}(x_2-t_2)^{\rho_2}\right]\right|^q dt_2\right)^{\frac{1}{q}} \left(\int_{a_2}^{x_2} \left|f_2^{(n_2+1)}(t_2)\right|^p dt_2\right)^{\frac{1}{p}}, \tag{13.35}$$

$\forall\, x_2 \in [a_2, b_2]$.

Hence we have (by (13.34), (13.35))

$$\left|D_{a_1*}^{n_1+\rho_1} f_1(x_1)\right| \left|D_{a_2*}^{n_2+\rho_2} f_2(x_2)\right| \leq \frac{|A_1(\rho_1)|\,|A_2(\rho_2)|}{(1-\rho_1)(1-\rho_2)}$$

$$\left(\int_{a_1}^{x_1} \left|E_{(\rho_1),\lambda_1}^{(\gamma_{j1})}\left[\frac{-\omega_1\rho_1}{1-\rho_1}(x_1-t_1)^{\rho_1}\right]\right|^p dt_1\right)^{\frac{1}{p}} \tag{13.36}$$

$$\left(\int_{a_2}^{x_2} \left|E_{(\rho_2),\lambda_2}^{(\gamma_{j2})}\left[\frac{-\omega_2\rho_2}{1-\rho_2}(x_2-t_2)^{\rho_2}\right]\right|^q dt_2\right)^{\frac{1}{q}}$$

$$\left\|f_1^{(n_1+1)}\right\|_q \left\|f_2^{(n_2+1)}\right\|_p \leq$$

(using Young's inequality for $a, b \geq 0$, $a^{\frac{1}{p}} b^{\frac{1}{q}} \leq \frac{a}{p} + \frac{b}{q}$)

$$\frac{|A_1(\rho_1)|\,|A_2(\rho_2)|}{(1-\rho_1)(1-\rho_2)} \left\{ \frac{\left(\int_{a_1}^{x_1} \left|E_{(\rho_1),\lambda_1}^{(\gamma_j 1)}\left[\frac{-\omega_1\rho_1}{1-\rho_1}(x_1-t_1)^{\rho_1}\right]\right|^p dt_1\right)}{p} + \right.$$

$$\left. \frac{\int_{a_2}^{x_2} \left|E_{(\rho_2),\lambda_2}^{(\gamma_j 2)}\left[\frac{-\omega_2\rho_2}{1-\rho_2}(x_2-t_2)^{\rho_2}\right]\right|^q dt_2}{q} \right\} \left\|f_1^{(n_1+1)}\right\|_q \left\|f_2^{(n_2+1)}\right\|_p, \tag{13.37}$$

$\forall\, x_\mu \in [a_\mu, b_\mu]$, $\mu = 1, 2$.

So far we have that

$$\frac{\left|D_{a_1*}^{n_1+\rho_1} f_1(x_1)\right|\left|D_{a_2*}^{n_2+\rho_2} f_2(x_2)\right|}{\left\{\frac{\left(\int_{a_1}^{x_1}\left|E_{(\rho_1),\lambda_1}^{(\gamma_j1)}\left[\frac{-\omega_1\rho_1}{1-\rho_1}(x_1-t_1)^{\rho_1}\right]\right|^p dt_1\right)}{p}+\frac{\int_{a_2}^{x_2}\left|E_{(\rho_2),\lambda_2}^{(\gamma_j2)}\left[\frac{-\omega_2\rho_2}{1-\rho_2}(x_2-t_2)^{\rho_2}\right]\right|^q dt_2}{q}\right\}}$$

$$\leq \frac{|A_1(\rho_1)||A_2(\rho_2)|}{(1-\rho_1)(1-\rho_2)}\left\|f_1^{(n_1+1)}\right\|_q\left\|f_2^{(n_2+1)}\right\|_p, \tag{13.38}$$

$\forall\, x_\mu \in [a_\mu, b_\mu]$, $\mu = 1, 2$.

The denominator in (13.38) can be zero only when $x_1 = a_1$ and $x_2 = a_2$.

Therefore we obtain (13.31) by integrating (13.38) over $[a_1, b_1] \times [a_2, b_2]$. ■

The counterpart of Theorem 13.5 follows (the right side inequality)

Theorem 13.6 *All as in Theorem 13.5, but now we denote*

$$\frac{-\omega_\mu\rho_\mu}{1-\rho_\mu}(t_\mu - x_\mu)^{\rho_\mu} := \left(\frac{-\omega_{1\mu}\rho_\mu}{1-\rho_\mu}(t_\mu - x_\mu)^{\rho_\mu}, \ldots, \frac{-\omega_{m\mu}\rho_\mu}{1-\rho_\mu}(t_\mu - x_\mu)^{\rho_\mu}\right), \tag{13.39}$$

$\mu = 1, 2.$

Then

$$\int_{a_1}^{b_1}\int_{a_2}^{b_2}\frac{\left|D_{b_1-}^{n_1+\rho_1} f_1(x_1)\right|\left|D_{b_2-}^{n_2+\rho_2} f_2(x_2)\right| dx_1 dx_2}{\left\{\frac{\left(\int_{x_1}^{b_1}\left|E_{(\rho_1),\lambda_1}^{(\gamma_j1)}\left[\frac{-\omega_1\rho_1}{1-\rho_1}(t_1-x_1)^{\rho_1}\right]\right|^p dt_1\right)}{p}+\frac{\int_{x_2}^{b_2}\left|E_{(\rho_2),\lambda_2}^{(\gamma_j2)}\left[\frac{-\omega_2\rho_2}{1-\rho_2}(t_2-x_2)^{\rho_2}\right]\right|^q dt_2}{q}\right\}}$$

$$\leq (b_1 - a_1)(b_2 - a_2)\frac{|A_1(\rho_1)||A_2(\rho_2)|}{(1-\rho_1)(1-\rho_2)}\left\|f_1^{(n_1+1)}\right\|_q\left\|f_2^{(n_2+1)}\right\|_p. \tag{13.40}$$

Proof As similar to the proof of Theorem 13.5 is omitted. ■

We need

Notation 13.7 Let $(\Omega_1, \Sigma_1, \mu_1)$ and $(\Omega_2, \Sigma_2, \mu_2)$ be measure spaces with positive σ-finite measures, and let $k : \Omega_1 \times \Omega_2 \to \mathbb{R}$ be nonnegative measurable functions, $k(x, \cdot)$ measurable on Ω_2, and

$$K(x) = \int_{\Omega_2} k(x, y)\, d\mu_2(y), \text{ for any } x \in \Omega_1. \tag{13.41}$$

We suppose that $K(x) > 0$ a.e. on Ω_1 and by a weight function u (shortly: a weight), we mean a nonnegative measurable function on the actual set. Let the measurable functions $g_j : \Omega_1 \to \mathbb{R}$, $j = 1, \ldots, r$, with the representation

$$g_j(x) = \int_{\Omega_2} k(x, y) f_j(y) d\mu_2(y), \tag{13.42}$$

where $f_j : \Omega_2 \to \mathbb{R}$ are measurable functions, $j = 1, ..., r$.

Denote by $\mathbf{x} = x := (x_1, ..., x_r) \in \mathbb{R}^r$, $\mathbf{g} := (g_1, ..., g_r)$ and $\mathbf{f} := (f_1, ..., f_r)$.

We consider here $\Phi : \mathbb{R}_+^r \to \mathbb{R}$ a convex function, which is increasing per coordinate, i.e. if $x_j \leq y_j$, $j = 1, ..., r$, then

$$\Phi(x_1, ..., x_r) \leq \Phi(y_1, ..., y_r).$$

In [2, p. 588], we proved that

Theorem 13.8 *Let u be a weight function on Ω_1, and k, K, g_j, f_j, $j = 1, ..., r \in \mathbb{N}$, and Φ defined as above. Assume that the function $x \to u(x) \frac{k(x,y)}{K(x)}$ is integrable on Ω_1 for each fixed $y \in \Omega_2$. Define v on Ω_2 by*

$$v(y) := \int_{\Omega_1} u(x) \frac{k(x, y)}{K(x)} d\mu_1(x) < \infty. \tag{13.43}$$

Then

$$\int_{\Omega_1} u(x) \Phi\left(\frac{|g_1(x)|}{K(x)}, ..., \frac{|g_r(x)|}{K(x)}\right) d\mu_1(x) \leq$$

$$\int_{\Omega_2} v(y) \Phi(|f_1(y)|, ..., |f_r(y)|) d\mu_2(y), \tag{13.44}$$

under the assumptions:

(i) f_j, $\Phi(|f_1|, ..., |f_r|)$, are $k(x, y) d\mu_2(y)$ -integrable, μ_1 -a.e. in $x \in \Omega_1$, for all $j = 1, ..., r$,

(ii) $v(y) \Phi(|f_1(y)|, ..., |f_r(y)|)$ is μ_2 -integrable.

We give under convexity a left fractional generalized Hardy type inequality.

Theorem 13.9 *Let u_1 be a weight function on $[a, b]$,*

$$k_{a*}(x, y) := \frac{A(\rho)}{1-\rho} \chi_{[a,x]}(y) E^{\gamma_j}_{(\rho),\lambda}\left[\frac{-\omega\rho}{1-\rho}(x-y)^{\rho}\right],$$

$$K_{a*}(x) := \frac{A(\rho)}{1-\rho} \int_a^b \chi_{[a,x]}(y) E^{\gamma_j}_{(\rho),\lambda}\left[\frac{-\omega\rho}{1-\rho}(x-y)^{\rho}\right] dy, \tag{13.45}$$

for any $x \in [a, b]$. Also consider $D^{n+\rho}_{a*} f_\mu$ and $f_\mu^{(n+1)} \in C([a, b])$, $\mu = 1, ..., r \in \mathbb{N}$, $n \in \mathbb{Z}_+$, and $\Phi : \mathbb{R}_+^r \to \mathbb{R}$ a convex function, which is increasing per coordinate. Assume that the function $x \to u_1(x) \frac{k_{a*}(x,y)}{K_{a*}(x)}$ is integrable on $[a, b]$ for each $y \in [a, b]$. Define v_1 on $[a, b]$ by

$$\nu_1(y) := \int_a^b u_1(x) \frac{k_{a*}(x,y)}{K_{a*}(x)} dx < \infty, \tag{13.46}$$

and ν_1 *is assumed to be integrable on* $[a,b]$.

Then

$$\int_a^b u_1(x) \Phi \left(\frac{\left| D_{a*}^{n+\rho} f_1(x) \right|}{K_{a*}(x)}, \ldots, \frac{\left| D_{a*}^{n+\rho} f_r(x) \right|}{K_{a*}(x)} \right) dx \leq$$

$$\int_a^b \nu_1(y) \Phi \left(\left| f_1^{(n+1)}(y) \right|, \ldots, \left| f_r^{(n+1)}(y) \right| \right) dy. \tag{13.47}$$

Proof See Notation 13.7 and apply Theorem 13.8. ■

Also we give the right side corresponding to (13.47) Hardy type inequality.

Theorem 13.10 *Let* u_2 *be a weight function on* $[a,b]$,

$$k_{b-}(x,y) := \frac{(-1)^{n+1} A(\rho)}{1-\rho} \chi_{[x,b]}(y) E_{(\rho),\lambda}^{\gamma_j} \left[\frac{-\omega\rho}{1-\rho} (y-x)^\rho \right],$$

$$K_{b-}(x) := \frac{(-1)^{n+1} A(\rho)}{1-\rho} \int_a^b \chi_{[x,b]}(y) E_{(\rho),\lambda}^{\gamma_j} \left[\frac{-\omega\rho}{1-\rho} (y-x)^\rho \right] dy, \tag{13.48}$$

for any $x \in [a,b]$. *Also consider* $D_{b-}^{n+\rho} f_\mu$ *and* $f_\mu^{(n+1)} \in C([a,b])$, $\mu = 1, \ldots, r \in \mathbb{N}$, $n \in \mathbb{Z}_+$, *and* $\Phi : \mathbb{R}_+^r \to \mathbb{R}$ *a convex function, which is increasing per coordinate. Assume that the function* $x \to u_2(x) \frac{k_{b-}(x,y)}{K_{b-}(x)}$ *is integrable on* $[a,b]$ *for each* $y \in [a,b]$. *Define* ν_2 *on* $[a,b]$ *by*

$$\nu_2(y) := \int_a^b u_2(x) \frac{k_{b-}(x,y)}{K_{b-}(x)} dx < \infty, \tag{13.49}$$

and ν_2 *is assumed to be integrable on* $[a,b]$.

Then

$$\int_a^b u_2(x) \Phi \left(\frac{\left| D_{b-}^{n+\rho} f_1(x) \right|}{K_{b-}(x)}, \ldots, \frac{\left| D_{b-}^{n+\rho} f_r(x) \right|}{K_{b-}(x)} \right) dx \leq$$

$$\int_a^b \nu_2(y) \Phi \left(\left| f_1^{(n+1)}(y) \right|, \ldots, \left| f_r^{(n+1)}(y) \right| \right) dy. \tag{13.50}$$

Proof See Notation 13.7 and apply Theorem 13.8. ■

Note 13.11 One can create a vast number of similar very interesting theorems using (13.3) and (13.5) and based on the author's monographs [1, 2]. To stay short we choose to skip this task.

Conclusion 13.12 The highlight of this chapter is the introduction of the fractional derivatives (13.2)–(13.7) with non singular kernel and many parameters, able to incorporate lots of data, so they can describe complex settings and situations in complete ways.

References

1. Anastassiou, G.: Fractional Differentiation Inequalities. Springer, Heidelberg, New York (2009)
2. Anastassiou, G.: Intelligent Comparisons: Analytic Inequalities. Springer, Heidelberg, New York (2016)
3. Anastassiou, G.: Multiparameter Fractional Differentiation with Non Singular Kernel (2021) (submitted)
4. Atangana, A., Baleanu, D.: New fractional derivatives with nonlocal and non-singular kernel: theory and application to heat transfer model. Therm. Sci. **20**(2), 763–769 (2016)
5. Caputo, M., Fabrizion, M.: A new definition of fractional derivative without singular kernel. Progr. Fract. Differ. Appl. **1**(2), 73–85 (2015)
6. Hewith, E., Stromberg, K.: Real and Abstract Analysis. Springer, New York, Berlin (1965)
7. Losada, J. Nieto, J.J.: Properties of a new fractional derivative without singular kernel. Progr. Fract. Differ. Appl. **1**(2), 87–92 (2015)
8. Saxena, R.K., Kalla, S.L., Saxena, R.: Multivariate analogue of generalized Mittag-Leffler function. Integ. Transforms Spec. Functions **22**(7), 533–548 (2011)
9. Srivastava, H.M., Daoust, M.C.: Certain generalized Newmann expansion associated with Kompe' de Feriet function. Nederl. Akad. Wetensch. Proc. Ser. A 72 (Indag. Math.) **31**, 449-457 (1969)
10. Srivastava, H.M., Daoust, M.C.: A note on the convergence of Kompe' de Feriet's double hypergeometric series. Math. Nachr. **53**, 151–159 (1972)
11. Srivastava, H.M., Karlsson, P.W.: Multiple Gaussian Hypergeometric Series. Ellis Horwood Limited, New York (1985)

Chapter 14
Advanced Hilfer Fractional Opial Inequalities

Here we present a detailed collection of Hilfer fractional left and right side Opial type inequalities. These cover forward, reverse and extreme cases, and involve one, two and several functions of various non-integer orders at various powers. Our estimates are very general covering many different settings. It follows [13].

14.1 Background

The original Opial inequality [17] (see also [16, p. 114]) states the following:

Theorem 14.1 *If $f \in C^1([0, a])$ with $f(0) = f(a) = 0$ and $f(x) > 0$ on $(0, a)$, then*

$$\int_0^a \left| f(x) f'(x) \right| dx \leq \frac{a}{4} \int_0^a \left(f'(x) \right)^2 dx. \tag{14.1}$$

The constant $\frac{a}{4}$ is the best possible constant.

This result for classical derivatives has been generalized in several directions (see, for instance, [1, 2]).

The author was the first to extend (14.1) for fractional derivatives, see [3], in 2000. For a complete treatment see his monograph [11] of 2009.

Here we present a complete collection of left and right side fractional Opial type inequalities involving the left and right side Hilfer fractional derivatives.

Let $-\infty < a < b < \infty$, the left and right Riemann-Liouville fractional integrals of order $\alpha \in \mathbb{C}$ ($\mathcal{R}(\alpha) > 0$) are defined by

$$\left(I_{a+}^{\alpha} f \right)(x) = \frac{1}{\Gamma(\alpha)} \int_a^x (x-t)^{\alpha-1} f(t) \, dt, \tag{14.2}$$

G. A. Anastassiou, *Unification of Fractional Calculi with Applications*, Studies in Systems, Decision and Control 398,
https://doi.org/10.1007/978-3-030-86920-5_14

$x > a$; where Γ stands for the gamma function,

and

$$\left(I_{b-}^{\alpha} f\right)(x) = \frac{1}{\Gamma(\alpha)} \int_x^b (t - x)^{\alpha-1} f(t)\, dt, \tag{14.3}$$

$x < b$.

The Riemann-Liouville left and right fractional derivatives of order $\alpha \in \mathbb{C}$ ($\mathcal{R}(\alpha) \geq 0$) are defined by

$$\left(\Delta_{a+}^{\alpha} y\right)(x) = \left(\frac{d}{dx}\right)^n \left(I_{a+}^{n-\alpha} y\right)(x) = \frac{1}{\Gamma(n-\alpha)} \left(\frac{d}{dx}\right)^n \int_a^x (x - t)^{n-\alpha-1} y(t)\, dt \tag{14.4}$$

($n = \lceil \mathcal{R}(\alpha) \rceil$, $\lceil \cdot \rceil$ means ceilling of the number; $x > a$)

$$\left(\Delta_{b-}^{\alpha} y\right)(x) = (-1)^n \left(\frac{d}{dx}\right)^n \left(I_{b-}^{n-\alpha} y\right)(x) =$$

$$\frac{(-1)^n}{\Gamma(n-\alpha)} \left(\frac{d}{dx}\right)^n \int_x^b (t - x)^{n-\alpha-1} y(t)\, dt \tag{14.5}$$

($n = \lceil \mathcal{R}(\alpha) \rceil$; $x < b$), respectively; where $\mathcal{R}(\alpha)$ is the real part of α.

In particular, when $\alpha = n \in \mathbb{Z}_+$, then

$$\left(\Delta_{a+}^{0} y\right)(x) = \left(\Delta_{b-}^{0} y\right)(x) = y(x);$$

$$\left(\Delta_{a+}^{n} y\right)(x) = y^{(n)}(x), \text{ and } \left(\Delta_{b-}^{n} y\right)(x) = (-1)^n y^{(n)}(x), \quad n \in \mathbb{N},$$

see [18].

We need to mention the left and right side Hilfer fractional derivatives.

Definition 14.2 ([19]) Let $\nu > 0$, $\nu \notin \mathbb{N}$, $\lceil \nu \rceil = n \in \mathbb{N}$, $f \in C([a, b])$. The Hilfer fractional derivative (left-sided and right-sided) ${}^{H}\mathbb{D}_{a+(b-)}^{\nu,\beta} f$ of order ν and type $0 \leq \beta \leq 1$, are defined by

$${}^{H}\mathbb{D}_{a+}^{\nu,\beta} f(x) = I_{a+}^{\beta(n-\nu)} \left(\frac{d}{dx}\right)^n I_{a+}^{(1-\beta)(n-\nu)} f(x), \tag{14.6}$$

and

$${}^{H}\mathbb{D}_{b-}^{\nu,\beta} f(x) = I_{b-}^{\beta(n-\nu)} \left(-\frac{d}{dx}\right)^n I_{b-}^{(1-\beta)(n-\nu)} f(x), \tag{14.7}$$

$x \in [a, b]$.

We make

Remark 14.3 When $\beta = 0$, we get the Riemann-Liouville fractional derivatives, while when $\beta = 1$ we obtain the Caputo type fractional derivatives as follows (with $f \in C^n([a,b])$):

$$D_{a+}^{\nu} f(x) = \frac{1}{\Gamma(n-\nu)} \int_a^x (x-t)^{n-\nu-1} f^{(n)}(t)\,dt, \tag{14.8}$$

and

$$D_{b-}^{\nu} f(x) = \frac{(-1)^n}{\Gamma(n-\nu)} \int_x^b (t-x)^{n-\nu-1} f^{(n)}(t)\,dt, \tag{14.9}$$

$\forall\, x \in [a,b]$.

We define $\delta := \nu + \beta(n-\nu)$. We notice that $n-1 < \nu \leq \nu + \beta(n-\nu) \leq \nu + n - \nu = n$, hence $\lceil \delta \rceil = n$. We can easily write that

$${}^{H}\mathbb{D}_{a+}^{\nu,\beta} f(x) = I_{a+}^{\delta-\nu} \Delta_{a+}^{\delta} f(x), \tag{14.10}$$

and

$${}^{H}\mathbb{D}_{b-}^{\nu,\beta} f(x) = I_{b-}^{\delta-\nu} \Delta_{b-}^{\delta} f(x), \tag{14.11}$$

$x \in [a,b]$.

We have that

$$\Delta_{a+}^{\delta} f(x) = \left(\frac{d}{dx}\right)^n I_{a+}^{(1-\beta)(n-\nu)} f(x), \tag{14.12}$$

and

$$\Delta_{b-}^{\delta} f(x) = \left(-\frac{d}{dx}\right)^n I_{b-}^{(1-\beta)(n-\nu)} f(x), \tag{14.13}$$

$x \in [a,b]$.

In particular, when $0 < \nu < 1$ and $0 \leq \beta \leq 1$; $\delta = \nu + \beta(1-\nu)$, we have that

$${}^{H}\mathbb{D}_{a+}^{\nu,\beta} f(x) = \frac{1}{\Gamma(\delta-\nu)} \int_a^x (x-t)^{\delta-\nu-1} \Delta_{a+}^{\delta} f(t)\,dt, \tag{14.14}$$

and

$${}^{H}\mathbb{D}_{b-}^{\nu,\beta} f(x) = \frac{1}{\Gamma(\delta-\nu)} \int_x^b (t-x)^{\delta-\nu-1} \Delta_{b-}^{\delta} f(t)\,dt, \tag{14.15}$$

$x \in [a,b]$.

Next we mention some Hilfer fractional representation formulae:

Theorem 14.4 ([12]) *Let $\nu > 0$, $\nu \notin \mathbb{N}$, $\lceil \nu \rceil = n$, $0 < \beta < 1$; $f \in C([a,b])$, $[a,b] \subset \mathbb{R}$; and set $\delta = \nu + \beta(n-\nu)$. Assume further that $\Delta_{a+}^{\delta} f \in C([a,b]) : \Delta_{a+}^{\delta-j} f(a) = 0$, for $j = 1, \ldots, n$. Let also $\gamma > 0 : \lceil \gamma \rceil = m$, with $\varepsilon = \gamma + \beta(m-\gamma)$, and assume that $\nu > \gamma$ and $\delta > \varepsilon$. Then*

$$ {}^{H}\mathbb{D}_{a+}^{\gamma,\beta} f(x) = \frac{1}{\Gamma(\nu-\gamma)} \int_a^x (x-t)^{\nu-\gamma-1} \, {}^{H}\mathbb{D}_{a+}^{\nu,\beta} f(t) \, dt, \tag{14.16} $$

$x \in [a, b]$, furthermore ${}^{H}\mathbb{D}_{a+}^{\gamma,\beta} f \in AC([a, b])$ (absolutely continuous functions) if $\nu - \gamma \geq 1$, and ${}^{H}\mathbb{D}_{a+}^{\gamma,\beta} f \in C([a, b])$ if $\nu - \gamma \in (0, 1)$.

We also mention

Theorem 14.5 ([12]) *Let $\nu > 0$, $\nu \notin \mathbb{N}$, $\lceil \nu \rceil = n$, $0 < \beta < 1$; $f \in C([a, b])$, $[a, b] \subset \mathbb{R}$; and set $\delta = \nu + \beta(n - \nu)$. Assume further that $\Delta_{b-}^{\delta} f \in C([a, b]) : \Delta_{b-}^{\delta-j} f(b) = 0$, for $j = 1, ..., n$. Let also $\gamma > 0 : \lceil \gamma \rceil = m$, with $\varepsilon = \gamma + \beta(m - \gamma)$, and assume that $\nu > \gamma$ and $\delta > \varepsilon$. Then*

$$ {}^{H}\mathbb{D}_{b-}^{\gamma,\beta} f(x) = \frac{1}{\Gamma(\nu-\gamma)} \int_x^b (t-x)^{\nu-\gamma-1} \, {}^{H}\mathbb{D}_{b-}^{\nu,\beta} f(t) \, dt, \tag{14.17} $$

$x \in [a, b]$, furthermore ${}^{H}\mathbb{D}_{b-}^{\gamma,\beta} f \in AC([a, b])$ if $\nu - \gamma \geq 1$, and ${}^{H}\mathbb{D}_{b-}^{\gamma,\beta} f \in C([a, b])$ if $\nu - \gamma \in (0, 1)$.

We continue with two extreme cases.

Theorem 14.6 ([12]) *When $\beta = 0$, Theorems 14.4, 14.5 are still valid. That is*

$$ \Delta_{a+}^{\gamma} f(x) = I_{a+}^{\nu-\gamma} \Delta_{a+}^{\nu} f(x), \tag{14.18} $$

and

$$ \Delta_{b-}^{\gamma} f(x) = I_{b-}^{\nu-\gamma} \Delta_{b-}^{\nu} f(x), \tag{14.19} $$

all $x \in [a, b]$, respectively.

The case of $\beta = 1$ collapses to Caputo type fractional derivatives and related representation formulae (see [12]), alreasy treted extensively in [11].

So, here we treat only $0 \leq \beta < 1$ case.

To motivate our chapter we mention the Opial type left side Hilfer fractional inequality:

Theorem 14.7 ([12]) *All as in Theorem 14.4, $0 \leq \beta < 1$. Let also $p, q > 1 : \frac{1}{p} + \frac{1}{q} = 1$, and assume that $\nu - \gamma > \frac{1}{q}$. Then*

$$ \int_a^x \left| {}^{H}\mathbb{D}_{a+}^{\gamma,\beta} f(w) \right| \left| {}^{H}\mathbb{D}_{a+}^{\nu,\beta} f(w) \right| dw \leq $$

$$ \frac{2^{-\frac{1}{q}} (x-a)^{(\nu-\gamma)+\left(\frac{1}{p}-\frac{1}{q}\right)}}{\Gamma(\nu-\gamma) \left[(p(\nu-\gamma-1)+1)(p(\nu-\gamma-1)+2) \right]^{\frac{1}{p}}} \left(\int_a^x \left| {}^{H}\mathbb{D}_{a+}^{\nu,\beta} f(w) \right|^q dw \right)^{\frac{2}{q}}, \tag{14.20} $$

$\forall \, x \in [a, b]$.

The corresponding Opial type right side Hilfer fractional inequality follows:

Theorem 14.8 ([12]) *All as in Theorem 14.5,* $0 \leq \beta < 1$. *Let also* $p, q > 1 : \frac{1}{p} + \frac{1}{q} = 1$, *and assume that* $\nu - \gamma > \frac{1}{q}$. *Then*

$$\int_x^b \left| {}^H\mathbb{D}_{b-}^{\gamma,\beta} f(w) \right| \left| {}^H\mathbb{D}_{b-}^{\nu,\beta} f(w) \right| dw \leq$$

$$\frac{2^{-\frac{1}{q}} (b-x)^{(\nu-\gamma)+\left(\frac{1}{p}-\frac{1}{q}\right)}}{\Gamma(\nu-\gamma)\left[(p(\nu-\gamma-1)+1)(p(\nu-\gamma-1)+2)\right]^{\frac{1}{p}}} \left(\int_x^b \left| {}^H\mathbb{D}_{b-}^{\nu,\beta} f(w) \right|^q dw \right)^{\frac{2}{q}}, \tag{14.21}$$

$\forall\, x \in [a, b]$.

14.2 Main Results

Next comes a complete collection of left and right side fractional Opial type inequalities engaging the left and right side Hilfer fractional derivatives.

14.2.1 Results involving one Function

We start with a left side one:

Theorem 14.9 *Here all as in Theorem 14.4,* $0 \leq \beta < 1$. *Then*

$$\int_a^x \left| {}^H\mathbb{D}_{a+}^{\gamma,\beta} f(w) \right| \left| {}^H\mathbb{D}_{a+}^{\nu,\beta} f(w) \right| dw \leq \frac{(x-a)^{(\nu-\gamma+1)}}{\Gamma(\nu-\gamma+2)} \left(\left\| {}^H\mathbb{D}_{a+}^{\nu,\beta} f \right\|_\infty \right)^2, \tag{14.22}$$

$x \in [a, b]$.

Proof Similar to Proposition 25.1, [4, p. 547]. ■

The right side one follows:

Theorem 14.10 *Here all as in Theorem 14.5,* $0 \leq \beta < 1$. *Then*

$$\int_x^b \left| {}^H\mathbb{D}_{b-}^{\gamma,\beta} f(w) \right| \left| {}^H\mathbb{D}_{b-}^{\nu,\beta} f(w) \right| dw \leq \frac{(b-x)^{\nu-\gamma+1}}{\Gamma(\nu-\gamma+2)} \left(\left\| {}^H\mathbb{D}_{b-}^{\nu,\beta} f \right\|_\infty \right)^2, \tag{14.23}$$

$x \in [a, b]$.

Proof Similar to (14.22). ■

We continue with converse results:

Theorem 14.11 *All as in Theorem 14.4,* $0 \leq \beta < 1$. *Suppose that* ${}^{H}\mathbb{D}_{a+}^{\nu,\beta} f(w) \not\equiv 0$ *and of fixed sign over* $[a, b]$. *Let* p, q *such that* $0 < p < 1$ *and* $\frac{1}{p} + \frac{1}{q} = 1$. *Then*

$$\int_a^x \left| {}^{H}\mathbb{D}_{a+}^{\gamma,\beta} f(w) \right| \left| {}^{H}\mathbb{D}_{a+}^{\nu,\beta} f(w) \right| dw \geq$$

$$\frac{2^{-\frac{1}{q}} (x-a)^{(\nu-\gamma)+\left(\frac{1}{p}-\frac{1}{q}\right)}}{\Gamma(\nu-\gamma) \left[(p(\nu-\gamma-1)+1)(p(\nu-\gamma-1)+2) \right]^{\frac{1}{p}}} \left(\int_a^x \left| {}^{H}\mathbb{D}_{a+}^{\nu,\beta} f(w) \right|^q dw \right)^{\frac{2}{q}}, \tag{14.24}$$

$\forall\, x \in [a, b]$.

Proof Similar to Theorem 25.3, [4, p. 547]. ■

Theorem 14.12 *All as in Theorem 14.5,* $0 \leq \beta < 1$. *Suppose that* ${}^{H}\mathbb{D}_{b-}^{\nu,\beta} f(w) \not\equiv 0$ *and of fixed sign over* $[a, b]$. *Let* p, q *such that* $0 < p < 1$ *and* $\frac{1}{p} + \frac{1}{q} = 1$. *Then*

$$\int_x^b \left| {}^{H}\mathbb{D}_{b-}^{\gamma,\beta} f(w) \right| \left| {}^{H}\mathbb{D}_{b-}^{\nu,\beta} f(w) \right| dw \geq$$

$$\frac{2^{-\frac{1}{q}} (b-x)^{(\nu-\gamma)+\left(\frac{1}{p}-\frac{1}{q}\right)}}{\Gamma(\nu-\gamma) \left[(p(\nu-\gamma-1)+1)(p(\nu-\gamma-1)+2) \right]^{\frac{1}{p}}} \left(\int_x^b \left| {}^{H}\mathbb{D}_{b-}^{\nu,\beta} f(w) \right|^q dw \right)^{\frac{2}{q}}, \tag{14.25}$$

$\forall\, x \in [a, b]$.

Proof Similar to (14.24). ■

We continue with

Theorem 14.13 *All as in Theorem 14.4,* $0 \leq \beta < 1$. *Additionally, let* $k > 0 : \lceil k \rceil = m_1$, *with* $\varepsilon_1 = k + \beta(m_1 - k)$, *and assume* $\delta > \varepsilon_1$. *Assume further that* $\nu - k > 1$ *and* $\gamma = k + 1$. *Let* $p, q > 1 : \frac{1}{p} + \frac{1}{q} = 1$. *Then*

$$\int_a^x \left| \left({}^{H}\mathbb{D}_{a+}^{k,\beta} f \right)(w) \right| \left| \left({}^{H}\mathbb{D}_{a+}^{\gamma,\beta} f \right)(w) \right| dw \leq$$

$$\frac{(x-a)^{\frac{2(p(\nu-k-1)+1)}{p}}}{2(\Gamma(\nu-k))^2 (p(\nu-k-1)+1)^{\frac{2}{p}}} \left(\int_a^x \left| \left({}^{H}\mathbb{D}_{a+}^{\nu,\beta} f \right)(w) \right|^q dw \right)^{\frac{2}{q}}, \tag{14.26}$$

$\forall\, x \in [a, b]$.

Proof Similar to Theorem 25.4, [4, p. 549]. ■

Theorem 14.14 *All as in Theorem 14.5, and the rest as in Theorem 14.13. Then*

$$\int_x^b \left|\left({}^H\mathbb{D}_{b-}^{k,\beta} f\right)(w)\right| \left|\left({}^H\mathbb{D}_{b-}^{\gamma,\beta} f\right)(w)\right| dw \leq$$

$$\frac{(b-x)^{\frac{2(p(\nu-k-1)+1)}{p}}}{2(\Gamma(\nu-k))^2 (p(\nu-k-1)+1)^{\frac{2}{p}}} \left(\int_x^b \left|\left({}^H\mathbb{D}_{b-}^{\nu,\beta} f\right)(w)\right|^q dw\right)^{\frac{2}{q}}, \quad (14.27)$$

$\forall\, x \in [a, b]$.

Proof Similar to (14.26). ■

We continue with extreme cases:

Theorem 14.15 *All as in Theorem 14.13, with $p = 1$, $q = \infty$. Then*

$$\int_a^x \left|\left({}^H\mathbb{D}_{a+}^{k,\beta} f\right)(w)\right| \left|\left({}^H\mathbb{D}_{a+}^{k+1,\beta} f\right)(w)\right| dw \leq$$

$$\frac{(x-a)^{2(\nu-k)} \left(\left\|{}^H\mathbb{D}_{a+}^{\nu,\beta} f\right\|_{\infty,(a,x)}\right)^2}{2(\Gamma(\nu-k+1))^2}, \quad (14.28)$$

$\forall\, x \in [a, b]$.

Proof Similar to Proposition 25.2, [4, p. 551]. ■

Theorem 14.16 *All as in Theorem 14.14; $p = 1$, $q = \infty$. Then*

$$\int_x^b \left|\left({}^H\mathbb{D}_{b-}^{k,\beta} f\right)(w)\right| \left|\left({}^H\mathbb{D}_{b-}^{k+1,\beta} f\right)(w)\right| dw \leq$$

$$\frac{(b-x)^{2(\nu-k)} \left(\left\|{}^H\mathbb{D}_{b-}^{\nu,\beta} f\right\|_{\infty,(x,b)}\right)^2}{2(\Gamma(\nu-k+1))^2}, \quad (14.29)$$

$\forall\, x \in [a, b]$.

Proof Similar to (14.28). ■

We continue with reverse inequalities:

Theorem 14.17 *All as in Theorem 14.13. Here $\left({}^H\mathbb{D}_{a+}^{\nu,\beta} f\right)(w) \not\equiv 0$ and of fixed sign over $[a, b]$, furthermore let $p, q : 0 < p < 1$ and $\frac{1}{p} + \frac{1}{q} = 1$. Then*

$$\int_a^x \left| {}^H\mathbb{D}_{a+}^{k,\beta} f(w) \right| \left| {}^H\mathbb{D}_{a+}^{\gamma,\beta} f(w) \right| dw \geq$$

$$\frac{(x-a)^{\frac{2(p(\nu-k-1)+1)}{p}}}{2(\Gamma(\nu-\gamma))^2 (p(\nu-k-1)+1)^{\frac{2}{p}}} \left(\int_a^x \left| \left({}^H\mathbb{D}_{a+}^{\nu,\beta} f \right)(w) \right|^q dw \right)^{\frac{2}{q}}, \quad (14.30)$$

$\forall\, x \in [a, b]$.

Proof Similar to Theorem 25.5, [4, p. 553]. ■

Theorem 14.18 *All as in Theorem 14.14. Here* $\left({}^H\mathbb{D}_{b-}^{\nu,\beta} f \right)(w) \not\equiv 0$ *and of fixed sign over* $[a, b]$, *furthermore let* $p, q : 0 < p < 1$ *and* $\frac{1}{p} + \frac{1}{q} = 1$. *Then*

$$\int_x^b \left| {}^H\mathbb{D}_{b-}^{k,\beta} f(w) \right| \left| {}^H\mathbb{D}_{b-}^{\gamma,\beta} f(w) \right| dw \geq$$

$$\frac{(b-x)^{\frac{2(p(\nu-k-1)+1)}{p}}}{2(\Gamma(\nu-k))^2 (p(\nu-k-1)+1)^{\frac{2}{p}}} \left(\int_x^b \left| \left({}^H\mathbb{D}_{b-}^{\nu,\beta} f \right)(w) \right|^q dw \right)^{\frac{2}{q}}, \quad (14.31)$$

$\forall\, x \in [a, b]$.

Proof Similar to (14.30). ■

Next we present

Theorem 14.19 *Let* $\nu > 0$, $\nu \notin \mathbb{N}$, $\lceil \nu \rceil = n$, $0 \leq \beta < 1$; $f \in C([a, b])$ *and set* $\delta = \nu + \beta(n - \nu)$. *Assume further that* $\Delta_{a+}^{\delta} f \in C([a, b]) : \Delta_{a+}^{\delta-j} f(a) = 0$, $j = 1, ..., n$. *Let also* $\gamma_i > 0 : \lceil \gamma_i \rceil = m_i$; $\nu - \gamma_i \geq 1$, *with* $\varepsilon_i = \gamma_i + \beta(m_i - \gamma_i)$ *and assume that* $\nu > \gamma_i$ *and* $\delta > \varepsilon_i$, $i = 1, ..., l$.

Here $a \leq x \leq b$; $q_1(x), q_2(x)$ *continuous functions on* $[a, b]$ *such that* $q_1(x) \geq 0$, $q_2(x) > 0$ *on* $[a, b]$, *and* $r_i > 0 : \sum_{i=1}^{l} r_i = r$. *Let* $s_1, s_1' > 1 : \frac{1}{s_1} + \frac{1}{s_1'} = 1$ *and* $s_2, s_2' > 1 : \frac{1}{s_2} + \frac{1}{s_2'} = 1$, *and* $p > s_2$.

Denote by

$$Q_1(x) := \left(\int_a^x (q_1(\omega))^{s_1'} d\omega \right)^{\frac{1}{s_1'}} \quad (14.32)$$

and

$$Q_2(x) := \left(\int_a^x (q_2(\omega))^{-\frac{s_2'}{p}} d\omega \right)^{\frac{r}{s_2'}}, \quad (14.33)$$

$$\sigma := \frac{p - s_2}{p s_2}.$$

Then

$$\int_a^x q_1(w)\prod_{i=1}^{l}\left|{}^H\mathbb{D}_{a+}^{\gamma_i,\beta}f(\omega)\right|^{r_i}d\omega\leq Q_1(x)\,Q_2(x)$$

$$\prod_{i=1}^{l}\left\{\frac{\sigma^{r_i\sigma}}{(\Gamma(\nu-\gamma_i))^{r_i}(\nu-\gamma_i-1+\sigma)^{r_i\sigma}}\right\} \tag{14.34}$$

$$\frac{(x-a)^{\left(\sum\limits_{i=1}^{l}(\nu-\gamma_i-1)r_i+\sigma r+\frac{1}{s_1}\right)}}{\left(\left(\sum\limits_{i=1}^{l}(\nu-\gamma_i-1)r_is_1\right)+rs_1\sigma+1\right)^{\frac{1}{s_1}}}\left(\int_a^x q_2(w)\left|{}^H\mathbb{D}_{a+}^{\nu,\beta}f(\omega)\right|^p d\omega\right)^{\frac{r}{p}},$$

$\forall\, x\in[a,b]$.

Proof Similar to Theorem 26.1, p. 567 of [4]. ■

Next we also present

Theorem 14.20 *Let* $\nu>0$, $\nu\notin\mathbb{N}$, $\lceil\nu\rceil=n$, $0\leq\beta<1$; $f\in C([a,b])$ *and set* $\delta=\nu+\beta(n-\nu)$. *Assume further that* $\Delta_{b-}^{\delta}f\in C([a,b]):\Delta_{b-}^{\delta-j}f(b)=0$, $j=1,...,n$. *Let also* $\gamma_i>0:\lceil\gamma_i\rceil=m_i$; $\nu-\gamma_i\geq1$, *with* $\varepsilon_i=\gamma_i+\beta(m_i-\gamma_i)$ *and assume that* $\nu>\gamma_i$ *and* $\delta>\varepsilon_i$, $i=1,...,l$.

Here $a\leq x\leq b$; $q_1(x)$, $q_2(x)$ *continuous functions on* $[a,b]$ *such that* $q_1(x)\geq0$, $q_2(x)>0$ *on* $[a,b]$, *and* $r_i>0:\sum\limits_{i=1}^{l}r_i=r$. *Let* $s_1,s_1'>1:\frac{1}{s_1}+\frac{1}{s_1'}=1$ *and* $s_2,s_2'>1:\frac{1}{s_2}+\frac{1}{s_2'}=1$, *and* $p>s_2$.

Denote by

$$\overline{Q}_1(x):=\left(\int_x^b(q_1(\omega))^{s_1'}d\omega\right)^{\frac{1}{s_1'}} \tag{14.35}$$

and

$$\overline{Q}_2(x):=\left(\int_x^b(q_2(\omega))^{-\frac{s_2'}{p}}d\omega\right)^{\frac{r}{s_2'}}, \tag{14.36}$$

$$\sigma:=\frac{p-s_2}{ps_2}.$$

Then

$$\int_x^b q_1(w)\prod_{i=1}^{l}\left|{}^H\mathbb{D}_{b-}^{\gamma_i,\beta}f(\omega)\right|^{r_i}d\omega\leq\overline{Q}_1(x)\,\overline{Q}_2(x)$$

$$\prod_{i=1}^{l}\left\{\frac{\sigma^{r_i\sigma}}{(\Gamma(\nu-\gamma_i))^{r_i}(\nu-\gamma_i-1+\sigma)^{r_i\sigma}}\right\} \tag{14.37}$$

$$\frac{(b-x)^{\left(\sum_{i=1}^{l}(\nu-\gamma_i-1)r_i+\sigma r+\frac{1}{s_1}\right)}}{\left(\left(\sum_{i=1}^{l}(\nu-\gamma_i-1)r_is_1\right)+rs_1\sigma+1\right)^{\frac{1}{s_1}}}\left(\int_x^b q_2(w)\left|{}^H\mathbb{D}_{b-}^{\nu,\beta}f(\omega)\right|^p d\omega\right)^{\frac{r}{p}},$$

$\forall\, x \in [a,b]$.

Proof Similar to Theorem 26.1, p. 567 of [4]. ∎

Converse theorems follow:

Theorem 14.21 *Let $\nu > 0$, $\nu \notin \mathbb{N}$, $\lceil \nu \rceil = n$, $0 \leq \beta < 1$; $f \in C([a,b])$ and set $\delta = \nu + \beta(n-\nu)$. Assume further that $\Delta_{a+}^{\delta} f \in C([a,b]) : \Delta_{a+}^{\delta-j} f(a) = 0$, $j = 1, \ldots, n$. Let also $\gamma_i > 0 : \lceil \gamma_i \rceil = m_i$; $\nu - \gamma_i \geq 1$, with $\varepsilon_i = \gamma_i + \beta(m_i - \gamma_i)$ and assume that $\nu > \gamma_i$ and $\delta > \varepsilon_i$, $i = 1, \ldots, l$; along with ${}^H\mathbb{D}_{a+}^{\nu,\beta} f > 0$.*

Here $a \leq x \leq b$; $q_1(x), q_2(x) > 0$ are continuous functions on $[a,b]$ and $r_i > 0 : \sum_{i=1}^{l} r_i = r$. Let $0 < s_1, s_2 < 1$ and $s_1', s_2' < 0$ such that $\frac{1}{s_1} + \frac{1}{s_1'} = 1$, $\frac{1}{s_2} + \frac{1}{s_2'} = 1$.

Denote by

$$Q_1(x) := \left(\int_a^x (q_1(\omega))^{s_1'} d\omega\right)^{\frac{1}{s_1'}} \tag{14.38}$$

and

$$Q_2(x) := \left(\int_a^x (q_2(\omega))^{-\frac{s_2'}{}} d\omega\right)^{\frac{r}{s_2'}}. \tag{14.39}$$

Set

$$\lambda := \frac{s_1 s_2}{s_1 s_2 - 1}.$$

Then

$$\int_a^x q_1(\omega)\left(\prod_{i=1}^{l}\left|{}^H\mathbb{D}_{a+}^{\gamma_i,\beta} f(\omega)\right|^{r_i}\right) d\omega \geq$$

$$\frac{Q_1(x)\,Q_2(x)}{\prod_{i=1}^{l}\left\{(\Gamma(\nu-\gamma_i))^{r_i}\left((\nu-\gamma_i-1)s_2^2 s_1+1\right)^{\left(\frac{r_i}{s_2^2 s_1}\right)}\right\}} \tag{14.40}$$

$$\frac{(x-a)^{\frac{\left\{\left(\sum_{i=1}^{l} r_i\left((\nu-\gamma_i-1)s_1+s_2^{-2}\right)\right)+1\right\}}{s_1}}}{\left\{\left(\sum_{i=1}^{l} r_i\left((\nu-\gamma_i-1)s_1+s_2^{-2}\right)\right)+1\right\}^{\frac{1}{s_1}}}\left(\int_a^x q_2^{\lambda s_2}(\omega)\left|{}^H\mathbb{D}_{a+}^{\nu,\beta} f(\omega)\right|^{\lambda s_2} d\omega\right)^{\frac{r}{\lambda s_2}},$$

$\forall\, x \in [a,b]$.

Proof Similar to Theorem 26.2, p. 570 of [4]. ■

Theorem 14.22 *Let $\nu > 0$, $\nu \notin \mathbb{N}$, $\lceil \nu \rceil = n$, $0 \leq \beta < 1$; $f \in C([a,b])$ and set $\delta = \nu + \beta(n - \nu)$. Assume further that $\Delta_{b-}^{\delta} f \in C([a,b]) : \Delta_{b-}^{\delta - j} f(b) = 0$, $j = 1, ..., n$. Let also $\gamma_i > 0 : \lceil \gamma_i \rceil = m_i$; $\nu - \gamma_i \geq 1$, with $\varepsilon_i = \gamma_i + \beta(m_i - \gamma_i)$ and assume that $\nu > \gamma_i$ and $\delta > \varepsilon_i$, $i = 1, ..., l$; along with ${}^{H}\mathbb{D}_{b-}^{\nu,\beta} f > 0$.*

Here $a \leq x \leq b$; $q_1(x), q_2(x) > 0$ are continuous functions on $[a,b]$ and $r_i > 0 : \sum_{i=1}^{l} r_i = r$. Let $0 < s_1, s_2 < 1$ and $s_1', s_2' < 0$ such that $\frac{1}{s_1} + \frac{1}{s_1'} = 1$, $\frac{1}{s_2} + \frac{1}{s_2'} = 1$.

Denote by

$$\overline{Q}_1(x) := \left(\int_x^b (q_1(\omega))^{s_1'} d\omega \right)^{\frac{1}{s_1'}} \tag{14.41}$$

and

$$\overline{Q}_2(x) := \left(\int_x^b (q_2(\omega))^{-s_2'} d\omega \right)^{\frac{r}{s_2'}}. \tag{14.42}$$

Set

$$\lambda := \frac{s_1 s_2}{s_1 s_2 - 1}.$$

Then

$$\int_x^b q_1(w) \prod_{i=1}^{l} \left| {}^{H}\mathbb{D}_{b-}^{\gamma_i,\beta} f(\omega) \right|^{r_i} d\omega \geq$$

$$\frac{\overline{Q}_1(x)\, \overline{Q}_2(x)}{\prod_{i=1}^{l} \left\{ (\Gamma(\nu - \gamma_i))^{r_i} \left((\nu - \gamma_i - 1) s_2^2 s_1 + 1 \right)^{\left(\frac{r_i}{s_2^2 s_1} \right)} \right\}} \tag{14.43}$$

$$\frac{(b - x)^{\frac{\left\{ \left(\sum_{i=1}^{l} r_i \left((\nu - \gamma_i - 1) s_1 + s_2^{-2} \right) \right) + 1 \right\}}{s_1}}}{\left\{ \left(\sum_{i=1}^{l} r_i \left((\nu - \gamma_i - 1) s_1 + s_2^{-2} \right) \right) + 1 \right\}^{\frac{1}{s_1}}} \left(\int_x^b q_2^{\lambda s_2}(\omega) \left| {}^{H}\mathbb{D}_{b-}^{\nu,\beta} f(\omega) \right|^{\lambda s_2} d\omega \right)^{\frac{r}{\lambda s_2}},$$

$\forall\, x \in [a,b]$.

Proof Similar to Theorem 26.2, p. 570 of [4]. ■

Extreme cases follow when $p = 1$ and $q = \infty$.

Theorem 14.23 *Let $\nu > 0$, $\nu \notin \mathbb{N}$, $\lceil \nu \rceil = n$, $0 \leq \beta < 1$; $f \in C([a,b])$ and set $\delta = \nu + \beta(n - \nu)$. Assume further that $\Delta_{a+}^{\delta} f \in C([a,b]) : \Delta_{a+}^{\delta - j} f(a) = 0$, $j = 1, ..., n$. Let also $\gamma_i > 0 : \lceil \gamma_i \rceil = m_i$; $\nu - \gamma_i \geq 1$, with $\varepsilon_i = \gamma_i + \beta(m_i - \gamma_i)$ and assume that $\nu > \gamma_i$ and $\delta > \varepsilon_i$, $i = 1, ..., l$. Here $a \leq x \leq b$, $0 \leq \overline{q} \in C([a,b])$ and*

$r_i > 0 : \sum_{i=1}^{l} r_i = r$. *Then*

$$\int_a^x \overline{q}(w) \prod_{i=1}^{l} \left| {}^H\mathbb{D}_{a+}^{\gamma_i,\beta} f(w) \right|^{r_i} d\omega \leq \tag{14.44}$$

$$\left\{ \frac{\|\overline{q}\|_{\infty,(a,x)} \left(\left\| {}^H\mathbb{D}_{a+}^{\nu,\beta} f \right\|_{\infty,(a,x)} \right)^r}{\prod_{i=1}^{l} \left(\Gamma(\nu - \gamma_i + 1) \right)^{r_i}} \right\} \left\{ \frac{(x-a)^{r\nu - \sum_{i=1}^{l} r_i\gamma_i + 1}}{\left(r\nu - \sum_{i=1}^{l} r_i\gamma_i + 1 \right)} \right\}.$$

Proof Similar to Proposition 26.1 of [4, p. 573]. ■

Theorem 14.24 *All as in Theorem 14.23, with now* $\Delta_{b-}^{\delta} f \in C([a,b]) : \Delta_{b-}^{\delta-j} f(b) = 0,\ j = 1, ..., n$. *Then*

$$\int_x^b \overline{q}(w) \prod_{i=1}^{l} \left| {}^H\mathbb{D}_{b-}^{\gamma_i,\beta} f(w) \right|^{r_i} dw \leq \tag{14.45}$$

$$\left\{ \frac{\|\overline{q}\|_{\infty,(x,b)} \left(\left\| {}^H\mathbb{D}_{b-}^{\nu,\beta} f \right\|_{\infty,(x,b)} \right)^r}{\prod_{i=1}^{l} \left(\Gamma(\nu - \gamma_i + 1) \right)^{r_i}} \right\} \left\{ \frac{(b-x)^{r\nu - \sum_{i=1}^{l} r_i\gamma_i + 1}}{\left(r\nu - \sum_{i=1}^{l} r_i\gamma_i + 1 \right)} \right\},$$

$\forall\, x \in [a,b]$.

Proof Similar to (14.44). ■

From now on we present only left side inequalities as the right side ones follow easily.

14.2.2 Results Involving Two Functions

We present

Theorem 14.25 *Let* $\nu > 0,\ \nu \notin \mathbb{N},\ \lceil \nu \rceil = n,\ 0 \leq \beta < 1;\ f_1, f_2 \in C([a,b])$ *and set* $\delta = \nu + \beta(n - \nu)$. *Assume further that* $\Delta_{a+}^{\delta} f_{j^*} \in C([a,b]) : \Delta_{a+}^{\delta-j} f_{j^*}(a) = 0$, $j = 1, ..., n;\ j^* = 1, 2$. *Let also* $\gamma_i > 0 : \lceil \gamma_i \rceil = m_i;\ \nu - \gamma_i \geq 1$, *with* $\varepsilon_i = \gamma_i + \beta(m_i - \gamma_i)$ *and assume that* $\nu > \gamma_i$ *and* $\delta > \varepsilon_i,\ i = 1, 2;\ a \leq x \leq b$. *Consider* $p(t) > 0$ *and* $q(t) \geq 0$, *with all* $p(t), \frac{1}{p(t)}, q(t) \in L_{\infty}(a,b)$. *Let* $\lambda_\nu > 0$ *and* $\lambda_\alpha, \lambda_{\overline{\beta}} \geq 0 : \lambda_\nu < p$, *where* $p > 1$. *Set*

$$P_k(\omega) := \int_a^{\omega} (\omega - t)^{\frac{(\nu-\gamma_k-1)p}{(p-1)}} (p(t))^{-\frac{1}{(p-1)}} dt, \tag{14.46}$$

$k = 1, 2,\ a \leq x \leq b;$

$$A(\omega) := \frac{q(\omega)(P_1(\omega))^{\lambda_\alpha\left(\frac{p-1}{p}\right)}(P_2(\omega))^{\lambda_{\overline{\beta}}\left(\frac{p-1}{p}\right)}(p(\omega))^{-\frac{\lambda_\nu}{p}}}{(\Gamma(\nu-\gamma_1))^{\lambda_\alpha}(\Gamma(\nu-\gamma_2))^{\lambda_{\overline{\beta}}}}, \tag{14.47}$$

$$A_0(x) := \left(\int_a^x (A(\omega))^{\frac{p}{p-\lambda_\nu}} d\omega\right)^{\frac{p-\lambda_\nu}{p}}, \tag{14.48}$$

and

$$\delta_1 := \begin{cases} 2^{1-\left(\frac{\lambda_\alpha+\lambda_\nu}{p}\right)}, & \text{if } \lambda_\alpha + \lambda_\nu \leq p, \\ 1, & \text{if } \lambda_\alpha + \lambda_\nu \geq p. \end{cases} \tag{14.49}$$

If $\lambda_{\overline{\beta}} = 0$, *we obtain that*

$$\int_a^x q(w)\left[\left|{}^H\mathbb{D}_{a+}^{\gamma_1,\beta} f_1(\omega)\right|^{\lambda_\alpha}\left|{}^H\mathbb{D}_{a+}^{\nu,\beta} f_1(\omega)\right|^{\lambda_\nu} + \right. \tag{14.50}$$

$$\left.\left|{}^H\mathbb{D}_{a+}^{\gamma_1,\beta} f_2(\omega)\right|^{\lambda_\alpha}\left|{}^H\mathbb{D}_{a+}^{\nu,\beta} f_2(\omega)\right|^{\lambda_\nu}\right] d\omega \leq$$

$$\left(A_0(x)\,|_{\lambda_{\overline{\beta}}=0}\right)\left(\frac{\lambda_\nu}{\lambda_\alpha+\lambda_\nu}\right)^{\frac{\lambda_\nu}{p}} \delta_1$$

$$\left[\int_a^x p(\omega)\left[\left|{}^H\mathbb{D}_{a+}^{\nu,\beta} f_1(\omega)\right|^p + \left|{}^H\mathbb{D}_{a+}^{\nu,\beta} f_2(\omega)\right|^p\right] d\omega\right]^{\frac{(\lambda_\alpha+\lambda_\nu)}{p}}.$$

Proof Similar toTheorem 2 of [5] and Theorem 4 of [9]. ■

The counterpart of the last theorem follows.

Theorem 14.26 *All here are as in Theorem 14.25. Denote*

$$\delta_3 := \begin{cases} 2^{\frac{\lambda_{\overline{\beta}}}{\lambda_\nu}} - 1, & \text{if } \lambda_{\overline{\beta}} \geq \lambda_\nu, \\ 1, & \text{if } \lambda_{\overline{\beta}} \leq \lambda_\nu. \end{cases} \tag{14.51}$$

If $\lambda_\alpha = 0$, *then*

$$\int_a^x q(w)\left[\left|{}^H\mathbb{D}_{a+}^{\gamma_2,\beta} f_2(\omega)\right|^{\lambda_{\overline{\beta}}}\left|{}^H\mathbb{D}_{a+}^{\nu,\beta} f_1(\omega)\right|^{\lambda_\nu} + \right. \tag{14.52}$$

$$\left| {}^{H}\mathbb{D}_{a+}^{\gamma_2,\beta} f_1(\omega) \right|^{\lambda_{\overline{\beta}}} \left| {}^{H}\mathbb{D}_{a+}^{\nu,\beta} f_2(\omega) \right|^{\lambda_\nu} \Bigg] d\omega \leq$$

$$\left(A_0(x)\,|_{\lambda_\alpha=0} \right) 2^{p-\frac{\lambda_\nu}{p}} \left(\frac{\lambda_\nu}{\lambda_{\overline{\beta}}+\lambda_\nu} \right)^{\frac{\lambda_\nu}{p}} \delta_3^{\frac{\lambda_\nu}{p}}$$

$$\left[\int_a^x p(\omega) \left[\left| {}^{H}\mathbb{D}_{a+}^{\nu,\beta} f_1(\omega) \right|^p + \left| {}^{H}\mathbb{D}_{a+}^{\nu,\beta} f_2(\omega) \right|^p \right] d\omega \right]^{\frac{\left(\lambda_\nu+\lambda_{\overline{\beta}}\right)}{p}},$$

all $a \leq x \leq b$.

Proof Similar toTheorem 3 of [5] and Theorem 5 of [9]. ■

The complete case $\lambda_\alpha, \lambda_{\overline{\beta}} \neq 0$ follows.

Theorem 14.27 *All here are as in Theorem 14.25. Denote*

$$\overline{\gamma}_1 := \begin{cases} 2^{\frac{\lambda_\alpha+\lambda_{\overline{\beta}}}{\lambda_\nu}} - 1, & \text{if } \lambda_\alpha + \lambda_{\overline{\beta}} \geq \lambda_\nu, \\ 1, & \text{if } \lambda_\alpha + \lambda_{\overline{\beta}} \leq \lambda_\nu, \end{cases} \tag{14.53}$$

and

$$\overline{\gamma}_2 := \begin{cases} 1, & \text{if } \lambda_\alpha + \lambda_{\overline{\beta}} + \lambda_\nu \geq p, \\ 2^{1-\left(\frac{\lambda_\alpha+\lambda_{\overline{\beta}}+\lambda_\nu}{p}\right)}, & \text{if } \lambda_\alpha + \lambda_{\overline{\beta}} + \lambda_\nu \leq p. \end{cases} \tag{14.54}$$

Then

$$\int_a^x q(w) \left[\left| {}^{H}\mathbb{D}_{a+}^{\gamma_1,\beta} f_1(\omega) \right|^{\lambda_\alpha} \left| {}^{H}\mathbb{D}_{a+}^{\gamma_2,\beta} f_2(\omega) \right|^{\lambda_{\overline{\beta}}} \left| {}^{H}\mathbb{D}_{a+}^{\nu,\beta} f_1(\omega) \right|^{\lambda_\nu} + \right. \tag{14.55}$$

$$\left. \left| {}^{H}\mathbb{D}_{a+}^{\gamma_2,\beta} f_1(\omega) \right|^{\lambda_{\overline{\beta}}} \left| {}^{H}\mathbb{D}_{a+}^{\gamma_1,\beta} f_2(\omega) \right|^{\lambda_\alpha} \left| {}^{H}\mathbb{D}_{a+}^{\nu,\beta} f_2(\omega) \right|^{\lambda_\nu} \right] d\omega \leq$$

$$A_0(x) \left(\frac{\lambda_\nu}{\left(\lambda_\alpha+\lambda_{\overline{\beta}}\right)\left(\lambda_\alpha+\lambda_{\overline{\beta}}+\lambda_\nu\right)} \right)^{\frac{\lambda_\nu}{p}} \left[\lambda_\alpha^{\frac{\lambda_\nu}{p}} \overline{\gamma}_2 + 2^{p-\frac{\lambda_\nu}{p}} \left(\overline{\gamma}_1 \lambda_{\overline{\beta}} \right)^{\frac{\lambda_\nu}{p}} \right]$$

$$\left(\int_a^x p(\omega) \left[\left| {}^{H}\mathbb{D}_{a+}^{\nu,\beta} f_1(\omega) \right|^p + \left| {}^{H}\mathbb{D}_{a+}^{\nu,\beta} f_2(\omega) \right|^p \right] d\omega \right)^{\frac{\left(\lambda_\alpha+\lambda_{\overline{\beta}}+\lambda_\nu\right)}{p}},$$

all $a \leq x \leq b$.

Proof As for Theorem 4 of [5] and Theorem 6 of [9]. ■

We continue with a special important case.

Theorem 14.28 *Let $\nu > 0$, $\nu \notin \mathbb{N}$, $\lceil \nu \rceil = n$, $0 \leq \beta < 1$; $f_1, f_2 \in C([a,b])$ and set $\delta = \nu + \beta(n-\nu)$. Assume further that $\Delta_{a+}^{\delta} f_{j^*} \in C([a,b]) : \Delta_{a+}^{\delta-j} f_{j^*}(a) = 0$, $j = 1, ..., n$; $j^* = 1, 2$. Let also $\gamma_i > 0 : \lceil \gamma_i \rceil = m_i$; $\nu - \gamma_i \geq 1$, with $\varepsilon_i = \gamma_i + \beta(m_i - \gamma_i)$ and assume that $\nu > \gamma_i$ and $\delta > \varepsilon_i$, $i = 1, 2$; but here it is $\gamma_2 = \gamma_1 + 1$; $a \leq x \leq b$. Consider $p(t) > 0$ and $q(t) \geq 0$, with all $p(t), \frac{1}{p(t)}, q(t) \in L_\infty(a,b)$. Let $\lambda_\alpha \geq 0$, $0 < \lambda_{\alpha+1} < 1$, and $p > 1$. Denote*

$$\theta_3 := \begin{cases} 2^{\frac{\lambda_\alpha}{\lambda_{\alpha+1}}} - 1, & \text{if } \lambda_\alpha \geq \lambda_{\alpha+1}, \\ 1, & \text{if } \lambda_\alpha \leq \lambda_{\alpha+1}, \end{cases} \tag{14.56}$$

$$L(x) := \left(2 \int_a^x (q(\omega))^{\frac{1}{1-\lambda_{\alpha+1}}} d\omega \right)^{1-\lambda_{\alpha+1}} \left(\frac{\theta_3 \lambda_{\alpha+1}}{\lambda_\alpha + \lambda_{\alpha+1}} \right)^{\lambda_{\alpha+1}}, \tag{14.57}$$

and

$$P_1(x) := \int_a^x (x-t)^{\frac{(\nu-\gamma_1-1)p}{p-1}} (p(t))^{-\frac{1}{p-1}} dt, \tag{14.58}$$

$$T(x) := L(x) \left(\frac{P_1(x)^{\frac{p-1}{p}}}{\Gamma(\nu-\gamma_1)} \right)^{\lambda_\alpha + \lambda_{\alpha+1}}, \tag{14.59}$$

$$\omega_1 := 2^{\frac{p-1}{p}(\lambda_\alpha + \lambda_{\alpha+1})}, \tag{14.60}$$

and

$$\Phi(x) := T(x)\, \omega_1. \tag{14.61}$$

Then

$$\int_a^x q(\omega) \Big[\left| {}^H\mathbb{D}_{a+}^{\gamma_1,\beta} f_1(\omega) \right|^{\lambda_\alpha} \left| {}^H\mathbb{D}_{a+}^{\gamma_1+1,\beta} f_2(\omega) \right|^{\lambda_{\alpha+1}} +$$

$$\left| {}^H\mathbb{D}_{a+}^{\gamma_1,\beta} f_2(\omega) \right|^{\lambda_\alpha} \left| {}^H\mathbb{D}_{a+}^{\gamma_1+1,\beta} f_1(\omega) \right|^{\lambda_{\alpha+1}} \Big] d\omega \leq \tag{14.62}$$

$$\Phi(x) \left[\int_a^x p(\omega) \left(\left| {}^H\mathbb{D}_{a+}^{\nu,\beta} f_1(\omega) \right|^p + \left| {}^H\mathbb{D}_{a+}^{\nu,\beta} f_2(\omega) \right|^p \right) d\omega \right]^{\frac{\lambda_\alpha + \lambda_{\alpha+1}}{p}},$$

all $a \leq x \leq b$.

Proof As in Theorem 5 of [5], and Theorem 8 of [9]. ∎

We give

Corollary 14.29 *All here are as in Theorem 14.25, with $\lambda_{\overline{\beta}} = 0$, $p(t) = q(t) = 1$. Then*

$$\int_a^x \left[\left| {}^H\mathbb{D}_{a+}^{\gamma_1,\beta} f_1(\omega) \right|^{\lambda_\alpha} \left| {}^H\mathbb{D}_{a+}^{\nu,\beta} f_1(\omega) \right|^{\lambda_\nu} + \left| {}^H\mathbb{D}_{a+}^{\gamma_1,\beta} f_2(\omega) \right|^{\lambda_\alpha} \left| {}^H\mathbb{D}_{a+}^{\nu,\beta} f_2(\omega) \right|^{\lambda_\nu} \right] d\omega \leq \tag{14.63}$$

$$C_1(x) \left(\int_a^x \left(\left| {}^H\mathbb{D}_{a+}^{\nu,\beta} f_1(\omega) \right|^p + \left| {}^H\mathbb{D}_{a+}^{\nu,\beta} f_2(\omega) \right|^p \right) d\omega \right)^{\frac{(\lambda_\alpha+\lambda_\nu)}{p}},$$

all $a \leq x \leq b$, *where*

$$C_1(x) := \left(A_0(x) |_{\lambda_{\overline{\beta}}=0} \right) \left(\frac{\lambda_\nu}{\lambda_\alpha + \lambda_\nu} \right)^{\frac{\lambda_\nu}{p}} \delta_1, \tag{14.64}$$

$$\delta_1 := \begin{cases} 2^{1-\frac{\lambda_\alpha+\lambda_\nu}{p}}, & \text{if } \lambda_\alpha + \lambda_\nu \leq p, \\ 1, & \text{if } \lambda_\alpha + \lambda_\nu \geq p. \end{cases} \tag{14.65}$$

We find that

$$\left(A_0(x) |_{\lambda_{\overline{\beta}}=0} \right) = \left\{ \left(\frac{(p-1)^{\frac{\lambda_\alpha p - \lambda_\alpha}{p}}}{(\Gamma(\nu-\gamma_1))^{\lambda_\alpha} (\nu p - \gamma_1 p - 1)^{\frac{\lambda_\alpha p - \lambda_\alpha}{p}}} \right) \cdot \right.$$

$$\left. \left(\frac{(p-\lambda_\nu)^{\frac{p-\lambda_\nu}{p}}}{(\lambda_\alpha \nu p - \lambda_\alpha \gamma_1 p - \lambda_\alpha + p - \lambda_\nu)^{\frac{p-\lambda_\nu}{p}}} \right) \right\} (x-a)^{\frac{\lambda_\alpha \nu p - \lambda_\alpha \gamma_1 p - \lambda_\alpha + p - \lambda_\nu}{p}}. \tag{14.66}$$

Proof As for Corollary 1 of [5], and Corollary 10 of [9]. ■

We continue with a related result regarding $\|\cdot\|_\infty$.

Theorem 14.30 *Let* $\nu > 0$, $\nu \notin \mathbb{N}$, $\lceil \nu \rceil = n$, $0 \leq \beta < 1$; $f_1, f_2 \in C([a,b])$ *and set* $\delta = \nu + \beta(n-\nu)$. *Assume further that* $\Delta_{a+}^{\delta} f_{j^*} \in C([a,b]) : \Delta_{a+}^{\delta-j} f_{j^*}(a) = 0$, $j = 1, ..., n$; $j^* = 1, 2$. *Let also* $\gamma_i > 0 : \lceil \gamma_i \rceil = m_i$; $\nu - \gamma_i \geq 1$, *with* $\varepsilon_i = \gamma_i + \beta(m_i - \gamma_i)$ *and assume that* $\nu > \gamma_i$ *and* $\delta > \varepsilon_i$, $i = 1, 2$; $a \leq x \leq b$. *Consider* $p(x) \geq 0$ *and* $p(x) \in L_\infty(a,b)$. *Let* $\lambda_\alpha, \lambda_{\overline{\beta}}, \lambda_\nu \geq 0$. *Set*

$$T(x) := \frac{(x-a)^{\left(\nu\lambda_\alpha - \gamma_1\lambda_\alpha + \nu\lambda_{\overline{\beta}} - \gamma_2\lambda_{\overline{\beta}} + 1\right)}}{\left(\nu\lambda_\alpha - \gamma_1\lambda_\alpha + \nu\lambda_{\overline{\beta}} - \gamma_2\lambda_{\overline{\beta}} + 1\right)} \cdot \tag{14.67}$$

$$\frac{\|p(s)\|_{\infty,(a,x)}}{(\Gamma(\nu-\gamma_1+1))^{\lambda_\alpha} (\Gamma(\nu-\gamma_2+1))^{\lambda_{\overline{\beta}}}}.$$

Then

$$\int_a^x p(w) \left[\left| {}^H\mathbb{D}_{a+}^{\gamma_1,\beta} f_1(\omega) \right|^{\lambda_\alpha} \left| {}^H\mathbb{D}_{a+}^{\gamma_2,\beta} f_2(\omega) \right|^{\lambda_{\overline{\beta}}} \left| {}^H\mathbb{D}_{a+}^{\nu,\beta} f_1(\omega) \right|^{\lambda_\nu} + \right.$$

$$\left. \left| {}^H\mathbb{D}_{a+}^{\gamma_2,\beta} f_1(\omega) \right|^{\lambda_{\overline{\beta}}} \left| {}^H\mathbb{D}_{a+}^{\gamma_1,\beta} f_2(\omega) \right|^{\lambda_\alpha} \left| {}^H\mathbb{D}_{a+}^{\nu,\beta} f_2(\omega) \right|^{\lambda_\nu} \right] d\omega \leq \tag{14.68}$$

$$\frac{T(x)}{2}\left[\left\|{}^{H}\mathbb{D}_{a+}^{\nu,\beta}f_1\right\|_{\infty,(a,x)}^{2(\lambda_\alpha+\lambda_\nu)}+\left\|{}^{H}\mathbb{D}_{a+}^{\nu,\beta}f_1\right\|_{\infty,(a,x)}^{2\lambda_{\overline{\beta}}}+\right.$$

$$\left.\left\|{}^{H}\mathbb{D}_{a+}^{\nu,\beta}f_2\right\|_{\infty,(a,x)}^{2\lambda_{\overline{\beta}}}+\left\|{}^{H}\mathbb{D}_{a+}^{\nu,\beta}f_2\right\|_{\infty,(a,x)}^{2(\lambda_\alpha+\lambda_\nu)}\right],$$

all $a \le x \le b$.

Proof Similar to Theorem 7 of [5], and Theorem 18 of [9]. ■

Next we give converse results involving two functions.

Theorem 14.31 *Let* $\nu > 0$, $\nu \notin \mathbb{N}$, $\lceil \nu \rceil = n$, $0 \le \beta < 1$; $f_1, f_2 \in C([a,b])$ *and set* $\delta = \nu + \beta(n-\nu)$. *Assume further that* $\Delta_{a+}^{\delta} f_{j^*} \in C([a,b]) : \Delta_{a+}^{\delta-j} f_{j^*}(a) = 0$, $j = 1, ..., n$; $j^* = 1, 2$. *Let also* $\gamma_i > 0 : \lceil \gamma_i \rceil = m_i$; $0 < p < 1$, $\frac{1}{p} > \nu - \gamma_i \ge 1$, *with* $\varepsilon_i = \gamma_i + \beta(m_i - \gamma_i)$ *and assume that* $\nu > \gamma_i$ *and* $\delta > \varepsilon_i$, $i = 1, 2$; $a \le x \le b$. *Consider* $p(t) > 0$ *and* $q(t) \ge 0$, *with all* $p(t), \frac{1}{p(t)}, q(t) \in L_\infty(a,b)$. *Further assume that* ${}^{H}\mathbb{D}_{a+}^{\nu,\beta} f_i$, $i = 1, 2$, *is of fixed sign a.e. on* $[a,b]$. *Let* $\lambda_\nu > 0$ *and* $\lambda_\alpha, \lambda_{\overline{\beta}} \ge 0$ *such that* $\lambda_\nu > p$. *Here* $P_k(\omega), A(\omega), A_0(x)$ *are as in (14.46), (14.47), and (14.48), respectively.*

Set

$$\delta_1 := 2^{1-\frac{(\lambda_\alpha+\lambda_\nu)}{p}}. \tag{14.69}$$

If $\lambda_{\overline{\beta}} = 0$, *then*

$$\int_a^x q(w)\left[\left|{}^{H}\mathbb{D}_{a+}^{\gamma_1,\beta}f_1(\omega)\right|^{\lambda_\alpha}\left|{}^{H}\mathbb{D}_{a+}^{\nu,\beta}f_1(\omega)\right|^{\lambda_\nu}+\right. \tag{14.70}$$

$$\left.\left|{}^{H}\mathbb{D}_{a+}^{\gamma_1,\beta}f_2(\omega)\right|^{\lambda_\alpha}\left|{}^{H}\mathbb{D}_{a+}^{\nu,\beta}f_2(\omega)\right|^{\lambda_\nu}\right]d\omega \ge$$

$$\left(A_0(x)\,|_{\lambda_{\overline{\beta}}=0}\right)\left(\frac{\lambda_\nu}{\lambda_\alpha+\lambda_\nu}\right)^{\frac{\lambda_\nu}{p}}\delta_1$$

$$\left[\int_a^x p(\omega)\left[\left|{}^{H}\mathbb{D}_{a+}^{\nu,\beta}f_1(\omega)\right|^p+\left|{}^{H}\mathbb{D}_{a+}^{\nu,\beta}f_2(\omega)\right|^p\right]d\omega\right]^{\frac{(\lambda_\alpha+\lambda_\nu)}{p}}.$$

Proof Similar toTheorem 5 of [10] and Theorem 4 of [8]. ■

We continue with

Theorem 14.32 *All here are as in Theorem 14.31. Further assume* $\lambda_{\overline{\beta}} \ge \lambda_\nu$. *Denote*

$$\delta_2 := 2^{1-\frac{\lambda_{\overline{\beta}}}{\lambda_\nu}}, \qquad \delta_3 := (\delta_2 - 1)\,2^{-\frac{\lambda_{\overline{\beta}}}{\lambda_\nu}}. \tag{14.71}$$

If $\lambda_\alpha = 0$, *then*

$$\int_a^x q(w)\left[\left|{}^H\mathbb{D}_{a+}^{\gamma_2,\beta} f_2(\omega)\right|^{\lambda_{\overline{\beta}}}\left|{}^H\mathbb{D}_{a+}^{\nu,\beta} f_1(\omega)\right|^{\lambda_\nu} + \right. \tag{14.72}$$

$$\left.\left|{}^H\mathbb{D}_{a+}^{\gamma_2,\beta} f_1(\omega)\right|^{\lambda_{\overline{\beta}}}\left|{}^H\mathbb{D}_{a+}^{\nu,\beta} f_2(\omega)\right|^{\lambda_\nu}\right] d\omega \geq$$

$$\left(A_0(x)\,|_{\lambda_\alpha=0}\right) 2^{p-\frac{\lambda_\nu}{p}}\left(\frac{\lambda_\nu}{\lambda_{\overline{\beta}}+\lambda_\nu}\right)^{\frac{\lambda_\nu}{p}} \delta_3^{\frac{\lambda_\nu}{p}}$$

$$\left(\int_a^x p(\omega)\left[\left|{}^H\mathbb{D}_{a+}^{\nu,\beta} f_1(\omega)\right|^p + \left|{}^H\mathbb{D}_{a+}^{\nu,\beta} f_2(\omega)\right|^p\right] d\omega\right)^{\frac{\left(\lambda_\alpha+\lambda_{\overline{\beta}}\right)}{p}},$$

all $a \leq x \leq b$.

Proof Similar toTheorem 6 of [10] and Theorem 5 of [8]. ■

Next we give a particular converse result involving two functions.

Theorem 14.33 *Let* $\nu > 0$, $\nu \notin \mathbb{N}$, $\lceil \nu \rceil = n$, $0 \leq \beta < 1$; $f_1, f_2 \in C([a,b])$ *and set* $\delta = \nu + \beta(n-\nu)$. *Assume further that* $\Delta_{a+}^{\delta} f_{j^*} \in C([a,b]) : \Delta_{a+}^{\delta-j} f_{j^*}(a) = 0$, $j = 1, ..., n$; $j^* = 1, 2$. *Let also* $\gamma_i > 0 : \lceil \gamma_i \rceil = m_i$; $0 < p < 1$, $\frac{1}{p} > \nu - \gamma_i \geq 1$, *with* $\varepsilon_i = \gamma_i + \beta(m_i - \gamma_i)$ *and assume that* $\nu > \gamma_i$ *and* $\delta > \varepsilon_i$, $i = 1, 2$; $\gamma_2 = \gamma_1 + 1$; $a \leq x \leq b$. *Consider* $p(t) > 0$ *and* $q(t) \geq 0$, *with all* $p(t), \frac{1}{p(t)}, q(t), \frac{1}{q(t)} \in L_\infty(a,b)$. *Further assume that* ${}^H\mathbb{D}_{a+}^{\nu,\beta} f_i$, $i = 1, 2$, *is of fixed sign a.e. on* $[a,b]$. *Let* $\lambda_\alpha \geq \lambda_{\alpha+1} > 1$.

Denote

$$\theta_3 := \left(2^{1-\frac{\lambda_\alpha}{\lambda_{\alpha+1}}} - 1\right) 2^{-\frac{\lambda_\alpha}{\lambda_{\alpha+1}}}, \tag{14.73}$$

$L(x)$ *is as in (14.57),* $P_1(x)$ *as in (14.58),* $T(x)$ *as in (14.59),* ω_1 *as in (14.60), and* Φ *as in (14.61). Then*

$$\int_a^x q(\omega)\left[\left|{}^H\mathbb{D}_{a+}^{\gamma_1,\beta} f_1(\omega)\right|^{\lambda_\alpha}\left|{}^H\mathbb{D}_{a+}^{\gamma_1+1,\beta} f_2(\omega)\right|^{\lambda_{\alpha+1}} + \right.$$

$$\left.\left|{}^H\mathbb{D}_{a+}^{\gamma_1,\beta} f_2(\omega)\right|^{\lambda_\alpha}\left|{}^H\mathbb{D}_{a+}^{\gamma_1+1,\beta} f_1(\omega)\right|^{\lambda_{\alpha+1}}\right] d\omega \geq \tag{14.74}$$

$$\Phi(x)\left[\int_a^x p(\omega)\left(\left|{}^H\mathbb{D}_{a+}^{\nu,\beta} f_1(\omega)\right|^p + \left|{}^H\mathbb{D}_{a+}^{\nu,\beta} f_2(\omega)\right|^p\right) d\omega\right]^{\frac{\lambda_\alpha+\lambda_{\alpha+1}}{p}},$$

all $a \leq x \leq b$.

Proof Similar to Theorem 7 of [10], and Theorem 7 of [8]. ■

14.2.3 Results Involving Several Functions

We present

Theorem 14.34 *Here all notations, terms and assumtions are as in Theorem 14.25, but for $f_j \in C([a,b])$, with $j = 1, ..., M \in \mathbb{N}$. Instead of δ_1 there, we define here*

$$\delta_1^* := \begin{cases} M^{1-\frac{\lambda_\alpha+\lambda_\nu}{p}}, & \text{if } \lambda_\alpha + \lambda_\nu \le p, \\ 2^{\frac{\lambda_\alpha+\lambda_\nu}{p}-1}, & \text{if } \lambda_\alpha + \lambda_\nu \ge p. \end{cases} \tag{14.75}$$

Call

$$\varphi_1(x) := \left(A_0(x)|_{\lambda_{\overline{\beta}}=0}\right)\left(\frac{\lambda_\nu}{\lambda_\alpha+\lambda_\nu}\right)^{\frac{\lambda_\nu}{p}}. \tag{14.76}$$

If $\lambda_{\overline{\beta}} = 0$, then

$$\int_a^x q(w)\left(\sum_{j=1}^M \left|{}^H\mathbb{D}_{a+}^{\gamma_1,\beta} f_j(\omega)\right|^{\lambda_\alpha}\left|{}^H\mathbb{D}_{a+}^{\nu,\beta} f_j(\omega)\right|^{\lambda_\nu}\right) d\omega \le$$

$$\delta_1^*\varphi_1(x)\left[\int_a^x p(\omega)\left(\sum_{j=1}^M \left|{}^H\mathbb{D}_{a+}^{\nu,\beta} f_j(\omega)\right|^p\right) d\omega\right]^{\frac{(\lambda_\alpha+\lambda_\nu)}{p}}, \tag{14.77}$$

all $a \le x \le b$.

Proof As in Theorem 2 of [6], and Theorem 4 of [7]. ■

We continue with

Theorem 14.35 *All here are as in Theorem 14.34. Denote*

$$\delta_3 := \begin{cases} 2^{\frac{\lambda_{\overline{\beta}}}{\lambda_\nu}} - 1, & \text{if } \lambda_{\overline{\beta}} \ge \lambda_\nu, \\ 1, & \text{if } \lambda_{\overline{\beta}} \le \lambda_\nu, \end{cases} \tag{14.78}$$

$$\varepsilon_2 := \begin{cases} 1, & \text{if } \lambda_\nu + \lambda_{\overline{\beta}} \ge p, \\ M^{1-\frac{\lambda_\nu+\lambda_{\overline{\beta}}}{p}}, & \text{if } \lambda_\nu + \lambda_{\overline{\beta}} \le p, \end{cases} \tag{14.79}$$

and

$$\varphi_2(x) := \left(A_0(x)\,|_{\lambda_\alpha=0}\right) 2^{\frac{p-\lambda_\nu}{p}} \left(\frac{\lambda_\nu}{\lambda_{\overline{\beta}}+\lambda_\nu}\right)^{\frac{\lambda_\nu}{p}} \delta_3^{\frac{\lambda_\nu}{p}}. \tag{14.80}$$

If $\lambda_\alpha = 0$, *then*

$$\int_a^x q(w) \left\{ \left\{ \sum_{j=1}^{M-1} \left[\left| {}^H\mathbb{D}_{a+}^{\gamma_2,\beta} f_{j+1}(\omega) \right|^{\lambda_{\overline{\beta}}} \left| {}^H\mathbb{D}_{a+}^{\nu,\beta} f_j(\omega) \right|^{\lambda_\nu} + \right. \right. \right.$$

$$\left. \left. \left| {}^H\mathbb{D}_{a+}^{\gamma_2,\beta} f_j(\omega) \right|^{\lambda_{\overline{\beta}}} \left| {}^H\mathbb{D}_{a+}^{\nu,\beta} f_{j+1}(\omega) \right|^{\lambda_\nu} \right] \right\} +$$

$$\left. \left[\left| {}^H\mathbb{D}_{a+}^{\gamma_2,\beta} f_M(\omega) \right|^{\lambda_{\overline{\beta}}} \left| {}^H\mathbb{D}_{a+}^{\nu,\beta} f_1(\omega) \right|^{\lambda_\nu} + \left| {}^H\mathbb{D}_{a+}^{\gamma_2,\beta} f_1(\omega) \right|^{\lambda_{\overline{\beta}}} \left| {}^H\mathbb{D}_{a+}^{\nu,\beta} f_M(\omega) \right|^{\lambda_\nu} \right] \right\} d\omega$$

$$\leq 2^{\frac{\lambda_\nu+\lambda_{\overline{\beta}}}{p}} \varepsilon_2 \varphi_2(x) \left\{ \int_a^x p(\omega) \left[\sum_{j=1}^{M} \left| {}^H\mathbb{D}_{a+}^{\nu,\beta} f_j(\omega) \right|^p \right] d\omega \right\}^{\frac{\lambda_\nu+\lambda_{\overline{\beta}}}{p}}, \tag{14.81}$$

all $a \leq x \leq b$.

Proof As for Theorem 3 of [6], and Theorem 5 of [7]. ■

We give the general case.

Theorem 14.36 *All are as in Theorem 14.34. Denote*

$$\overline{\gamma}_1 := \begin{cases} 2^{\frac{\lambda_\alpha+\lambda_{\overline{\beta}}}{\lambda_\nu}} - 1, & \text{if } \lambda_\alpha + \lambda_{\overline{\beta}} \geq \lambda_\nu, \\ 1, & \text{if } \lambda_\alpha + \lambda_{\overline{\beta}} \leq \lambda_\nu, \end{cases} \tag{14.82}$$

and

$$\overline{\gamma}_2 := \begin{cases} 1, & \text{if } \lambda_\alpha + \lambda_{\overline{\beta}} + \lambda_\nu \geq p, \\ 2^{1-\left(\frac{\lambda_\alpha+\lambda_{\overline{\beta}}+\lambda_\nu}{p}\right)}, & \text{if } \lambda_\alpha + \lambda_{\overline{\beta}} + \lambda_\nu \leq p. \end{cases} \tag{14.83}$$

Set

$$\varphi_3(x) := A_0(x) \left(\frac{\lambda_\nu}{\left(\lambda_\alpha+\lambda_{\overline{\beta}}\right)\left(\lambda_\alpha+\lambda_{\overline{\beta}}+\lambda_\nu\right)}\right)^{\frac{\lambda_\nu}{p}} \left[\lambda_\alpha^{\frac{\lambda_\nu}{p}} \overline{\gamma}_2 + 2^{\frac{p-\lambda_\nu}{p}} \left(\overline{\gamma}_1 \lambda_{\overline{\beta}}\right)^{\frac{\lambda_\nu}{p}}\right], \tag{14.84}$$

and

$$\varepsilon_3 := \begin{cases} 1, & \text{if } \lambda_\alpha + \lambda_{\overline{\beta}} + \lambda_\nu \geq p, \\ M^{1-\left(\frac{\lambda_\alpha+\lambda_{\overline{\beta}}+\lambda_\nu}{p}\right)}, & \text{if } \lambda_\alpha + \lambda_{\overline{\beta}} + \lambda_\nu \leq p. \end{cases} \tag{14.85}$$

Then

$$\int_a^x q(w)\left[\sum_{j=1}^{M-1}\left[\left|{}^H\mathbb{D}_{a+}^{\gamma_1,\beta} f_j(\omega)\right|^{\lambda_\alpha}\left|{}^H\mathbb{D}_{a+}^{\gamma_2,\beta} f_{j+1}(\omega)\right|^{\lambda_{\overline{\beta}}}\left|{}^H\mathbb{D}_{a+}^{\nu,\beta} f_j(\omega)\right|^{\lambda_\nu}+\right.\right.$$

$$\left.\left|{}^H\mathbb{D}_{a+}^{\gamma_2,\beta} f_j(\omega)\right|^{\lambda_{\overline{\beta}}}\left|{}^H\mathbb{D}_{a+}^{\gamma_1,\beta} f_{j+1}(\omega)\right|^{\lambda_\alpha}\left|{}^H\mathbb{D}_{a+}^{\nu,\beta} f_{j+1}(\omega)\right|^{\lambda_\nu}\right]+$$

$$\left[\left|{}^H\mathbb{D}_{a+}^{\gamma_1,\beta} f_1(\omega)\right|^{\lambda_\alpha}\left|{}^H\mathbb{D}_{a+}^{\gamma_2,\beta} f_M(\omega)\right|^{\lambda_{\overline{\beta}}}\left|{}^H\mathbb{D}_{a+}^{\nu,\beta} f_1(\omega)\right|^{\lambda_\nu}+\right.$$

$$\left.\left.\left|{}^H\mathbb{D}_{a+}^{\gamma_2,\beta} f_1(\omega)\right|^{\lambda_{\overline{\beta}}}\left|{}^H\mathbb{D}_{a+}^{\gamma_1,\beta} f_M(\omega)\right|^{\lambda_\alpha}\left|{}^H\mathbb{D}_{a+}^{\nu,\beta} f_M(\omega)\right|^{\lambda_\nu}\right]\right]d\omega \leq$$

$$2^{\frac{\lambda_\alpha+\lambda_{\overline{\beta}}+\lambda_\nu}{p}}\varepsilon_3\varphi_3(x)\left\{\int_a^x p(\omega)\left[\sum_{j=1}^{M}\left|{}^H\mathbb{D}_{a+}^{\nu,\beta} f_j(\omega)\right|^p\right]d\omega\right\}^{\frac{\left(\lambda_\alpha+\lambda_{\overline{\beta}}+\lambda_\nu\right)}{p}}, \quad (14.86)$$

all $a \leq x \leq b$.

Proof As for Theorem 4 of [6] and Theorem 6 of [7]. ■

We give

Theorem 14.37 *All here are as in Theorem 14.28, but for* $f_j \in C([a,b])$, $j = 1, \ldots, M \in \mathbb{N}$. *Also put*

$$\varepsilon_4 := \begin{cases} 1, & \text{if } \lambda_\alpha + \lambda_{\alpha+1} \geq p, \\ M^{1-\left(\frac{\lambda_\alpha+\lambda_{\alpha+1}}{p}\right)}, & \text{if } \lambda_\alpha + \lambda_{\alpha+1} \leq p. \end{cases} \quad (14.87)$$

Then

$$\int_a^x q(w)\left\{\{\sum_{j=1}^{M-1}\left[\left|{}^H\mathbb{D}_{a+}^{\gamma_1,\beta} f_j(\omega)\right|^{\lambda_\alpha}\left|{}^H\mathbb{D}_{a+}^{\gamma_1+1,\beta} f_{j+1}(\omega)\right|^{\lambda_{\alpha+1}}+\right.\right.$$

$$\left.\left|{}^H\mathbb{D}_{a+}^{\gamma_1,\beta} f_{j+1}(\omega)\right|^{\lambda_\alpha}\left|{}^H\mathbb{D}_{a+}^{\gamma_1+1,\beta} f_j(\omega)\right|^{\lambda_{\alpha+1}}\right]\}+\left[\left|{}^H\mathbb{D}_{a+}^{\gamma_1,\beta} f_1(\omega)\right|^{\lambda_\alpha}\left|{}^H\mathbb{D}_{a+}^{\gamma_1+1,\beta} f_M(\omega)\right|^{\lambda_{\alpha+1}}+\right.$$

$$\left.\left.\left|{}^H\mathbb{D}_{a+}^{\gamma_1,\beta} f_M(\omega)\right|^{\lambda_\alpha}\left|{}^H\mathbb{D}_{a+}^{\gamma_1+1,\beta} f_1(\omega)\right|^{\lambda_{\alpha+1}}\right]\right\}d\omega \leq$$

$$2^{\frac{\lambda_\alpha+\lambda_{\alpha+1}}{p}}\varepsilon_4\Phi(x)\left\{\int_a^x p(\omega)\left[\sum_{j=1}^{M}\left|{}^H\mathbb{D}_{a+}^{\nu,\beta} f_j(\omega)\right|^p\right]d\omega\right\}^{\frac{\left(\lambda_\alpha+\lambda_{\alpha+1}\right)}{p}}, \quad (14.88)$$

all $a \leq x \leq b$.

Proof As for Theorem 5 of [6] and Theorem 7 of [7]. ■

We continue with results regarding $\|\cdot\|_{\infty}$.

Theorem 14.38 *All are as in Theorem 14.30, but for* $f_j \in C([a,b])$, $j = 1, ..., M \in \mathbb{N}$. *Then*

$$\int_a^x p(w) \left\{ \left[\sum_{j=1}^{M-1} [\left| {}^H\mathbb{D}_{a+}^{\gamma_1,\beta} f_j(\omega) \right|^{\lambda_\alpha} \left| {}^H\mathbb{D}_{a+}^{\gamma_2,\beta} f_{j+1}(\omega) \right|^{\lambda_{\overline{\beta}}} \left| {}^H\mathbb{D}_{a+}^{\nu,\beta} f_j(\omega) \right|^{\lambda_\nu} + \right.\right.$$

$$\left. \left| {}^H\mathbb{D}_{a+}^{\gamma_2,\beta} f_j(\omega) \right|^{\lambda_{\overline{\beta}}} \left| {}^H\mathbb{D}_{a+}^{\gamma_1,\beta} f_{j+1}(\omega) \right|^{\lambda_\alpha} \left| {}^H\mathbb{D}_{a+}^{\nu,\beta} f_{j+1}(\omega) \right|^{\lambda_\nu}] \right] +$$

$$\left[\left| {}^H\mathbb{D}_{a+}^{\gamma_1,\beta} f_1(\omega) \right|^{\lambda_\alpha} \left| {}^H\mathbb{D}_{a+}^{\gamma_2,\beta} f_M(\omega) \right|^{\lambda_{\overline{\beta}}} \left| {}^H\mathbb{D}_{a+}^{\nu,\beta} f_1(\omega) \right|^{\lambda_\nu} + \right.$$

$$\left.\left. \left| {}^H\mathbb{D}_{a+}^{\gamma_2,\beta} f_1(\omega) \right|^{\lambda_{\overline{\beta}}} \left| {}^H\mathbb{D}_{a+}^{\gamma_1,\beta} f_M(\omega) \right|^{\lambda_\alpha} \left| {}^H\mathbb{D}_{a+}^{\nu,\beta} f_M(\omega) \right|^{\lambda_\nu} \right] \right\} d\omega \leq \tag{14.89}$$

$$T(x) \left\{ \sum_{j=1}^{M} \left[\left\| {}^H\mathbb{D}_{a+}^{\nu,\beta} f_j \right\|_{\infty,(a,x)}^{2(\lambda_\alpha+\lambda_\nu)} + \left\| {}^H\mathbb{D}_{a+}^{\nu,\beta} f_j \right\|_{\infty,(a,x)}^{2\lambda_{\overline{\beta}}} \right] \right\},$$

all $a \leq x \leq b$.

Proof Based on Theorem 14.30. ■

One based on Sect. 14.2.2 can write down converse results for several functions. Also to write down right side inequalities for two or several functions and to cover interesting specific examples.

All these results can have great applications to related fractional differential equations proving uniqueness of solutions or giving upper bounds to their solutions. The reader is referred to author's monograph of 2009, [11], to continue such studies. We choose to stop here.

References

1. Agarwal, R.P.: Sharp Opial-type inequalities involving r-derivatives and their applications. Tohoku Math. J. **47**, 567–593 (1995)
2. Agarwal, R.P., Pang, P.Y.H.: Opial Inequalities with Applications in Differential and Difference Equations. Kluwer Academic Publishers, Dordrecht, Boston, London (1995)
3. Anastassiou, G.A.: Opial type inequalities involving functions and their ordinary and fractional derivatives. Commun. Appl. Anal. **4**(4), 547–560 (2000)

4. Anastassiou, G.A.: Quantitative Approximation. Chapman & Hall/CRC, Boca Rarton, New York (2001)
5. Anastassiou, G.A.: Opial type inequalities involving fractional derivatives of two functions and applications. Comput. Math. Appl. **48**, 1701–1731 (2004)
6. Anastassiou, G.A.: Fractional Opial inequalities for several functions with applications. J. Comput. Anal. Appl. **7**(3), 233–259 (2005)
7. Anastassiou, G.A.: Riemann-Liouville fractional Opial inequalities for several functions with applications. Commun. Appl. Anal. **12**(4), 377–398 (2008)
8. Anastassiou, G.A.: Reverse Riemann-Liouville fractional Opial inequalities for several functions. Complex Variables Elliptic Eqs. **53**(6), 523–544 (2008)
9. Anastassiou, G.A.: Opial type inequalities involving Riemann-Liouville fractional derivatives of two functions with applications. Math. Comput. Model. **48**, 344–374 (2008)
10. Anastassiou, G.A.: Converse fractional Opial inequalities for several functions. CUBO **10**(1), 117–142 (2008)
11. Anastassiou, G.A.: Fractional Differentiation Inequalities. Springer, Heidelberg, New York (2009)
12. Anastassiou, G.: Advancements on ψ-Hilfer fractional calculus and fractional integral inequalities. In: Discontinuity, Nonlinear and Complexity (2021) (accepted)
13. Anastassiou, G.A.: A collection of Hilfer fractional Opial inequalities (2021) (submitted)
14. Anastassiou, G.A., Koliha, J., Pecaric, J.: Opial inequalities for fractional derivatives. Dyn. Syst. Appl. **10**, 395–406 (2001)
15. Anastassiou, G.A., Koliha, J., Pecaric, J.: Opial type L_p-inequalities for fractional derivatives. Intern. J. Math. Math. Sci. **31**(2), 85–95 (2002)
16. Mitrinovic, D.S., Pecaric, J.E., Fink, A.M.: Classical and New Inequalities in Analysis. Kluwer Academic Publishers, Dordrecht (1993)
17. Opial, Z.: Sur une inégalité. Ann. Polon. Math. **8**, 29–32 (1960)
18. Samko, S.G., Kilbas, A.A., Marichev, O.I.: Fractional Integrals and Derivatives: Theory and Applications. Gordon and Breach Science Publishers, Switzerland (1993)
19. Tomovski, Ž, Hilfer, R., Srivastava, H.M.: Fractional and operational calculus with generalized fractional derivative operators and Mittag-Leffler type functions. Integral Transforms Spec Funct. **21**(11), 797–814 (2010)

Chapter 15
Exotic Fractional Integral Inequalities

Here we present a thorough collection of Opial and Hardy type fractional inequalities involving also convexity, based on Luchko's generalized fractional calculus, and Prabhakar's partial and mixed of variable degree multivariate fractional integrals. It follows [7].

15.1 Background

In this chapter we have been motivated by the deep work of Luchko ([11]) and the author's work ([4]). We derive a collection of fractional integral Opial and Hardy type inequalities, basic and general, covering Luchko's fractional calculus ([11]) and other extreme fractional integral cases ([4]).

15.1.1. All the following come from [11]. We need

Definition 15.1 By $C_{-1}(0,+\infty)$, we denote the space of functions continuous on the positive real semi-axis that can have an integrable singularity of a power function type at the point zero:

$$C_{-1}(0,+\infty) := \left\{ f : f(t) = t^{p-1} f_1(t),\ t > 0,\ p > 0,\ f_1 \in C([0,+\infty)) \right\}. \tag{15.1}$$

The Laplace convolution is denoted by

$$(f * g)(t) := \int_0^t f(t-\tau)\, g(\tau)\, d\tau. \tag{15.2}$$

By $\{1\}$ we denote the function which is identical equal to 1 on $[0,+\infty)$.

A function $f = f(t)$, $t > 0$ is said to belong to the space $C_{-1}^m(0,+\infty)$ if and only if $f^{(m)} \in C_{-1}(0,+\infty)$, where $m \in \mathbb{Z}_+$.

Clearly $C_{-1}^m(0,+\infty) \subset C_{-1}(0,+\infty)$.

G. A. Anastassiou, *Unification of Fractional Calculi with Applications*,
Studies in Systems, Decision and Control 398,
https://doi.org/10.1007/978-3-030-86920-5_15

Sonine (1884) ([12]) considered the functions k and K such that

$$(k * K)(t) = \{1\}, \ t > 0. \tag{15.3}$$

The functions κ, k are called Sonine kernels and (15.3) is called Sonine condition. The kernel k is called the dual or associate kernel to Sonine kernel κ. Clearly then κ is the dual kernel of k.

We denote the set of the Sonine kernels by S. Define

$$C_{-1,0}(0, +\infty) := \left\{f : f(t) = t^{p-1} f_1(t), \ t > 0, 0 < p < 1, \ f_1 \in C([0, +\infty))\right\}. \tag{15.4}$$

Obviously it is $C_{-1,0}(0, +\infty) \subset C_{-1}(0, +\infty)$.

Here we consider the following subset of S, denoted by S_{-1}:

$$S_{-1} := \left\{\kappa, k \in C_{-1,0}(0, +\infty) \text{ and } (\kappa * k)(t) = \{1\}\right\}. \tag{15.5}$$

Let $\kappa \in S_{-1}$, the (GFI) generalized fractional integral with the kernel κ is defined by

$$\left(I_{(\kappa)} f\right)(t) := (\kappa * f)(t) = \int_0^t \kappa(t - \tau) f(\tau) d\tau, \tag{15.6}$$

$t > 0$, where $f \in C_{-1}(0, +\infty)$.

We need

Definition 15.2 Let $\kappa \in S_{-1}$ and k be its associate Sonine kernel. The (GFDs) generalized fractional derivatives of the Riemann-Liouville and Caputo types associate to the GFI (15.6) are defined as follows:

$$\left(\mathbb{D}_{(k)} f\right)(t) := \frac{d}{dt}(k * f)(t), \ t > 0, \tag{15.7}$$

and

$$\left({}_*\mathbb{D}_{(k)} f\right)(t) := \left(\mathbb{D}_{(k)} f\right)(t) - f(0) k(t), \ t > 0. \tag{15.8}$$

It is easy to see that

$$\left({}_*\mathbb{D}_{(k)} f\right)(t) = \left(\mathbb{D}_{(k)} [f(\cdot) - f(0)]\right)(t), \ t > 0. \tag{15.9}$$

We mention

Theorem 15.3 ([11]) *(Second Fundamental Theorem for the GFD). Let $\kappa \in S_{-1}$ and k be its associate Sonine kernel. Then, the relations*

$$I_{(\kappa)}\left({}_*\mathbb{D}_{(k)} f\right)(t) = f(t) - f(0), \ t > 0, \tag{15.10}$$

and

$$I_{(\kappa)}\left(\mathbb{D}_{(k)} f\right)(t) = f(t), \; t > 0, \tag{15.11}$$

hold valid for functions $f \in C^1_{-1}(0, +\infty)$.

Above it is

$$C^1_{-1}(0, +\infty) := \left\{ f = f(t), \; t > 0 : f' \in C_{-1}(0, +\infty) \right\}. \tag{15.12}$$

For $f \in C^1_{-1}(0, +\infty)$, we obtain

$$\left(\mathbb{D}_{(k)} f\right)(t) = \left(k * f'\right)(t) + f(0)\, k(t), \;\; t > 0, \tag{15.13}$$

and

$$\left({}_*\mathbb{D}_{(k)} f\right)(t) = \left(k * f'\right)(t), \;\; t > 0. \tag{15.14}$$

We can rewrite (15.10) and (15.11) ($f \in C^1_{-1}(0, +\infty)$) as

$$\int_0^t \kappa(t-\tau) \left({}_*\mathbb{D}_{(k)} f\right)(\tau)\, d\tau = f(t) - f(0), \;\; t > 0, \tag{15.15}$$

and

$$\int_0^t \kappa(t-\tau) \left(D_{(k)} f\right)(\tau)\, d\tau = f(t), \;\; t > 0, \tag{15.16}$$

respectively.

We need

Definition 15.4 Let $\kappa \in S_{-1}$. The n-fold GFI ($n \in \mathbb{N}$) is defined as a composition of n GFIs with the kernel κ:

$$\left(I^n_{(\kappa)} f\right)(t) := \left(I_{(\kappa)} ... I_{(\kappa)} f\right)(x) = \left(\kappa^n * f\right)(t), \tag{15.17}$$

$\kappa^n := \underbrace{\kappa * ... * \kappa}_{n}$, $t > 0$. The kernel κ^n, $n \in \mathbb{N}$, is from $C_{-1}(0, +\infty)$, but it is not always a Sonine kernel.

Definition 15.5 Let $\kappa \in S_{-1}$ and k be its associate Sonine kernel. The n-fold GFD ($n \in \mathbb{N}$) in the Riemann-Liouville sense is defined by

$$\left(\mathbb{D}^n_{(k)} f\right)(t) := \frac{d^n}{dt^n}\left(k^n * f\right)(t), \tag{15.18}$$

$k^n := \underbrace{k * ... * k}_{n}$, $t > 0$.

We mention

Theorem 15.6 ([11]) *Let $\kappa \in S_{-1}$ and k be its associate Sonine kernel. Then*

$$\left(\mathbb{D}^n_{(k)} f\right)(t) = \left(k^n * f^{(n)}\right)(t) + \sum_{j=0}^{n-1} f^{(j)}(0) \frac{d^{n-j-1}}{dt^{n-j-1}} k^n(t), \ t > 0, \tag{15.19}$$

for $f \in C^n_{-1}(0, +\infty)$.

We need

Definition 15.7 Let $\kappa \in S_{-1}$ and k be its associate Sonine kernel. The n-fold GFD ($n \in \mathbb{N}$) in the Caputo sense is defined as follows:

$$\left({}_*\mathbb{D}^n_{(k)} f\right)(t) := \left(\mathbb{D}^n_{(k)} f\right)(t) - \sum_{j=0}^{n-1} f^{(j)}(0) \frac{d^{n-j-1}}{dt^{n-j-1}} k^n(t), \ \ t > 0. \tag{15.20}$$

By [11], on the space $C^n_{-1}(0, +\infty)$, we get the simpler form

$$\left({}_*\mathbb{D}^n_{(k)} f\right)(t) = \left(k^n * f^{(n)}\right)(t), \ t > 0. \tag{15.21}$$

We mainly need

Theorem 15.8 ([11]) *(Second Fundamental Theorem for the n-fold GFD) Let $\kappa \in S_{-1}$ and k be its associate Sonine kernel. Then*

(i)

$$\left(I^n_{(\kappa)}\left({}_*\mathbb{D}^n_{(k)} f\right)\right)(t) = f(t) - \sum_{j=0}^{n-1} f^{(j)}(0) \frac{t^j}{j!}, \ \ t > 0, \tag{15.22}$$

for $f \in C^n_{-1}(0, +\infty)$,

and

(ii)

$$\left(I^n_{(\kappa)}\left(\mathbb{D}^n_{(k)} f\right)\right)(t) = f(t), \tag{15.23}$$

for $f \in C^n_{-1}(0, +\infty)$, $t > 0$; under the conditions

$$\left(\kappa^n(\tau) * \frac{d^{n-j-1}}{d\tau^{n-j-1}} k^n(\tau)\right)(t) = \frac{d^{n-j-1}}{dt^{n-j-1}} \left(\kappa^n(\tau) * k^n(\tau)\right)(t), \ \ j = 0, ..., n-1. \tag{15.24}$$

We make

Remark 15.9 ([11]) The properties:

$$I_{(\kappa)} : C_{-1}(0, +\infty) \to C_{-1}(0, +\infty), \tag{15.25}$$

the commutative law

$$I_{(\kappa_1)} I_{(\kappa_2)} = I_{(\kappa_2)} I_{(\kappa_1)}, \ \kappa_1, \kappa_2 \in S_{-1}, \tag{15.26}$$

and the index law

$$I_{(\kappa_1)} I_{(\kappa_2)} = I_{(\kappa_1 * \kappa_2)}, \kappa_1, \kappa_2 \in S_{-1}, \tag{15.27}$$

are valid over the space $C_{-1}(0, +\infty)$.

15.1.2. Here we consider the Prabhakar function (also known as the three parameter Mittag-Leffler function, an entire function if $z \in \mathbb{C}$), (see [8], [9, p. 97])

$$E_{\alpha,\beta}^{\gamma}(z) = \sum_{k=0}^{\infty} \frac{(\gamma)_k}{k! \Gamma(\alpha k + \beta)} z^k,$$

where Γ is the gamma function; $\alpha, \beta, \gamma \in \mathbb{R} : \alpha, \beta > 0$, $z \in \mathbb{R}$, and $(\gamma)_k = \gamma(\gamma+1) \ldots (\gamma + k - 1)$. It is $E_{\alpha,\beta}^{0}(z) = \frac{1}{\Gamma(\beta)}$.

Let $\prod_{i=1}^{N} [a_i, b_i] \subset \mathbb{R}^N$, $N > 1$, $a_i < b_i$, $a_i, b_i \in \mathbb{R}$; $f \in C\left(\prod_{i=1}^{N} [a_i, b_i]\right)$. Let also $x_i, t_i \in [a_i, b_i]$; $\rho_i, \mu_i > 0$; $\gamma_i, \omega_i \in \mathbb{R}$; $i = 1, \ldots, N$. Here $\psi_i \in C^1([a_i, b_i])$ which is increasing, $i=1, \ldots, N$. We set $a = (a_1, \ldots, a_N)$, $b = (b_1, \ldots, b_N)$, $x = (x_1, \ldots, x_N)$, $t = (t_1, \ldots, t_N)$; $\rho = (\rho_1, \ldots, \rho_N)$, $\mu = (\mu_1, \ldots, \mu_N)$, $\gamma = (\gamma_1, \ldots, \gamma_N)$, $\omega = (\omega_1, \ldots, \omega_N)$, $\psi := (\psi_1, \ldots, \psi_N)$.

We need

Definition 15.10 We define the left and right partial Prabhakar fractional integrals with respect to ψ_k, respectively, for $k = 1, \ldots, N$, as follows:

$$\left({}^{P}e_{\rho_k,\mu_k,\omega_k,a_k+}^{\gamma_k;\psi_k} f\right)(x) := \int_{a_k}^{x_k} \psi_k'(t_k) (\psi_k(x_k) - \psi_k(t_k))^{\mu_k - 1} \tag{15.28}$$

$$E_{\rho_k,\mu_k}^{\gamma_k} \left[\omega_k (\psi_k(x_k) - \psi_k(t_k))^{\rho_k}\right] f(x_1, \ldots, x_{k-1}, t_k, x_{k+1}, \ldots, x_N) \, dt_k,$$

with $a_k \leq x_k \leq b_k$;

and

$$\left({}^{P}e_{\rho_k,\mu_k,\omega_k,b_k-}^{\gamma_k;\psi_k} f\right)(x) := \int_{x_k}^{b_k} \psi_k'(t_k) (\psi_k(t_k) - \psi_k(x_k))^{\mu_k - 1} \tag{15.29}$$

$$E_{\rho_k,\mu_k}^{\gamma_k} \left[\omega_k (\psi_k(t_k) - \psi_k(x_k))^{\rho_k}\right] f(x_1, \ldots, x_{k-1}, t_k, x_{k+1}, \ldots, x_N) \, dt_k,$$

with $a_k \leq x_k \leq b_k$.

By [6], (15.28), (15.29) are continuous functions in x_k, $k = 1, \ldots, N$.

We mention

Definition 15.11 ([13, p. 35]) Here we consider the Prabhakar function wsith respect to another function $\alpha(x) > 0$, $x \in [a, b]$, $\alpha \in C([a, b])$, as follows:

$$E_{\alpha(x),\beta}^{\gamma}(z) := \sum_{k=0}^{\infty} \frac{(\gamma)_k}{k!\Gamma(k\alpha(x) + \beta)} z^k, \tag{15.30}$$

where $\beta > 0$, $\gamma \in \mathbb{R}$, $z \in \mathbb{R}$.

We need

Definition 15.12 All as above, however now instead of ρ we have $\rho(x) = (\rho_1(x_1), ..., \rho_N(x_N))$, where $0 < \rho_i \in C([a_i, b_i])$, for $i = 1, ..., N$. We define the left and right mixed Prabhakar multiple fractional integrals of variable degree with respect to ψ, respectively, as follows:

$$\left({}_{v}^{M}e_{\rho,\mu,\omega,a+}^{\gamma;\psi} f\right)(x) := \int_{a_1}^{x_1} ... \int_{a_N}^{x_N} \prod_{i=1}^{N} \left[\psi_i'(t_i)(\psi_i(x_i) - \psi_i(t_i))^{\mu_i - 1}\right.$$

$$\left. E_{\rho_i(t_i),\mu_i}^{\gamma_i}\left[\omega_i(\psi_i(x_i) - \psi_i(t_i))^{\rho_i(t_i)}\right]\right] f(t_1, ..., t_N)\, dt_1...dt_N, \tag{15.31}$$

with $x_i \geq a_i$, $i = 1, ..., N$;

$$\left({}_{v}^{M}e_{\rho,\mu,\omega,b-}^{\gamma;\psi} f\right)(x) := \int_{x_1}^{b_1} ... \int_{x_N}^{b_N} \prod_{i=1}^{N} \left[\psi_i'(t_i)(\psi_i(t_i) - \psi_i(x_i))^{\mu_i - 1}\right.$$

$$\left. E_{\rho_i(t_i),\mu_i}^{\gamma_i}\left[\omega_i(\psi_i(t_i) - \psi_i(x_i))^{\rho_i(t_i)}\right]\right] f(t_1, ..., t_N)\, dt_1...dt_N, \tag{15.32}$$

with $x_i \leq b_i$, $i = 1, ..., N$.

In this chapter based on Luchko ([11]) fractional calculus, namely on (15.15), (15.16), (15.22), (15.23) we derive several fractional Opial type inequalities, basic fractional Hardy type inequalities and generalized fractional Hardy type inequalities engaging convexity. Furthermore using the exotic fractional integrals (15.28), (15.29), (15.31), (15.32), we obtain also generalized fractional Hardy type inequalities involving convexity.

15.2 Prerequisites

15.2.1. Let $(\Omega_1, \Sigma_1, \mu_1)$ and $(\Omega_2, \Sigma_2, \mu_2)$ be measure spaces with positive σ-finite measures, and let $k_i : \Omega_1 \times \Omega_2 \to \mathbb{R}$ be nonnegative measurable functions, $k_i(x, \cdot)$

measurable on Ω_2, and

$$K_i(x) = \int_{\Omega_2} k_i(x, y)\, d\mu_2(y), \text{ for any } x \in \Omega_1, \tag{15.33}$$

$i = 1, ..., m \in \mathbb{N}$. We assume that $K_i(x) > 0$ a.e. on Ω_1 and the weight functions are nonnegative measurable functions on the related set.

We consider measurable functions $g_i : \Omega_1 \to \mathbb{R}$ with the representation

$$g_i(x) = \int_{\Omega_2} k_i(x, y)\, f_i(y)\, d\mu_2(y), \tag{15.34}$$

where $f_i : \Omega_2 \to \mathbb{R}$ are measurable functions, $i = 1, ..., m$. Here u stands for a weight function on Ω_1 (nonnegative measurable function).

We mention

Theorem 15.13 ([2, p. 29], p. 29) *Let $j \in \{1, ..., m\}$ be fixed. Assume that the function $x \to \left(\frac{u(x) \prod_{i=1}^{m} k_i(x,y)}{\prod_{i=1}^{m} K_i(x)} \right)$ is integrable on Ω_1, for each $y \in \Omega_2$. Define λ_m on Ω_2 by*

$$\lambda_m(y) := \int_{\Omega_1} \left(\frac{u(x) \prod_{i=1}^{m} k_i(x, y)}{\prod_{i=1}^{m} K_i(x)} \right) d\mu_1(x) < \infty. \tag{15.35}$$

Here $\Phi_i : \mathbb{R}_+ \to \mathbb{R}_+$, $i = 1, ..., m$, are convex and increasing functions.

Then

$$\int_{\Omega_1} u(x) \prod_{i=1}^{m} \Phi_i \left(\left| \frac{g_i(x)}{K_i(x)} \right| \right) d\mu_1(x) \leq$$

$$\left(\prod_{\substack{i=1 \\ i \neq j}}^{m} \int_{\Omega_2} \Phi_i(|f_i(y)|)\, d\mu_2(y) \right) \left(\int_{\Omega_2} \Phi_j\left(\left|f_j(y)\right|\right) \lambda_m(y)\, d\mu_2(y) \right), \tag{15.36}$$

true for all measurable functions, $i = 1, ..., m$, $f_i : \Omega_2 \to \mathbb{R}$ such that:

(i) f_i, $\Phi_i(|f_i|)$, are both $k_i(x, y)\, d\mu_2(y)$ - integrable, μ_1-a.e. in $x \in \Omega_1$,
(ii) $\lambda_m \Phi_j\left(\left|f_j\right|\right)$; $\Phi_1(|f_1|)$, $\Phi_2(|f_2|)$, $\Phi_3(|f_3|)$, ..., $\widehat{\Phi_j\left(\left|f_j\right|\right)}$, ..., $\Phi_m(|f_m|)$, are all μ_2-integrable,

and for all corresponding functions g_i given by (15.34). Above $\widehat{\Phi_j\left(\left|f_j\right|\right)}$ means missing item.

15.2.2. We need

Remark 15.14 In the next we consider for $i = 1, ..., m$, the measurable functions $f_{1i}, f_{2i} : \Omega_2 \to \mathbb{R}$, and

$$g_{1i}(x) = \int_{\Omega_2} k_i(x, y) f_{1i}(y) d\mu_2(y), \tag{15.37}$$

and

$$g_{2i}(x) = \int_{\Omega_2} k_i(x, y) f_{2i}(y) d\mu_2(y), \tag{15.38}$$

these are now instead of (15.34).

Again here $u \geq 0$ is a weight measurable function on Ω_1.

We will use the following rational result:

Theorem 15.15 ([3, p. 405]) *Here $0 < f_{2i}(y) < \infty$, a.e., $i = 1, ..., m$. Let $\overline{j} \in \{1, ..., m\}$ be fixed. Assume that the function $x \to \left(\frac{u(x) \prod_{i=1}^{m} k_i(x,y) f_{2i}(y)}{\prod_{i=1}^{m} g_{2i}(x)} \right)$ is integrable on Ω_1, for each $y \in \Omega_2$. Define λ_m^{**} on Ω_2 by*

$$\lambda_m^{**}(y) := \left(\prod_{i=1}^{m} f_{2i}(y) \right) \int_{\Omega_1} \left(\frac{u(x) \prod_{i=1}^{m} k_i(x, y)}{\prod_{i=1}^{m} g_{2i}(x)} \right) d\mu_1(x) < \infty. \tag{15.39}$$

Here $\Phi_i : \mathbb{R}_+ \to \mathbb{R}_+$, $i = 1, ..., m$, are convex and increasing functions. Then

$$\int_{\Omega_1} u(x) \prod_{i=1}^{m} \Phi_i \left(\left| \frac{g_{1i}(x)}{g_{2i}(x)} \right| \right) d\mu_1(x) \leq$$

$$\left(\prod_{\substack{i=1 \\ i \neq \overline{j}}}^{m} \int_{\Omega_2} \Phi_i \left(\left| \frac{f_{1i}(y)}{f_{2i}(y)} \right| \right) d\mu_2(y) \right) \left(\int_{\Omega_2} \Phi_{\overline{j}} \left(\left| \frac{f_{1\overline{j}}(y)}{f_{2\overline{j}}(y)} \right| \right) \lambda_m^{**}(y) d\mu_2(y) \right), \tag{15.40}$$

true for all measurable functions, $i = 1, ..., m$, $f_{1i}, f_{2i} : \Omega_2 \to \mathbb{R}$ such that:

(i) $\frac{f_{1i}(y)}{f_{2i}(y)}$, $\Phi_i \left(\left| \frac{f_{1i}(y)}{f_{2i}(y)} \right| \right)$, are both $k_i(x, y) f_{2i}(y) d\mu_2(y)$ - integrable, μ_1-a.e. in $x \in \Omega_1$,

(ii) $\lambda_m^{**} \Phi_{\overline{j}} \left(\left| \frac{f_{1\overline{j}}(y)}{f_{2\overline{j}}(y)} \right| \right)$, and $\Phi_i \left(\left| \frac{f_{1i}(y)}{f_{2i}(y)} \right| \right)$, for $i = \{1, ..., m\} - \{\overline{j}\}$, are all μ_2-integrable;

and for all corresponding g_{1i} given by (15.37), and g_{2i} given by (15.38).

15.2.3. Here we follow [3, p. 441], see Chap. 22.

Let $(\Omega_1, \Sigma_1, \mu_1)$ and $(\Omega_2, \Sigma_2, \mu_2)$ be measure spaces with positive σ-finite measures, and let $k_j : \Omega_1 \times \Omega_2 \to \mathbb{R}$ be nonnegative measurable functions, $k_j(x, \cdot)$ measurable on Ω_2, and

$$K_j(x) = \int_{\Omega_2} k_j(x, y)\, d\mu_2(y), \text{ for any } x \in \Omega_1, \tag{15.41}$$

$j = 1, ..., r \in \mathbb{N}$. We assume that $K_j(x) > 0$ a.e. on Ω_1 and the weight functions are nonnegative measurable functions on the related set.

We consider measurable functions $g_j : \Omega_1 \to \mathbb{R}$ with the representation

$$g_j(x) = \int_{\Omega_2} k_j(x, y) f_j(y)\, d\mu_2(y), \tag{15.42}$$

where $f_j : \Omega_2 \to \mathbb{R}$ are measurable functions, $j = 1, ..., r$.

Here u stands for a weight function on Ω_1 ($u \geq 0$, which is measurable).

We will use the following general result:

Theorem 15.16 ([3, p. 442]) *Assume that the functions ($j = 1, 2, ..., r \in \mathbb{N}$) $x \to \left(u(x) \frac{k_j(x,y)}{K_j(x)}\right)$ are integrable on Ω_1, for each fixed $y \in \Omega_2$. Define u_j on Ω_2 by*

$$u_j(y) := \int_{\Omega_1} u(x) \frac{k_j(x, y)}{K_j(x)} d\mu_1(x) < \infty. \tag{15.43}$$

Let $p_j > 1 : \sum_{j=1}^{r} \frac{1}{p_j} = 1$. Let the functions $\Phi_j : \mathbb{R}_+ \to \mathbb{R}_+$, $j = 1, ..., r$, be convex and increasing.

Then

$$\int_{\Omega_1} u(x) \prod_{j=1}^{r} \Phi j \left(\left| \frac{g_j(x)}{K_j(x)} \right| \right) d\mu_1(x) \leq$$

$$\prod_{j=1}^{r} \left(\int_{\Omega_2} u_j(y) \Phi_j \left(\left| f_j(y) \right| \right)^{p_j} d\mu_2(y) \right)^{\frac{1}{p_j}}, \tag{15.44}$$

for all measurable functions $f_j : \Omega_2 \to \mathbb{R}$ ($j = 1, ..., r$) such that

(i) f_j, $\Phi_j \left(\left| f_j \right| \right)^{p_j}$, are both $k_j(x, y)\, d\mu_2(y)$ - integrable, μ_1-a.e. in $x \in \Omega_1$, $j = 1, ..., r$,

(ii) $u_j \Phi_j \left(\left| f_j \right| \right)^{p_j}$ is μ_2-integrable, $j = 1, ..., r$,
and for all corresponding functions g_j ($j = 1, ..., r$) given by (15.42).

15.2.4. Let $(\Omega_1, \Sigma_1, \mu_1)$ and $(\Omega_2, \Sigma_2, \mu_2)$ be measure spaces with positive σ-finite measures, and let $k : \Omega_1 \times \Omega_2 \to \mathbb{R}$ be nonnegative measurable functions, $k(x, \cdot)$ measurable on Ω_2, and

$$K(x) = \int_{\Omega_2} k(x, y)\, d\mu_2(y), \text{ for any } x \in \Omega_1. \tag{15.45}$$

We suppose that $K(x) > 0$ a.e. on Ω_1 and by a weight function u (shortly: a weight), we mean a nonnegative measurable function on the actual set. Let the measurable functions $g_j : \Omega_1 \to \mathbb{R}$, $j = 1, ..., r$, with the representation

$$g_j(x) = \int_{\Omega_2} k(x, y)\, f_j(y)\, d\mu_2(y), \tag{15.46}$$

where $f_j : \Omega_2 \to \mathbb{R}$ are measurable functions, $j = 1, ..., r$.

Denote by $\mathbf{x} = x := (x_1, ..., x_r) \in \mathbb{R}^r$, $\mathbf{g} := (g_1, ..., g_r)$ and $\mathbf{f} := (f_1, ..., f_r)$.

We consider here $\Phi : \mathbb{R}^r_+ \to \mathbb{R}$ a convex function, which is increasing per coordinate, i.e. if $x_j \leq y_j$, $j = 1, ..., r$, then

$$\Phi(x_1, ..., x_r) \leq \Phi(y_1, ..., y_r).$$

In [3, p. 588], we proved that

Theorem 15.17 *Let u be a weight function on Ω_1, and k, K, g_j, f_j, $j = 1, ..., r \in \mathbb{N}$, and Φ defined as above. Assume that the function $x \to u(x) \frac{k(x,y)}{K(x)}$ is integrable on Ω_1 for each fixed $y \in \Omega_2$. Define v on Ω_2 by*

$$v(y) := \int_{\Omega_1} u(x) \frac{k(x, y)}{K(x)} d\mu_1(x) < \infty. \tag{15.47}$$

Then

$$\int_{\Omega_1} u(x)\, \Phi\left(\frac{|g_1(x)|}{K(x)}, ..., \frac{|g_r(x)|}{K(x)}\right) d\mu_1(x) \leq$$

$$\int_{\Omega_2} v(y)\, \Phi(|f_1(y)|, ..., |f_r(y)|)\, d\mu_2(y), \tag{15.48}$$

under the assumptions:

(i) f_j, $\Phi(|f_1|, ..., |f_r|)$, are $k(x, y)\, d\mu_2(y)$ -integrable, μ_1 -a.e. in $x \in \Omega_1$, for all $j = 1, ..., r$,
(ii) $v(y)\, \Phi(|f_1(y)|, ..., |f_r(y)|)$ is μ_2 -integrable.

15.3 Main Results

We start with some fractional Opial type inequalities based on 1.1.

Theorem 15.18 *Let $\overline{p}, q > 1 : \frac{1}{\overline{p}} + \frac{1}{q} = 1$, $p > \frac{1}{q}$. Let $f \in C^1_{-1}(0, +\infty)$, $\kappa \in S_{-1}$ and k be its associate Sonine kernel. Assume that $\mathbb{D}_{(k)} f \in L_q((0, +\infty))$. Then*

$$\int_0^t |f(w)| \left|\left(\mathbb{D}_{(k)} f\right)(w)\right| dw \leq$$

$$2^{-\frac{1}{q}} \left(\int_0^t \left(\int_0^w |\kappa(w-\tau)|^{\overline{p}} d\tau \right) dw \right)^{\frac{1}{\overline{p}}} \left(\int_0^t \left|\left(\mathbb{D}_{(k)} f\right)(w)\right|^q dw \right)^{\frac{2}{q}}, \quad (15.49)$$

$\forall\, t > 0$.

Proof From (15.16) by Hölder's inequality we have

$$|f(t)| \leq \int_0^t |\kappa(t-\tau)| \left|\left(\mathbb{D}_{(k)} f\right)(\tau)\right| d\tau \leq$$

$$\left(\int_0^t |\kappa(t-\tau)|^{\overline{p}} d\tau \right)^{\frac{1}{\overline{p}}} \left(\int_0^t \left|\left(\mathbb{D}_{(k)} f\right)(\tau)\right|^q d\tau \right)^{\frac{1}{q}}, \quad (15.50)$$

$\forall\, t \in (0, +\infty)$.

Call

$$\rho(t) := \int_0^t \left|\left(\mathbb{D}_{(k)} f\right)(\tau)\right|^q d\tau, \quad \rho(0) = 0. \quad (15.51)$$

Thus

$$\rho'(t) = \left|\left(\mathbb{D}_{(k)} f\right)(t)\right|^q \geq 0, \text{ exists a.e. on } (0, +\infty), \quad (15.52)$$

and

$$\left(\rho'(t)\right)^{\frac{1}{q}} = \left|\left(\mathbb{D}_{(k)} f\right)(t)\right| \geq 0, \text{ exists a.e. over } (0, +\infty). \quad (15.53)$$

Consequently we get

$$|f(w)| \left|\left(\mathbb{D}_{(k)} f\right)(w)\right| \leq \left(\int_0^w |\kappa(w-\tau)|^{\overline{p}} d\tau \right)^{\frac{1}{\overline{p}}} \left(\rho(w)\, \rho'(w)\right)^{\frac{1}{q}}, \quad (15.54)$$

a.e. for $w > 0$.

Then, by applying again Hölder's inequality:

$$\int_0^t |f(w)| \left|\left(\mathbb{D}_{(k)} f\right)(w)\right| dw \leq \int_0^t \left(\int_0^w |\kappa(w-\tau)|^{\overline{p}} d\tau \right)^{\frac{1}{\overline{p}}} \left(\rho(w)\, \rho'(w)\right)^{\frac{1}{q}} dw \leq$$

$$\left(\int_0^t \left(\int_0^w |\kappa(w-\tau)|^{\overline{p}} d\tau \right) dw \right)^{\frac{1}{\overline{p}}} \left(\int_0^t \rho(w)\, \rho'(w)\, dw \right)^{\frac{1}{q}} =$$

$$\left(\int_0^t \left(\int_0^w |\kappa(w-\tau)|^{\overline{p}} d\tau \right) dw \right)^{\frac{1}{\overline{p}}} \left(\frac{\rho^2(t)}{2} \right)^{\frac{1}{q}} = \quad (15.55)$$

$$2^{-\frac{1}{q}}\left(\int_0^t\left(\int_0^w|\kappa(w-\tau)|^{\overline{p}}\,d\tau\right)dw\right)^{\frac{1}{\overline{p}}}\left(\int_0^t\left|\left(\mathbb{D}_{(k)}f\right)(w)\right|^q dw\right)^{\frac{2}{q}}.$$

The theorem is proved. ■

We continue with

Theorem 15.19 *Let* $\overline{p}, q > 1 : \frac{1}{\overline{p}} + \frac{1}{q} = 1$, $p > \frac{1}{q}$. *Let* $f \in C^1_{-1}(0, +\infty)$, $\kappa \in S_{-1}$ *and* k *be its associate Sonine kernel. Assume that* ${}_*\mathbb{D}_{(k)}f \in L_q((0, +\infty))$. *Then*

$$\int_0^t|f(w)-f(0)|\left|\left({}_*\mathbb{D}_{(k)}f\right)(w)\right|dw \le$$

$$2^{-\frac{1}{q}}\left(\int_0^t\left(\int_0^w|\kappa(w-\tau)|^{\overline{p}}\,d\tau\right)dw\right)^{\frac{1}{\overline{p}}}\left(\int_0^t\left|\left({}_*\mathbb{D}_{(k)}f\right)(w)\right|^q dw\right)^{\frac{2}{q}}, \quad (15.56)$$

$\forall\, t > 0$.

Proof Similar to (15.49) based on (15.15). ■

We present

Theorem 15.20 *Let* $\overline{p}, q > 1 : \frac{1}{\overline{p}} + \frac{1}{q} = 1$, $p > \frac{1}{q}$. *Let* $f \in C^n_{-1}(0, +\infty)$, $n \in \mathbb{N}$; $\kappa \in S_{-1}$ *and* k *be its associate Sonine kernel. Assume that* $\left(\kappa^n(\tau) * \frac{d^{n-j-1}}{d\tau^{n-j-1}}k^n(\tau)\right)(t) = \frac{d^{n-j-1}}{d\tau^{n-j-1}}\left(\kappa^n(\tau) * k^n(\tau)\right)(t)$, $j = 0, ..., n-1$; *and* $\left(I^{n-1}_{(\kappa)}\left(\mathbb{D}^n_{(k)}f\right)\right) \in L_q((0, +\infty))$. *Then*

$$\int_0^t|f(w)|\left|\left(I^{n-1}_{(\kappa)}\left(\mathbb{D}^n_{(k)}f\right)\right)(w)\right|dw \le$$

$$2^{-\frac{1}{q}}\left(\int_0^t\left(\int_0^w|\kappa(w-\tau)|^{\overline{p}}\,d\tau\right)dw\right)^{\frac{1}{\overline{p}}}\left(\int_0^t\left|\left(I^{n-1}_{(\kappa)}\left(\mathbb{D}^n_{(k)}f\right)\right)(w)\right|^q dw\right)^{\frac{2}{q}}, \quad (15.57)$$

$\forall\, t > 0$.

Proof Similar to (15.49) using (15.23). ■

The counterpart of the last result follows.

Theorem 15.21 *Let* $\overline{p}, q > 1 : \frac{1}{\overline{p}} + \frac{1}{q} = 1$, $p > \frac{1}{q}$. *Let* $f \in C^n_{-1}(0, +\infty)$, $n \in \mathbb{N}$; $\kappa \in S_{-1}$ *and* k *be its associate Sonine kernel. Assume that* $\left(I^{n-1}_{(\kappa)}\left({}_*\mathbb{D}^n_{(k)}f\right)\right) \in L_q((0, +\infty))$. *Then*

$$\int_0^t\left|f(w)-\sum_{j=0}^{n-1}f^{(j)}(0)\frac{w^j}{j!}\right|\left|\left(I^{n-1}_{(\kappa)}\left({}_*\mathbb{D}^n_{(k)}f\right)\right)(w)\right|dw \le$$

$$2^{-\frac{1}{q}}\left(\int_0^t\left(\int_0^w|\kappa(w-\tau)|^{\overline{p}}d\tau\right)dw\right)^{\frac{1}{\overline{p}}}\left(\int_0^t\left|\left(I_{(\kappa)}^{n-1}\left({}_*\mathbb{D}_{(k)}^n f\right)\right)(w)\right|^q dw\right)^{\frac{2}{q}}, \tag{15.58}$$

$\forall\, t>0$.

Proof Similar to (15.49) using (15.22). ■

We continue with some converse fractional Opial type inequalities based on 1.1.

Theorem 15.22 *Let* $0<\overline{p}<1$, $q<0$: $\frac{1}{\overline{p}}+\frac{1}{q}=1$. *Let* $f\in C_{-1}^1(0,+\infty)$, $0\le\kappa\in S_{-1}$ *and* k *be its associate Sonine kernel. Assume that* $\mathbb{D}_{(k)}f\in L_\infty((0,+\infty))$ *that is of the same sign a.e. in* $(0,+\infty)$, *with* $\left(\mathbb{D}_{(k)}f\right)^{-1}\in L_\infty((0,+\infty))$. *Then*

$$\int_0^t|f(w)|\left|\left(\mathbb{D}_{(k)}f\right)(w)\right|dw\ge$$

$$2^{-\frac{1}{q}}\left(\int_0^t\left(\int_0^w|\kappa(w-\tau)|^{\overline{p}}d\tau\right)dw\right)^{\frac{1}{\overline{p}}}\left(\int_0^t\left|\left(\mathbb{D}_{(k)}f\right)(w)\right|^q dw\right)^{\frac{2}{q}}, \tag{15.59}$$

$\forall\, t>0$.

Proof From (15.16) and by reverse Hölder's inequality, see [10, p. 191], we have:

$$|f(t)|=\int_0^t k(t-\tau)\left|\left(\mathbb{D}_{(k)}f\right)(\tau)\right|d\tau\ge$$

$$\left(\int_0^t|\kappa(t-\tau)|^{\overline{p}}d\tau\right)^{\frac{1}{\overline{p}}}\left(\int_0^t\left|\left(\mathbb{D}_{(k)}f\right)(\tau)\right|^q d\tau\right)^{\frac{1}{q}}, \tag{15.60}$$

$\forall\, t\in(0,+\infty)$.

Call

$$\rho(t):=\int_0^t\left|\left(\mathbb{D}_{(k)}f\right)(\tau)\right|^q d\tau,\ \ \rho(0)=0. \tag{15.61}$$

Thus

$$\rho'(t)=\left|\left(\mathbb{D}_{(k)}f\right)(t)\right|^q\ge0,\ \text{exists a.e. on }(0,+\infty), \tag{15.62}$$

and

$$\left(\rho'(t)\right)^{\frac{1}{q}}=\left|\left(\mathbb{D}_{(k)}f\right)(t)\right|\ge0,\ \text{exists a.e. over }(0,+\infty). \tag{15.63}$$

Consequently we get

$$|f(w)|\left|\left(\mathbb{D}_{(k)}f\right)(w)\right|\ge\left(\int_0^w|\kappa(w-\tau)|^{\overline{p}}d\tau\right)^{\frac{1}{\overline{p}}}\left(\rho(w)\rho'(w)\right)^{\frac{1}{q}}, \tag{15.64}$$

a.e. for $w>0$.

Then, by applying again the reverse Hölder's inequality:

$$\int_0^t |f(w)| \left|\left(\mathbb{D}_{(k)} f\right)(w)\right| dw \geq \int_0^t \left(\int_0^w |\kappa(w-\tau)|^{\overline{p}} d\tau\right)^{\frac{1}{\overline{p}}} \left(\rho(w)\rho'(w)\right)^{\frac{1}{q}} dw \geq$$

$$\left(\int_0^t \left(\int_0^w |\kappa(w-\tau)|^{\overline{p}} d\tau\right) dw\right)^{\frac{1}{\overline{p}}} \left(\int_0^t \rho(w)\rho'(w)\,dw\right)^{\frac{1}{q}} =$$

$$\left(\int_0^t \left(\int_0^w |\kappa(w-\tau)|^{\overline{p}} d\tau\right) dw\right)^{\frac{1}{\overline{p}}} \left(\frac{\rho^2(t)}{2}\right)^{\frac{1}{q}} = \tag{15.65}$$

$$2^{-\frac{1}{q}} \left(\int_0^t \left(\int_0^w |\kappa(w-\tau)|^{\overline{p}} d\tau\right) dw\right)^{\frac{1}{\overline{p}}} \left(\int_0^t \left|\left(\mathbb{D}_{(k)} f\right)(w)\right|^q dw\right)^{\frac{2}{q}}.$$

The theorem is proved. ■

We continue with

Theorem 15.23 *Let* $0 < \overline{p} < 1$, $q < 0 : \frac{1}{\overline{p}} + \frac{1}{q} = 1$. *Let* $f \in C_{-1}^1(0, +\infty)$, $0 \leq \kappa \in S_{-1}$ *and* k *be its associate Sonine kernel. Assume that* ${}_*\mathbb{D}_{(k)} f \in L_\infty((0, +\infty))$ *that is of the same sign a.e. in* $(0, +\infty)$, *with* $\left({}_*\mathbb{D}_{(k)} f\right)^{-1} \in L_\infty((0, +\infty))$.

Then

$$\int_0^t |f(w) - f(0)| \left|\left({}_*\mathbb{D}_{(k)} f\right)(w)\right| dw \geq$$

$$2^{-\frac{1}{q}} \left(\int_0^t \left(\int_0^w |\kappa(w-\tau)|^{\overline{p}} d\tau\right) dw\right)^{\frac{1}{\overline{p}}} \left(\int_0^t \left|\left({}_*\mathbb{D}_{(k)} f\right)(w)\right|^q dw\right)^{\frac{2}{q}}, \tag{15.66}$$

$\forall\, t > 0$.

Proof Similar to (15.59) based on (15.15). ■

We present

Theorem 15.24 *Let* $0 < \overline{p} < 1$, $q < 0 : \frac{1}{\overline{p}} + \frac{1}{q} = 1$. *Let* $f \in C_{-1}^n(0, +\infty)$, $n \in \mathbb{N}$; $0 \leq \kappa \in S_{-1}$ *and* k *be its associate Sonine kernel. Assume that* $\left(\kappa^n(\tau) * \frac{d^{n-j-1}}{d\tau^{n-j-1}} k^n(\tau)\right)(t) = \frac{d^{n-j-1}}{dt^{n-j-1}} \left(\kappa^n(\tau) * k^n(\tau)\right)(t)$, $j = 0, ..., n-1$; *and* $\left(I_{(\kappa)}^{n-1}\left(\mathbb{D}_{(k)}^n f\right)\right) \in L_\infty((0, +\infty))$ *that is of the same sign a.e. in* $(0, +\infty)$, *with* $\left(I_{(\kappa)}^{n-1}\left(\mathbb{D}_{(k)}^n f\right)\right)^{-1} \in L_\infty((0, +\infty))$.

Then

$$\int_0^t |f(w)| \left|\left(I_{(\kappa)}^{n-1}\left(\mathbb{D}_{(k)}^n f\right)\right)(w)\right| dw \geq$$

$$2^{-\frac{1}{q}}\left(\int_0^t\left(\int_0^w |\kappa(w-\tau)|^{\overline{p}}\,d\tau\right)dw\right)^{\frac{1}{\overline{p}}}\left(\int_0^t\left|\left(I_{(\kappa)}^{n-1}\left(\mathbb{D}_{(k)}^n f\right)\right)(w)\right|^q dw\right)^{\frac{2}{q}}, \tag{15.67}$$

$\forall\, t>0$.

Proof Similar to (15.59) using (15.23). ■

The counterpart of the last result follows.

Theorem 15.25 *Let* $0<\overline{p}<1$, $q<0$: $\frac{1}{\overline{p}}+\frac{1}{q}=1$. *Let* $f\in C_{-1}^n(0,+\infty)$, $n\in\mathbb{N}$; $0\leq\kappa\in S_{-1}$ *and* k *be its associate Sonine kernel. Assume that* $\left(I_{(\kappa)}^{n-1}\left({}_*\mathbb{D}_{(k)}^n f\right)\right)\in L_\infty((0,+\infty))$ *that is of the same sign a.e. in* $(0,+\infty)$, *with* $\left(I_{(\kappa)}^{n-1}\left({}_*\mathbb{D}_{(k)}^n f\right)\right)^{-1}\in L_\infty((0,+\infty))$.

Then

$$\int_0^t\left|f(w)-\sum_{j=0}^{n-1}f^{(j)}(0)\frac{w^j}{j!}\right|\left|\left(I_{(\kappa)}^{n-1}\left({}_*\mathbb{D}_{(k)}^n f\right)\right)(w)\right|dw\geq$$

$$2^{-\frac{1}{q}}\left(\int_0^t\left(\int_0^w |\kappa(w-\tau)|^{\overline{p}}\,d\tau\right)dw\right)^{\frac{1}{\overline{p}}}\left(\int_0^t\left|\left(I_{(\kappa)}^{n-1}\left({}_*\mathbb{D}_{(k)}^n f\right)\right)(w)\right|^q dw\right)^{\frac{2}{q}}, \tag{15.68}$$

$\forall\, t>0$.

Proof Similar to (15.59) using (15.22). ■

We continue with some basic fractional L_p Hardy type inequalities.

Theorem 15.26 *Let* $\overline{p},q>1$: $\frac{1}{\overline{p}}+\frac{1}{q}=1$, $p>\frac{1}{q}$. *Let* $f_i\in C_{-1}^1(0,+\infty)$, $i=1,\ldots,N\in\mathbb{N}$, $\kappa\in S_{-1}$ *and* k *be its associate Sonine kernel. Assume that* $\mathbb{D}_{(k)}f_i\in L_q((0,+\infty))$, $i=1,\ldots,N$. *Then*

$$\left\|\prod_{i=1}^N f_i\right\|_{\overline{p},(0,a)}\leq\left(\int_0^a\left(\int_0^t|\kappa(t-\tau)|^{\overline{p}}\,d\tau\right)^N dt\right)^{\frac{1}{\overline{p}}}\left(\prod_{i=1}^N\left\|\mathbb{D}_{(k)}f_i\right\|_{q,(0,+\infty)}\right), \tag{15.69}$$

$\forall\, a>0$.

Proof From (15.16) by Hölder's inequality we have

$$|f_i(t)|\leq\int_0^t|\kappa(t-\tau)|\left|\left(\mathbb{D}_{(k)}f_i\right)(\tau)\right|d\tau\leq$$

$$\left(\int_0^t|\kappa(t-\tau)|^{\overline{p}}\,d\tau\right)^{\frac{1}{\overline{p}}}\left(\int_0^t\left|\left(\mathbb{D}_{(k)}f_i\right)(\tau)\right|^q d\tau\right)^{\frac{1}{q}}\leq$$

$$\left(\int_0^t |\kappa (t-\tau)|^{\overline{p}}\, d\tau\right)^{\frac{1}{\overline{p}}} \left\| \mathbb{D}_{(k)} f_i \right\|_{q,(0,+\infty)}, \tag{15.70}$$

$\forall\, t > 0, i = 1, ..., N.$

Therefore we get

$$\prod_{i=1}^{N} |f_i(t)| \leq \left(\int_0^t |\kappa (t-\tau)|^{\overline{p}}\, d\tau\right)^{\frac{N}{\overline{p}}} \prod_{i=1}^{N} \left\| \mathbb{D}_{(k)} f_i \right\|_{q,(0,+\infty)}, \tag{15.71}$$

$\forall\, t > 0.$

Hence it holds

$$\left(\prod_{i=1}^{N} |f_i(t)|\right)^{\overline{p}} \leq \left(\int_0^t |\kappa (t-\tau)|^{\overline{p}}\, d\tau\right)^{N} \left(\prod_{i=1}^{N} \left\| \mathbb{D}_{(k)} f_i \right\|_{q,(0,+\infty)}\right)^{\overline{p}}, \tag{15.72}$$

$\forall\, t > 0.$

Consequently we derive $(a > 0)$

$$\int_0^a \left(\prod_{i=1}^{N} |f_i(t)|\right)^{\overline{p}} dt \leq \left(\int_0^a \left(\int_0^t |\kappa (t-\tau)|^{\overline{p}}\, d\tau\right)^{N} dt\right) \left(\prod_{i=1}^{N} \left\| \mathbb{D}_{(k)} f_i \right\|_{q,(0,+\infty)}\right)^{\overline{p}}, \tag{15.73}$$

proving the claim. ■

We continue with

Theorem 15.27 *Let $\overline{p}, q > 1 : \frac{1}{\overline{p}} + \frac{1}{q} = 1$, $p > \frac{1}{q}$. Let $f_i \in C_{-1}^{1}(0, +\infty)$, $i = 1, ...,$ $N \in \mathbb{N}$, $\kappa \in S_{-1}$ and k be its associate Sonine kernel. Assume that ${}_*\mathbb{D}_{(k)} f_i \in$ $L_q((0, +\infty))$, $i = 1, ..., N$. Then*

$$\left\| \prod_{i=1}^{N} (f_i(\cdot) - f_i(0)) \right\|_{\overline{p},(0,a)} \leq$$

$$\left(\int_0^a \left(\int_0^t |\kappa (t-\tau)|^{\overline{p}}\, d\tau\right)^{N} dt\right)^{\frac{1}{\overline{p}}} \left(\prod_{i=1}^{N} \left\| {}_*\mathbb{D}_{(k)} f_i \right\|_{q,(0,+\infty)}\right), \tag{15.74}$$

$\forall\, a > 0.$

Proof Similar to (15.69) using (15.15). ■

It follows:

Theorem 15.28 *Let $\overline{p}, q > 1 : \frac{1}{\overline{p}} + \frac{1}{q} = 1$, $p > \frac{1}{q}$. Let $f_i \in C_{-1}^{n}(0, +\infty)$, $n \in \mathbb{N}$, $i = 1, ..., N \in \mathbb{N}$, $\kappa \in S_{-1}$ and k be its associate Sonine kernel. Assume that*

$\left(\kappa^n(\tau) * \frac{d^{n-j-1}}{d\tau^{n-j-1}} k^n(\tau)\right)(t) = \frac{d^{n-j-1}}{d\tau^{n-j-1}} \left(\kappa^n(\tau) * k^n(\tau)\right)(t)$, $j = 0, ..., n-1$; *and* $\left(I_{(\kappa)}^{n-1}\left(\mathbb{D}_{(k)}^n f_i\right)\right) \in L_q((0, +\infty))$, $i = 1, ..., N$. *Then*

$$\left\| \prod_{i=1}^N f_i \right\|_{\overline{p},(0,a)} \leq \left(\int_0^a \left(\int_0^t |\kappa(t-\tau)|^{\overline{p}} d\tau \right)^N dt \right)^{\frac{1}{\overline{p}}} \left(\prod_{i=1}^N \left\| I_{(\kappa)}^{n-1}\left(\mathbb{D}_{(k)}^n f_i\right) \right\|_{q,(0,+\infty)} \right), \tag{15.75}$$

$\forall\, a > 0$.

Proof Similar to (15.69) using (15.23). ■

It follows:

Theorem 15.29 *Let* $\overline{p}, q > 1 : \frac{1}{\overline{p}} + \frac{1}{q} = 1$, $p > \frac{1}{q}$. *Let* $f_i \in C_{-1}^n(0, +\infty)$, $n \in \mathbb{N}$, $i = 1, ..., N \in \mathbb{N}$, $\kappa \in S_{-1}$ *and* k *be its associate Sonine kernel. Assume that* $\left(I_{(\kappa)}^{n-1}\left({}_*\mathbb{D}_{(k)}^n f_i\right)\right) \in L_q((0, +\infty))$, $i = 1, ..., N$. *Then*

$$\left\| \prod_{i=1}^N \left(f_i(w) - \sum_{j=0}^{n-1} f_i^{(j)}(0) \frac{w^j}{j!} \right) \right\|_{\overline{p},(0,a)} \leq$$

$$\left(\int_0^a \left(\int_0^t |\kappa(t-\tau)|^{\overline{p}} d\tau \right)^N dt \right)^{\frac{1}{\overline{p}}} \left(\prod_{i=1}^N \left\| I_{(\kappa)}^{n-1}\left({}_*\mathbb{D}_{(k)}^n f_i\right) \right\|_{q,(0,+\infty)} \right), \tag{15.76}$$

$\forall\, a > 0$.

Proof Similar to (15.69) using (15.22). ■

Next we present very general fractional Hardy type inequalities under convexity. We make

Remark 15.30 Here we would like to apply Sect. 15.2.4 and Theorem 15.17 to Sect. 15.1.1.

Let $0 \leq \kappa \in S_{-1}$, by (15.6) we have that

$$\left(I_{(\kappa)} 1\right)(x) = \int_0^\infty \chi_{(0,x)}(y)\, \kappa(x-y)\, dy, \quad x > 0, \tag{15.77}$$

where χ is the characteristic function.

Let $f_j \in C_{-1}^1(0, +\infty)$, with $\mathbb{D}_{(k)} f_j \in L_1((0, +\infty))$, $j = 1, ..., r$. By (15.16) we obtain

$$f_j(x) = \int_0^{+\infty} \chi_{(0,x)}(y)\, \kappa(x-y) \left(\mathbb{D}_{(k)} f_j\right)(y)\, dy, \tag{15.78}$$

$j = 1, ..., r$.

Let $\Phi : \mathbb{R}_+^r \to \mathbb{R}$ a convex function, which is increasing per coordinate.

We obtain

Theorem 15.31 *Let u be a weight function on $(0, +\infty)$, and all as in Remark 15.30. Assume that the function $x \to u(x) \frac{\chi_{(0,x)}(y)\kappa(x-y)}{(I_{(\kappa)}1)(x)}$ is integrable on $(0, +\infty)$ for each fixed $y \in (0, +\infty)$. Define υ on $(0, +\infty)$ by*

$$\upsilon(y) := \int_0^{+\infty} u(x) \frac{\chi_{(0,x)}(y)\kappa(x-y)}{(I_{(\kappa)}1)(x)} dx < \infty. \tag{15.79}$$

Then

$$\int_0^{+\infty} u(x)\Phi\left(\frac{|f_1(x)|}{(I_{(\kappa)}1)(x)}, ..., \frac{|f_r(x)|}{(I_{(\kappa)}1)(x)}\right) dx \leq$$

$$\int_0^{+\infty} \upsilon(y)\Phi\left(\left|\mathbb{D}_{(k)} f_1(y)\right|, ..., \left|\mathbb{D}_{(k)} f_r(y)\right|\right) dy, \tag{15.80}$$

under the assumptions:

(i) $\mathbb{D}_{(k)} f_j$, $\Phi\left(\left|\mathbb{D}_{(k)} f_1\right|, ..., \left|\mathbb{D}_{(k)} f_r\right|\right)$, *are* $\chi_{(0,x)}(y)\kappa(x-y)\,dy$*-integrable,* dx*-a,e, in* $x \in (0, +\infty)$, *for all* $j = 1, ..., r$,

(ii) $\upsilon(y)\Phi\left(\left|\mathbb{D}_{(k)} f_1(y)\right|, ..., \left|\mathbb{D}_{(k)} f_r(y)\right|\right)$ *is* dy*-integrable.*

We continue with

Remark 15.32 Here again we would like to apply Sect. 15.2.4 and Theorem 15.17 to Sect. 15.1.1.

Let $0 \leq \kappa \in S_{-1}$, by (15.6) we have that

$$\left(I_{(\kappa)}1\right)(x) = \int_0^{\infty} \chi_{(0,x)}(y)\kappa(x-y)\,dy, \quad x > 0, \tag{15.81}$$

where χ is the characteristic function.

Let $f_j \in C_{-1}^1(0, +\infty)$, with ${}_*\mathbb{D}_{(k)} f_j \in L_1((0, +\infty))$, $j = 1, ..., r$. By (15.15) we obtain

$$f_j(x) - f_j(0) = \int_0^{+\infty} \chi_{(0,x)}(y)\kappa(x-y)\left({}_*\mathbb{D}_{(k)} f_j\right)(y)\,dy, \tag{15.82}$$

$j = 1, ..., r$.

Let $\Phi : \mathbb{R}_+^r \to \mathbb{R}$ a convex function, which is increasing per coordinate.

We obtain

Theorem 15.33 *Let u be a weight function on $(0, +\infty)$, and all as in Remark 15.32. Assume that the function $x \to u(x) \frac{\chi_{(0,x)}(y)\,\kappa(x-y)}{(I_{(\kappa)}1)(x)}$ is integrable on $(0, +\infty)$ for each fixed $y \in (0, +\infty)$. Define υ on $(0, +\infty)$ by*

$$\upsilon(y) := \int_0^{+\infty} u(x) \frac{\chi_{(0,x)}(y) \kappa(x-y)}{\left(I_{(\kappa)}1\right)(x)} dx < \infty. \tag{15.83}$$

Then

$$\int_0^{+\infty} u(x) \Phi\left(\frac{|f_1(x) - f_1(0)|}{\left(I_{(\kappa)}1\right)(x)}, ..., \frac{|f_r(x) - f_r(0)|}{\left(I_{(\kappa)}1\right)(x)}\right) dx \leq \tag{15.84}$$

$$\int_0^{+\infty} \upsilon(y) \Phi\left(\left|{}_*\mathbb{D}_{(k)} f_1(y)\right|, ..., \left|{}_*\mathbb{D}_{(k)} f_r(y)\right|\right) dy,$$

under the assumptions:

(i) ${}_*\mathbb{D}_{(k)} f_j$, $\Phi\left(\left|{}_*\mathbb{D}_{(k)} f_1\right|, ..., \left|{}_*\mathbb{D}_{(k)} f_r\right|\right)$, *are* $\chi_{(0,x)}(y) \kappa(x-y) dy$*-integrable,* dx*-a,e, in* $x \in (0, +\infty)$, *for all* $j = 1, ..., r$,

(ii) $\upsilon(y) \Phi\left(\left|{}_*\mathbb{D}_{(k)} f_1(y)\right|, ..., \left|{}_*\mathbb{D}_{(k)} f_r(y)\right|\right)$ *is* dy*-integrable.*

We make

Remark 15.34 Here we apply againTheorem 15.17.

Let $0 \leq \kappa \in S_{-1}$ and κ be its associate Sonine kernel such that (15.24) is valid, and $\left(I_{(\kappa)}1\right)(x)$ is as in (15.77). Let $f_j \in C_{-1}^n(0, +\infty)$, $n \in \mathbb{N}$, with $\left(I_{(\kappa)}^{n-1}\left(\mathbb{D}_{(k)}^n f_j\right)\right) \in L_1((0, +\infty))$, $j = 1, ..., r$. By (15.23) we obtain

$$f_j(x) = \int_0^{+\infty} \chi_{(0,x)}(y) \kappa(x-y) \left(I_{(\kappa)}^{n-1}\left(\mathbb{D}_{(k)}^n f_j\right)\right)(y) dy, \tag{15.85}$$

$j = 1, ..., r$.

Let $\Phi : \mathbb{R}_+^r \to \mathbb{R}$ a convex function, which is increasing per coordinate.

We derive

Theorem 15.35 *Let* u *be a weight function on* $(0, +\infty)$, *and all as in Remark 15.34. Assume that the function* $x \to u(x) \frac{\chi_{(0,x)}(y)\kappa(x-y)}{\left(I_{(\kappa)}1\right)(x)}$ *is integrable on* $(0, +\infty)$ *for each fixed* $y \in (0, +\infty)$. *Define* υ *on* $(0, +\infty)$ *by*

$$\upsilon(y) := \int_0^{+\infty} u(x) \frac{\chi_{(0,x)}(y) \kappa(x-y)}{\left(I_{(\kappa)}1\right)(x)} dx < \infty. \tag{15.86}$$

Then

$$\int_0^{+\infty} u(x) \Phi\left(\frac{|f_1(x)|}{\left(I_{(\kappa)}1\right)(x)}, ..., \frac{|f_r(x)|}{\left(I_{(\kappa)}1\right)(x)}\right) dx \leq \tag{15.87}$$

$$\int_0^{+\infty} \upsilon(y) \Phi\left(\left|\left(I_{(\kappa)}^{n-1}\left(\mathbb{D}_{(k)}^n f_1\right)\right)(y)\right|, ..., \left|\left(I_{(\kappa)}^{n-1}\left(\mathbb{D}_{(k)}^n f_r\right)\right)(y)\right|\right) dy,$$

under the assumptions:

(i) $\left(I_{(\kappa)}^{n-1}\left(\mathbb{D}_{(k)}^{n} f_j\right)\right)$, $\Phi\left(\left|\left(I_{(\kappa)}^{n-1}\left(\mathbb{D}_{(k)}^{n} f_1\right)\right)\right|, ..., \left|\left(I_{(\kappa)}^{n-1}\left(\mathbb{D}_{(k)}^{n} f_r\right)\right)\right|\right)$, *are* $\chi_{(0,x)}$ $(y)\,\kappa\,(x-y)$ *dy-integrable, dx-a.e. in* $x \in (0,+\infty)$, *for all* $j = 1, ..., r$,

(ii) $\upsilon(y)\,\Phi\left(\left|\left(I_{(\kappa)}^{n-1}\left(\mathbb{D}_{(k)}^{n} f_1\right)\right)(y)\right|, ..., \left|\left(I_{(\kappa)}^{n-1}\left(\mathbb{D}_{(k)}^{n} f_r\right)\right)(y)\right|\right)$ *is dy-integrable.*

It follows:

Remark 15.36 Here we apply again Theorem 15.17.

Let $0 \le \kappa \in S_{-1}$ and k be its associate Sonine kernel, and $\left(I_{(\kappa)}1\right)(x)$ is as in (15.77). Let $f_j \in C_{-1}^{n}(0,+\infty)$, $n \in \mathbb{N}$, with $\left(I_{(\kappa)}^{n-1}\left({}_*\mathbb{D}_{(k)}^{n} f_j\right)\right) \in L_1((0,+\infty))$, $j = 1, ..., r$.

Here we denote

$$\overline{f}_j(x) := f_j(x) - \sum_{\overline{j}=0}^{n-1} f_j^{(\overline{j})}(0)\,\frac{x^{\overline{j}}}{\overline{j}!}, \quad x \in (0,+\infty),\ j = 1, ..., r.$$

By (15.22) we obtain

$$\overline{f}_j(x) = \int_0^{+\infty} \chi_{(0,x)}(y)\,\kappa(x-y)\left(I_{(\kappa)}^{n-1}\left({}_*\mathbb{D}_{(k)}^{n} f_j\right)\right)(y)\,dy, \tag{15.88}$$

$j = 1, ..., r$.

Let $\Phi : \mathbb{R}_+^r \to \mathbb{R}$ a convex function, which is increasing per coordinate.

We derive

Theorem 15.37 *Let u be a weight function on* $(0,+\infty)$, *and all as in Remark 15.36. Assume that the function* $x \to u(x)\,\frac{\chi_{(0,x)}(y)\kappa(x-y)}{\left(I_{(\kappa)}1\right)(x)}$ *is integrable on* $(0,+\infty)$ *for each fixed* $y \in (0,+\infty)$. *Define υ on* $(0,+\infty)$ *by*

$$\upsilon(y) := \int_0^{+\infty} u(x)\,\frac{\chi_{(0,x)}(y)\,\kappa(x-y)}{\left(I_{(\kappa)}1\right)(x)}\,dx < \infty. \tag{15.89}$$

Then

$$\int_0^{+\infty} u(x)\,\Phi\left(\frac{\left|\overline{f}_1(x)\right|}{\left(I_{(\kappa)}1\right)(x)}, ..., \frac{\left|\overline{f}_r(x)\right|}{\left(I_{(\kappa)}1\right)(x)}\right) dx \le \tag{15.90}$$

$$\int_0^{+\infty} \upsilon(y)\,\Phi\left(\left|\left(I_{(\kappa)}^{n-1}\left({}_*\mathbb{D}_{(k)}^{n} f_1\right)\right)(y)\right|, ..., \left|\left(I_{(\kappa)}^{n-1}\left({}_*\mathbb{D}_{(k)}^{n} f_r\right)\right)(y)\right|\right) dy,$$

under the assumptions:

(i) $\left(I_{(\kappa)}^{n-1}\left({}_*\mathbb{D}_{(k)}^{n} f_j\right)\right)$, $\Phi\left(\left|\left(I_{(\kappa)}^{n-1}\left({}_*\mathbb{D}_{(k)}^{n} f_1\right)\right)\right|, ..., \left|\left(I_{(\kappa)}^{n-1}\left({}_*\mathbb{D}_{(k)}^{n} f_r\right)\right)\right|\right)$, *are* $\chi_{(0,x)}(y)\,\kappa(x-y)$ *dy-integrable, dx-a.e. in* $x \in (0,+\infty)$, *for all* $j = 1, ..., r$,

(ii) $\upsilon(y)\Phi\left(\left|\left(I_{(\kappa)}^{n-1}\left({}_{*}\mathbb{D}_{(k)}^{n}f_1\right)\right)(y)\right|,\ldots,\left|\left(I_{(\kappa)}^{n-1}\left({}_{*}\mathbb{D}_{(k)}^{n}f_r\right)\right)(y)\right|\right)$ *is dy-integrable.*

To keep our chapter short we choose to skip applications of Sect. 15.1.1 to Theorems 15.13–15.16.

We make

Remark 15.38 Our goal here is to apply Theorem 15.13 to generalized left and right partial Prabhakar fractional integrals, see Definition 15.10. Our setting now is as follows:

Let $\prod_{i=1}^{N}[a_i,b_i]\subset\mathbb{R}^N$, $N>1$, $a_i<b_i$, $a_i,b_i\in\mathbb{R}$; $f_j\in C\left(\prod_{i=1}^{N}[a_i,b_i]\right)$, $j=1,\ldots,m$. Let also $x_i,t_i\in[a_i,b_i]$; $\rho_{ij},\mu_{ij},\gamma_{ij},\omega_{ij}>0$, $i=1,\ldots,N$; $j=1,\ldots,m$. Here $\psi_i\in C^1([a_i,b_i])$ which is increasing, $i=1,\ldots,N$. We set $a=(a_1,\ldots,a_N)$, $b=(b_1,\ldots,b_N)$, $x=(x_1,\ldots,x_N)$, $t=(t_1,\ldots,t_N)$; $\rho_j=(\rho_{1j},\ldots,\rho_{Nj})$, $\mu_j=(\mu_{1j},\ldots,\mu_{Nj})$, $\gamma_j=(\gamma_{1j},\ldots,\gamma_{Nj})$, $\omega_j=(\omega_{1j},\ldots,\omega_{Nj})$, $\psi:=(\psi_1,\ldots,\psi_N)$, $j=1,\ldots,m$.

The corresponding left and right partial Prabhakar fractional integrals with respect to ψ_k, respectively, for $k=1,\ldots,N$, came as follows (see Definition 15.10):

$$\left({}^{P}e^{\gamma_{kj};\psi_k}_{\rho_{kj},\mu_{kj},\omega_{kj},a_k+}f_j\right)(x):=\int_{a_k}^{x_k}\psi_k'(t_k)\left(\psi_k(x_k)-\psi_k(t_k)\right)^{\mu_{kj}-1}$$

$$E^{\gamma_{kj}}_{\rho_{kj},\mu_{kj}}\left[\omega_{kj}\left(\psi_k(x_k)-\psi_k(t_k)\right)^{\rho_{kj}}\right]f_j(x_1,\ldots,x_{k-1},t_k,x_{k+1},\ldots,x_N)\,dt_k, \tag{15.91}$$

with $a_k\leq x_k\leq b_k$;

and

$$\left({}^{P}e^{\gamma_{kj};\psi_k}_{\rho_{kj},\mu_{kj},\omega_{kj},b_k-}f_j\right)(x):=\int_{x_k}^{b_k}\psi_k'(t_k)\left(\psi_k(t_k)-\psi_k(x_k)\right)^{\mu_{kj}-1} \tag{15.92}$$

$$E^{\gamma_{kj}}_{\rho_{kj},\mu_{kj}}\left[\omega_{kj}\left(\psi_k(t_k)-\psi_k(x_k)\right)^{\rho_{kj}}\right]f(x_1,\ldots,x_{k-1},t_k,x_{k+1},\ldots,x_N)\,dt_k,$$

with $a_k\leq x_k\leq b_k$;

for all $j=1,\ldots,m$. The functions (15.91) and (15.92) are continuous functions in x_k, $k=1,\ldots,N$.

The corresponding to (15.35) functions are defined as follows for appropiate weight $u\geq 0$:

$$A^{\psi_k}_{m+}(y_k)=\left(\psi_k'(y_k)\right)^m\int_{y_k}^{b_k}u(x_k)\left(\frac{\left(\psi_k(x_k)-\psi_k(y_k)\right)^{\sum_{j=1}^{m}\mu_{kj}-m}}{\left(\psi_k(x_k)-\psi_k(a_k)\right)^{\sum_{j=1}^{m}\mu_{kj}}}\right)$$

$$\prod_{j=1}^{m}\left(\frac{E^{\gamma_{kj}}_{\rho_{kj},\mu_{kj}}\left[\omega_{kj}\left(\psi_k(x_k)-\psi_k(y_k)\right)^{\rho_{kj}}\right]}{E^{\gamma_{kj}}_{\rho_{kj},\mu_{kj}+1}\left[\omega_{kj}\left(\psi_k(x_k)-\psi_k(a_k)\right)^{\rho_{kj}}\right]}\right)dx_k<\infty, \tag{15.93}$$

$a_k \leq y_k \leq b_k$, and it is assumed to be integrable, similarly define

$$A_{m-}^{\psi_k}(y_k) = \left(\psi_k'(y_k)\right)^m \int_{a_k}^{y_k} u(x_k) \left(\frac{(\psi_k(y_k) - \psi_k(x_k))^{\sum_{j=1}^{m} \mu_{kj} - m}}{(\psi_k(b_k) - \psi_k(x_k))^{\sum_{j=1}^{m} \mu_{kj}}} \right)$$

$$\prod_{j=1}^{m} \left(\frac{E_{\rho_{kj},\mu_{kj}}^{\gamma_{kj}} \left[\omega_{kj} (\psi_k(y_k) - \psi_k(x_k))^{\rho_{kj}} \right]}{E_{\rho_{kj},\mu_{kj}+1}^{\gamma_{k}j} \left[\omega_{kj} (\psi_k(b_k) - \psi_k(x_k))^{\rho_{kj}} \right]} \right) dx_k < \infty, \tag{15.94}$$

$a_k \leq y_k \leq b_k$, and it is assumed to be integrable.

Based on Theorem 15.13 and as in the proof of Theorem 18 of [5], we present the following generalized partial Prabhakar left fractional Hardy type inequality:

Theorem 15.39 *All as in Remark 15.38. Here $\Phi_j : \mathbb{R}_+ \to \mathbb{R}_+$, $j = 1, ..., m$, are convex and increasing functions. Then*

$$\int_{a_k}^{b_k} u(x_k) \prod_{j=1}^{m} \Phi_j \left(\frac{\left| \left({}^{P}e_{\rho_{kj},\mu_{kj},\omega_{kj},a_k+}^{\gamma_{kj};\psi_k} f_j \right)(x) \right|}{(\psi_k(x_k) - \psi_k(a_k))^{\mu_{kj}} E_{\rho_{kj},\mu_{kj+1}}^{\gamma_{kj}} \left[\omega_{kj} (\psi_k(x_k) - \psi_k(a_k))^{\rho_{kj}} \right]} \right) dx_k$$

$$\leq \left(\prod_{\substack{j=1 \\ j \neq \bar{j}}}^{m} \int_{a_k}^{b_k} \Phi_j \left(\left| f_j (x_1, ..., x_{k-1}, y_k, x_{k+1}, ..., x_N) \right| \right) dy_k \right)$$

$$\left(\int_{a_k}^{b_k} \Phi_{\bar{j}} \left(\left| f_{\bar{j}} (x_1, ..., x_{k-1}, y_k, x_{k+1}, ..., x_N) \right| \right) A_{m+}^{\psi_k}(y_k) \, dy_k \right). \tag{15.95}$$

Based on Theorem 15.13 and as in the proof of Theorem 20 of [5], we present the following generalized partial Prabhakar right fractional Hardy type inequality:

Theorem 15.40 *All as in Remark 15.38. Here $\Phi_j : \mathbb{R}_+ \to \mathbb{R}_+$, $j = 1, ..., m$, are convex and increasing functions. Then*

$$\int_{a_k}^{b_k} u(x_k) \prod_{j=1}^{m} \Phi_j \left(\frac{\left| \left({}^{P}e_{\rho_{kj},\mu_{k}j,\omega_{kj},b_k-}^{\gamma_{kj};\psi_k} f_j \right)(x) \right|}{(\psi_k(b_k) - \psi_k(x_k))^{\mu_{kj}} E_{\rho_{kj},\mu_{kj+1}}^{\gamma_{kj}} \left[\omega_{kj} (\psi_k(b_k) - \psi_k(x_k))^{\rho_{kj}} \right]} \right) dx_k$$

$$\leq \left(\prod_{\substack{j=1 \\ j \neq \bar{j}}}^{m} \int_{a_k}^{b_k} \Phi_j \left(\left| f_j (x_1, ..., x_{k-1}, y_k, x_{k+1}, ..., x_N) \right| \right) dy_k \right)$$

$$\left(\int_{a_k}^{b_k} \Phi_{\overline{j}} \left(\left| f_{\overline{j}} \left(x_1, ..., x_{k-1}, y_k, x_{k+1}, ..., x_N \right) \right| \right) A_{m-}^{\psi_k} \left(y_k \right) dy_k \right). \tag{15.96}$$

We make

Remark 15.41 Our goal here is to apply Theorem 15.15 to generalized left and right partial Prabhakar fractional integrals, see Definition 15.10. Our setting now is as follows:

Let $\prod_{i=1}^{N} [a_i, b_i] \subset \mathbb{R}^N$, $N > 1$, $a_i < b_i$, $a_i, b_i \in \mathbb{R}$; $f_{\lambda j} \in C\left(\prod_{i=1}^{N} [a_i, b_i] \right)$, $\lambda = 1, 2$; $j = 1, ..., m$. Let also $x_i, t_i \in [a_i, b_i]$; $\rho_{ij}, \mu_{ij}, \gamma_{ij}, \omega_{ij} > 0$, $i = 1, ..., N$; $j = 1, ..., m$. Here $\psi_i \in C^1([a_i, b_i])$ which is increasing, $i = 1, ..., N$. We set $a = (a_1, ..., a_N), b = (b_1, ..., b_N), x = (x_1, ..., x_N), t = (t_1, ..., t_N)$; $\rho_j = (\rho_{1j}, ..., \rho_{Nj})$, $\mu_j = (\mu_{1j}, ..., \mu_{Nj})$, $\gamma_j = (\gamma_{1j}, ..., \gamma_{Nj})$, $\omega_j = (\omega_{1j}, ..., \omega_{Nj})$, $\psi := (\psi_1, ..., \psi_N)$, $j = 1, ..., m$.

The corresponding left and right partial Prabhakar fractional integrals with respect to ψ_k, respectively, for $k = 1, ..., N$, here came as follows: $\left({}^{P}e_{\rho_{kj}, \mu_{k} j, \omega_{kj}, a_k+}^{\gamma_{kj}; \psi_k} f_{\lambda j} \right)(x)$ and $\left({}^{P}e_{\rho_{kj}, \mu_{kj}, \omega_{kj}, b_k-}^{\gamma_{kj}; \psi_k} f_{\lambda j} \right)(x)$, with $a_k \leq x_k \leq b_k$; as in (15.91), (15.92), for all $\lambda = 1, 2$; $j = 1, ..., m$.

Denote by $\overline{f}_{2j}(y_k) := f_{2j}(x_1, ..., x_{k-1}, y_k, x_{k+1}, ..., x_N)$, $j = 1, ..., m$.

Here we assume that $0 < \overline{f}_{2j}(y_k) < \infty$, $j = 1, ..., m$; $\overline{j} \in \{1, ..., m\}$ is fixed.

Define $B_{m+}^{\psi_k}$ so that $(a_k \leq y_k \leq b_k)$

$$B_{m+}^{\psi_k}(y_k) := \left(\prod_{j=1}^{m} \overline{f}_{2j}(y_k) \right) \left(\psi_k'(y_k) \right)^m \tag{15.97}$$

$$\int_{y_k}^{b_k} \left(\frac{u(x_k) \left(\psi_k(x_k) - \psi_k(y_k) \right)^{\sum_{j=1}^{m} \mu_{kj} - m} \prod_{j=1}^{m} E_{\rho_{kj}, \mu_{kj}}^{\gamma_{kj}} \left[\omega_{kj} \left(\psi_k(x_k) - \psi_k(y_k) \right)^{\rho_{kj}} \right]}{\prod_{j=1}^{m} \left({}^{P}e_{\rho_{kj}, \mu_{kj}, \omega_{kj}, a_k+}^{\gamma_{kj}; \psi_k} f_{2j} \right)(x)} \right) dx_k < \infty,$$

and $B_{m+}^{\psi_k}$ is assumed to be integrable over $[a_k, b_k]$.

Similarly we define $B_{m-}^{\psi_k}$ so that $(a_k \leq y_k \leq b_k)$

$$B_{m-}^{\psi_k}(y_k) := \left(\prod_{j=1}^{m} \overline{f}_{2j}(y_k) \right) \left(\psi_k'(y_k) \right)^m \tag{15.98}$$

$$\int_{a_k}^{y_k}\left(\frac{u(x_k)\left(\psi_k(y_k)-\psi_k(x_k)\right)^{\sum_{j=1}^{m}\mu_{kj}-m}\prod_{j=1}^{m}E_{\rho_{kj},\mu_{kj}}^{\gamma_{kj}}\left[\omega_{kj}\left(\psi_k(y_k)-\psi_k(x_k)\right)^{\rho_{kj}}\right]}{\prod_{j=1}^{m}\left({}^{P}e_{\rho_{kj},\mu_{kj},\omega_{kj},b_k-}^{\gamma_{kj};\psi_k}f_{2j}\right)(x)}\right)dx_k<\infty,$$

and $B_{m-}^{\psi_k}$ is assumed to be integrable over $[a_k, b_k]$.

Here $\Phi_j : \mathbb{R}_+ \to \mathbb{R}_+$, $j = 1, ..., m$, are convex and increasing functions.

Next come the ψ-Prabhakar-Riemann-Liouville left and right side Hardy type rational fractional inequalities, respectively:

Theorem 15.42 *It holds*

$$\int_{a_k}^{b_k}u(x_k)\prod_{j=1}^{m}\Phi_j\left(\frac{\left|\left({}^{P}e_{\rho_{kj},\mu_{kj},\omega_{kj},a_k+}^{\gamma_{kj};\psi_k}f_{1j}\right)(x)\right|}{\left({}^{P}e_{\rho_{kj},\mu_{kj},\omega_{kj},a_k+}^{\gamma_{kj};\psi_k}f_{2j}\right)(x)}\right)dx_k\leq$$

$$\left(\prod_{\substack{j=1\\ j\neq\overline{j}}}^{m}\int_{a_k}^{b_k}\Phi_j\left(\frac{\left|\overline{f}_{1j}(y_k)\right|}{\overline{f}_{2j}(y_k)}\right)dy_k\right)\left(\int_{a_k}^{b_k}\Phi_{\overline{j}}\left(\frac{\left|\overline{f}_{1\overline{j}}(y_k)\right|}{\overline{f}_{2\overline{j}}(y_k)}\right)B_{m+}^{\psi_k}(y_k)\,dy_k\right),\tag{15.99}$$

where $\overline{f}_{\lambda j}(y_k) := f_{\lambda j}(x_1, ..., x_{k-1}, y_k, x_{k+1}, ..., x_N)$, $\lambda = 1, 2;\ j = 1, ..., m$.

Proof Based on Theorem 15.15, Remarks 15.14, 15.41, and similar to the proof of Theorem 29 in [5]. ■

Theorem 15.43 *It holds*

$$\int_{a_k}^{b_k}u(x_k)\prod_{j=1}^{m}\Phi_j\left(\frac{\left|\left({}^{P}e_{\rho_{kj},\mu_{kj},\omega_{kj},b_k-}^{\gamma_{kj};\psi_k}f_{1j}\right)(x)\right|}{\left({}^{P}e_{\rho_{kj},\mu_{kj},\omega_{kj},b_k-}^{\gamma_{kj};\psi_k}f_{2j}\right)(x)}\right)dx_k\leq$$

$$\left(\prod_{\substack{j=1\\ j\neq\overline{j}}}^{m}\int_{a_k}^{b_k}\Phi_j\left(\frac{\left|\overline{f}_{1j}(y_k)\right|}{\overline{f}_{2j}(y_k)}\right)dy_k\right)\left(\int_{a_k}^{b_k}\Phi_{\overline{j}}\left(\frac{\left|\overline{f}_{1\overline{j}}(y_k)\right|}{\overline{f}_{2\overline{j}}(y_k)}\right)B_{m-}^{\psi_k}(y_k)\,dy_k\right),\tag{15.100}$$

where $\overline{f}_{\lambda j}(y_k) := f_{\lambda j}(x_1, ..., x_{k-1}, y_k, x_{k+1}, ..., x_N)$, $\lambda = 1, 2;\ j = 1, ..., m$.

Proof Based on Theorem 15.15, Remarks 15.14, 15.41, and similar to the proof of Theorem 30 in [5]. ■

We make

Remark 15.44 Let $\prod_{i=1}^{N}[a_i,b_i]\subset\mathbb{R}^N, N>1, a_i<b_i, a_i,b_i\in\mathbb{R}$; $f_j\in C\left(\prod_{i=1}^{N}[a_i,b_i]\right)$, $j=1,...,r$. Let also $x_i,t_i\in[a_i,b_i]$; $0<\rho_i\in C([a_i,b_i])$; $\mu_i,\gamma_i,\omega_i>0$, for $i=1,...,N$. Here $\psi_i\in C^1([a_i,b_i])$ which is increasing, $i=1,...,N$. We set $a=(a_1,...,a_N)$, $b=(b_1,...,b_N)$, $x=(x_1,...,x_N)$, $t=(t_1,...,t_N)$; $\rho(t)=(\rho_1(t_1),...,\rho_N(t_N)), \mu_j=(\mu_{1j},...,\mu_{Nj}), \gamma=(\gamma_1,...,\gamma_N), \omega=(\omega_1,...,\omega_n), \psi:=(\psi_1,...,\psi_N)$.

We define the left and right mixed Prabhakar multiple fractional integrals for f_j of variable degree with respect to ψ, respectively (see Definition 15.12 and (15.31), (15.32)), as $\left({}_{\upsilon}^{M}e_{\rho,\mu,\omega,a+}^{\gamma;\psi}f_j\right)$, $\left({}_{\upsilon}^{M}e_{\rho,\mu,\omega,b-}^{\gamma;\psi}f_j\right)$, $j=1,...,r$; which integrals are considered here as measurable functions.

The purpose it is to apply Theorem 15.16.

Here $p_j>1:\sum_{j=1}^{r}\frac{1}{p_j}=1$, and $\Phi_j:\mathbb{R}_+\to\mathbb{R}_+$, $j=1,...,r$, are convex and increasing functions, where $u\geq 0$ is a weight function.

We define the kernels $(y=(y_1,...,y_N)\in\prod_{i=1}^{N}(a_i,b_i))$

$$k_{a+}(x,y):=\chi_{\prod_{i=1}^{N}(a_i,x_i]}(y)$$

$$\prod_{i=1}^{N}\left[\psi_i'(y_i)(\psi_i(x_i)-\psi_i(y_i))^{\mu_i-1}E_{\rho_i(y_i),\mu_i}^{\gamma_i}\left[\omega_i(\psi_i(x_i)-\psi_i(y_i))^{\rho_i(y_i)}\right]\right], \tag{15.101}$$

where χ is the characteristic function,

and

$$k_{b-}(x,y):=\chi_{\prod_{i=1}^{N}[x_i,b_i)}(y)$$

$$\prod_{i=1}^{N}\left[\psi_i'(y_i)(\psi_i(y_i)-\psi_i(x_i))^{\mu_i-1}E_{\rho_i(y_i),\mu_i}^{\gamma_i}\left[\omega_i(\psi_i(y_i)-\psi_i(x_i))^{\rho_i(y_i)}\right]\right], \tag{15.102}$$

$\forall\, x,y\in\prod_{i=1}^{N}(a_i,b_i)$.

We also define the integrals

$$K_{a+}(x):=\int_{\prod_{i=1}^{N}(a_i,b_i)}k_{a+}(x,y)\,dy, \tag{15.103}$$

and

$$K_{b-}(x) := \int_{\prod_{i=1}^{N}(a_i,b_i)} k_{b-}(x,y)\,dy, \qquad (15.104)$$

$\forall\, x \in \prod_{i=1}^{N}(a_i, b_i)$.

We present the following variable generalized fractional Hardy type inequalities under convexity:

Theorem 15.45 *All as in Remark 15.44. Assume that the function* $x \to \left(u(x)\frac{k_{a+}(x,y)}{K_{a+}(x)}\right)$ *is integrable on* $\prod_{i=1}^{N}(a_i, b_i)$, *for each fixed* $y \in \prod_{i=1}^{N}(a_i, b_i)$. *Define* φ^{+} *on* $\prod_{i=1}^{N}(a_i, b_i)$ *by*

$$\varphi^{+}(y) := \int_{\prod_{i=1}^{N}(a_i,b_i)} u(x)\frac{k_{a+}(x,y)}{K_{a+}(x)}dx < \infty, \qquad (15.105)$$

for appropiate weight $u \geq 0$, *so that* φ^{+} *is integrable on* $\prod_{i=1}^{N}(a_i, b_i)$. *Then*

$$\int_{\prod_{i=1}^{N}(a_i,b_i)} u(x)\prod_{j=1}^{r}\Phi_j\left(\frac{\left|{}^{M}_{\upsilon}e^{\gamma;\psi}_{\rho,\mu,\omega,a+}f_j(x)\right|}{K_{a+}(x)}\right)dx \leq \qquad (15.106)$$

$$\prod_{j=1}^{r}\left(\int_{\prod_{i=1}^{N}(a_i,b_i)} \varphi^{+}(y)\,\Phi_j\left(|f_j(y)|\right)^{p_j}dy\right)^{\frac{1}{p_j}}.$$

We finish with

Theorem 15.46 *All as in Remark 15.44. Assume that the function* $x \to \left(u(x)\frac{k_{b-}(x,y)}{K_{b-}(x)}\right)$ *is integrable on* $\prod_{i=1}^{N}(a_i, b_i)$, *for each fixed* $y \in \prod_{i=1}^{N}(a_i, b_i)$. *Define* φ^{-} *on* $\prod_{i=1}^{N}(a_i, b_i)$ *by*

$$\varphi^{-}(y) := \int_{\prod_{i=1}^{N}(a_i,b_i)} u(x)\frac{k_{b-}(x,y)}{K_{b-}(x)}dx < \infty, \qquad (15.107)$$

for appropiate weight $u \geq 0$, *so that* φ^{-} *is integrable on* $\prod_{i=1}^{N}(a_i, b_i)$. *Then*

$$\int_{\prod_{i=1}^{N}(a_i,b_i)} u(x)\prod_{j=1}^{r}\Phi_j\left(\frac{\left|{}^{M}_{\upsilon}e^{\gamma;\psi}_{\rho,\mu,\omega,b-}f_j(x)\right|}{K_{b-}(x)}\right)dx \leq \qquad (15.108)$$

$$\prod_{j=1}^{r}\left(\int_{\prod_{i=1}^{N}(a_i,b_i)}\varphi^{-}(y)\,\Phi_j\left(\left|f_j(y)\right|\right)^{p_j}dy\right)^{\frac{1}{p_j}}.$$

The interested researcher can produce a wealth of similar results by surveying Chaps. 21–27 of [1, 3] and . We choose to stop here.

References

1. Anastassiou, G.A.: Fractional Differentiation Inequalities. Springer, Heidelberg, New York (2009)
2. Anastassiou, G.A.: Univariate Hardy-type fractional inequalities, Chapter 2. In: Anastassiou, G., Duman, O. (eds.) "Advances in Applied Mathematics and Approximation Theory", Contributions from AMAT 2012, pp. 21–56. Springer, New York (2013)
3. Anastassiou, G.A.: Intelligent Comparisons: Analytic Inequalities. Springer, Heidelberg, New York (2016)
4. Anastassiou, G.A.: Generalized Multivariate Prabhakar Type Fractional Integrals and Inequalities. (2021) (submitted)
5. Anastassiou, G.A.: Advanced Hilfer-Prabhakar-Hardy Type Fractional Inequalities (2021) (submitted)
6. Anastassiou, G.A.: Foundations of Generalized Prabhakar-Hilfer Fractional Calculus with Applications (2021) (submitted)
7. Anastassiou, G.A.: Exotic Fractional Inequalities Revisited (2021) (submitted)
8. Giusti, A., et al.: A practical guide to Prabhakar fractional calculus. Fract. Calculus Appl. Anal. **23**(1), 9–54 (2020)
9. Gorenflo, R., Kilbas, A., Mainardi, F., Rogosin, S.: Mittag-Leffler Functions. Related Topics and Applications. Springer, Heidelberg, New York (2014)
10. Hewith, E., Stromberg, K.: Real and Abstract Analysis. Springer, New York, Heidelberg, Berlin (1965)
11. Luchko, Y.: General fractional integrals and derivatives with the Sonine kernels. Mathematics **9**, 594 (2021). https://doi.org/10.3390/math9060594
12. Sonine, N.: Sur la généralisation d'une formule d'Abel. Acta Math. **4**, 171–176 (1884)
13. Yang, X.-J.: General Fractional Derivatives. Theory, Methods and Applications. Taylor & Francis Group, CRC Press, Boca Raton, FL (2019)

Chapter 16
Fuzzy Fractional Calculus

Here we study very general fuzzy fractional integral-differential operators, give their basic properties and derive related Opial and Hardy type inequalities. Then we describe a list of specific such operators of singular and non-singular kernels that fulfill this general theory. It follows [10].

16.1 Fuzzy Mathematical Analysis Needed

This chapter is a continuation of author's [5], Chap. 34.

We need the following basic fuzzy background.

Definition 16.1 (see [25]) Let $\mu : \mathbb{R} \to [0, 1]$ with the following properties

(i) is normal, i.e., $\exists\, x_0 \in \mathbb{R};\ \mu(x_0) = 1$.

(ii) $\mu(\lambda x + (1-\lambda) y) \geq \min\{\mu(x), \mu(y)\}, \forall\, x, y \in \mathbb{R}, \forall\, \lambda \in [0, 1]$ (μ is called a convex fuzzy subset).

(iii) μ is upper semicontinuous on $\mathbb{R}$, i.e. $\forall\, x_0 \in \mathbb{R}$ and $\forall\, \varepsilon > 0$, $\exists$ neighborhood $V(x_0) : \mu(x) \leq \mu(x_0) + \varepsilon, \forall\, x \in V(x_0)$.

(iv) The set $\overline{\sup p(\mu)}$ is compact in $\mathbb{R}$, (where $\sup p(\mu) := \{x \in \mathbb{R} : \mu(x) > 0\}$).

We call μ a fuzzy real number. Denote the set of all μ with $\mathbb{R}_{\mathcal{F}}$.

E.g. $\chi_{\{x_0\}} \in \mathbb{R}_{\mathcal{F}}$, for any $x_0 \in \mathbb{R}$, where $\chi_{\{x_0\}}$ is the characteristic function at x_0.

For $0 < r \leq 1$ and $\mu \in \mathbb{R}_{\mathcal{F}}$ define

$$[\mu]^r := \{x \in \mathbb{R} : \mu(x) \geq r\}$$

and

$$[\mu]^0 := \overline{\{x \in \mathbb{R} : \mu(x) \geq 0\}}.$$

Then it is well known that for each $r \in [0, 1]$, $[\mu]^r$ is a closed and bounded interval on $\mathbb{R}$ ([16]).

G. A. Anastassiou, *Unification of Fractional Calculi with Applications*,
Studies in Systems, Decision and Control 398,
https://doi.org/10.1007/978-3-030-86920-5_16

For $u, v \in \mathbb{R}_{\mathcal{F}}$ and $\lambda \in \mathbb{R}$, we define uniquely the sum $u \oplus v$ and the product $\lambda \odot u$ by

$$[u \oplus v]^r = [u]^r + [v]^r, \quad [\lambda \odot u]^r = \lambda [u]^r, \ \forall r \in [0, 1],$$

where

- $[u]^r + [v]^r$ means the usual addition of two intervals (as substes of $\mathbb{R}$) and
- $\lambda [u]^r$ means the usual product between a scalar and a subset of $\mathbb{R}$ (see, e.g. [25]).

Notice $1 \odot u = u$ and it holds

$$u \oplus v = v \oplus u, \lambda \odot u = u \odot \lambda.$$

If $0 \le r_1 \le r_2 \le 1$ then $[u]^{r_2} \subseteq [u]^{r_1}$. Actually $[u]^r = \left[u_-^{(r)}, u_+^{(r)}\right]$, where $u_-^{(r)} \le u_+^{(r)}$, $u_-^{(r)}, u_+^{(r)} \in \mathbb{R}$, $\forall r \in [0, 1]$.

For $\lambda > 0$ one has $\lambda u_\pm^{(r)} = (\lambda \odot u)_\pm^{(r)}$, respectively.

Define

$$D : \mathbb{R}_{\mathcal{F}} \times \mathbb{R}_{\mathcal{F}} \to \mathbb{R}_{\mathcal{F}}$$

by

$$D(u, v) := \sup_{r \in [0,1]} \max \left\{ \left| u_-^{(r)} - v_-^{(r)} \right|, \left| u_+^{(r)} - v_+^{(r)} \right| \right\}$$

$$= \sup_{r \in [0,1]} \text{Hausdorff distance} \left([u]^r, [v]^r \right),$$

where $[v]^r = \left[v_-^{(r)}, v_+^{(r)}\right]$; $u, v \in \mathbb{R}_{\mathcal{F}}$. We have that D is a metric on $\mathbb{R}_{\mathcal{F}}$.

Then $(\mathbb{R}_{\mathcal{F}}, D)$ is a complete metric space, see [25], [26], with the properties

$$D(u \oplus w, v \oplus w) = D(u, v), \quad \forall u, v, w \in \mathbb{R}_{\mathcal{F}},$$
$$D(k \odot u, k \odot v) = |k| \, D(u, v), \quad \forall u, v \in \mathbb{R}_{\mathcal{F}}, \forall k \in \mathbb{R},$$
$$D(u \oplus v, w \oplus e) \le D(u, w) + D(v, e), \quad \forall u, v, w, e \in \mathbb{R}_{\mathcal{F}}.$$

Here Σ^* stands for fuzzy summation and $\widetilde{0} := \chi_{\{0\}} \in \mathbb{R}_{\mathcal{F}}$ is the neutral element with respect to $\oplus$, i.e.,

$$u \oplus \widetilde{0} = \widetilde{0} \oplus u = u, \ \forall u \in \mathbb{R}_{\mathcal{F}}.$$

Denote

$$D^*(f, g) := \sup_{x \in [a,b]} D(f, g),$$

where $f, g : [a, b] \to \mathbb{R}_{\mathcal{F}}$.

We mention

Definition 16.2 Let $f : [a, b] \subseteq \mathbb{R} \rightarrow \mathbb{R}_{\mathcal{F}}$, we define the (first) fuzzy modulus of continuity of f by

$$\omega_1^{(\mathcal{F})}(f, \delta) = \sup_{x,y \in [a,b],\ |x-y| \leq \delta} D(f(x), f(y)), \ \delta > 0.$$

We define by $C_{\mathcal{F}}^U([a, b])$ the space of fuzzy uniformly continuous functions from $[a, b] \rightarrow \mathbb{R}_{\mathcal{F}}$, also $C_{\mathcal{F}}([a, b])$ is the space of fuzzy continuous functions on $[a, b]$.

It is clear that

$$C_{\mathcal{F}}^U([a, b]) = C_{\mathcal{F}}([a, b]).$$

We mention

Proposition 16.3 *([3]) Let* $f \in C_{\mathcal{F}}^U([a, b])$. *Then* $\omega_1^{(\mathcal{F})}(f, \delta) < \infty$, *for any* $\delta > 0$.

Proposition 16.4 *([3]) It holds*

$$\lim_{\delta \rightarrow 0} \omega_1^{(\mathcal{F})}(f, \delta) = \omega_1^{(\mathcal{F})}(f, 0) = 0,$$

iff $f \in C_{\mathcal{F}}^U([a, b])$.

Proposition 16.5 *([3]) Here* $[f]^r = \left[f_-^{(r)}, f_+^{(r)}\right]$, $r \in [0, 1]$. *If* $f \in C_{\mathcal{F}}([a, b])$ *then* $f_{\pm}^{(r)} \in C([a, b])$, *for* $r \in [0, 1]$, *in fact these are equicontinuous families, respectively in* $\pm$. *Furthermore* f *is a fuzzy bounded function.*

We need

Definition 16.6 Let $x, y \in \mathbb{R}_{\mathcal{F}}$. If there exists $z \in \mathbb{R}_{\mathcal{F}} : x = y \oplus z$, then we call z the H-difference on x and y, denoted $x - y$.

Definition 16.7 ([25]) Let $T := [x_0, x_0 + \beta] \subset \mathbb{R}$, with $\beta > 0$. A function $f : T \rightarrow \mathbb{R}_{\mathcal{F}}$ is H-differentiable at $x \in T$ if there exists an $f'(x) \in \mathbb{R}_{\mathcal{F}}$ such that the limits (with respect to D)

$$\lim_{h \rightarrow 0+} \frac{f(x+h) - f(x)}{h}, \quad \lim_{h \rightarrow 0+} \frac{f(x) - f(x-h)}{h}$$

exist and are equal to $f'(x)$.

We call f' the H-derivative or fuzzy derivative of f at x.

Above is assumed that the H-differences $f(x+h) - f(x)$, $f(x) - f(x-h)$ exist in $\mathbb{R}_{\mathcal{F}}$ in a neighborhood of x.

We denote by $C_{\mathcal{F}}^N([a, b])$, $N \geq 1$, the space of all N-times continuously fuzzy differentiable functions from $[a, b]$ into $\mathbb{R}_{\mathcal{F}}$.

We mention

Theorem 16.8 *([18]) Let $f : [a, b] \to \mathbb{R}_{\mathcal{F}}$ be H-fuzzy differentiable. Let $t \in [a, b]$, $0 \le r \le 1$. Clearly*

$$[f(t)]^r = \left[f(t)_-^{(r)}, f(t)_+^{(r)} \right] \subseteq \mathbb{R}.$$

Then $(f(t))_\pm^{(r)}$ are differentiable and

$$[f'(t)]^r = \left[\left(f(t)_-^{(r)} \right)', \left(f(t)_+^{(r)} \right)' \right].$$

I.e.

$$(f')_\pm^{(r)} = \left(f_\pm^{(r)} \right)', \quad \forall r \in [0, 1].$$

Remark 16.9 ([2]) Let $f \in C_{\mathcal{F}}^N([a, b])$, $N \ge 1$. Then by Theorem 16.8 we obtain

$$\left[f^{(i)}(t) \right]^r = \left[\left(f(t)_-^{(r)} \right)^{(i)}, \left(f(t)_+^{(r)} \right)^{(i)} \right],$$

for $i = 0, 1, 2, ..., N$, and in particular we have that

$$\left(f^{(i)} \right)_\pm^{(r)} = \left(f_\pm^{(r)} \right)^{(i)},$$

for any $r \in [0, 1]$, all $i = 0, 1, 2, ..., N$.

Note 16.10 ([2]) Let $f \in C_{\mathcal{F}}^N([a, b])$, $N \ge 1$. Then by Theorem 16.8 we have $f_\pm^{(r)} \in C^N([a, b])$, for any $r \in [0, 1]$.

We need also a particular case of the Fuzzy Henstock integral ($\delta(x) = \frac{\delta}{2}$), see [25].

Definition 16.11 ([14], p. 644) Let $f : [a, b] \to \mathbb{R}_{\mathcal{F}}$. We say that f is Fuzzy-Riemann integrable to $I \in \mathbb{R}_{\mathcal{F}}$ if for any $\varepsilon > 0$, there exists $\delta > 0$ such that for any division $P = \{[u, v]; \xi\}$ of $[a, b]$ with the norms $\Delta(P) < \delta$, we have

$$D\left(\sum_P^{*} (v - u) \odot f(\xi), I \right) < \varepsilon.$$

We write

$$I := (FR) \int_a^b f(x)\, dx.$$

We mention

Theorem 16.12 *([16]) Let $f : [a, b] \to \mathbb{R}_{\mathcal{F}}$ be fuzzy continuous. Then*

$$(FR) \int_a^b f(x)\, dx$$

exists and belongs to $\mathbb{R}_{\mathcal{F}}$, furthermore it holds

$$\left[(FR) \int_a^b f(x)\, dx \right]^r = \left[\int_a^b (f)_-^{(r)}(x)\, dx, \int_a^b (f)_+^{(r)}(x)\, dx \right],$$

$\forall\, r \in [0, 1]$.

Theorem 16.13 *([25]) Let $f \in C_{\mathcal{F}}([a, b])$ and $c \in [a, b]$. Then*

$$(FR) \int_a^b f(x)\, dx = (FR) \int_a^c f(x)\, dx + (FR) \int_c^b f(x)\, dx.$$

Theorem 16.14 *([16]) Let $f, g \in C_{\mathcal{F}}([a, b])$ and $c_1, c_2 \in \mathbb{R}$. Then*

$$(FR) \int_a^b (c_1 f(x) + c_2 g(x))\, dx = c_1 (FR) \int_a^b f(x)\, dx + c_2 (FR) \int_a^b g(x)\, dx.$$

Also we need

Lemma 16.15 *([1]) If $f, g : [a, b] \subseteq \mathbb{R} \to \mathbb{R}_{\mathcal{F}}$ are fuzzy continuous functions, then the function $F : [a, b] \to \mathbb{R}_+$ defined by $F(x) := D(f(x), g(x))$ is continuous on $[a, b]$, and*

$$D\left((FR) \int_a^b f(x)\, dx, (FR) \int_a^b g(x)\, dx \right) \leq (FR) \int_a^b D(f(x), g(x))\, dx.$$

For the definition of general fuzzy integral we follow [19] next.

Definition 16.16 Let (Ω, Σ, μ) be a complete σ-finite measure space. We call $F : \Omega \to R_{\mathcal{F}}$ measurable iff $\forall$ closed $B \subseteq \mathbb{R}$ the function $F^{-1}(B) : \Omega \to [0, 1]$ defined by

$$F^{-1}(B)(w) := \sup_{x \in B} F(w)(x), \text{ all } w \in \Omega$$

is measurable, see [19].

Theorem 16.17 *([19]) For $F : \Omega \to \mathbb{R}_{\mathcal{F}}$,*

$$F(w) = \left\{ \left(F_-^{(r)}(w), F_+^{(r)}(w) \right) | 0 \leq r \leq 1 \right\},$$

the following are equivalent

(1) F is measurable,

(2) $\forall\, r \in [0, 1]$, $F_{-}^{(r)}$, $F_{+}^{(r)}$ are measurable.

Following [19], given that for each $r \in [0, 1]$, $F_{-}^{(r)}$, $F_{+}^{(r)}$ are integrable we have that the parametrized representation

$$\left\{ \left(\int_A F_{-}^{(r)} d\mu, \int_A F_{+}^{(r)} d\mu \right) |0 \leq r \leq 1 \right\}$$

is a fuzzy real number for each $A \in \Sigma$.

The last fact leads to

Definition 16.18 ([19]) A measurable function $F : \Omega \to \mathbb{R}_{\mathcal{F}}$,

$$F(w) = \left\{ \left(F_{-}^{(r)}(w), F_{+}^{(r)}(w) \right) |0 \leq r \leq 1 \right\}$$

is integrable if for each $r \in [0, 1]$, $F_{\pm}^{(r)}$ are integrable, or equivalently, if $F_{\pm}^{(0)}$ are integrable.

In this case, the fuzzy integral of F over $A \in \Sigma$ is defined by

$$\int_A F d\mu := \left\{ \left(\int_A F_{-}^{(r)} d\mu, \int_A F_{+}^{(r)} d\mu \right) |0 \leq r \leq 1 \right\}.$$

By [19], F is integrable iff $w \to \|F(w)\|_{\mathcal{F}}$ is real-valued integrable.

Here denote

$$\|u\|_{\mathcal{F}} := D\left(u, \widetilde{0}\right), \ \forall\, u \in \mathbb{R}_{\mathcal{F}}.$$

We need also

Theorem 16.19 *([19]) Let $F, G : \Omega \to \mathbb{R}_{\mathcal{F}}$ be integrable. Then*

(1) Let $a, b \in \mathbb{R}$, then $aF + bG$ is integrable and for each $A \in \Sigma$,

$$\int_A (aF + bG) d\mu = a \int_A F d\mu + b \int_A G d\mu;$$

(2) $D(F, G)$ is a real- valued integrable function and for each $A \in \Sigma$,

$$D\left(\int_A F d\mu, \int_A G d\mu \right) \leq \int_A D(F, G) d\mu.$$

In particular,

$$\left\| \int_A F d\mu \right\|_{\mathcal{F}} \leq \int_A \|F\|_{\mathcal{F}} d\mu.$$

Above μ could be the Lebesgue measure, with all the basic properties valid here too.

Basically here we have

$$\left[\int_A F d\mu\right]^r = \left[\int_A F_-^{(r)} d\mu, \int_A F_+^{(r)} d\mu\right],$$

i.e.

$$\left(\int_A F d\mu\right)_\pm^{(r)} = \int_A F_\pm^{(r)} d\mu, \quad \forall r \in [0, 1],$$

respectively.

16.2 Fractional Calculus Background

Definition 16.20 We consider the Prabhakar function (also known as the three parameter Mittag-Leffler function), (see [17], p. 97; [15])

$$E_{\alpha,\beta}^{\gamma}(z) = \sum_{k=0}^{\infty} \frac{(\gamma)_k}{k!\Gamma(\alpha k + \beta)} z^k, \tag{16.1}$$

where Γ is the gamma function; $\alpha, \beta, \gamma \in \mathbb{R} : \alpha, \beta > 0,\ z \in \mathbb{R}$, and $(\gamma)_k = \gamma(\gamma + 1) \dots (\gamma + k - 1)$. It is $E_{\alpha,\beta}^{0}(z) = \frac{1}{\Gamma(\beta)}$.

Let $a, b \in \mathbb{R}$, $a < b$ and $x \in [a, b]$; $f \in C([a, b])$. Let also $\psi \in C^1([a, b])$ which is increasing. The left and right Prabhakar fractional integrals with respect to ψ are defined ([9]) as follows:

$$\left(e_{\rho,\mu,\omega,a+}^{\gamma,\psi} f\right)(x) = \int_a^x \psi'(t)(\psi(x) - \psi(t))^{\mu-1} E_{\rho,\mu}^{\gamma}\left[\omega(\psi(x) - \psi(t))^{\rho}\right] f(t)\, dt, \tag{16.2}$$

and

$$\left(e_{\rho,\mu,\omega,b-}^{\gamma;\psi} f\right)(x) = \int_x^b \psi'(t)(\psi(t) - \psi(x))^{\mu-1} E_{\rho,\mu}^{\gamma}\left[\omega(\psi(t) - \psi(x))^{\rho}\right] f(t)\, dt, \tag{16.3}$$

where $\rho, \mu > 0$; $\gamma, \omega \in \mathbb{R}$.

Functions (16.2) and (16.3) are continuous, see [9].

Next, additionally, assume that $\psi'(x) \neq 0$ over $[a, b]$.

Let $\psi, f \in C^N([a, b])$, where $N = \lceil\mu\rceil$, ($\lceil\cdot\rceil$ is the ceiling of the number), $0 < \mu \notin \mathbb{N}$. We define the ψ-Prabhakar-Caputo left and right fractional derivatives of order μ ([9]) as follows ($x \in [a, b]$):

$$\left({}^{C}D^{\gamma;\psi}_{\rho,\mu,\omega,a+}f\right)(x)=\int_a^x \psi'(t)\left(\psi(x)-\psi(t)\right)^{N-\mu-1}$$

$$E^{-\gamma}_{\rho,N-\mu}\left[\omega\left(\psi(x)-\psi(t)\right)^{\rho}\right]\left(\frac{1}{\psi'(t)}\frac{d}{dt}\right)^{N}f(t)\,dt, \tag{16.4}$$

and

$$\left({}^{C}D^{\gamma;\psi}_{\rho,\mu,\omega,b-}f\right)(x)=(-1)^N\int_x^b \psi'(t)\left(\psi(t)-\psi(x)\right)^{N-\mu-1}$$

$$E^{-\gamma}_{\rho,N-\mu}\left[\omega\left(\psi(t)-\psi(x)\right)^{\rho}\right]\left(\frac{1}{\psi'(t)}\frac{d}{dt}\right)^{N}f(t)\,dt. \tag{16.5}$$

One can write these (see (16.4), (16.5)) as

$$\left({}^{C}D^{\gamma;\psi}_{\rho,\mu,\omega,a+}f\right)(x)=\left(e^{-\gamma;\psi}_{\rho,N-\mu,\omega,a+}f^{[N]}_{\psi}\right)(x), \tag{16.6}$$

and

$$\left({}^{C}D^{\gamma;\psi}_{\rho,\mu,\omega,b-}f\right)(x)=(-1)^N\left(e^{-\gamma;\psi}_{\rho,N-\mu,\omega,b-}f^{[N]}_{\psi}\right)(x), \tag{16.7}$$

where

$$f^{[N]}_{\psi}(x)=f^{(N)}_{\psi}f(x):=\left(\frac{1}{\psi'(x)}\frac{d}{dx}\right)^{N}f(x), \tag{16.8}$$

$\forall\, x\in[a,b]$.

Functions (16.6) and (16.7) are continuous on $[a,b]$. We will apply (16.2)-(16.8) for $\psi(x)=x$.

We need

Definition 16.21 Here we use the multivariate analogue of generalized Mittag-Leffler function, see [21], defined for $\lambda,\gamma_j,\rho_j,z_j\in\mathbb{C}$, $Re\left(\rho_j\right)>0\ (j=1,...,m)$ in terms of a multiple series of the form:

$$E^{(\gamma_j)}_{(\rho_j),\lambda}(z_1,...,z_m)=E^{(\gamma_1,...,\gamma_m)}_{(\rho_1,...,\rho_m),\lambda}(z_1,...,z_m)=$$

$$\sum_{k_1,...,k_m=0}^{\infty}\frac{(\gamma_1)_{k_1}\cdots(\gamma_m)_{k_m}}{\Gamma\left(\lambda+\sum_{j=1}^{m}k_j\rho_j\right)}\frac{z_1^{k_1}...z_m^{k_m}}{k_1!...k_m!}, \tag{16.9}$$

where $\left(\gamma_j\right)_{k_j}$ is the Pochhammer symbol, Γ is the gamma function. This is a special case of the generalized Lauricella series in several variables, see [22], p. 454 and [24].

By [23], p. 157, (16.9) converges for $Re\left(\rho_j\right)>0$, $j=1,...,m$.

In what follows we will use the particular case of $E_{(\rho,\ldots,\rho),\lambda}^{(\gamma_1,\ldots,\gamma_m)}[\omega_1 t^\rho, \ldots, \omega_m t^\rho]$, denoted by $E_{(\rho),\lambda}^{(\gamma_j)}[\omega_1 t^\rho, \ldots, \omega_m t^\rho]$, where $0 < \rho < 1$, $t \geq 0$, $\lambda > 0$, $\gamma_j \in \mathbb{R}$ with $(\gamma_j)_{k_j} := \gamma_j(\gamma_j + 1)\ldots(\gamma_j + k_j - 1)$, $\omega_j \in \mathbb{R} - \{0\}$, for $j = 1, \ldots, m$. Let $f \in C^1([a, b])$, we define the following Caputo type generalized left fractional derivative with non singular kernel of order ρ, as

$$D_{a*}^{\rho} f(x) := {}_{(\gamma_j)(\omega_j)}^{CA} D_{a*}^{\rho,\lambda} f(x) :=$$

$$\frac{A(\rho)}{1-\rho} \int_a^x E_{(\rho),\lambda}^{(\gamma_j)} \left[\frac{-\omega_1 \rho}{1-\rho}(x-t)^\rho, \ldots, \frac{-\omega_m \rho}{1-\rho}(x-t)^\rho \right] f'(t)\,dt, \quad x \in [a, b], \tag{16.10}$$

where $A(\rho)$ is a normalizing constant.

Let now $f \in C^{n+1}([a, b])$, $n \in \mathbb{Z}_+$.

We define the Caputo type generalized left fractional derivative with non singular kernel of order $n + \rho$, as

$$D_{a*}^{n+\rho} f(x) := {}_{(\gamma_j)(\omega_j)}^{CA} D_{a*}^{n+\rho,\lambda} f(x) :=$$

$$\frac{A(\rho)}{1-\rho} \int_a^x E_{(\rho),\lambda}^{(\gamma_j)} \left[\frac{-\omega_1 \rho}{1-\rho}(x-t)^\rho, \ldots, \frac{-\omega_m \rho}{1-\rho}(x-t)^\rho \right] f^{(n+1)}(t)\,dt, \quad x \in [a, b]. \tag{16.11}$$

Similarly, we define the Caputo type generalized right fractional derivative with non singular kernel of order ρ, as

$$D_{b-}^{\rho} f(x) := {}_{(\gamma_j)(\omega_j)}^{CA} D_{b-}^{\rho,\lambda} f(x) :=$$

$$\frac{-A(\rho)}{1-\rho} \int_x^b E_{(\rho),\lambda}^{(\gamma_j)} \left[\frac{-\omega_1 \rho}{1-\rho}(t-x)^\rho, \ldots, \frac{-\omega_m \rho}{1-\rho}(t-x)^\rho \right] f'(t)\,dt, \quad x \in [a, b]. \tag{16.12}$$

And, for $f \in C^{n+1}([a, b])$, $n \in \mathbb{Z}_+$, we define

$$D_{b-}^{n+\rho} f(x) := {}_{(\gamma_j)(\omega_j)}^{CA} D_{b-}^{n+\rho,\lambda} f(x) :=$$

$$(-1)^{n+1} \frac{A(\rho)}{1-\rho} \int_x^b E_{(\rho),\lambda}^{(\gamma_j)} \left[\frac{-\omega_1 \rho}{1-\rho}(t-x)^\rho, \ldots, \frac{-\omega_m \rho}{1-\rho}(t-x)^\rho \right] f^{(n+1)}(t)\,dt, \tag{16.13}$$

$x \in [a, b]$.

The above derivatives generalize the Atangana-Baleanu fractional derivative, see [11], also see next.

We mention the following non-singular kernel fractional derivatives.

Definition 16.22 Let $f \in C^1([a,b])$, $a < b$, $\alpha \in (0,1)$, the Atangana-Baleanu left fractional derivative ([11]) is given by

$$\left({}^{ABC}D_{a+}^{\alpha} f\right)(x) = \frac{B(\alpha)}{1-\alpha} \int_a^x E_\alpha\left(-\alpha \frac{(x-t)^\alpha}{1-\alpha}\right) f'(t)\, dt, \tag{16.14}$$

$\forall\, x \in [a,b]$, where $B(\alpha)$ is a normalizing constant and E_α is the one parameter Mittag-Leffler function:

$$E_\alpha(x) = \sum_{k=0}^{\infty} \frac{x^k}{\Gamma(\alpha k + 1)}, \quad x \in \mathbb{R}, \tag{16.15}$$

which is an entire function converging for all $x \in \mathbb{R}$.

The right side fractional derivative is defined by

$$\left({}^{ABC}D_{b-}^{\alpha} f\right)(x) = -\frac{B(\alpha)}{1-\alpha} \int_x^b E_\alpha\left(-\alpha \frac{(t-x)^\alpha}{1-\alpha}\right) f'(t)\, dt, \tag{16.16}$$

$\forall\, x \in [a,b]$.

Let now $f \in C^{n+1}([a,b])$, $n \in \mathbb{Z}_+$.

We define more generally, the left side fractional derivative:

$$\left({}^{ABC}D_{a+}^{n+\alpha} f\right)(x) = \frac{B(\alpha)}{1-\alpha} \int_a^x E_\alpha\left(-\alpha \frac{(x-t)^\alpha}{1-\alpha}\right) f^{(n+1)}(t)\, dt, \tag{16.17}$$

$\forall\, x \in [a,b]$,

and the right side one:

$$\left({}^{ABC}D_{b-}^{n+\alpha} f\right)(x) = (-1)^{n+1} \frac{B(\alpha)}{1-\alpha} \int_x^b E_\alpha\left(-\alpha \frac{(t-x)^\alpha}{1-\alpha}\right) f^{(n+1)}(t)\, dt, \tag{16.18}$$

$\forall\, x \in [a,b]$.

Fractional derivatives (16.14), (16.16), (16.17), (16.18) generalize the following Caputo-Fabrizio non-singular kernel fractional derivatives, see [13], [20].

Definition 16.23 Let $f \in C^1([a,b])$, $\alpha \in (0,1)$, the Caputo-Fabrizio ([13], [20]) corresponding fractional derivatives have as follows:

$${}^{CF}D_{a+}^{\alpha} f(x) = \frac{1}{1-\alpha} \int_a^x \exp\left(-\frac{\alpha}{1-\alpha}(x-t)\right) f'(t)\, dt, \tag{16.19}$$

and

$${}^{CF}D_{b-}^{\alpha} f(x) = -\frac{1}{1-\alpha} \int_x^b \exp\left(-\frac{\alpha}{1-\alpha}(t-x)\right) f'(t)\, dt, \tag{16.20}$$

$\forall\, x \in [a, b]$.

When $f \in C^{n+1}([a, b])$, $n \in \mathbb{Z}_+$, we have

$$^{CF}D_{a+}^{n+\alpha} f(x) = \frac{1}{1-\alpha} \int_a^x \exp\left(-\frac{\alpha}{1-\alpha}(x-t)\right) f^{(n+1)}(t)\,dt, \tag{16.21}$$

and

$$^{CF}D_{b-}^{n+\alpha} f(x) = \frac{(-1)^{n+1}}{1-\alpha} \int_x^b \exp\left(-\frac{\alpha}{1-\alpha}(t-x)\right) f^{(n+1)}(t)\,dt, \tag{16.22}$$

$\forall\, x \in [a, b]$.

Finally we give

Definition 16.24 Let $f \in C^1([a, b])$, $\alpha \in (0, 1)$, the Atangana-Gomez ([12]) corresponding fractional derivatives have as follows:

$$^{AG}D_{a+}^{\alpha} f(x) = \frac{AG(\alpha)}{1-\alpha} \int_a^x \exp\left(-\frac{\alpha}{1-\alpha}(x-t)^2\right) f'(t)\,dt, \tag{16.23}$$

and

$$^{AG}D_{b-}^{\alpha} f(x) = -\frac{AG(\alpha)}{1-\alpha} \int_x^b \exp\left(-\frac{\alpha}{1-\alpha}(x-t)^2\right) f'(t)\,dt, \tag{16.24}$$

$\forall\, x \in [a, b]$.

When $f \in C^{n+1}([a, b])$, $n \in \mathbb{Z}_+$, we have

$$^{AG}D_{a+}^{n+\alpha} f(x) = \frac{AG(\alpha)}{1-\alpha} \int_a^x \exp\left(-\frac{\alpha}{1-\alpha}(x-t)^2\right) f^{(n+1)}(t)\,dt, \tag{16.25}$$

and

$$^{AG}D_{b-}^{n+\alpha} f(x) = \frac{(-1)^{n+1} AG(\alpha)}{1-\alpha} \int_x^b \exp\left(-\frac{\alpha}{1-\alpha}(x-t)^2\right) f^{(n+1)}(t)\,dt, \tag{16.26}$$

$\forall\, x \in [a, b]$, where $AG(\alpha)$ is a normalization constant.

In [6], Chap. 34, we had introduced and studied the left and right fuzzy fractional Riemann-Liouville integrals and the left and right fuzzy fractional Caputo derivatives.

In this chapter we define the corresponding to Sect. 16.2, fuzzy fractional entities, and we study some of their basic properties. Furthermore we derive some related fuzzy fractional inequalities of Opial and Hardy types.

16.3 Main Results

We need

Definition 16.25 Let the kernel $k : [0, b - a] \to \mathbb{R}_+$; $a < b$, be an integrable function, and let $\nu \geq 0$ with $n = \lceil \nu \rceil \in \mathbb{Z}_+$, and $f \in C^n_{\mathcal{F}}([a, b])$.

We define the generalized fuzzy fractional derivatives of order ν (or integrals $\nu = 0$) left and right; respectively, as follows:

$$ {}^k D^{\mathcal{F}}_{a+} f(x) := \int_a^x k(x - t) \odot f^{(n)}(t)\, dt, \tag{16.27} $$

and

$$ {}^k D^{\mathcal{F}}_{b-} f(x) := (-1)^n \odot \int_x^b k(t - x) \odot f^{(n)}(t)\, dt, \tag{16.28} $$

$\forall\, x \in [a, b]$.

Furthermore, when $g \in C^n([a, b])$ we define

$$ {}^k D_{a+} g(x) := \int_a^x k(x - t)\, g^{(n)}(t)\, dt, \tag{16.29} $$

and

$$ {}^k D_{b-} g(x) := (-1)^n \int_x^b k(t - x)\, g^{(n)}(t)\, dt, \tag{16.30} $$

$\forall\, x \in [a, b]$, $n \in \mathbb{Z}_+$, the corresponding real fractional derivatives and integrals.

We make

Remark 16.26 We have that (by Definition 16.18)

$$ {}^k D^{\mathcal{F}}_{a+} f(x) = \int_a^x k(x - t) \odot f^{(n)}(t)\, dt = $$

$$ \left\{ \left(\int_a^x k(x - t) \left(f^{(n)}\right)^{(r)}_{-}(t)\, dt, \int_a^x k(x - t) \left(f^{(n)}\right)^{(r)}_{+}(t)\, dt \right) | 0 \leq r \leq 1 \right\} = \tag{16.31} $$
$$ \left\{ \left(\int_a^x k(x - t) \left(f^{(r)}_{-}\right)^{(n)}(t)\, dt, \int_a^x k(x - t) \left(f^{(r)}_{+}\right)^{(n)}(t)\, dt \right) | 0 \leq r \leq 1 \right\}. $$

So, we obtain

$$ \left[{}^k D^{\mathcal{F}}_{a+} f(x) \right]^r = \left[\int_a^x k(x - t) \left(f^{(r)}_{-}\right)^{(n)}(t)\, dt, \int_a^x k(x - t) \left(f^{(r)}_{+}\right)^{(n)}(t)\, dt \right], \tag{16.32} $$

$0 \leq r \leq 1$.

That is

$$\left({}^{k}D_{a+}^{\mathcal{F}} f(x)\right)_{\pm}^{(r)} = \int_{a}^{x} k(x-t) \left(f_{\pm}^{(r)}\right)^{(n)}(t)\, dt = \left({}^{k}D_{a+}\left(f_{\pm}^{(r)}\right)\right)(x), \quad (16.33)$$

i.e. we get

$$\left({}^{k}D_{a+}^{\mathcal{F}} f\right)_{\pm}^{(r)} = {}^{k}D_{a+}\left(f_{\pm}^{(r)}\right), \quad \forall\, r \in [0,1]. \quad (16.34)$$

Similarly, we obtain that

$${}^{k}D_{b-}^{\mathcal{F}} f(x) = (-1)^{n} \odot \int_{x}^{b} k(t-x) \odot f^{(n)}(t)\, dt =$$

$$\left\{\left((-1)^{n} \int_{x}^{b} k(t-x) \left(f^{(n)}\right)_{-}^{(r)}(t)\, dt, (-1)^{n} \int_{x}^{b} k(t-x) \left(f^{(n)}\right)_{+}^{(r)}(t)\, dt\right)\right. \quad (16.35)$$

$$\left. | 0 \leq r \leq 1\right\} =$$

$$\left\{\left((-1)^{n} \int_{x}^{b} k(t-x) \left(f_{-}^{(r)}\right)^{(n)}(t)\, dt, (-1)^{n} \int_{x}^{b} k(t-x) \left(f_{+}^{(r)}\right)^{(n)}(t)\, dt\right)\right.$$

$$\left. | 0 \leq r \leq 1\right\}.$$

So, we obtain

$$\left[{}^{k}D_{b-}^{\mathcal{F}} f(x)\right]^{r} =$$

$$\left[(-1)^{n} \int_{x}^{b} k(t-x) \left(f_{-}^{(r)}\right)^{(n)}(t)\, dt, (-1)^{n} \int_{x}^{b} k(t-x) \left(f_{+}^{(r)}\right)^{(n)}(t)\, dt\right], \quad (16.36)$$

$0 \leq r \leq 1$

That is

$$\left({}^{k}D_{b-}^{\mathcal{F}} f(x)\right)_{\pm}^{(r)} = (-1)^{n} \int_{x}^{b} k(t-x) \left(f_{\pm}^{(r)}\right)^{(n)}(t)\, dt = \left({}^{k}D_{b-}\left(f_{\pm}^{(r)}\right)\right)(x), \quad (16.37)$$

i.e. we get

$$\left({}^{k}D_{b-}^{\mathcal{F}} f\right)_{\pm}^{(r)} = {}^{k}D_{b-}\left(f_{\pm}^{(r)}\right), \quad \forall\, r \in [0,1]. \quad (16.38)$$

Another important matter follows:

It holds that $\lim_{y \to x} \int_{x-a}^{y-a} k(z)\, dz = \lim_{y \to x} \int_{0}^{b-a} \chi_{[x-a, y-a]}(z)\, k(z)\, dz = 0$, by the integrability of k and dominated convergence theorem, where $x, y \in [a,b]$, $y \geq x$; χ is the characteristic function.

Similarly, it holds that $\lim_{y \to x} \int_{b-x}^{b-y} k(z)\, dz = 0$, where $x, y \in [a,b]$, $y \leq x$.

We give

Theorem 16.27 *All as in Definition 16.25.* ${}^kD_{a+}^{\mathcal{F}}f(x)$ *is fuzzy continuous in* $x \in [a,b]$.

Proof Without loss of generality we may assume $a \leq x \leq y \leq b$, that is $0 \leq x - a \leq y - a \leq b - a$. So, we have

$$D\left({}^kD_{a+}^{\mathcal{F}}f(x), {}^kD_{a+}^{\mathcal{F}}f(y)\right) =$$

$$D\left(\int_a^x k(x-t) \odot f^{(n)}(t)\,dt, \int_a^y k(y-t) \odot f^{(n)}(t)\,dt\right) =$$

$$D\left(\int_0^{x-a} k(z) \odot f^{(n)}(x-z)\,dz, \int_0^{y-a} k(z) \odot f^{(n)}(y-z)\,dz\right) = \tag{16.39}$$

$$D\left(\int_0^{x-a} k(z) \odot f^{(n)}(x-z)\,dz\,,\right.$$

$$\left.\int_0^{x-a} k(z) \odot f^{(n)}(y-z)\,dz \oplus \int_{x-a}^{y-a} k(z) \odot f^{(n)}(y-z)\,dz\right) \leq$$

$$D\left(\int_0^{x-a} k(z) \odot f^{(n)}(x-z)\,dz, \int_0^{x-a} k(z) \odot f^{(n)}(y-z)\,dz\right) +$$

$$D\left(\int_{x-a}^{y-a} k(z) \odot f^{(n)}(y-z)\,dz, \widetilde{0}\right) \leq$$

$$\int_0^{x-a} k(z)\,D\left(f^{(n)}(x-z), f^{(n)}(y-z)\right)dz + \int_{x-a}^{y-a} k(z)\,D\left(f^{(n)}(y-z), \widetilde{0}\right)dz \leq \tag{16.40}$$

$$\left(\int_0^{x-a} k(z)\,dz\right)\omega_1^{(\mathcal{F})}\left(f^{(n)}, y-x\right) + D^*\left(f^{(n)}, \widetilde{0}\right)\int_{x-a}^{y-a} k(z)\,dz \leq$$

$$\left\{\left(\int_0^{b-a} k(z)\,dz\right)\omega_1^{(\mathcal{F})}\left(f^{(n)}, y-x\right) + D^*\left(f^{(n)}, \widetilde{0}\right)\int_{x-a}^{y-a} k(z)\,dz\right\} \to 0,$$

as $y \to x$, by noticing that $f^{(n)} \in C_{\mathcal{F}}^U([a,b])$. ■

We also present

Theorem 16.28 *All as in Definition 16.25.* ${}^kD_{b-}^{\mathcal{F}}f(x)$ *is fuzzy continuous in* $x \in [a,b]$.

Proof Without loss of generality we suppose $a \leq y \leq x \leq b$, that is $0 \leq b - x \leq b - y \leq b - a$. So, we have

$$D\left({}^{k}D_{b-}^{\mathcal{F}} f(x), {}^{k}D_{b-}^{\mathcal{F}} f(y)\right) =$$

$$D\left((-1)^n \odot \int_x^b k(t-x) \odot f^{(n)}(t)\,dt, (-1)^n \odot \int_y^b k(t-y) \odot f^{(n)}(t)\,dt\right) =$$

$$D\left(\int_x^b k(t-x) \odot f^{(n)}(t)\,dt, \int_y^b k(t-y) \odot f^{(n)}(t)\,dt\right) =$$

$$D\left(\int_0^{b-x} k(z) \odot f^{(n)}(z+x)\,dz, \int_0^{b-y} k(z) \odot f^{(n)}(z+y)\,dz\right) = \tag{16.41}$$

$$D\left(\int_0^{b-x} k(z) \odot f^{(n)}(z+x)\,dz,\right.$$

$$\left.\int_0^{b-x} k(z) \odot f^{(n)}(z+y)\,dz \oplus \int_{b-x}^{b-y} k(z) \odot f^{(n)}(z+y)\,dz\right) \leq$$

$$D\left(\int_0^{b-x} k(z) \odot f^{(n)}(z+x)\,dz, \int_0^{b-x} k(z) \odot f^{(n)}(z+y)\,dz\right) +$$

$$D\left(\int_{b-x}^{b-y} k(z) \odot f^{(n)}(z+y)\,dz, \widetilde{0}\right) \leq$$

$$\int_0^{b-x} k(z)\, D\left(f^{(n)}(z+x), f^{(n)}(z+y)\right) dz + \int_{b-x}^{b-y} k(z)\, D\left(f^{(n)}(z+y), \widetilde{0}\right) dz \leq \tag{16.42}$$

$$\left(\int_0^{b-x} k(z)\,dz\right) \omega_1^{(\mathcal{F})}\left(f^{(n)}, x-y\right) + D^*\left(f^{(n)}, \widetilde{0}\right) \int_{b-x}^{b-y} k(z)\,dz \leq$$

$$\left\{\left(\int_0^{b-a} k(z)\,dz\right) \omega_1^{(\mathcal{F})}\left(f^{(n)}, x-y\right) + D^*\left(f^{(n)}, \widetilde{0}\right) \int_{b-x}^{b-y} k(z)\,dz\right\} \to 0,$$

as $y \to x$, by noticing that $f^{(n)} \in C_{\mathcal{F}}^{U}([a,b])$. ■

We present the following left Opial type general inequality.

Theorem 16.29 *Let $p, q > 1 : \frac{1}{p} + \frac{1}{q} = 1$. All here as in Definition 16.25. Assume that $k \in L_p(0, b-a)$. Then*

$$\int_a^x D\left({}^{k}D_{a+}^{\mathcal{F}} f(w), \widetilde{0}\right) D\left(f^{(n)}(w), \widetilde{0}\right) dw \leq$$

$$2^{-\frac{1}{q}} \left(\int_a^x \left(\int_a^w k(w-t)^p\,dt\right) dw\right)^{\frac{1}{p}} \left(\int_a^x \left(D\left(f^{(n)}(w), \widetilde{0}\right)\right)^q dw\right)^{\frac{2}{q}}, \tag{16.43}$$

$\forall\, x \in [a,b]$.

Proof We have that

$$D\left({}^{k}D_{a+}^{\mathcal{F}} f(w), \widetilde{o}\right) \overset{(16.27)}{=} D\left(\int_a^x k(x-t) \odot f^{(n)}(t)\, dt, \widetilde{o}\right) \leq$$

$$\int_a^x k(x-t)\, D\left(f^{(n)}(t), \widetilde{o}\right) dt \leq \tag{16.44}$$

(by Hölder's inequality)

$$\left(\int_a^x k(x-t)^p\, dt\right)^{\frac{1}{p}} \left(\int_a^x \left(D\left(f^{(n)}(t), \widetilde{o}\right)\right)^q dt\right)^{\frac{1}{q}}.$$

Call

$$\rho(x) := \int_a^x \left(D\left(f^{(n)}(t), \widetilde{o}\right)\right)^q dt, \ \ \rho(a) = 0. \tag{16.45}$$

Then

$$\rho'(x) = \left(D\left(f^{(n)}(x), \widetilde{o}\right)\right)^q \geq 0,$$

and

$$\left(\rho'(x)\right)^{\frac{1}{q}} = \left(D\left(f^{(n)}(x), \widetilde{o}\right)\right) \geq 0, \ \ \forall\, x \in [a,b]. \tag{16.46}$$

Consequently, we get

$$\left({}^{k}D_{a+}^{\mathcal{F}} f(w), \widetilde{o}\right) D\left(f^{(n)}(w), \widetilde{o}\right) \leq \left(\int_a^w k(w-t)^p\, dt\right)^{\frac{1}{p}} \left(\rho(w)\, \rho'(w)\right)^{\frac{1}{q}}, \tag{16.47}$$

$\forall\, w \in [a,b]$.

Then, by applying again Hölder's inequality:

$$\int_a^x D\left({}^{k}D_{a+}^{\mathcal{F}} f(w), \widetilde{o}\right) D\left(f^{(n)}(w), \widetilde{o}\right) dw \leq$$

$$\int_a^x \left[\left(\int_a^w k(w-t)^p\, dt\right)^{\frac{1}{p}} \left(\rho(w)\, \rho'(w)\right)^{\frac{1}{q}}\right] dw \leq$$

$$\left(\int_a^x \left(\int_a^w k(w-t)^p\, dt\right) dw\right)^{\frac{1}{p}} \left(\int_a^x \rho(w)\, d\rho(w)\right)^{\frac{1}{q}} = \tag{16.48}$$

$$\left(\int_a^x \left(\int_a^w k(w-t)^p\, dt\right) dw\right)^{\frac{1}{p}} \left(\frac{\rho^2(x)}{2}\right)^{\frac{1}{q}} =$$

$$2^{-\frac{1}{q}}\left(\int_a^x\left(\int_a^w k(w-t)^p\,dt\right)dw\right)^{\frac{1}{p}}\left(\int_a^x\left(D\left(f^{(n)}(w),\widetilde{0}\right)\right)^q dw\right)^{\frac{2}{q}}, \quad (16.49)$$

proving the claim. ■

The right side corresponding fractional Opial type general inequality follows:

Theorem 16.30 *All as in Theorem 16.29. Then*

$$\int_x^b D\left({}^kD_{b-}^{\mathcal{F}}f(w),\widetilde{0}\right)D\left(f^{(n)}(w),\widetilde{0}\right)dw \leq$$

$$2^{-\frac{1}{q}}\left(\int_x^b\left(\int_w^b k(t-w)^p\,dt\right)dw\right)^{\frac{1}{p}}\left(\int_x^b\left(D\left(f^{(n)}(w),\widetilde{0}\right)\right)^q dw\right)^{\frac{2}{q}}, \quad (16.50)$$

$\forall\, x \in [a,b]$.

Proof As similar to the proof of Theorem 16.29 is omitted. ■

Next follow general left and right fractional Hardy type inequalities.

Theorem 16.31 *All as in Theorem 16.29. Then*
(1)

$$\left\|D\left({}^kD_{a+}^{\mathcal{F}}f(\cdot),\widetilde{0}\right)\right\|_{q,[a,b]} \leq$$

$$\left(\int_a^b\left(\int_a^x k(x-t)^p\,dt\right)^{\frac{q}{p}}dx\right)^{\frac{1}{q}}\left\|D\left(f^{(n)}(\cdot),\widetilde{0}\right)\right\|_{q,[a,b]}. \quad (16.51)$$

and
(2)

$$\left\|D\left({}^kD_{b-}^{\mathcal{F}}f(\cdot),\widetilde{0}\right)\right\|_{q,[a,b]} \leq$$

$$\left(\int_a^b\left(\int_x^b k(t-x)^p\,dt\right)^{\frac{q}{p}}dx\right)^{\frac{1}{q}}\left\|D\left(f^{(n)}(\cdot),\widetilde{0}\right)\right\|_{q,[a,b]}. \quad (16.52)$$

Proof (1) By (16.44) we obtain

$$D\left({}^kD_{a+}^{\mathcal{F}}f(x),\widetilde{0}\right) \leq \left(\int_a^x k(x-t)^p\,dt\right)^{\frac{1}{p}}\left\|D\left(f^{(n)}(\cdot),\widetilde{0}\right)\right\|_{q,[a,b]}. \quad (16.53)$$

Hence we have

$$D\left({}^kD_{a+}^{\mathcal{F}}f(x),\widetilde{0}\right)^q \leq \left(\int_a^x k(x-t)^p\,dt\right)^{\frac{q}{p}}\left\|D\left(f^{(n)}(\cdot),\widetilde{0}\right)\right\|_{q,[a,b]}^q, \quad (16.54)$$

and

$$\int_a^b D\left({}^k D_{a+}^{\mathcal{F}} f(x), \widetilde{0}\right)^q dx \leq \left(\int_a^b \left(\int_a^x k(x-t)^p dt\right)^{\frac{q}{p}} dx\right) \left\| D\left(f^{(n)}(\cdot), \widetilde{0}\right)\right\|_{q,[a,b]}^q . \tag{16.55}$$

Therefore it holds

$$\left(\int_a^b D\left({}^k D_{a+}^{\mathcal{F}} f(x), \widetilde{0}\right)^q dx\right)^{\frac{1}{q}} \leq$$

$$\left(\int_a^b \left(\int_a^x k(x-t)^p dt\right)^{\frac{q}{p}} dx\right)^{\frac{1}{q}} \left\| D\left(f^{(n)}(\cdot), \widetilde{0}\right)\right\|_{q,[a,b]}, \tag{16.56}$$

proving the claim.

(2) the proof of (16.52) is similar and thus it is omitted. ■

One can derive a vast wealth of fuzzy fractional inequalities based on author's monographs [4–8].

Next we write down Definition 16.20, when $\psi(x) = x$.

Definition 16.32 Let $f \in C([a,b])$. The left and right Prabhakar fractional integrals are defined as follows:

$$\left(e_{\rho,\mu,\omega,a+}^{\gamma} f\right)(x) = \int_a^x (x-t)^{\mu-1} E_{\rho,\mu}^{\gamma}\left[\omega(x-t)^{\rho}\right] f(t)\, dt, \tag{16.57}$$

and

$$\left(e_{\rho,\mu,\omega,b-}^{\gamma} f\right)(x) = \int_x^b (t-x)^{\mu-1} E_{\rho,\mu}^{\gamma}\left[\omega(t-x)^{\rho}\right] f(t)\, dt, \tag{16.58}$$

where $\rho, \mu > 0$; $\gamma, \omega \in \mathbb{R}$; $\forall\, x \in [a,b]$.

Functions (16.57) and (16.58) are continuous, see [8].

Let $\lceil \mu \rceil = N$, $f \in C^N([a,b])$, $0 < \mu \notin \mathbb{N}$. We define the Prabhakar-Caputo left and right fractional derivatives of order μ as follows ($x \in [a,b]$):

$$\left({}^C D_{\rho,\mu,\omega,a+}^{\gamma} f\right)(x) = \int_a^x (x-t)^{N-\mu-1} E_{\rho,N-\mu}^{-\gamma}\left[\omega(x-t)^{\rho}\right] f^{(N)}(t)\, dt, \tag{16.59}$$

and

$$\left({}^C D_{\rho,\mu,\omega,b-}^{\gamma} f\right)(x) = (-1)^N \int_x^b (t-x)^{N-\mu-1} E_{\rho,N-\mu}^{-\gamma}\left[\omega(t-x)^{\rho}\right] f^{(N)}(t)\, dt, \tag{16.60}$$

that is the continuous functions on $[a,b]$:

$$\left({}^C D_{\rho,\mu,\omega,a+}^{\gamma} f\right)(x) = \left(e_{\rho,N-\mu,\omega,a+}^{-\gamma} f^{(N)}\right)(x), \tag{16.61}$$

and

$$\left({}^{C}D^{\gamma}_{\rho,\mu,\omega,b-}f\right)(x) = (-1)^N \left(e^{-\gamma}_{\rho,N-\mu,\omega,b-}f^{(N)}\right)(x), \tag{16.62}$$

$\forall\, x \in [a, b]$.

Based on Sect. 16.2 real valued fractional integrals and derivatives (see also Definition 16.32) we define the corresponding fuzzy valued fractional integrals and derivatives.

We give

Definition 16.33 Let $f \in C_{\mathcal{F}}([a, b])$. The fuzzy left and right Prabhakar fractional integrals are defined as follows:

$$\left(e^{\gamma,\mathcal{F}}_{\rho,\mu,\omega,a+}f\right)(x) = \int_a^x (x-t)^{\mu-1} E^{\gamma}_{\rho,\mu}\left[\omega(x-t)^{\rho}\right] \odot f(t)\,dt, \tag{16.63}$$

and

$$\left(e^{\gamma,\mathcal{F}}_{\rho,\mu,\omega,b-}f\right)(x) = \int_x^b (t-x)^{\mu-1} E^{\gamma}_{\rho,\mu}\left[\omega(t-x)^{\rho}\right] \odot f(t)\,dt, \tag{16.64}$$

where $\rho, \mu > 0$; $\gamma, \omega \in \mathbb{R}_+$; $\forall\, x \in [a, b]$.

Functions (16.63) and (16.64) are fuzzy continuous, see Theorems 16.27, 16.28.

Let $\lceil \mu \rceil = N$, $f \in C^N_{\mathcal{F}}([a, b])$, $0 < \mu \notin \mathbb{N}$. We define the fuzzy Prabhakar-Caputo left and right fractional derivatives of order μ as follows ($x \in [a, b]$):

$$\left({}^{C}D^{\gamma,\mathcal{F}}_{\rho,\mu,\omega,a+}f\right)(x) = \int_a^x (x-t)^{N-\mu-1} E^{-\gamma}_{\rho,N-\mu}\left[\omega(x-t)^{\rho}\right] \odot f^{(N)}(t)\,dt, \tag{16.65}$$

and

$$\left({}^{C}D^{\gamma,\mathcal{F}}_{\rho,\mu,\omega,b-}f\right)(x) = (-1)^N \odot \int_x^b (t-x)^{N-\mu-1} E^{-\gamma}_{\rho,N-\mu}\left[\omega(t-x)^{\rho}\right] \odot f^{(N)}(t)\,dt, \tag{16.66}$$

that is the fuzzy continuous functions on $[a, b]$:

$$\left({}^{C}D^{\gamma,\mathcal{F}}_{\rho,\mu,\omega,a+}f\right)(x) = \left(e^{-\gamma,\mathcal{F}}_{\rho,N-\mu,\omega,a+}f^{(N)}\right)(x), \tag{16.67}$$

and

$$\left({}^{C}D^{\gamma,\mathcal{F}}_{\rho,\mu,\omega,b-}f\right)(x) = (-1)^N \odot \left(e^{-\gamma,\mathcal{F}}_{\rho,N-\mu,\omega,b-}f^{(N)}\right)(x), \tag{16.68}$$

$\forall\, x \in [a, b]$.

We need

Definition 16.34 Here we use the multivariate analogue of generalized Mittag-Leffler function, see [21], defined for $\lambda, \gamma_j, \rho_j, z_j \in \mathbb{C}$, $Re\left(\rho_j\right) > 0$ $(j = 1, ..., m)$ in terms of a multiple series of the form:

$$E_{(\rho_j),\lambda}^{(\gamma_j)}(z_1, ..., z_m) = E_{(\rho_1,...,\rho_m),\lambda}^{(\gamma_1,...,\gamma_m)}(z_1, ..., z_m) =$$

$$\sum_{k_1,...,k_m=0}^{\infty} \frac{(\gamma_1)_{k_1} \cdots (\gamma_m)_{k_m}}{\Gamma\left(\lambda + \sum_{j=1}^{m} k_j \rho_j\right)} \frac{z_1^{k_1} ... z_m^{k_m}}{k_1! ... k_m!}, \tag{16.69}$$

where $\left(\gamma_j\right)_{k_j}$ is the Pochhammer symbol and Γ is the gamma function.

In what follows we will use the particular case of $E_{(\rho,...,\rho),\lambda}^{(\gamma_1,...,\gamma_m)}[\omega_1 t^\rho, ..., \omega_m t^\rho]$, denoted by $E_{(\rho),\lambda}^{(\gamma_j)}[\omega_1 t^\rho, ..., \omega_m t^\rho]$, where $0 < \rho < 1$, $t \geq 0$, $\lambda > 0$, $\gamma_j \in \mathbb{R}_+$ with $\left(\gamma_j\right)_{k_j} := \gamma_j\left(\gamma_j + 1\right) ... \left(\gamma_j + k_j - 1\right)$, $\omega_j \in \mathbb{R}_+ - \{0\}$, for $j = 1, ..., m$. Let $f \in C_{\mathcal{F}}^1([a, b])$, we define the following fuzzy Caputo type generalized left fractional derivative with non singular kernel of order ρ, as

$$D_{a*}^{\rho,\mathcal{F}} f(x) := {}_{(\gamma_j)(\omega_j)}^{CA} D_{a*}^{\rho,\lambda,\mathcal{F}} f(x) :=$$

$$\frac{A(\rho)}{1-\rho} \odot \int_a^x E_{(\rho),\lambda}^{(\gamma_j)}\left[\frac{-\omega_1 \rho}{1-\rho}(x-t)^\rho, ..., \frac{-\omega_m \rho}{1-\rho}(x-t)^\rho\right] \odot f'(t)\, dt, \quad x \in [a, b], \tag{16.70}$$

where $A(\rho)$ is a normalizing constant.

Let now $f \in C_{\mathcal{F}}^{n+1}([a, b])$, $n \in \mathbb{Z}_+$.

We define the fuzzy Caputo type generalized left fractional derivative with non singular kernel of order $n + \rho$, as

$$D_{a*}^{n+\rho,\mathcal{F}} f(x) := {}_{(\gamma_j)(\omega_j)}^{CA} D_{a*}^{n+\rho,\lambda,\mathcal{F}} f(x) :=$$

$$\frac{A(\rho)}{1-\rho} \odot \int_a^x E_{(\rho),\lambda}^{(\gamma_j)}\left[\frac{-\omega_1 \rho}{1-\rho}(x-t)^\rho, ..., \frac{-\omega_m \rho}{1-\rho}(x-t)^\rho\right] \odot f^{(n+1)}(t)\, dt, \quad x \in [a, b]. \tag{16.71}$$

Similarly, we define the fuzzy Caputo type generalized right fractional derivative with non singular kernel of order ρ, as

$$D_{b-}^{\rho,\mathcal{F}} f(x) := {}_{(\gamma_j)(\omega_j)}^{CA} D_{b-}^{\rho,\lambda,\mathcal{F}} f(x) :=$$

$$\frac{-A(\rho)}{1-\rho} \odot \int_x^b E_{(\rho),\lambda}^{(\gamma_j)}\left[\frac{-\omega_1 \rho}{1-\rho}(t-x)^\rho, ..., \frac{-\omega_m \rho}{1-\rho}(t-x)^\rho\right] \odot f'(t)\, dt, \quad x \in [a, b]. \tag{16.72}$$

And, for $f \in C_{\mathcal{F}}^{n+1}([a, b])$, $n \in \mathbb{Z}_+$, we define

$$D_{b-}^{n+\rho,\mathcal{F}} f(x) := {}_{(\gamma_j)(\omega_j)}^{CA} D_{b-}^{n+\rho,\lambda,\mathcal{F}} f(x) :=$$

$$(-1)^{n+1} \frac{A(\rho)}{1-\rho} \odot \int_x^b E_{(\rho),\lambda}^{(\gamma_j)} \left[\frac{-\omega_1 \rho}{1-\rho} (t-x)^{\rho}, \ldots, \frac{-\omega_m \rho}{1-\rho} (t-x)^{\rho} \right] \odot f^{(n+1)}(t)\, dt, \tag{16.73}$$

$x \in [a, b]$.

The above fuzzy fractional derivatives generalize the fuzzy Atangana-Baleanu type fractional derivatives, see next.

We mention the following non-singular kernel fuzzy fractional derivatives.

Definition 16.35 Let $f \in C_{\mathcal{F}}^{1}([a, b])$, $a < b$, $\alpha \in (0, 1)$, the fuzzy Atangana-Baleanu type left fractional derivative is given by

$$\left({}^{ABC}D_{a+}^{\alpha,\mathcal{F}} f \right)(x) = \frac{B(\alpha)}{1-\alpha} \odot \int_a^x E_{\alpha} \left(-\alpha \frac{(x-t)^{\alpha}}{1-\alpha} \right) \odot f'(t)\, dt, \tag{16.74}$$

$\forall\, x \in [a, b]$, where $B(\alpha)$ is a normalizing constant and E_{α} is the one parameter Mittag-Leffler function:

$$E_{\alpha}(x) = \sum_{k=0}^{\infty} \frac{x^k}{\Gamma(\alpha k + 1)}, \quad x \in \mathbb{R}, \tag{16.75}$$

which is an entire function converging for all $x \in \mathbb{R}$.

The right side fuzzy fractional derivative is defined by

$$\left({}^{ABC}D_{b-}^{\alpha,\mathcal{F}} f \right)(x) = -\frac{B(\alpha)}{1-\alpha} \odot \int_x^b E_{\alpha} \left(-\alpha \frac{(t-x)^{\alpha}}{1-\alpha} \right) \odot f'(t)\, dt, \tag{16.76}$$

$\forall\, x \in [a, b]$.

Let now $f \in C_{\mathcal{F}}^{n+1}([a, b])$, $n \in \mathbb{Z}_{+}$.

We define more generally, the fuzzy left side fractional derivative:

$$\left({}^{ABC}D_{a+}^{n+\alpha,\mathcal{F}} f \right)(x) = \frac{B(\alpha)}{1-\alpha} \odot \int_a^x E_{\alpha} \left(-\alpha \frac{(x-t)^{\alpha}}{1-\alpha} \right) \odot f^{(n+1)}(t)\, dt, \tag{16.77}$$

$\forall\, x \in [a, b]$,

and the fuzzy right side fractional derivative:

$$\left({}^{ABC}D_{b-}^{n+\alpha,\mathcal{F}} f \right)(x) = (-1)^{n+1} \frac{B(\alpha)}{1-\alpha} \odot \int_x^b E_{\alpha} \left(-\alpha \frac{(t-x)^{\alpha}}{1-\alpha} \right) \odot f^{(n+1)}(t)\, dt, \tag{16.78}$$

$\forall\, x \in [a, b]$.

Fuzzy fractional derivatives (16.74), (16.76), (16.77), (16.78) generalize the following fuzzy Caputo-Fabrizio type non-singular kernel fractional derivatives.

Definition 16.36 Let $f \in C^1_{\mathcal{F}}([a,b])$, $\alpha \in (0,1)$, the fuzzy Caputo-Fabrizio type fractional derivatives have as follows:

$$ {}^{CF}D^{\alpha,\mathcal{F}}_{a+} f(x) = \frac{1}{1-\alpha} \odot \int_a^x \exp\left(-\frac{\alpha}{1-\alpha}(x-t)\right) \odot f'(t)\,dt, \tag{16.79} $$

and

$$ {}^{CF}D^{\alpha,\mathcal{F}}_{b-} f(x) = -\frac{1}{1-\alpha} \odot \int_x^b \exp\left(-\frac{\alpha}{1-\alpha}(t-x)\right) \odot f'(t)\,dt, \tag{16.80} $$

$\forall\, x \in [a,b]$.

When $f \in C^{n+1}_{\mathcal{F}}([a,b])$, $n \in \mathbb{Z}_+$, we have

$$ {}^{CF}D^{n+\alpha,\mathcal{F}}_{a+} f(x) = \frac{1}{1-\alpha} \odot \int_a^x \exp\left(-\frac{\alpha}{1-\alpha}(x-t)\right) \odot f^{(n+1)}(t)\,dt, \tag{16.81} $$

and

$$ {}^{CF}D^{n+\alpha,\mathcal{F}}_{b-} f(x) = \frac{(-1)^{n+1}}{1-\alpha} \odot \int_x^b \exp\left(-\frac{\alpha}{1-\alpha}(t-x)\right) \odot f^{(n+1)}(t)\,dt, \tag{16.82} $$

$\forall\, x \in [a,b]$.

Finally we give

Definition 16.37 Let $f \in C^1_{\mathcal{F}}([a,b])$, $\alpha \in (0,1)$, the fuzzy Atangana-Gomez type fractional derivatives have as follows:

$$ {}^{AG}D^{\alpha,\mathcal{F}}_{a+} f(x) = \frac{AG(\alpha)}{1-\alpha} \odot \int_a^x \exp\left(-\frac{\alpha}{1-\alpha}(x-t)^2\right) \odot f'(t)\,dt, \tag{16.83} $$

and

$$ {}^{AG}D^{\alpha,\mathcal{F}}_{b-} f(x) = -\frac{AG(\alpha)}{1-\alpha} \odot \int_x^b \exp\left(-\frac{\alpha}{1-\alpha}(x-t)^2\right) \odot f'(t)\,dt, \tag{16.84} $$

$\forall\, x \in [a,b]$.

When $f \in C^{n+1}_{\mathcal{F}}([a,b])$, $n \in \mathbb{Z}_+$, we have

$$ {}^{AG}D^{n+\alpha,\mathcal{F}}_{a+} f(x) = \frac{AG(\alpha)}{1-\alpha} \odot \int_a^x \exp\left(-\frac{\alpha}{1-\alpha}(x-t)^2\right) \odot f^{(n+1)}(t)\,dt, \tag{16.85} $$

and

$$ {}^{AG}D^{n+\alpha,\mathcal{F}}_{b-} f(x) = \frac{(-1)^{n+1}AG(\alpha)}{1-\alpha} \odot \int_x^b \exp\left(-\frac{\alpha}{1-\alpha}(x-t)^2\right) \odot f^{(n+1)}(t)\,dt, \tag{16.86} $$

$\forall\, x \in [a, b]$, where $AG(\alpha)$ is a normalization constant.

We make

Remark 16.38 In this chapter we presented a number of specific new fuzzy fractional integrals and derivatives mostly of non-singular kernels, see Definitions 16.33-16.37. All these are special cases of the fuzzy fractional integrals (16.27), (16.28), respectively, we introduced in Definition 16.25, and we studied in Remark 16.26 and Theorems 16.27-16.31. Therefore all these special cases satisfy all the established general properties of (16.27), (16.28), respectively. That is our specific fuzzy fractional entities satisfy (16.34), (16.38), they are all fuzzy continuous and satisfy the fuzzy fractional inequalities (16.43), (16.50)-(16.52), respectively.

To keep our presentation short as a sample we mention only the following:
The Opial inequalities follow:

Corollary 16.39 *Let* $p, q > 1 : \frac{1}{p} + \frac{1}{q} = 1$. *All as in Definition 16.33. Then*
(1)

$$\int_a^x D\left(\left(e_{\rho,\mu,\omega,a+}^{\gamma,\mathcal{F}} f\right)(w), \widetilde{0}\right) D\left(f(w), \widetilde{0}\right) dw \leq$$

$$2^{-\frac{1}{q}} \left(\int_a^x \left(\int_a^w \left[(w-t)^{\mu-1} E_{\rho,\mu}^{\gamma}\left[\omega (w-t)^{\rho}\right]\right]^p dt\right) dw\right)^{\frac{1}{p}} \tag{16.87}$$

$$\left(\int_a^x \left(D\left(f(w), \widetilde{0}\right)\right)^q dw\right)^{\frac{2}{q}}, \quad \forall\, x \in [a, b],$$

and
(2)

$$\int_x^b D\left(\left(e_{\rho,\mu,\omega,b-}^{\gamma,\mathcal{F}} f\right)(w), \widetilde{0}\right) D\left(f(w), \widetilde{0}\right) dw \leq$$

$$2^{-\frac{1}{q}} \left(\int_x^b \left(\int_w^b \left[(t-w)^{\mu-1} E_{\rho,\mu}^{\gamma}\left[\omega (t-w)^{\rho}\right]\right]^p dt\right) dw\right)^{\frac{1}{p}} \tag{16.88}$$

$$\left(\int_x^b \left(D\left(f(w), \widetilde{0}\right)\right)^q dw\right)^{\frac{2}{q}}, \quad \forall\, x \in [a, b].$$

Proof Apply (16.43) and (16.50), respectively. ■

The Hardy inequalities follow:

Corollary 16.40 *Let* $p, q > 1 : \frac{1}{p} + \frac{1}{q} = 1$. *All as in Definition 16.35. Then*
(1)

$$\left\| D\left({}^{ABC}D_{a+}^{n+\alpha,\mathcal{F}} f(\cdot), \widetilde{0} \right) \right\|_{q,[a,b]} \leq \frac{|B(\alpha)|}{1-\alpha}$$

$$\left(\int_a^b \left(\int_a^x \left(E_\alpha \left(-\alpha \frac{(x-t)^\alpha}{1-\alpha} \right) \right)^p dt \right)^{\frac{q}{p}} dx \right)^{\frac{1}{q}} \left\| D\left(f^{(n+1)}(\cdot), \widetilde{0} \right) \right\|_{q,[a,b]}, \tag{16.89}$$

and
(2)

$$\left\| D\left({}^{ABC}D_{b-}^{n+\alpha,\mathcal{F}} f(\cdot), \widetilde{0} \right) \right\|_{q,[a,b]} \leq \frac{|B(\alpha)|}{1-\alpha}$$

$$\left(\int_a^b \left(\int_x^b \left(E_\alpha \left(-\alpha \frac{(t-x)^\alpha}{1-\alpha} \right) \right)^p dt \right)^{\frac{q}{p}} dx \right)^{\frac{1}{q}} \left\| D\left(f^{(n+1)}(\cdot), \widetilde{0} \right) \right\|_{q,[a,b]}, \tag{16.90}$$

Proof By Theorem 16.31. ■

References

1. Anastassiou, G.A.: Rate of convergence of Fuzzy neural network operators, univariate case. J. Fuzzy Math. **10**(3), 755–780 (2002)
2. Anastassiou, G.A.: Higher order fuzzy Korovkin theory via inequalities. Commun. Appl. Anal. **10** (2), 359–392 (2006)
3. Anastassiou, G.A.: Fuzzy Korovkin theorems and inequalities. J. Fuzzy Math. **15**(1), 169–205 (2007)
4. Anastassiou, G.A.: Fractional Differentiation Inequalities. Research Monograph, Springer, New York (2009)
5. Anastassiou, G.A.: Fuzzy Mathematics: Approximation Theory. Springer, Heidelberg, New York (2010)
6. Anastassiou, G.A.: Intelligent Mathematics: Computational Analysis. Springer, Heidelberg, New York (2011)
7. Anastassiou, G.: Advances on Fractional Inequalities. Springer, Heidelberg, New York (2011)
8. Anastassiou, G.A.: Intelligent Comparisons: Analytic Inequalities. Springer, Heidelberg, New York (2016)
9. Anastassiou, G.A.: Foundations of Generalized Prabhakar-Hilfer fractional Calculus with Applications. [2021, Submitted]
10. Anastassiou, G.A.: Fuzzy Fractional Calculus Revisited (2021). [Submitted]
11. Atangana, A., Baleanu, D.: New fractional derivatives with nonlocal and non-singular kernel: theory and application to heat transfer model. Therm. Sci. **20**(2), 763–769 (2016)
12. Atangana, A., Gomez-Aguilar, J.F.: Numerical approximation of Riemann-Liouville definition of fractional derivative: from Riemann-Liouville to Atangana-Baleanu. Numer. Methods. Partial Differ. Equ. **34**(5), 1502–1523 (2018)

13. Caputo, M., Fabrizio, M.: A new definition of fractional derivative without singular kernel. Progr. Fract. Differ. Appl. **1**(2), 73–85 (2015)
14. Gal, S.: Approximation Theory in Fuzzy Setting. In: Anastassiou G (ed) Chapter 13 in Handbook of Analytic-Computational Methods in Applied Mathematics, pp. 617–666. Chapman & Hall/CRC, 2000, Boca Raton, New York
15. Giusti, A., et al.: A practical Guide to Prabhakar fractional calculus. Frac. Calc. Appl. Anal. **23**(1), 9–54 (2020)
16. Goetschel, R., Jr., Voxman, W.: Elementary fuzzy calculus. Fuzzy Sets Syst. **18**, 31–43 (1986)
17. Gorenflo, R., Kilbas, A., Mainardi, F., Rogosin, S.: Mittag-Leffler functions. Related Topics and Applications. Springer, Heidelberg, New York (2014)
18. Kaleva, O.: Fuzzy differential equations. Fuzzy Sets Syst. **24**, 301–317 (1987)
19. Kim, Y.K., Ghil, B.M.: Integrals of fuzzy-number-valued functions. Fuzzy Sets Syst. **86**, 213–222 (1997)
20. Losada, J., Nieto, J.J.: Properties of a new fractional derivative without singular kernel. Progr. Fract. Differ. Appl. /textbf1 (2), 87–92 (2015)
21. Saxena, R.K., Kalla, S.L.: Ravi Saxena. Multivariate analogue of generalized Mittag-Leffler function, integral transforms and special functions **22**(7), 533–548 (2011)
22. Srivastava, H.M., Daoust, M.C.: Certain generalized Newmann expansion associated with Kompe' de Feriet function. Nederl. Akad. Wetensch. Proc. Ser. A 72 (Indag. Math.) 31, 449-457 (1969)
23. Srivastava, H.M., Daoust, M.C.: A note on the convergence of Kompe' de Feriet's double hypergeometric series. Math. Nachr. **53**, 151–159 (1972)
24. Srivastava, H.M., Karlsson, P.W.: Multiple Gaussian Hypergeometric Series. Ellis Horwood Limited, New York (1985)
25. Wu, C., Gong, Z.: On Henstock integral of fuzzy-number-valued functions (I), Fuzzy Sets and Systems, 120. No. **3**, 523–532 (2001)
26. Wu, C., Ma, M.: On embedding problem of fuzzy number spaces: Part 1. Fuzzy Sets Syst. **44**, 33–38 (1991)

Chapter 17
Conclusion

During the last 50 years fractional calculus due to its wide applications to many applied sciences has become a main trend in mathematics. Its predominant kinds are the old Riemann-Liouville fractional calculus and the newer one of Caputo type.

Around these two versions have been built a plethora of other variants and all of these involve singular kernels. More recently researchers presented also new fractional calculi of non singular kernels.

The recent Hilfer fractional calculus unifies the Riemann-Liouville and Caputo fractional Calculi and the Prabhakar fractional calculus unifies both singular and non-singular kernel fractional calculi.

Finally the newer Hilfer-Prabhakar fractional calculus is the most general one unifying all trends and for different values of its parameters we get the particular fractional calculi.

In this monograph we presented and employed unifying advanced and generalized versions of Hilfer, Prabhakar and Hilfer-Prabhakar fractional calculi and we established related unifying fractional integral inequalities of the following types: Iyengar, Landau, Polya, Ostrowski, Hilbert-Pachpatte, Hardy, Opial, Csiszar's f-Divergence, self-adjoint operator and related to fuzziness. Our results are univariate and multivariate. This monograph's results are expected to find applications in many areas of pure and applied mathematics, especially in fractional inequalities and fractional differential equations. Other interesting applications can be in applied sciences like geophysics, physics, chemistry, economics and engineering, etc.

The advantage of this unification is the uniform action taken in describing the various natural phenomena.

G. A. Anastassiou, *Unification of Fractional Calculi with Applications*,
Studies in Systems, Decision and Control 398,
https://doi.org/10.1007/978-3-030-86920-5_17